ɔn No. 150 of the International Lithosphere Programme

Geological Society Special Publications
Series Editor K. COE

Publicati

Inve

GEOLOGICAL SOCIETY SPECIAL PUBLICATION NO. 44

Inversion Tectonics

EDITED BY

M. A. COOPER
BP Canada Inc.
Oil and Gas Division
333 Fifth Avenue SW
Calgary
Alberta T2P 3B6
Canada

G. D. WILLIAMS
Department of Geology
University of Keele
Keele
Staffs ST5 5BG, UK

1989

Published for
The Geological Society by
Blackwell Scientific Publications
OXFORD LONDON EDINBURGH
BOSTON MELBOURNE

Published for
The Geological Society by
Blackwell Scientific Publications
Osney Mead, Oxford OX2 0EL
(*Orders*: Tel. 0865 240201)
8 John Street, London WC1N 2ES
23 Ainslie Place, Edinburgh EH3 6AJ
3 Cambridge Center, Suite 208, Cambridge
Massachusetts 02142, USA
107 Barry Street, Carlton, Victoria 3053
Australia

First published 1989

Set by Setrite Typesetters, Hong Kong
Printed and bound in Great Britain
at the University Press, Cambridge

DISTRIBUTORS

USA
Publishers' Business Services
PO Box 447
Brookline Village
Massachusetts 02147
(*Orders*: Tel. (617) 524–7678)

Canada
Oxford University Press
70 Wynford Drive
Don Mills
Ontario M3C 1J9
(*Orders*: Tel. (416) 441–2941)

Australia
Blackwell Scientific Publications
(Australia) Pty Ltd
107 Barry Street, Carlton
Victoria 3053
(*Orders*: Tel. (03) 347–0300)

British Library
Cataloguing in Publication Data

Inversion Tectonics
1. Tectonics
I. Cooper, M. A. II. Williams, G. D.
III. Series
551.1.'36

ISBN 0–632–02502–6

Library of Congress
Cataloguing-in-Publication Data

Inversion tectonics. (Geological Society special publication; no. 44)
Includes index.
1. Inversions (Geology)—Europe—Congresses.
2. Inversions (Geology)—Congresses.
I. Cooper, M. A. II. Williams, G. D.
III. Series.
QE606.5.E85I59 1989 551.8′6 88–34276

ISBN 0–632–02502–6

Contents

Introduction

M. A. Cooper & G. D. Williams

This volume contains a selection of papers presented at the Inversion Tectonics Meeting held at the Royal Society on 3–4 March 1987. Two of the papers, those by Hayward & Graham, and Morley, were not actually presented at the meeting as the contributors were unable to attend. Six of the papers presented were not submitted in manuscript form for inclusion; however, we have included the abstracts at the end of the volume.

The volume has been divided into four sections: Modelling and Theoretical Concepts, Inversion in the Alps and Alpine Foreland, Inversion on the European Continental Shelf, and Inversion in Other Geological Environments. At the end we have included a general discussion on the usage and definition of inversion that comprises some editorial thoughts and discussion contributions made at the meeting.

In the first section, Modelling and Theoretical Concepts, Williams *et al.* discuss the analysis of inversion structures using the concept of the null point. Hayward & Graham introduce many of the characteristic features of inversion structures illustrated by examples from W Europe. The section is concluded by McClay who presents the results of modelling inversion using sand/clay interlayered models.

The second section, Inversion in the Alps and Alpine Foreland, contains strongly contrasting views on the usage of the concept of inversion. Ziegler presents a comprehensive review of inversion in NW Europe which is applied to the Alpine orogenic belt by de Graciansky *et al.* and by Butler. The paper by Simpson *et al.* briefly discusses inversion in the Wessex Basin based on extensive exploration data. The final paper by Roberts extends the effect of the inversion far further to the west and north than is traditionally accepted.

The third section, Inversion on the European Continental Shelf, deals with 'traditional' inversion illustrated by seismic data from the Danish section of the North Sea by Cartwright, the Western Approaches by Chapman and the southern North Sea by Badley *et al.*

The fourth and final section, Inversion in Other Geological Environments, discusses the application of the concepts of inversion to areas in which the reactivation of pre-existing structures is demonstrable but complex. Powell & Williams describe the extensional collapse of the Lewis Thrust sheet in the Rockies of NW Montana. McClay *et al.* present evidence for inversion of Devonian half-grabens in the Canadian Rockies. An example of inversion in the Scandinavian Caledonides is documented by Morley, and Coward *et al.* discuss inversion related to movements on the Great Glen Fault in the Devonian of N Scotland. In the final paper, Daly *et al.* describe how reactivation of fundamental basement lineaments controlled the geometry of the East African Rift system.

This collection of papers thus provides a very broad spectrum of examples of inversion tectonics which goes beyond the traditional usage of the term. This reflects the current range of research on inversion problems and whilst this generates some controversy (see discussion) we believe that there is value in illustrating the whole of current usage.

M. A. Cooper, BP Canada Inc., Oil and Gas Division, 333 Fifth Avenue SW, Calgary, Alberta T2P 3B6, Canada.
G. D. Williams, Department of Geology, University of Keele, Keele, Staffordshire ST5 5BG, UK.

Modelling and Theoretical Concepts

Geometry and kinematics of inversion tectonics

G. D. Williams, C. M. Powell & M. A. Cooper

SUMMARY: Positive inversion tectonics involves the reversal of extensional fault movement during contractional tectonics. Basin stratigraphy developed before, during and after extensional fault movements may be described as pre-, syn- and postrift sequences. Growth fault activity may be graphically displayed using thickness changes in stratigraphic intervals from fault footwall to hanging wall. Alternatively, it may be recorded using a hanging wall displacement/distance plot. Contractional reactivation of extensional faults puts progressively older synrift markers into net contraction. The point of change from net extension to net contraction is the null point. Its position in the synrift stratigraphy may be used to quantify the inversion ratio, which is defined as the ratio of contractional to extensional movement. Negative inversion is the reactivation in extension of a significant portion of an existing contractional system. Stratigraphic separation diagrams constructed from geological maps may be used to define the null point of individual faults and to quantify their inversion ratio.

Basic nomenclature

Structural inversion occurs when basin-controlling extensional faults reverse their movement during compressional tectonics, and, to varying degrees, basins are turned inside out to become positive features. The result is that individual faults may retain net extension at depth and show net contraction associated with anticline growth in their upper portions (Fig. 1). The converse is also possible. Thus inversion can be considered in terms of both positive (uplift) and negative (subsidence) senses relative to the immediately preceeding history of a fault system (Glennie & Boegner 1981).

In this paper, a fault is considered to be of extensional mode if it involves the displacement of an approximately horizontal marker horizon in the fault hanging wall below its pre-deformational regional elevation, and is of contractional mode if a hanging wall marker is displaced above its regional elevation (Fig. 2). A marker horizon may consist of an originally horizontal or subhorizontal stratigraphic interface, a marker bed on a geological cross-section, or a sequence boundary on a seismic reflection section. This definition is simple; it does not involve discussions of net slip of the fault and closely resembles fault dip separation.

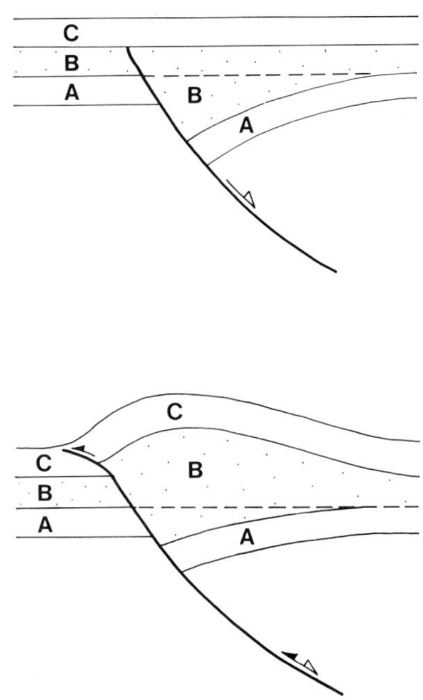

FIG. 1. Schematic diagram of a classical positive inversion structure. A, B and C are stratigraphic sequences. A, prerift; B, synrift; C, postrift sequence.

In our definition we consider apparent dip as portrayed in a cross-section or seismic section, rather than true fault dips. Cross-sections through a fault in any direction will reveal its extensional or contractional mode if our definition is applied. Only a perfectly bed-parallel fault (a flat on flat) and a perfectly strike-slip fault is neither extensional nor contractional and may be considered a neutral mode fault. A fault which appears vertical in the plane of a section cannot be classified in this scheme. Because net slip is not considered in this definition we may apply the contractional/extensional nomenclature to all faults including those

From COOPER, M. A. & WILLIAMS, G. D. (eds), *Inversion Tectonics*, Geological Society Special Publication No. 44, 3–15

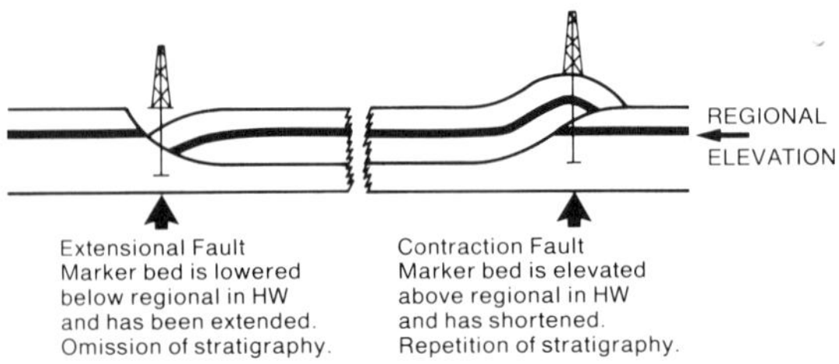

FIG. 2. Definition of A, extensional and B, contractional faults.

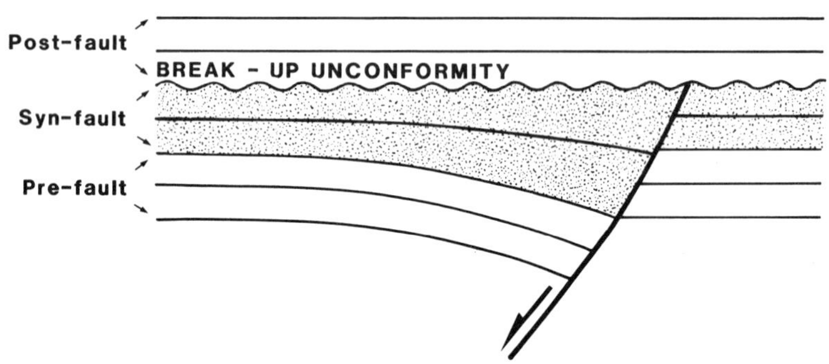

FIG. 3. Schematic diagram of stratigraphy accumulated before, during and after extensional fault movement. The break-up unconformity is shown as a wavy line at the top of the synrift sequence.

generated in a strike slip system at transpressive and transtensional fault bends.

Structural inversion involves fault movements giving rise to a switching of fault mode. A perfect reversal of fault net slip is unlikely and most inversion structures probably result from superimposed oblique slip movements. Our definition may be used to separate individual fault modes and to analyse structural inversion and is preferable to existing, complex, fault nomenclature.

In this paper we shall consider both positive inversion of a stratigraphy accumulated in extensional regimes and the negative inversion of thrust systems.

Extensional faulting

Stratigraphy built up on continental margins or in continental extensional basins, in its simplest form, may exhibit three distinct sequences (Fig. 3).

1 A prerift sequence is deposited prior to any extensional fault movement.

2 A synrift sequence is deposited during extensional faulting. Marked stratigraphic thickness changes from fault footwall to hanging wall are indicative of growth faulting.

3 A postrift sequence is deposited after the cessation of extensional faulting. The postrift sequence may be deposited after a period of non-deposition and/or erosion marked by a break-up unconformity which may remove part of the synrift sequence.

Percentage interval change

A listric extensional fault active during sediment deposition gives rise to a rollover anticline in the hanging wall above the curved portion of the fault (Hamblin 1965; Gibbs 1984). A rotational fault block model dominated by planar faults gives a similar effect. A wedge of synrift sediments thickens towards the fault in either case (Fig. 3). The fault movement history is recorded in stratigraphic thickness changes between footwall and hanging wall. This may be graphically displayed by plotting changes in stratigraphic thickness (footwall to hanging wall) against chronostratigraphy (Fig. 4A) (Beach 1984). Percentage interval change (I) is given by:

$$I = \frac{I_h - I_f}{I_f} \times 100 \qquad (1)$$

where I_f is the thickness of a stratigraphic interval in the footwall of an extension fault and I_h is the thickness of the same interval in the hanging wall. Both I_h and I_f are measured perpendicular to bedding in the plane of the section. They should be measured immediately adjacent to the fault.

Peaks in the resulting curve correspond to periods of maximum fault activity in the synrift sequence. A break-up unconformity will remove part of the synrift sequence. In this case, the fault activity curve will be incomplete and will not return to the ordinate (Fig. 4B). The unconformity will remove a thickness of section which represents a greater time interval in the footwall than in the hanging wall. This is because the footwall represents a condensed stratigraphic sequence.

Displacement/distance

A synrift sedimentary sequence accumulated during growth faulting will show increasing extensional displacements of stratigraphic markers with depth. This may be graphically represented using a modification of the displacement/distance plot of Williams & Chapman (1983). Using the top of the synrift sequence as a reference point, distance along a fault to individual hanging wall markers can be measured. This effectively records hanging wall stratigraphic thicknesses. Hanging wall distance is plotted against displacement (or more correctly, dip separation) of individual markers. Both parameters are measured in the plane of the section and therefore represent *apparent hanging wall distance* and *apparent extensional displacement* (Fig. 5). A curve with a negative gradient results. Changes in gradient relate to variations through time of fault activity. In this respect, the curve is similar to the percentage interval change plot (Figs 4A & B). Because the prerift and postrift sequences were not accumulated during growth faulting, any fault dislocation of these sequences should show no dislocation changes. This yields a curve with an infinite gradient or a slope of 90° on the displacement/distance plot.

In the preceding discussion we have not considered effects such as sediment compaction or displacement gradients along individual faults. These important effects are quantifiable.

Positive inversion of extensional fault systems

Our definition of extensional, contractional and neutral mode faults relies upon the geometry of faults and of marker horizons relative to regional elevation in the plane of a cross-section or seismic reflection section. The contractional reactivation of an extensional mode fault leads to an apparent reversal of movement when considered in the plane of the section. Using a typical pre-, syn- and postrift stratigraphy accumulated before, during and after extensional fault movements, we now consider the geometry of the system during contraction. In this analysis we assume that the stratigraphy is complete and that no break-up unconformity is present.

Contractional fault movement immediately displaces the postrift sequence in the fault hanging wall above its pre-deformational regional elevation. Continued contractional movement on the fault progressively puts synrift marker horizons into contraction (Fig. 6A, B & C). This process occurs sequentially. Markers at the top of the synrift sequence regain their regional elevations and move into contraction before lower synrift markers. If contractional faulting is arrested during this process, the upper part of the synrift sequence may be in net contraction whilst the lower part remains in net extension. The net extensional and net contractional portions are separated by a synrift marker that has regained its regional elevation by contractional movement. This marker now appears unfaulted and this point on the fault is defined as the *null point*. Progressive contractional inversion of an extensional synrift sequence causes the null point to move down the synrift sequence.

Inversion ratio

It is possible to quantify the relative magnitudes of contractional and extensional movement measured in the plane of the section using the position of the null point in a synrift sequence. The inversion ratio (R_i) is defined as the ratio of contractional to extensional displacement. It may be calculated by considering the thickness of the hanging wall synrift sequence measured parallel to the fault and the position of the null point in this sequence (Fig. 7).

$$R_i = \frac{d_c}{d_h} \qquad (2)$$

where d_h is the thickness of synrift sequence

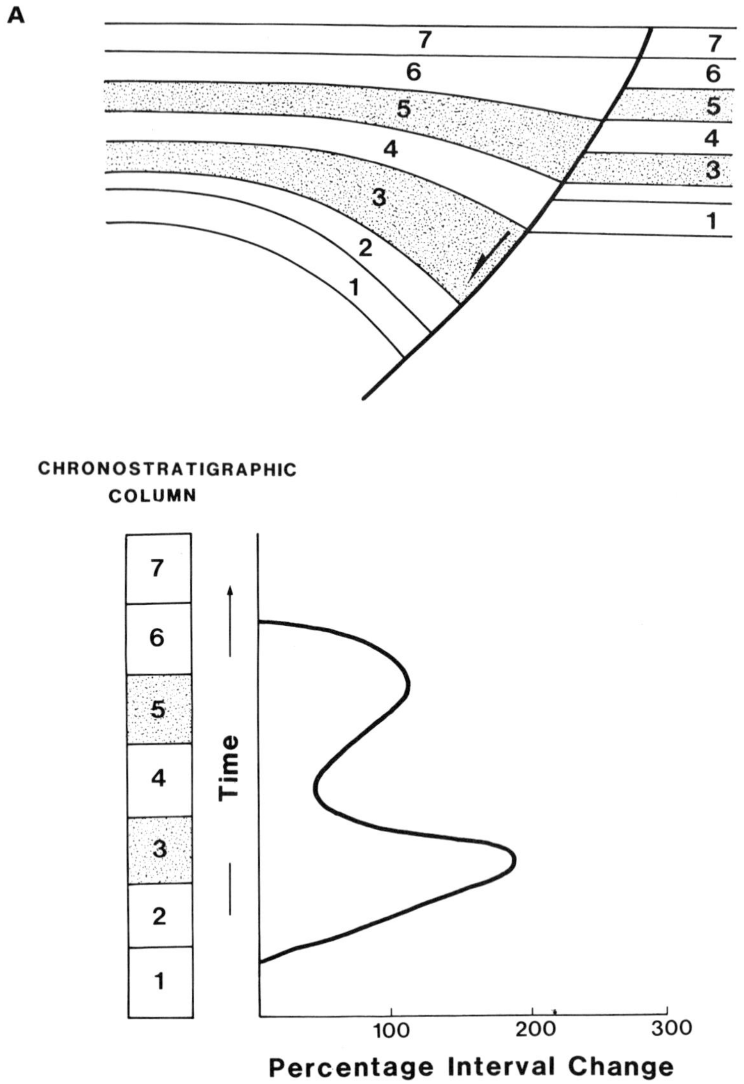

FIG. 4. 'Percentage interval change' plots; A, with no break-up unconformity; B, with break-up unconformity.

parallel to the fault and d_c is the thickness of synrift sequence in contraction (above the null point). We may rewrite this as:

$$R_i = 1 - \frac{d_e}{d_h} \quad (3)$$

where d_e is the thickness of synrift sequence parallel to the fault below the null point.

From equations (2) and (3), it is seen that if the null point lies at the top of the synrift sequence, $R_i = 0$ ($d_c = 0$ and $d_e = d_h$). That is, no contractional inversion has occurred. If the null point is at the base of the synrift sequence, $R_i = 1$ ($d_c = d_h$ and $d_e = 0$). Total inversion of the synrift sequence has occurred and all markers in the prerift sequence have regained their pre-deformational regional elevations.

Displacement/distance

The displacement/distance plot (Fig. 5) may be further modified to graphically display inversion structures. Using the top of the synrift sequence as a reference datum, distance to

B

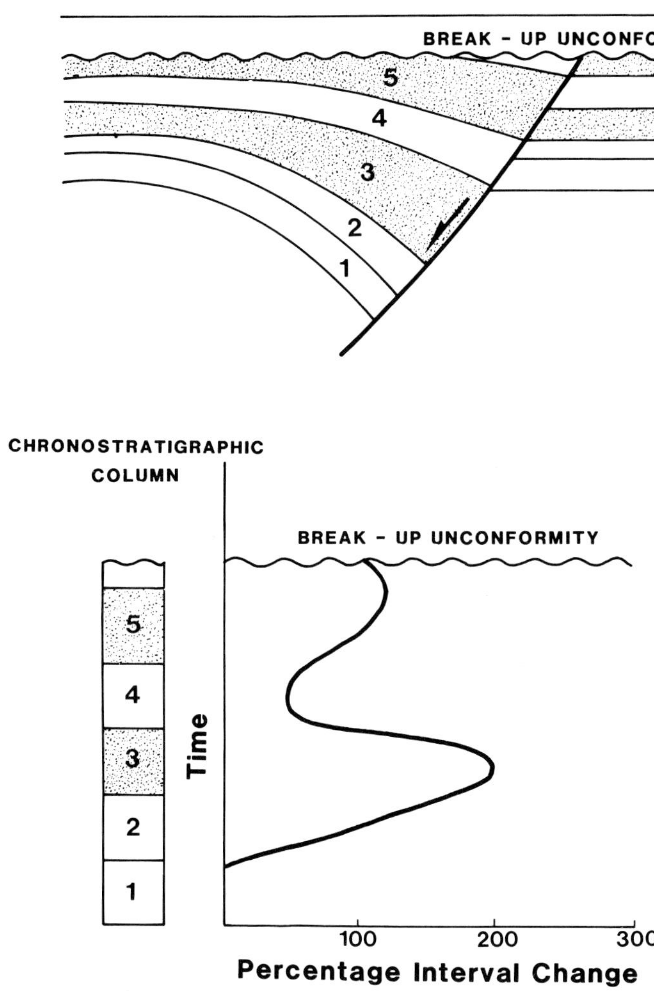

marker horizons is measured parallel to the fault in its hanging wall. Contractional displacement of markers is plotted to the left and extensional displacement is plotted to the right of the ordinate (Fig. 8). A curve with a negative gradient results, and the intercept of the curve with the ordinate defines the null point. The shape of this curve is a result of extensional tectonics, and gradient changes record variations in extensional fault growth. During progressive contractional inversion, the whole curve retains its form, but moves to the left. This causes the null point to move down the ordinate and effectively down section. The parameters d_h, d_c and d_e used in equations (2) and (3) may be read directly from this diagram (Fig. 8). An example of an inverted fault from the Gippsland Basin of SE Australia (Fig. 9) shows inversion of only one fault in an array of extensional faults (Davis 1983). The apparent inversion ratio of this structure is $R_i = 0.7$. However, because this figure is a time section the appropriate velocity function must be applied to convert the inversion ratio to its true value.

Negative inversion of contractional fault systems

The extensional reactivation of originally contractional systems appears to be less common than positive inversion. However, the reactivation may be only partial, and only evident in certain portions of the system. This can produce

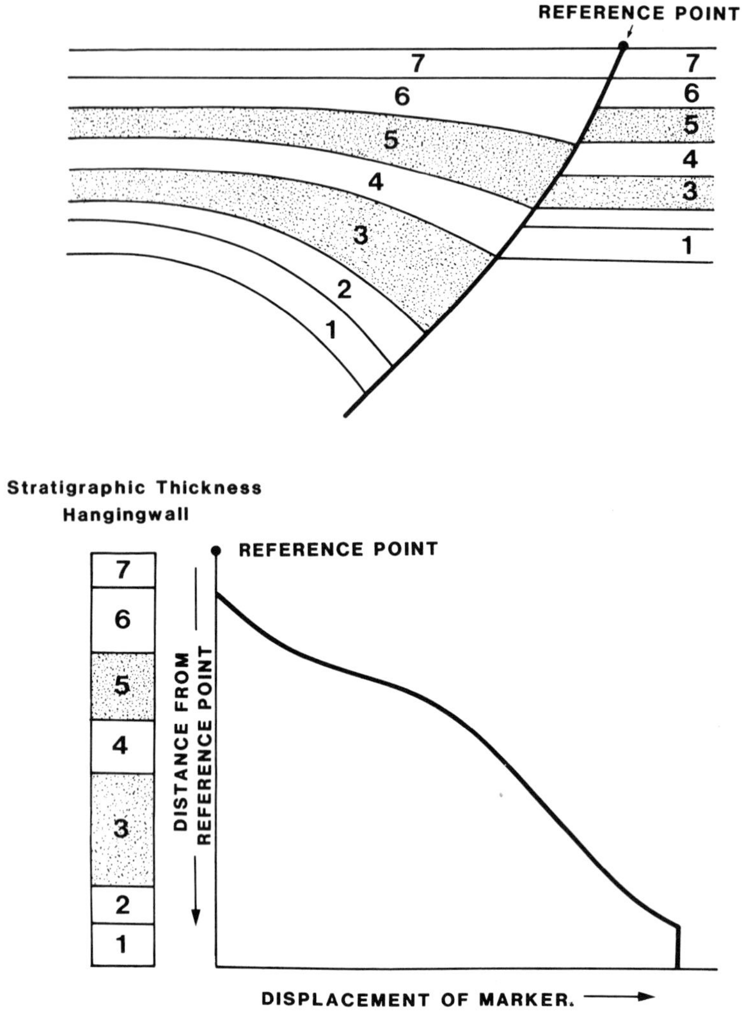

FIG. 5. Displacement/distance plot for the hanging wall of an extensional growth fault.

complex relationships between the portions of faults that are in net extension and those in net contraction. This can be conveniently analysed using the concept of null points and null zones (or neutral faults). These are fault segments which show neither contraction nor extension.

No re-use of the thrust system (Fig. 10)

In this case, an extension fault cuts through a thrust system and may detach at a deeper level. If the extension fault cuts through the thrust system without intersecting repeated stratigraphy across a thrust (which is possible though unlikely) then no null point exists. If the extension fault cross cuts thrust-repeated stratigraphy, then in the early stages of extension the footwall cutoff of the thrust which produces repetition will be the null point. The newly formed extension fault will show contraction below this point. A lower null point which marks the return to extension will be located at the footwall cutoff of the stratigraphically lowest thrusted bed. Both null points will disappear when the hanging wall cutoff of the stratigraphically highest thrusted bed coincides with the upper null point. At this point the extensional fault will be entirely in net extension with no local apparent contraction.

Partial re-use of single thrust (Fig. 11)

In this case an extensional fault has developed with a branch point to the thrust system at the

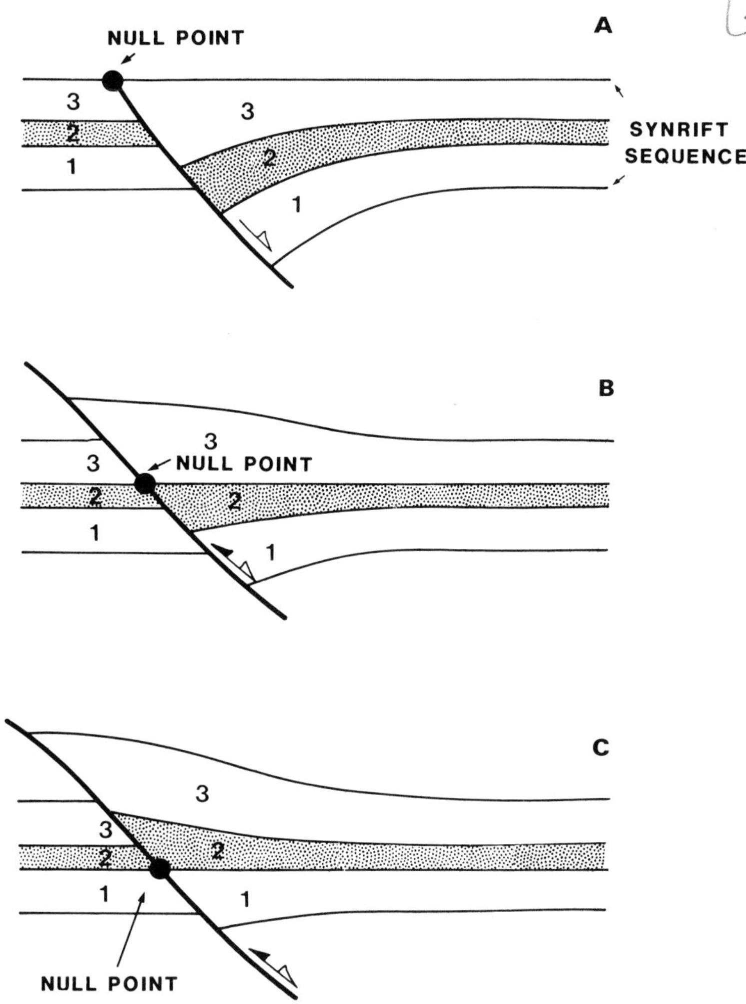

FIG. 6. Sequential diagrams to show the contractional inversion of an extensional fault. The null point moves progressively down the synrift sequence with increased contractional inversion.

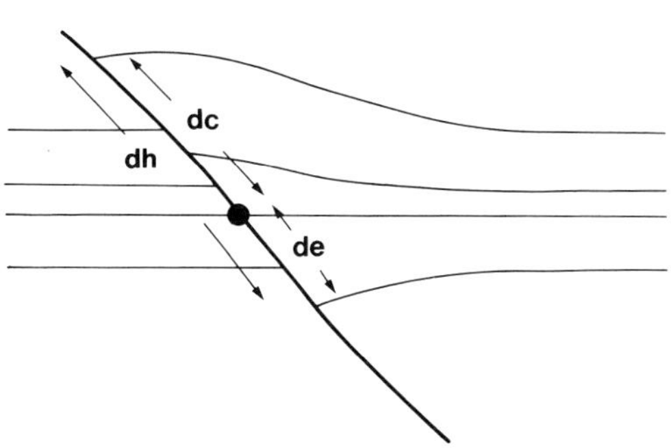

FIG. 7. Parameters used in equations (2) and (3) for the calculation of inversion ratio.

top of the thrust ramp. The null point of the system is this branch point. However, with continued extensional movement, the null point will disappear once the hanging wall cutoff of the stratigraphically highest repeated bed coincides with the branch point. At this position, the null point becomes a null zone which extends down the fault. Any additional extension causes this zone to disappear and the whole thrust ramp becomes extensional.

This does not mean that all the contractional slip has been reversed because the null point is derived from stratigraphic relationships. The extensional fault has shortcut the thrust hanging wall leaving behind a thrusted portion in its

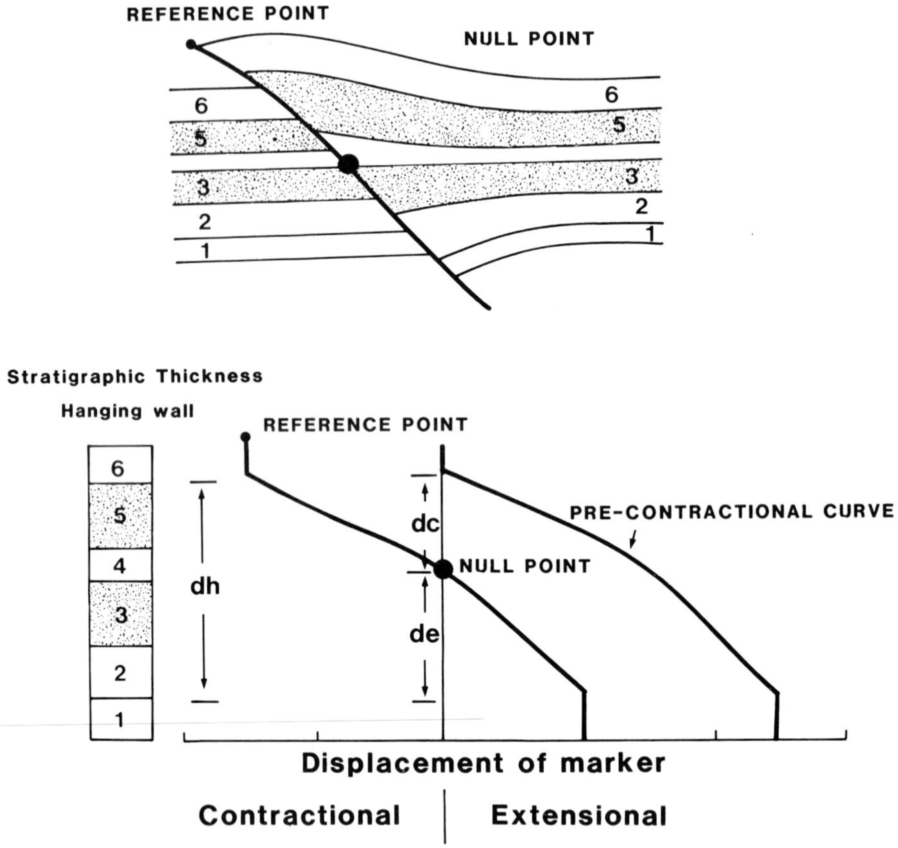

Fig. 8. Displacement/distance plot for a contractionally inverted extensional fault. Curve on right represents the situation prior to inversion. Parameters for equations (2) and (3) are shown.

footwall. If the erosion level was below that of the upper thrust plane then it would be impossible to detect the early thrust phase. The thrust/extension fault plane is a neutral fault showing neither contraction nor extension.

Partial reactivation of a complex thrust system (Fig. 12)

This thrust system comprises a major thrust (thrust 2) with a hanging wall imbricate (thrust 3). Thrust 1 has developed in the footwall of thrust 2 with a branch point at the base of a footwall ramp on thrust 2. Fig. 12 shows two stages in the progressive negative inversion of the thrust system. Thrust 1 is in net extension where it is a ramp, but turns into a neutral fault (a null zone) where it is parallel to bedding. When extensional reactivation has removed all the initial contraction of thrust 1, so that the hanging wall and footwall cutoffs re-match, then a null zone will exist for the whole length of thrust 1. Thrust 2 shows no extensional reactivation and thus has no null point. It remains in net contraction.

After the first increment of extension thrust 3 is still in net contraction until the null point is reached at the base of the extensional synrift package. With greater extension the null point migrates down the ramp until the hanging wall and footwall cutoffs of the repeated section re-match. At this point a null zone is created from the base of the synrift sequence to the next hanging wall ramp cutoff (Fig. 12B). This null zone will remain fixed until the cutoffs of the deeper thrust ramp are re-matched. At this stage, the contractional displacement on thrust 2, which was shortcut by extensional reactivation, has been completely removed by extension. With any additional extension, the whole

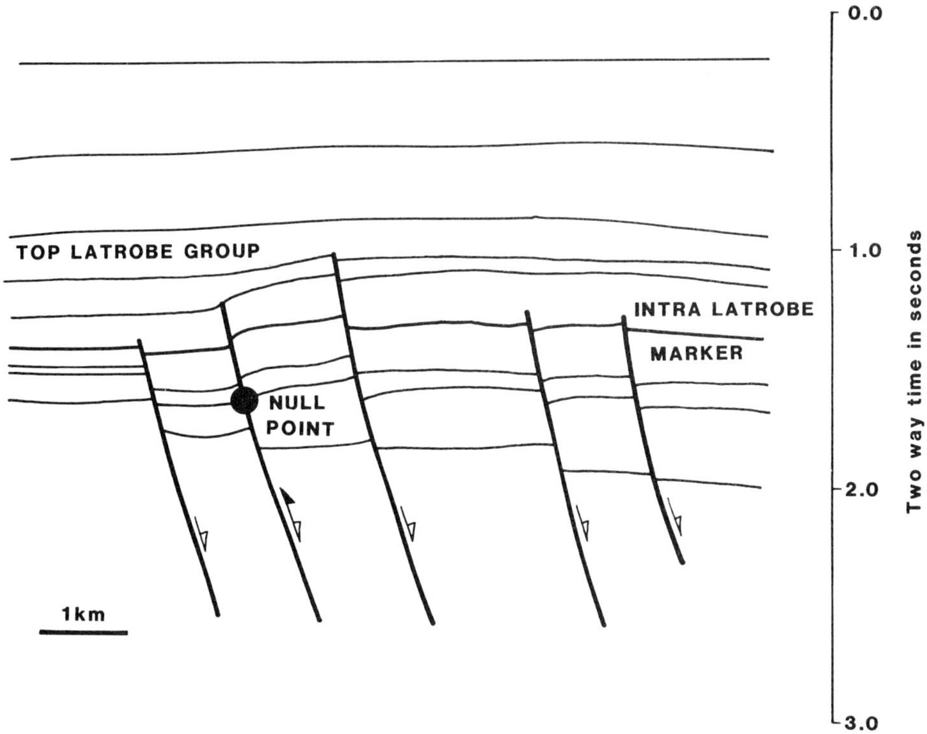

FIG. 9. Line drawing of an interpreted seismic section across the Snapper Field, Gippsland Basin, SE Australia (after Davis 1983).

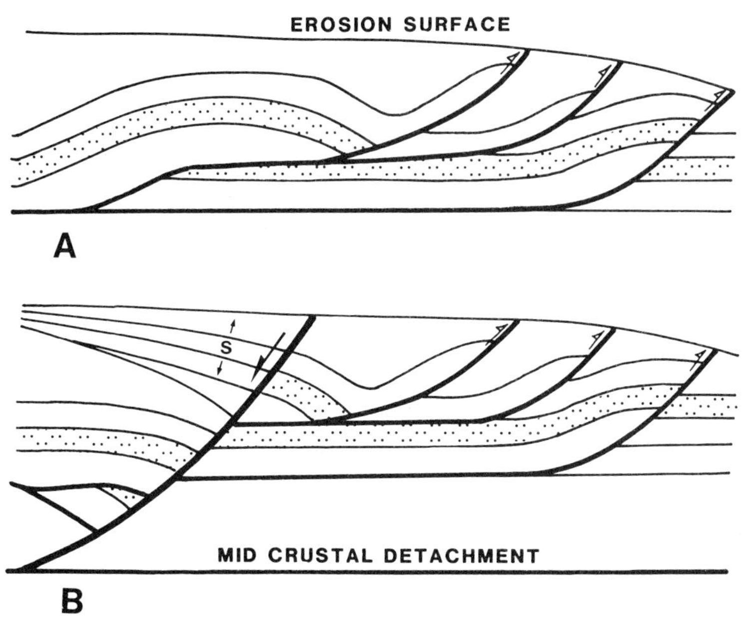

FIG. 10. A, Eroded linked thrust system. B, Extensional fault cross cutting the pre-existing thrust system with no re-use of thrust faults. S represents a synrift extensional sequence.

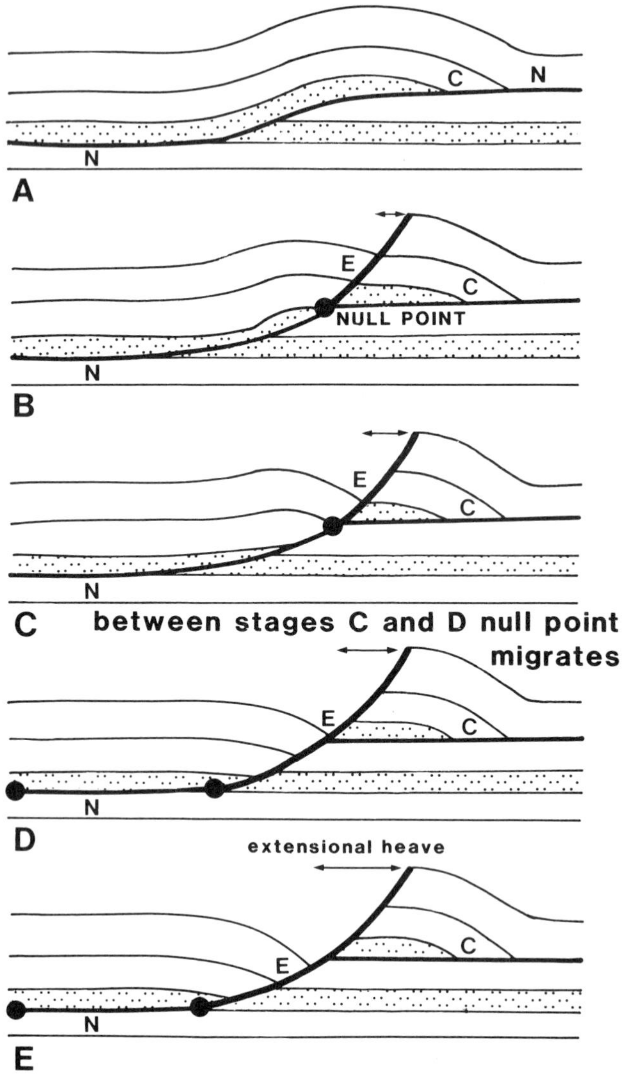

FIG. 11. Sequential diagrams to show the shortcutting and partial reactivation of a single thrust fault by extensional movements. Faults are labelled C, contractional; N, neutral; and E; extensional (represented by thicker lines).

system would be represented by extensional and neutral faults with only the shortcut remnant of thrust 2 showing net contraction.

Recognizing inversion on geological maps

The recognition of inversion tectonics and the quantification of inversion ratio may be performed using geological maps where structures have some degree of plunge. A good stratigraphic control is essential. A region that has undergone positive inversion of a synrift (extensional) stratigraphy will contain faults that apparently change from normal to reverse across the map. In the simplest case, an extensional fault displaces younger material in its hanging wall above older rocks in its footwall. A contractional fault displaces older material on younger rocks. Considerations of net slip are unimportant. A simplified geological map of the Weymouth District (BGS 19) illustrates the

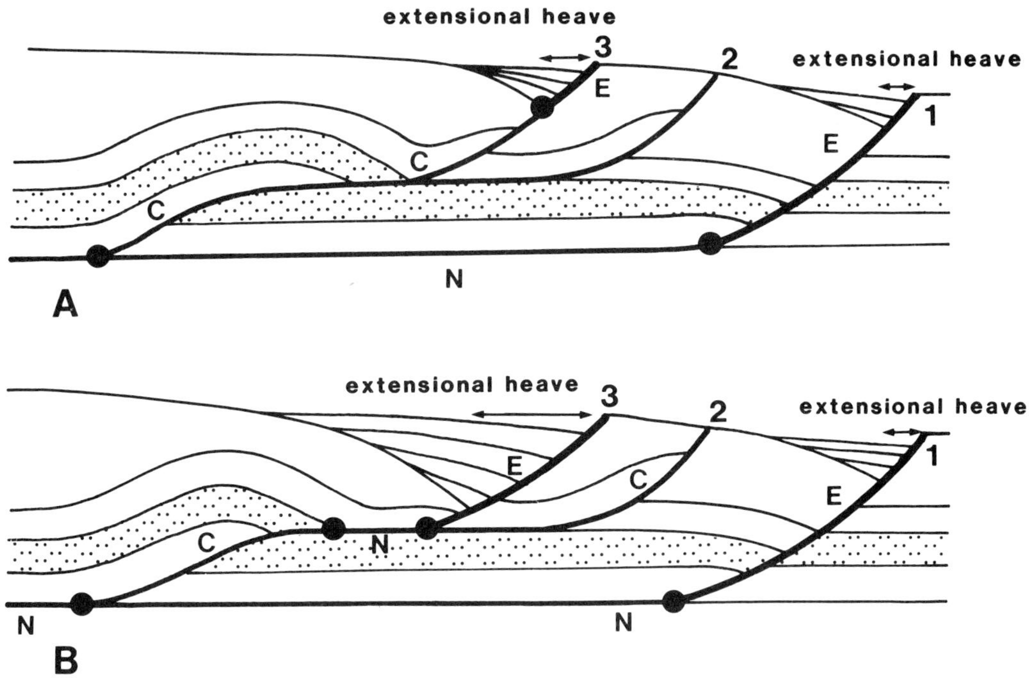

FIG. 12. Thrust system as in Fig. 10, partially reactivated in extension. A and B show two stages in reactivation. Faults are labelled C, contractional; N, neutral, and E, extensional (represented by thicker lines).

concept of changing mode along a single fault (Fig. 13). This effect results from positive inversion tectonics. In this region, the Jurassic and Lower Cretaceous sequences are considered to be synrift. The Upper Cretaceous is a postrift sequence deposited above a significant unconformity (Stoneley 1982). The point of change along the fault from net normal to net reverse displacement is the null point. In this case it occurs at the Upper Cretaceous unconformity.

The geometry of inversion may be analysed and the inversion ratio may be calculated using stratigraphic separation diagrams (Elliott & Johnson 1980) for individual faults. Stratigraphic separation is the displacement of a stratigraphic marker by a fault. Displacement is measured perpendicular to bedding. Stratigraphic separation may be measured in map view along a fault at positions of known stratigraphy. Curves of hanging wall and footwall stratigraphy cut by the fault may be plotted against fault map length. In normal faults, the hanging wall curve plots above that of the footwall. In reverse faults the footwall curve plots above that of the hanging wall. Where footwall and hanging wall stratigraphy differs, both stratigraphic sequences may be incorporated in the stratigraphic separation diagram by plotting individual curves using the appropriate stratigraphy (Fig. 10).

Where faults have undergone positive inversion, the curves cross at the null point. Using the hanging wall synrift stratigraphy, parameters d_h, d_e and d_c may be measured for use in equations (2) and (3) for the calculation of inversion ratio. However, this is not possible in the Weymouth example (Fig. 10) because of erosion and partial omission of Jurassic and Lower Cretaceous stratigraphy in the fault hanging walls. A phase of contractional inversion coupled with erosion took place before the deposition of the Upper Cretaceous/Tertiary section.

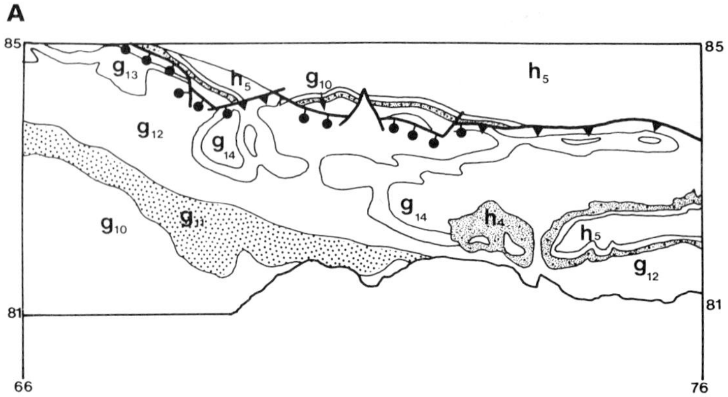

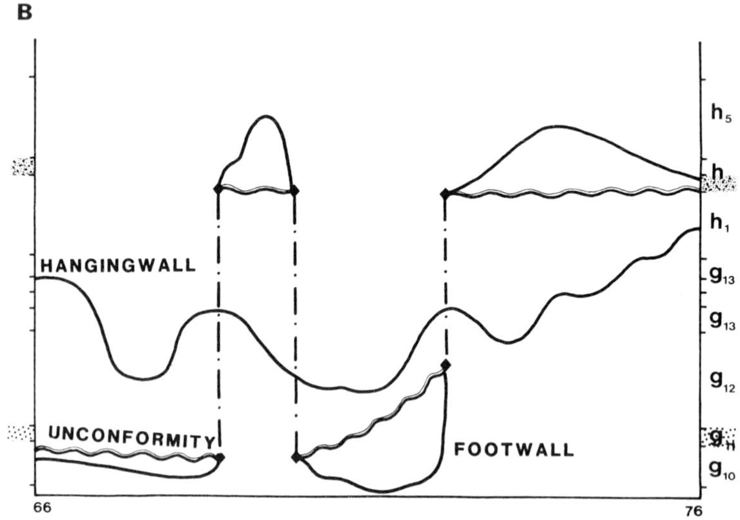

Fig. 13. A, Simplified geological map of the area N of Weymouth (from BGS sheet Weymouth) to show switching in fault mode from normal to reverse as a result of positive inversion. ▼▼, reverse fault; ●●, normal faults. B, Stratigraphic separation diagram constructed from Weymouth sheet. Note that this diagram is plotted on a full stratigraphic frame. Certain parts of the Jurassic/Lower Cretaceous stratigraphy is missing below the Albian (h4) unconformity as a result of early contractional inversion coupled with erosion. The missing stratigraphy is represented by dashed lines in the foot wall curve.

References and further reading

Beach, A. 1984. Structural evolution of the Witch Ground Graben. *Journal of the Geological Society of London*, **141**, 621–628.

Davis, P. N. 1983. Gippsland Basin, SE Australia. *In*: Bally, A. W. (ed.) *Seismic Expression of Structural Styles. American Association of Petrology Geologists Special Publication*, **3**, 3.3.19–3.3.24.

Elliott, D. & Johnson, M. R. W. 1980. Structural evolution in the northern part of the Moine Thrust Belt, Northwest Scotland. *Transactions of the Royal Society of Edinburgh: Earth Sciences*, **71**, 69–96.

Gibbs, A. D. 1984. Structural evolution of extensional basin margins. *Journal of the Geological Society of London*, **141**, 609–619.

Glennie, K. W. & Boegner, P. L. E. 1981. Sole Pit inversion tectonics. *In*: Illing, L. V. & Hobson, G. D. (eds) *Petroleum Geology of the Continental Shelf of NW Europe*. Institute of Petroleum, London, pp. 110–120.

Hamblin, W. K. 1965. Origin of "reverse drag" on

the downthrown side of normal faults. *Geological Society of America Bulletin*, **76**, 1145–1164.
STONELEY, R. 1982. The structural development of the Wessex Basin. *Journal of the Geological Society of London*, **139**, 545–554.
WILLIAMS, G. D. and CHAPMAN, T. J. 1983. Strains developed in the hanging walls of thrusts due to their slip/propagation rate: a dislocation mode. *Journal of Structural Geology*, **5**, 563–571.

G. D. WILLIAMS & C. M. POWELL, Department of Geology, University of Keele, Keele, Staffordshire, ST5 5BG, UK.
M. A. COOPER, BP Canada Inc., Oil and Gas Division, 333 Fifth Avenue SW, Calgary, Alberta T2P 3B6, Canada.

Some geometrical characteristics of inversion

A. B. Hayward & R. H. Graham

SUMMARY: The inversion of extensional fault systems results in the reversal of slip on the faults and expulsion of the synrift fill. During inversion the beds in the cover sequence shorten before the net extension at the basement level has been cancelled. Shortening of the sedimentary cover generates folding and backthrusting in the still downthrown hanging wall block. Intracratonic inverted basins in different parts of the Alpine Foreland show similar structural geometries with the major extensional faults which controlled basin development reactivated during subsequent compression. We use examples from the Western Approaches and offshore Holland (Broad Fourteens Basin) to illustrate the structural styles developed during inversion. The fundamental control on compressional structural geometry exerted by pre-existing extensional structures is also visible in more complexly deformed orogenic belts, like the Western Alps and the Pyrenees. In these areas inversion also occurs, but more commonly extensional faults which may not have inverted act as an indirect control on the location of ramps, and/or thrust orientation. Seismic data are normally required to establish these effects with certainty. However, as the body of knowledge builds up, it is possible to recognize certain geometrical characteristics which suggest the control of extensional faults in thrust belts. These include footwall shortcuts, out of sequence structures and arcuate thrust–fold traces.

Introduction

The Inversion Tectonics meeting at the Royal Society, London on 3–4 March 1987, provoked lively discussion about the meaning and use of the term inversion. Should it be restricted to 'intracratonic' deformation in the otherwise undeformed foreland regions of orogenic belts, or should it be extended to include the restacking of highly stretched continental margins within the orogenic zones themselves? In this paper we present a series of examples which we believe demonstrate the progressive intensity of inversion ranging from the mild deformational structures of the Western Approaches Basin through to more intense deformation seen in the areas in offshore Holland and the structural styles associated with inversion in the Alpine and Pyrenean areas. In our discussion of inversion within orogenic areas we also review some of the geometrical phenomena which seem to be commonly associated with the presence of pre-existing extensional faults in thrust belts. These may eventually prove to be empirical criteria by which we may come to recognize the presence of extensional faults and half-graben expulsion without seismic, well or complete stratigraphic data.

Theoretical considerations and general principles

In recent years the nature and geometry of extensional fault systems have become increasingly well understood, due both to the much improved resolution of industrial seismic data, and the shooting of a number of deep crustal seismic lines. Most extensional basins are seen to be linkages of asymmetrical half-grabens. The impression of overall full graben symmetry in rift systems is created either by antithetic faults linking in to a controlling fault, or by periodic changes of dip of the controlling faults themselves. This is achieved either by extension parallel transfer faults, (Bally 1981; Gibbs 1983) or by more complex transfer zones (Rosendahl *et al.* 1986).

The 'positive' inversion of a previously formed sedimentary basin has been modelled by Bally (1984) using a simple half-graben geometry which undergoes slip reversal on the main boundary fault. The slip reversal produces volumetric problems in the synrift sediments as they become elevated. The problems are overcome by expelling the synrift sequence both forwards out of the graben towards the

From Cooper, M. A. & Williams, G. D. (eds), 1989, *Inversion Tectonics*
Geological Society Special Publications No. 44, 17–39.

footwall and backwards on to the old rollover (Figs 1 and 2). In some examples the forward expulsion produces a ramp anticline which is underlain by a flat detachment lying above the footwall (Fig. 1). Backwards expulsion may occur either on a new set of backthrusts (Fig. 2) or on older antithetic extensional faults which become reactivated as reverse faults.

During inversion the synrift sequence is shortened above the prerift surface (Fig. 2). Before the net extension at prerift level has been cancelled shortening of the sedimentary cover may generate backthrusting and folding in the still downthrown hanging wall block. In the case of mild inversion (e.g. the Western Approaches Basin) where there is simple dip reversal of a major extensional fault the contents of the half-graben are extruded as a major fold. In some cases of more severe inversion (e.g. the Broad Fourteens Basin) backthrusts are generated.

In other areas, for example the half-grabens in the Pelvoux Massif of the Western French Alps, there has been rotation of the bounding faults and intense internal deformation of the contents of the half-grabens (Gillcrist *et al. in press*; Gillcrist, Coward *et al. this volume*). Fig. 3 shows possible geometries for half-graben extrusion and deformation. In all of them the basement in either the hanging wall or footwall of the major fault may act as a butress against thrust translation so that a backward propagating deformation sequence may develop (Fig. 3A), an out of sequence thrust may cut the inversion (Fig. 3B) or a shortcut thrust might cut through the extensional fault footwall. Footwall shortcut thrusts are discussed later.

In the following section we use two examples of both mild and moderate inversion in an intracratonic setting to demonstrate some of the geometrical characteristics that we have outlined above.

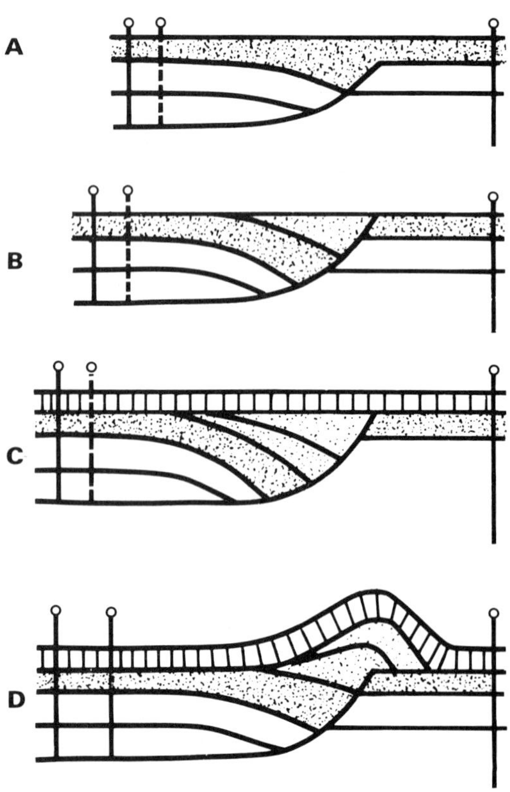

FIG. 1. Graphical construction of a simple listric growth fault subsequently inverted. A, B and C show progressive extension and deposition of a synrift fill (shaded); D shows the subsequent inversion geometry.

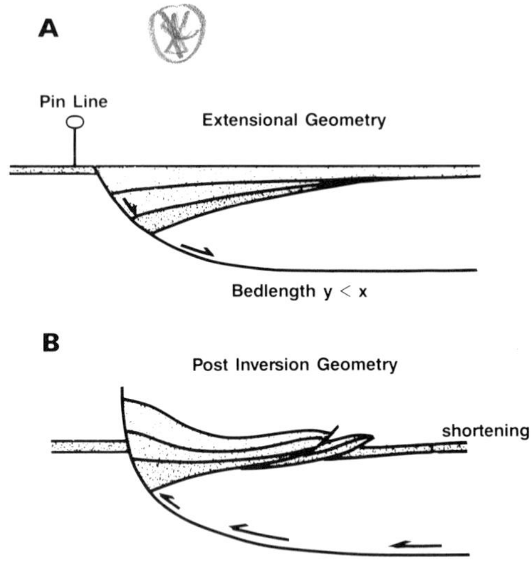

FIG. 2. A, The excessive bed length in the cover sequence to an asymmetric half-graben and B, the accommodation structures that develop during inversion of the half-graben (in part after Bally 1984).

The Western Approaches Basin: an example of mild inversion

The Western Approaches Basin (Fig. 4) forms a ENE–WSW trending Mesozoic extensional basin of Early Jurassic to Late Cretaceous age. It was inverted during the late Oligocene–Early Miocene.

The basin can be divided into three sub-basins

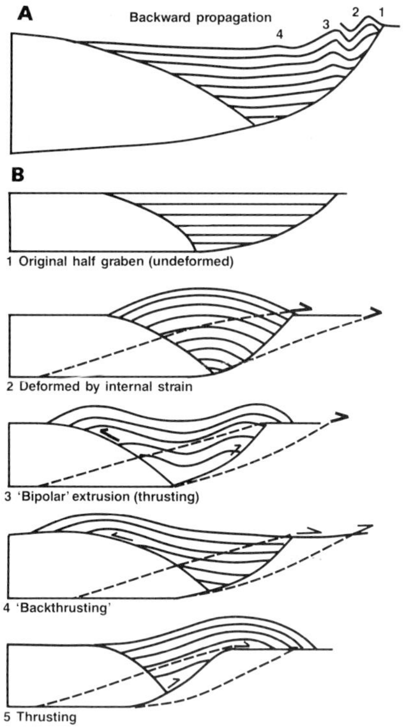

FIG. 3. A, The development of a backward propogating thrust sequence produced by the buttress effect of the footwall fault block during inversion. B, Theoretical models for the extrusion of half-graben synrift fills. Heavy line indicates potential for late out of sequence thrusting.

(Fig. 4) which comprise three inverted half-grabens. In the SW sub-basin a major reverse fault (Fig. 5) along the southern margin of the basin marks the boundary. The NW margin is defined by a number of minor faults (Fig. 5).

The geometry of the sedimentary fill of the half-graben indicates that during the development of the basin different faults were dominant at different times. The Jurassic sequence shows marked thickening to the north into what was clearly an extensional fault of Jurassic age. However, the Lower and Upper Cretaceous section thickens to the S into a major listric fault which controlled the development of the basin throughout the Cretaceous. Restoration of the cross-section in Fig. 5 demonstrates that throughout the Cretaceous the S sub-basin formed an asymmetric half-graben approximately 50 km wide controlled by a major listric extensional fault in the S. The N margin is defined by an extensional fault antithetic to the major basin controlling fault.

During Late Oligocene/Early Miocene inversion both faults were reactivated as reverse faults. The reversal of movement on the major basement faults resulted in uplift, the formation of a series of broad anticlinal culminations in the hanging wall and mild basin-wide inversion. Minor antithetic and synthetic faults were also reactivated.

In contrast to the S sub-basin, the present day geometry of the central sub-basin is controlled by a major reverse fault along the N boundary of the basin (Fig. 6). The Mesozoic stratigraphy

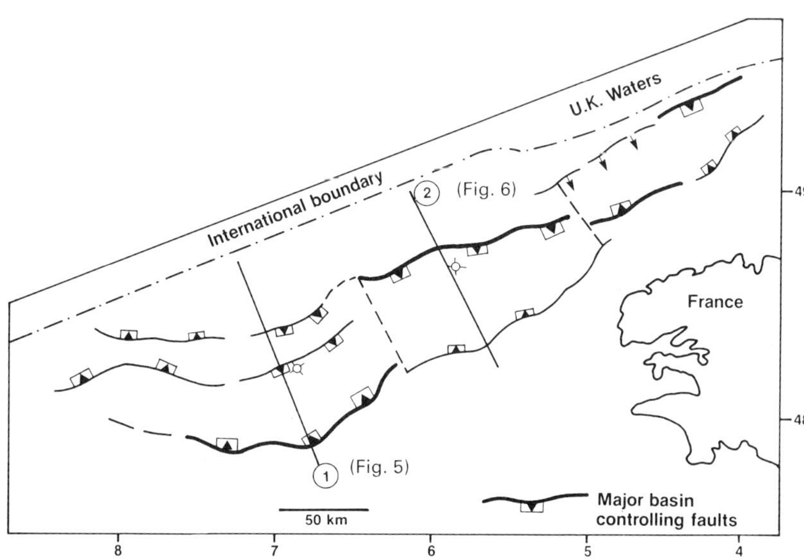

FIG. 4. Structural outline map of the Western Approaches Basin.

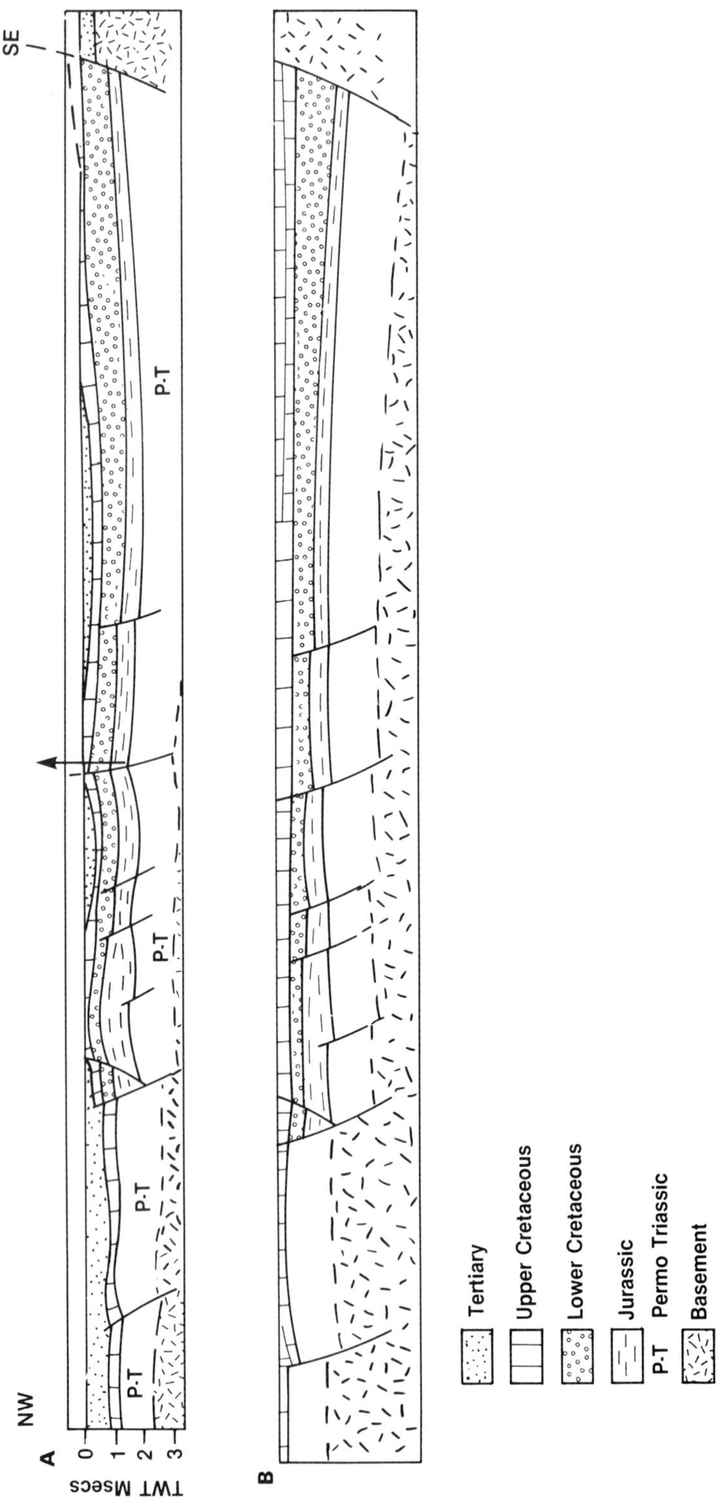

FIG. 5. A, Present day cross-section 1 across the southern sub-basin of the Western Approaches, based on a regional seismic line. Note vertical scale in two way time, $V = H$. B, The same section schematically restored to top Upper Cretaceous.

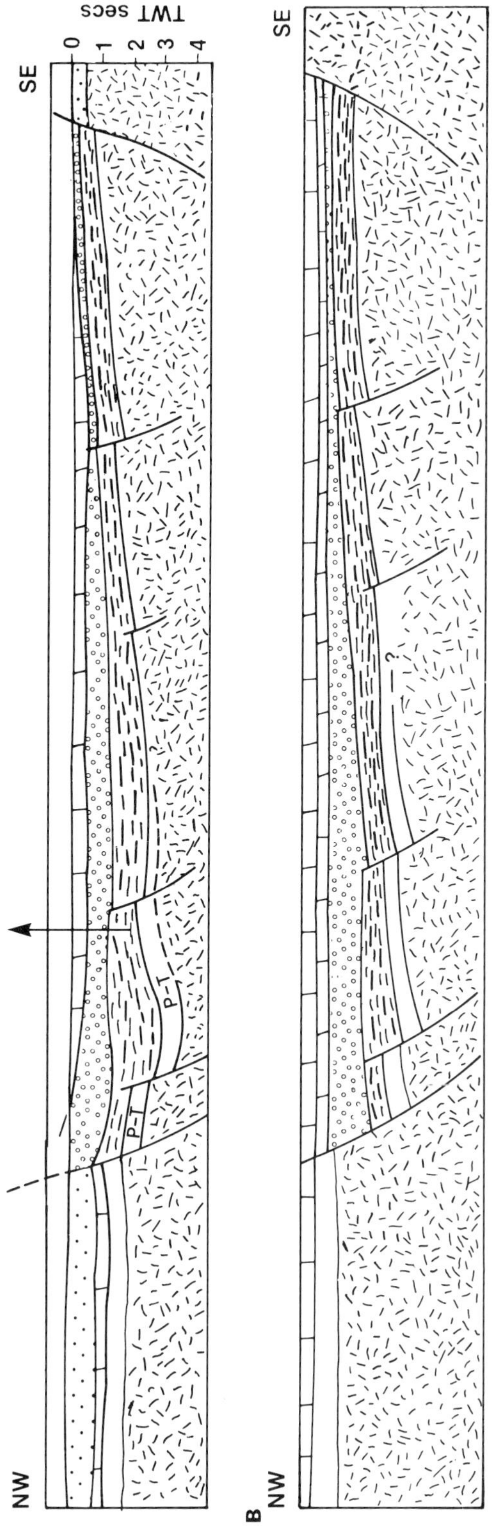

FIG. 6. A, Present day cross-section 2 across the central sub-basin of the Western Approaches, based on a regional seismic line. Note vertical scale in two way time, $V = H$. B, The same section schematically restored to top Upper Cretaceous.

shows a consistent thickening to the north into what was previously a major listric extensional fault. Restoration of the cross-section to the pre-inversion geometry demonstrates that the central sub-basin formed a half-graben approximately 50 km wide controlled by a major listric extensional fault in the N. The S margin of the basin was marked by minor antithetic faults (Fig. 6). During the Late Oligocene/Early Miocene inversion the major extensional fault was reactivated as a reverse fault. Most of the deformation was taken up on the major basement fault producing mild basin-wide inversion.

The N sub-basin also shows a change in asymmetry; the major extensional fault and subsequent reverse faults lie along the S margin of the basin. A change of asymmetry or polarity along strike is typical of many extensional basins (Bally 1981). In this example it clearly influences the subsequent inversion geometry.

The Broad Fourteens Basin: an example of moderate inversion

The Broad Fourteens Basin located offshore of Holland (Fig. 7) forms a NW–SE trending Mesozoic extensional basin of Triassic to Late Cretaceous age. It underwent moderate inversion during the Campanian (Late Cretaceous) (Oele *et al.* 1981).

The present day sub-Tertiary structure (Fig. 8) is controlled by a major bounding fault along the W margin of the basin. Across this fault the Jurassic sediments of the basin are juxtaposed with Cretaceous sediments of the basin margin. In contrast, along the E margin of the basin there is no major basin boundary fault and the Jurassic–Cretaceous sedimentary sequence thins progressively onto the Texel–Ijsselmeer High. Restoration of the cross-section (Fig. 8) demonstrates that the Broad Fourteens Basin formed a large asymmetric half-graben approximately 60 km wide throughout the Jurassic and Cretaceous. It was controlled by a major listric extensional fault along the W margin of the basin. The Jurassic–Cretaceous sedimentary sequence shows a dramatic thickening into the fault and thins onto the hinge area along the E margin of the basin. Minor Jurassic–Cretaceous extensional faults in the hinge area of the basin represent antithetic faults to the major listric fault.

During the Late Cretaceous inversion, the major listric extensional fault and minor antithetic faults were reactivated as thrusts or reverse faults. As in the Western Approaches Basin most of the deformation was taken up on the major basement fault, resulting in the basin-wide inversion of the Broad Fourteens Basin. In contrast to the Western Approaches Basin where there is very little shortening, restored cross-sections across the Broad Fourteens Basin show a total shortening of around ten per cent. The pre-inversion rollover geometry of the basin (Fig. 8) results in the cover sequence of Mesozoic sediments having a greater bed length than the underlying basement surface. The additional shortening required in the cover sequence is accommodated by the backward expulsion of the half-graben fill on a series of reverse faults and listric thrusts. These are detached mostly in the Permian Zechstein evaporite sequence and are developed principally in the hinge area of the graben. A geoseismic section over the rollover (NE) margin of the Broad Fourteens Basin is shown in Fig. 9. Imbricate slices which are detached in Zechstein evaporites end in a triangle zone on the crest of rollover. They were formed during the Late Cretaceous inversion and expulsion of the graben fill towards the rollover shoulder. In areas where there are no Zechstein evaporites to form a suitable detachment, the additional shortening required in the cover sequence is accommodated by the development of a series of small pop-up structures.

Inversion in orogenic belts

Recent years have seen the widespread application of thin skinned structural models, first developed by petroleum geologists working in the E foothills of the Canadian Rockies (e.g. Bally *et al.*, 1966; Dahlstrom 1969), to broad areas of the Alpine fold belt (e.g. Boyer & Elliot 1982; Butler 1983; Butler *et al.* 1986). The application of this type of model has met with varying degrees of success. In the Rocky Mountain foothills the stratigraphy and basement surface is essentially uniform and there are no pre-thrusting extensional faults. Deposition was controlled by the progressive downward flexure of the former passive margin. On the other hand, much of the Alpine area underwent Mesozoic fault-controlled extension (e.g. Trümpy 1980; Lemoine *et al.* 1986). The control exerted by early extensional faults on the compressional geometry during later orogenic deformation is now being recognized from a large number of mountain belts (Jackson 1980; Dewey 1982; Zeigler 1983; Harding 1985; Bally 1984; Etheridge 1986 and Lemoine *et al.* 1986). In the next section we show that in the Western

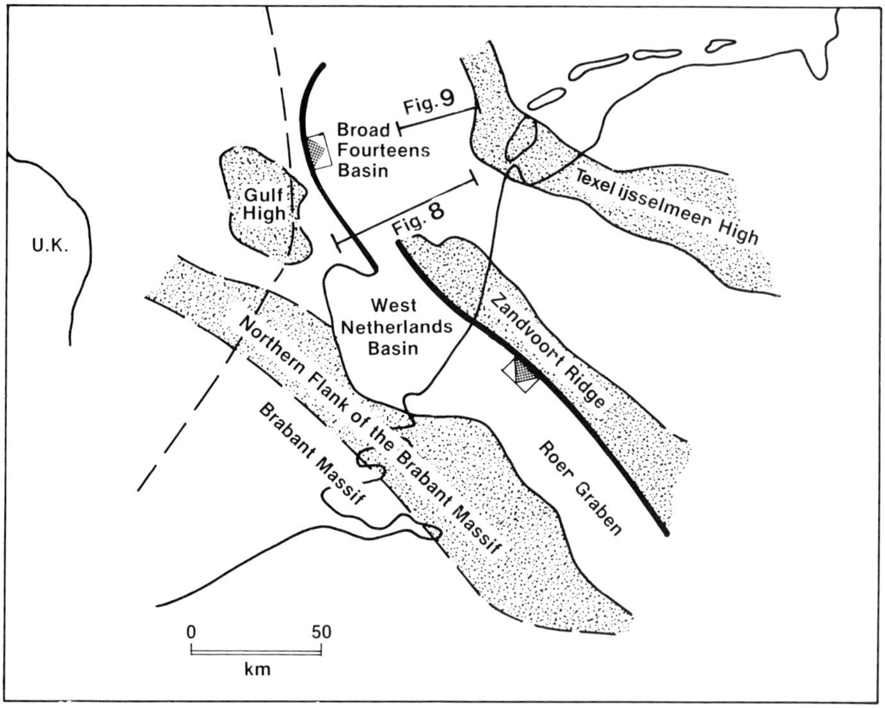

FIG. 7. Structural outline map of offshore Holland.

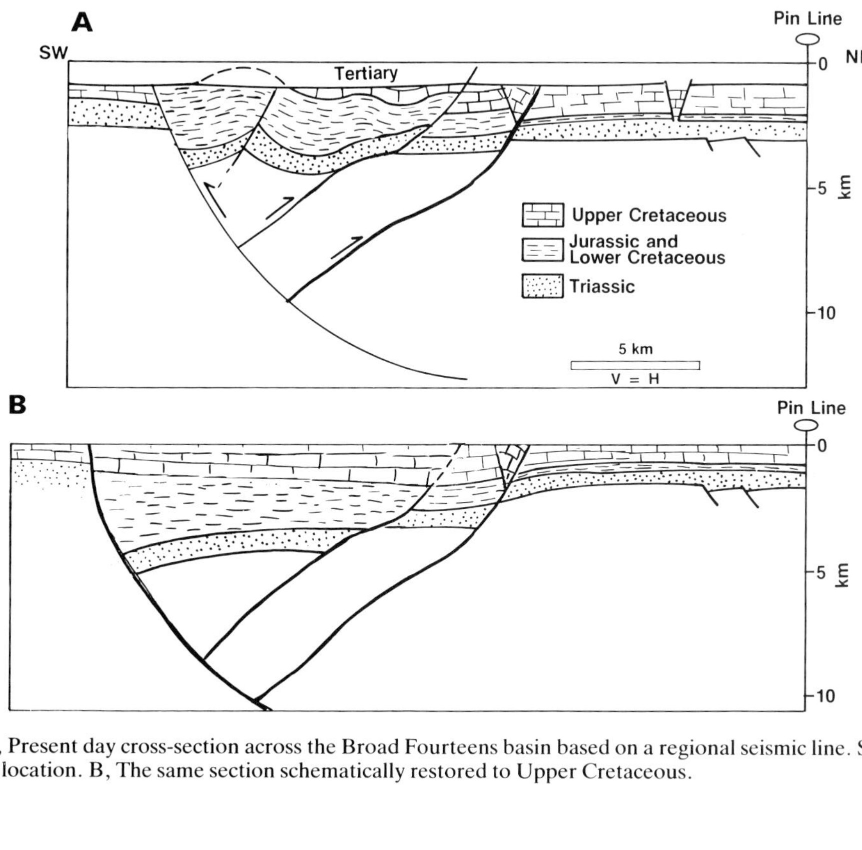

FIG. 8. A, Present day cross-section across the Broad Fourteens basin based on a regional seismic line. See Fig. 7 for location. B, The same section schematically restored to Upper Cretaceous.

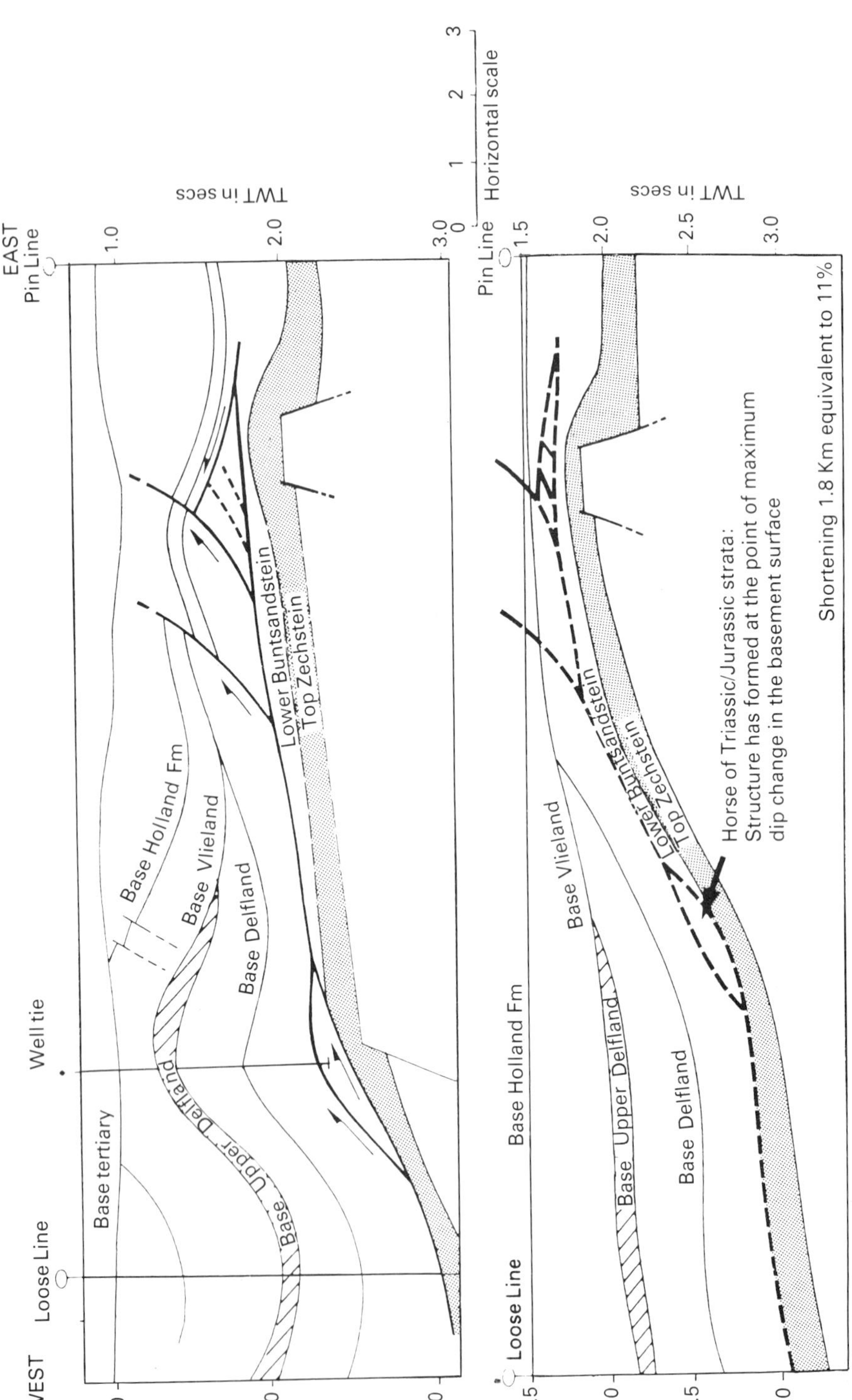

FIG. 9. A, Geoseismic section across the rollover margin of the Broad Fourteens Basin. Vertical scale in two way time, $V = H$. B, section restored in time at Base Chalk. For location see Fig. 7.

Alps and Pyrenees the influence of pre-existing extensional structures in these areas goes beyond the simple inversion seen in the Alpine foreland.

Western Alps

It has long been recognized that the stratigraphic distinctiveness of individual thrust sheets in the Alpine system (the tectono-stratigraphic units of classic Alpine stratigraphy) is related to pre-tectonic palaeogeography. Many authors (Trümpy 1980, Trümpy & Lemoine 1987, Lemoine *et al.* 1986, de Graciansky *et al. this volume* and others) have all demonstrated that this palaeogeography is a reflection of Mesozoic extensional geometry associated with the stretching events which preceeded the development of the Ligurian and Tethyan oceans.

Is this to say that every major thrust sheet is an extruded half-graben and every major thrust an inverted extensional fault? Trumpy and Lemoine (1987) have argued this case for the Helvetic Nappes, though the argument is refuted by Butler, *this volume*.

We briefly outline two Alpine examples ot the influence extensional faults on later contractional structures. Both show geometrical relationships that contrast with the inversion described in early sections. The first example is the section across the Belldonne and associated external crystalline massifs (also discussed by Butler, *this volume*; Gillcrist, *et al.* 1988 *this volume* and de Graciansky *et al.*, *this volume*). The second is the Digne thrust in the external Alpine zones of Haute Provence.

Belledonne section

A section from the external zones of the Alps to the Frontal Pennine thrust (leading internal thrust) (Figs 10 and 11) crosses the shelf edge of the Tethyan margin of Europe. The Mesozoic rocks of the Jura are shelf carbonates. The rocks thrust over them in the hanging wall of the leading thrust of the sub-Alpine zone are thicker and more shaley. They are derived from the continental slope. These rocks are deformed into a major syncline W of the elevated crystalline basement which forms the Belledonne and

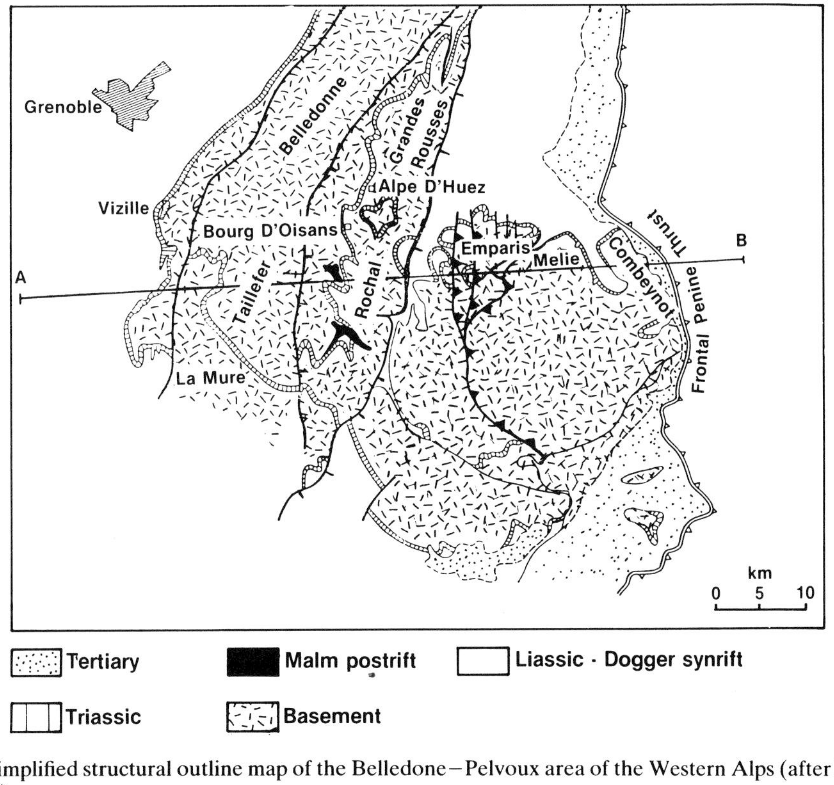

FIG. 10. Simplified structural outline map of the Belledone–Pelvoux area of the Western Alps (after Lemoine et al. 1981).

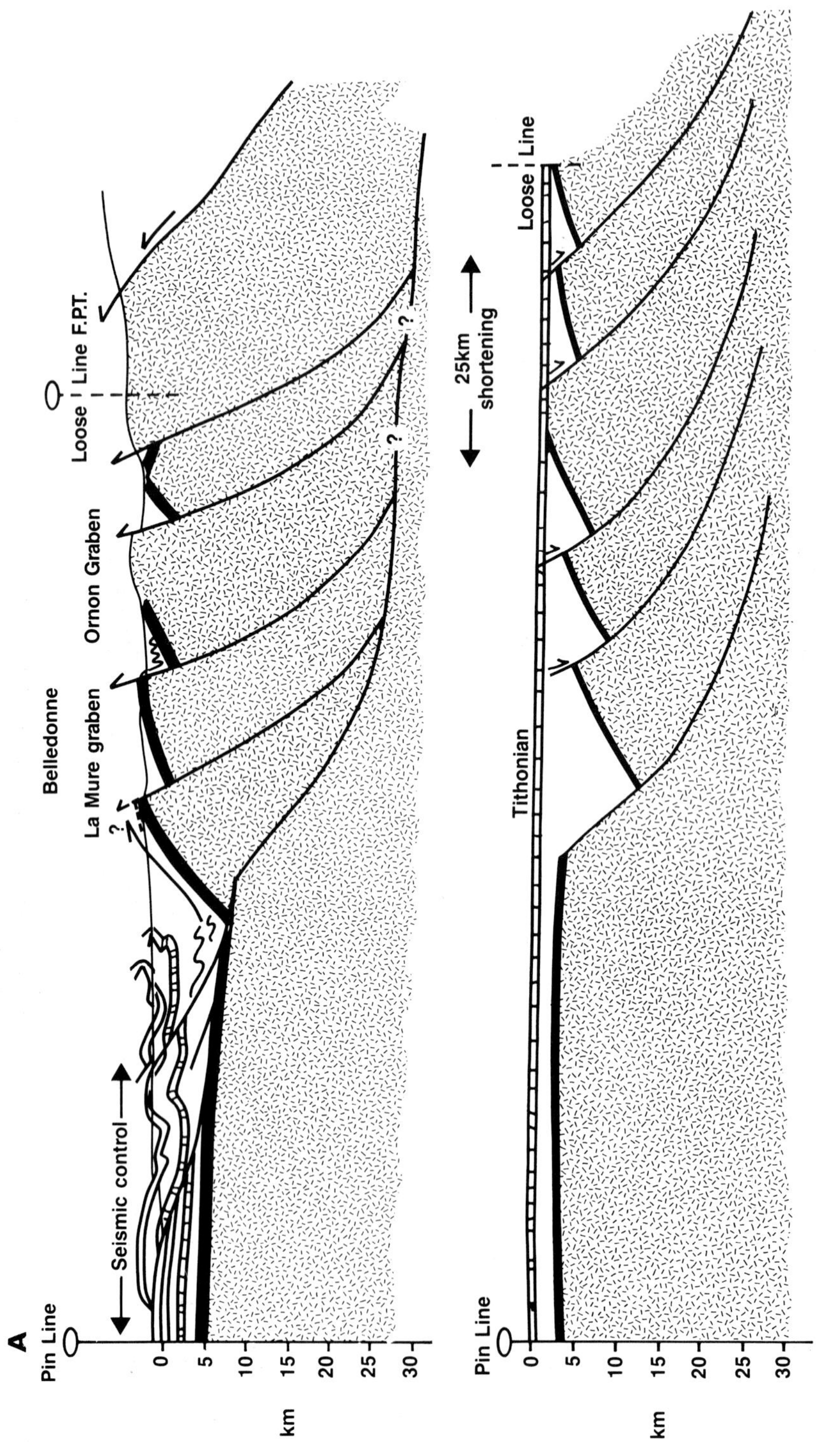
A
Pin Line
Seismic control
La Mure graben
Belledonne
Ornon Graben
Loose Line F.P.T.
km
0
5
10
15
20
25
30
Pin Line
Tithonian
25km shortening
Loose Line
km
0
5
10
15
20
25
30

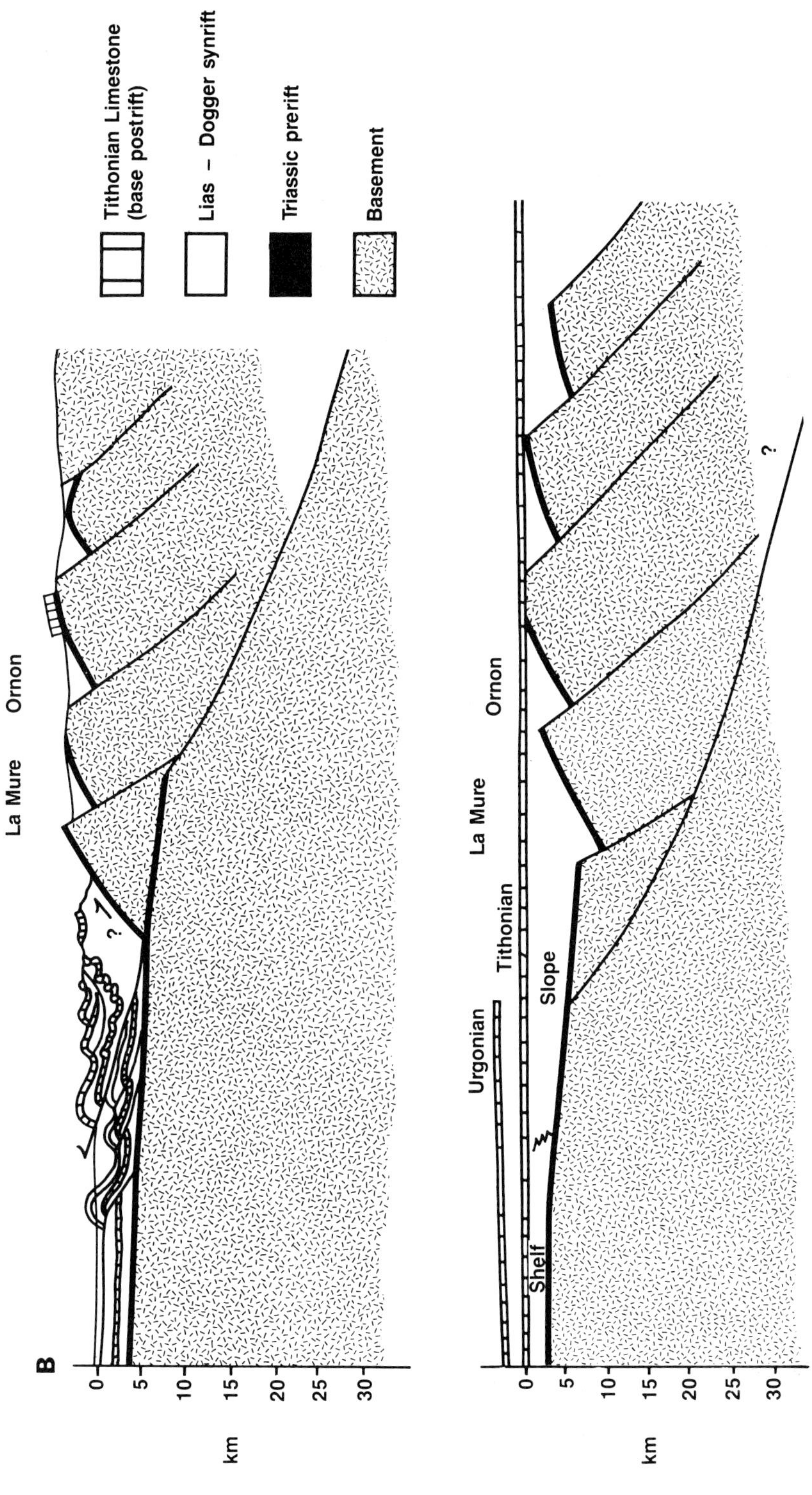

FIG. 11. A, Structural cross-section across the external zones of the Western Alps (in part after Lemoine *et al.* 1981), with restoration (see Fig. 10 for location). Based on the inversion model. B, The same cross-section based on the shortcut model.

associated 'external' massifs. The basement is elevated 12 km above its regional level. Seismic refraction studies (Menard 1976) have revealed a major low velocity zone dipping 30° or so eastwards beneath the Belledonne massif. It seems reasonable to suggest that this is the thrust which brought about the uplift of the massif.

The recent detailed studies of Lemoine, de Graciansky and their colleagues (Lemoine *et al.* 1981; 1985; de Graciansky *et al., this volume*) have convincingly demonstrated the presence of extensional fault blocks and half-grabens within these external crystalline massifs, the detailed structure of which is described by Gillcrist *et al., in press* and Gillcrist *et al., this volume*. Though intensely internally deformed and transformed into nipped-in synclines, the half-grabens are clearly recognizable. The bounding faults of the more W half-graben (La Mure, Bourg d'Oisans) may have been rotated and over steepened, but there is little evidence of actual reversal of movement on these bounding faults, i.e. no demonstrable 'inversion' in the sense in which we have used the term in this paper.

Adjacent to the old extensional faults the synrift sediments exhibit an intensification of internal deformation (folds and cleavage). The implication is that detachment at the base of the cover sequence is transformed into strain in the vicinity of the blockage represented by the old extensional faults. In addition, backthrusts are developed. These represent expulsion of the synrift sequence backwards over the next rotated fault block. Lemoine has used the term 'buffer stop' to describe the buttress effect that produces this deformation.

Only in the E part of this section, not far below the leading internal zone thrust (Frontal Penine Thrust) is there any major 'forward' thrusting of the half-graben sediments (La Grave thrust). Davies (1982) has described small scale inversion in this region, but whether the large scale La Grave structure is genuine inversion of an old extensional fault or a shortcut phenomenon remains unclear.

The Alpine elevation of the deformed half-grabens *en masse* might itself be inversion. Fig. 11 is a section and restoration which allows this possibility. However, an equally viable alternative is the shortcut model (Fig. 11) which assumes that the most westerly original extensional fault was too steep to be reactivated as a thrust first acted as a buttress and was then cut through by the thrust on which the Belledonne massif was elevated.

Digne Thrust System

The Digne Thrust is one of the most important structures in the external Alpine zones of France (Fig. 12). It is 60 or more kilometres long from tip to tip. It transports a thick sequence of basinal Jurassic and Cretaceous black shales (Vocontian facies) southwestwards over thin Mesozoic platform carbonates of the Provencal facies (Figs 13 and 14). South of the S tip, the thrust is replaced by a stack of imbricate thrusts and associated folds whose traces curve through 90° around the so-called 'Arc de Castellane'. The platform facies occurs everywhere S and W of the arc and the thrust system telescopes the facies transition.

The lithological contrast between the footwall and hanging wall is dramatically seen in a window into the footwall of the thrust sheet located in the Gorges de Barles area, 15 km N of the town of Digne (Fig. 12). The Lias and Middle Jurassic is 500 m thick in the footwall compared with 1500 m thick in the hanging wall. Traced southwestwards in the transport direction the hanging wall facies gradually changes and becomes thinner and less shaley until, in the frontal part of the thrust sheet, the facies is not dissimilar to the footwall facies. This gradual change suggests an original structural position on the rollover margin of an extensional basin. Our interpretation of the Digne thrust is that it represents the contents of a basin thrust out over the original extensional fault rollover in a similar style to the Broad Fourteens Basin. In this case the rollover lies below the Valensole plateau and forms the footwall block of the Durance fault, one of the most important palaeofaults of Southeastern France.

The fault or faults which controlled the basin in which the rocks of the Digne thrust sheet were deposited cannot be identified with certainty but are inferred to lie along the margin of the Argentera massif. The structural evolution of the Digne thrust sheet is schematically summarized in Fig. 15. The figure makes the point that although the Digne thrust is an example of an extensional basin 'extruded' as a thrust sheet, there is no simple reactivation of the controlling extensional fault. Indeed, complex internal deformation makes it impossible to recognize the controlling fault with any certainty. It may actually be folded, as indicated in Fig. 15.

The North Pyrenees

The Pyrenean Mountains comprise a central axial zone of Palaeozoic rocks flanked by folded and thrusted Mesozoic and Tertiary strata. They

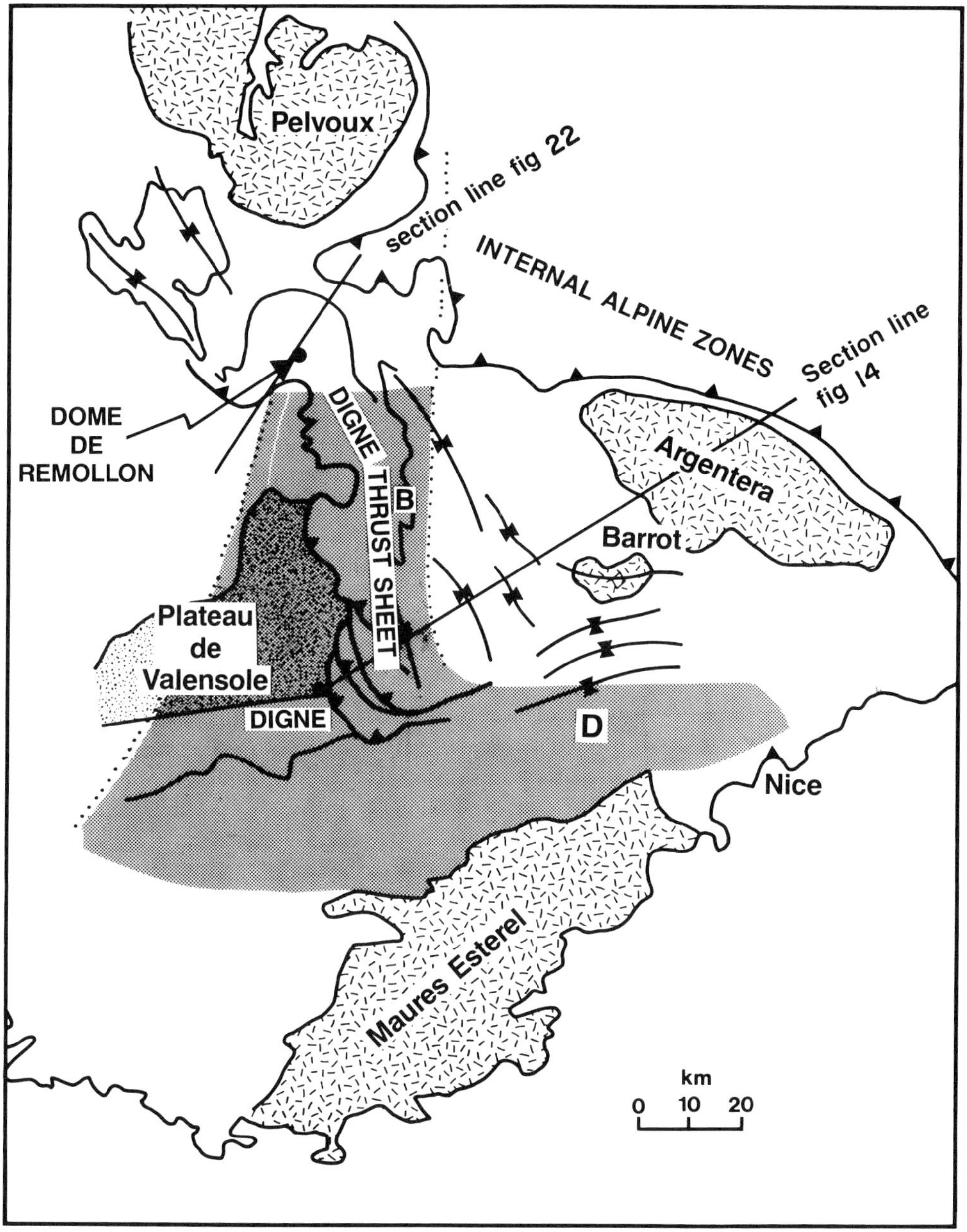

FIG. 12. Map of SE France showing locations of section lines. Stipple is Molasse, hackle ornament is pre-Triassic outcrop. Shaded area is possible extent of Mesozic platform carbonates beneath thrusted basinal rocks with dotted lines as limit (definite fault in west, possible fault in east). The letters A–D refer to contrasted stratigraphies shown in Fig. 13.

have been variously interpreted as both a strike-slip fold belt dominated by steep upright structures (e.g. Choukroune & Matauer 1978) and as a thin skinned fold and thrust belt (Williams & Fischer 1985). Neither interpretation fully takes into account the observed stratigraphic variations or structural style. Recently, improved industrial seismic resolution in the North Pyrenean thrust belt and within the structured foreland zone of the Aquitaine Basin demonstrates convincingly that the Pyrenees formed a major Mesozoic extensional basin controlled by a complex extensional fault system that exerted a major influence during the subsequent Eocene compression.

The major period of crustal extension was related to the opening of the Bay of Biscay in the Lower Cretaceous (Boillot 1984). The northward movement of Iberia throughout the Tertiary resulted in the continental collision and the development of the Pyrenean fold and thrust belt.

We describe below two sections from the central and W area of the North Pyrenees which document both basin inversion and half-graben extrusion. We also discuss the way that earlier extensional faults controlled the location of subsequent thrust ramps.

The first section (Fig. 16) in the extreme W runs from the margin of the Mt Ursuya massif northwards across the North Pyrenean thrust belt and Aquitaine basin. The section is well constrained by seismic and well data. North of the Mt Ursuya massif the Ursuya thrust sheet comprises a Triassic to Upper Cretaceous sequence that has a pronounced wedge shaped geometry. A thick Lower Cretaceous sequence in the north of the thrust sheet passes southwards into a much thinner Lower Cretaceous sequence. Along the N edge of the Ursuya massif the Upper Cretaceous rests directly on basement and is overthrust by a backthrust which carries a complete Mesozoic sequence southwards. This is presumed to represent a half graben that has been completely inverted and extruded both forwards and backwards.

To the north the Came Thrust Sheet drilled

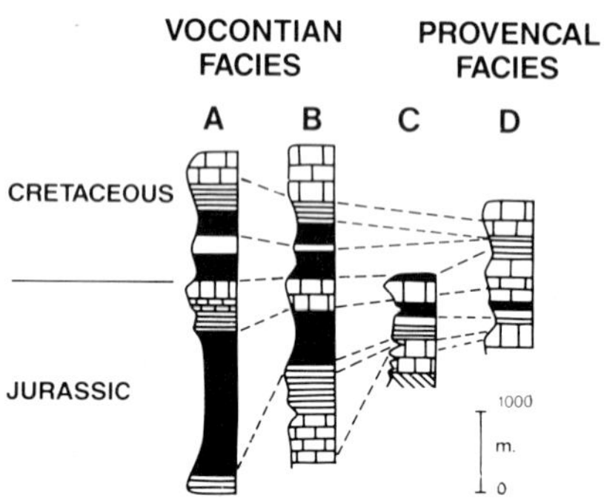

FIG. 13. Contrasting stratigraphy in the basins of SE France (after Debelmas 1974).

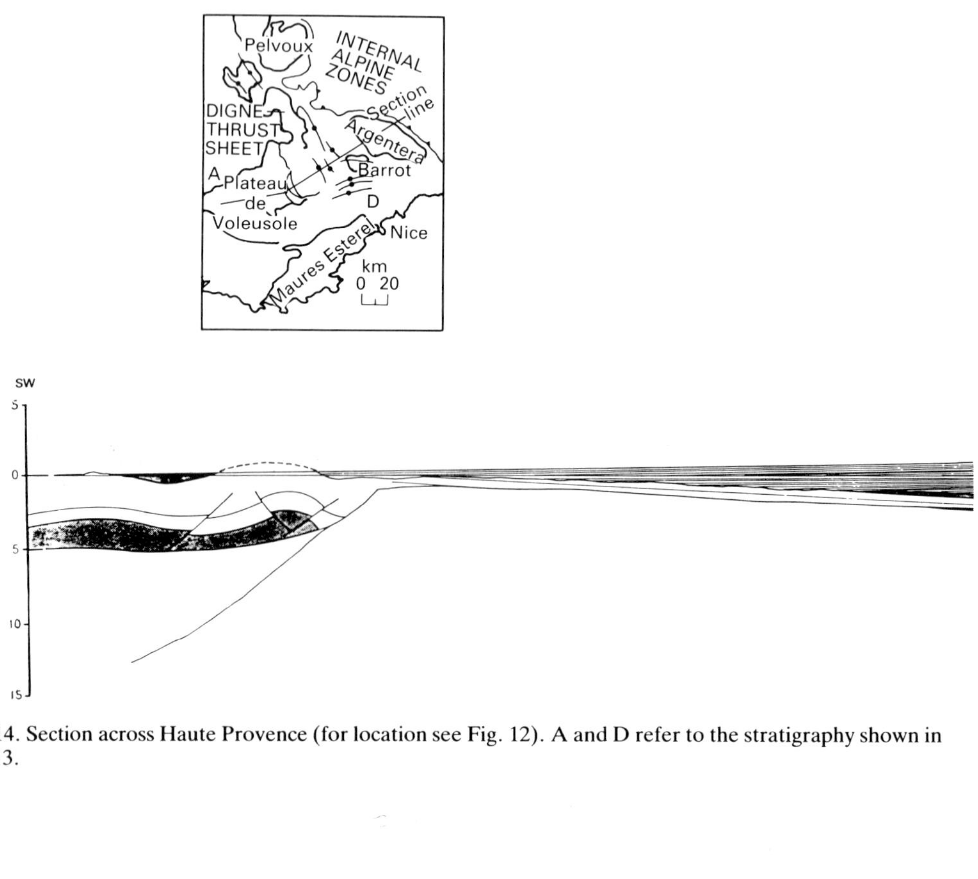

FIG. 14. Section across Haute Provence (for location see Fig. 12). A and D refer to the stratigraphy shown in Fig. 13.

by the Came well (Fig. 16) has a much thicker and more basinal Lower Cretaceous sequence when compared with the Lower Cretaceous carbonate platform sequence penetrated in the St Lons well in the footwall of the thrust sheet. Restoration of this section suggests the presence of a series of extensional half grabens of Lower Cretaceous age. In contrast to those in the Western Alps described earlier, these half grabens have been completely inverted. The second section (Fig. 17) crosses the St Suzanne anticline which is the leading structure in the central W area of the North Pyreenes. The St Suzanne structure is thin skinned and detached within the Upper Triassic evaporite sequence. However, the Upper Cretaceous sequence in the footwall to the structure comprises a thin ~500 m thick carbonate platform sequence that is represented in the hanging wall by a thick basinal shale sequence ~1500 m thick (Fig. 17). Restoration of this section implies the presence of an extensional fault of Upper Cretaceous age that has controlled the location of the subsequent thin skinned thrust ramp.

In the Southern Pyrenees there was no significant Mesozoic extension. The sedimentary sequence comprising a thin Upper Cretaceous carbonate sequence rests unconformably on Triassic evaporites and Palaeozoic basement. The boundary between the North and South Pyrenean zones is marked by a major thrust to the present day, which is likely to be the W continuation of the North Pyrenean fault (J. R. Hossack, personal communication). Restoration of cross-sections implies that this fault was a major extensional fault throughout the Cretaceous. It is tempting to speculate that the reactivation of this fault is responsible for much of the elevation of the axial zone basement massif in this area.

The Southern Pyrenees are a major basement duplex locally with a roof thrust in Triassic evaporites. This represents thrust deformation on the footwall of the old extensional fault. In contrast, in the North Pyrenees much of the deformation can be related to the earlier extensional structures. In some areas reactivation of extensional faults as reverse faults or thrust has resulted in major inversion and graben extrusion. Elsewhere steps in the basement surface control the location of subsequent thin skinned thrust ramps.

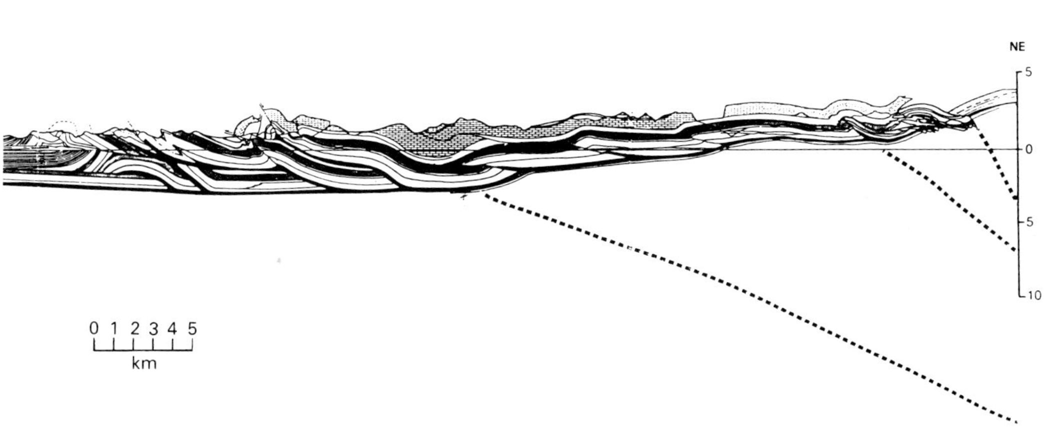

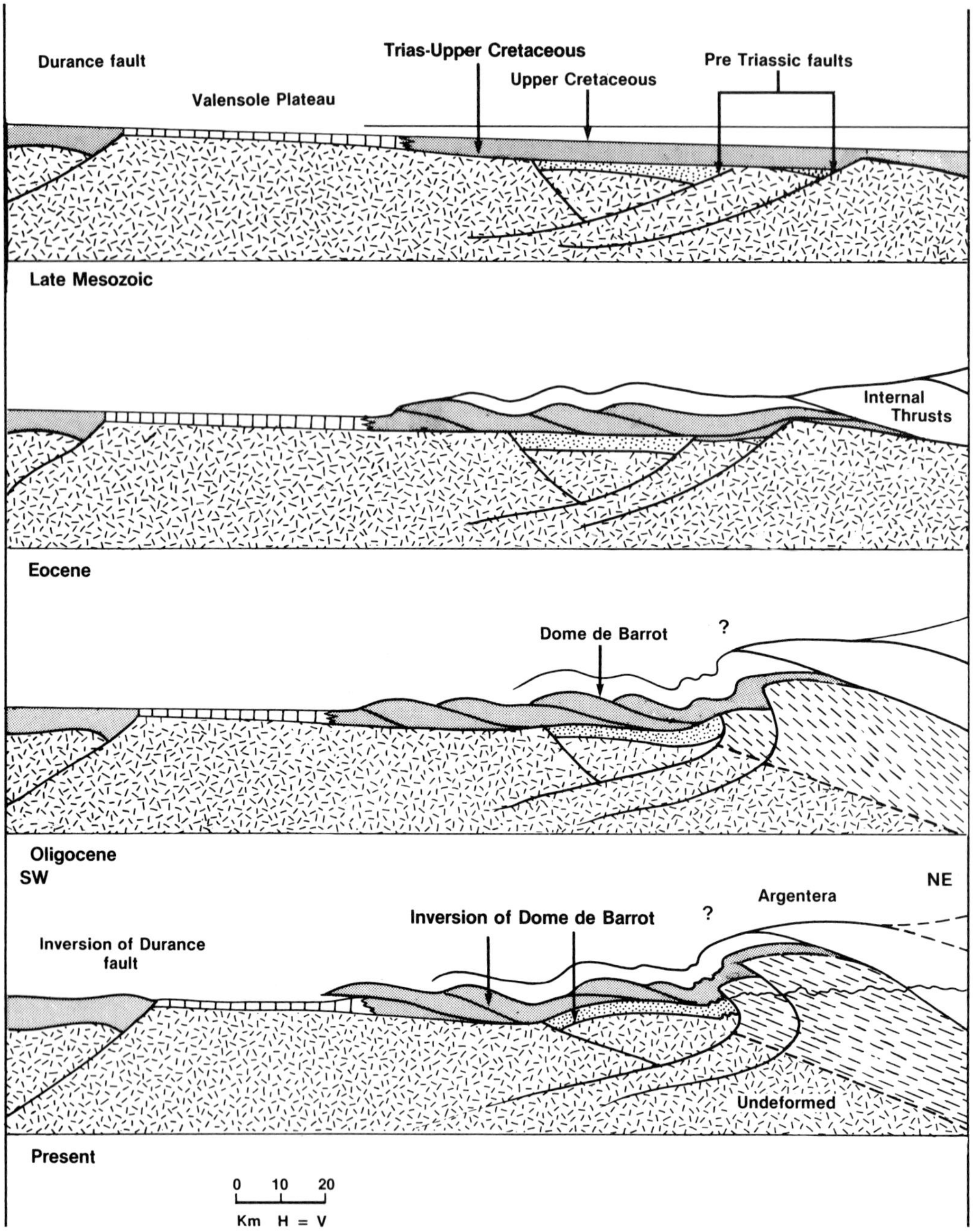

FIG. 15. Possible structural evolution of the Digne Basin.

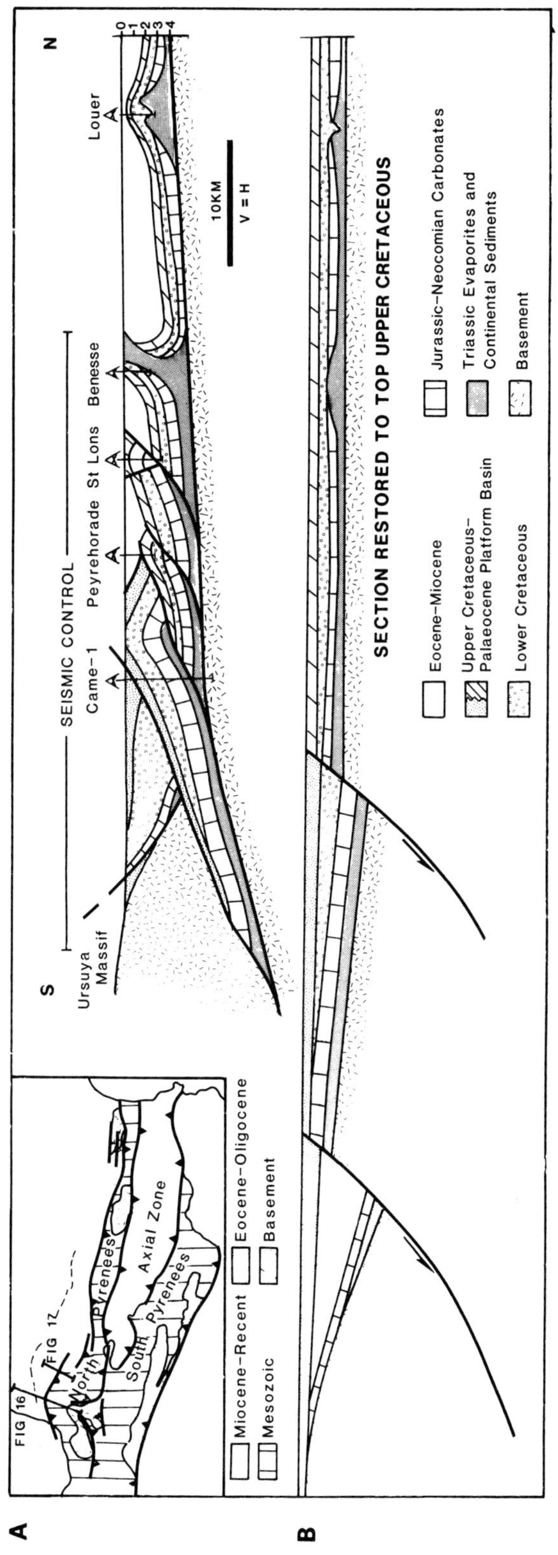

FIG. 16. A, Structural cross-section 1 North Pyrenees. Inset shows location. B, The same section restored to top Upper Cretaceous.

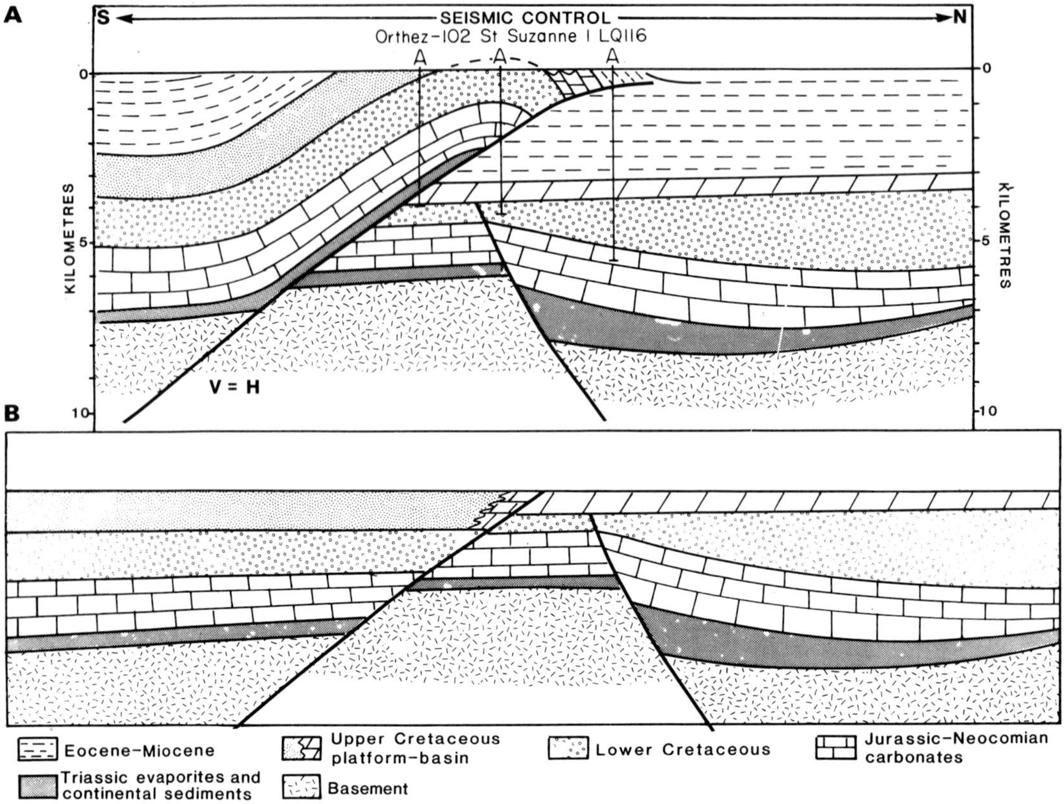

FIG. 17. A, Structural cross-section across the St Suzanne Anticline, North Pyrenees. See Fig. 16 for location. B, The same section restored to top Upper Cretaceous showing control on ramp location by earlier extensional fault. See Fig. 16 for the regional context of this structure.

Indications of the influence of extensional fault systems in orogenic belts

In this section we examine a number of features which are common in the many orogenic belts and which we believe indicate the influence of pre-existing extensional fault systems on the subsequent thrust geometry.

Out of sequence thrusts and arcuate thrust traces

These features seem to be indications of the influence of sub-thrust extensional patterns and sequences. The Digne thrust system described above shows clear evidence of both (Figs 12 and 14). The Castellane arc is almost certainly a response of southwesterly directed thrusts breaking as oblique thrust ramps as they climb sub-thrust obstacles. The trend of the ramps roughly reflects the trend of the sub-thrust structures (Fig. 18) which are most reasonably interpreted as part of the original extensional geometry of this area of the Alpine margin.

The leading part of the Digne thrust shows a complex movement sequence. The thrust is out of sequence in the sense that it cuts both limbs of folds in both its footwall and hanging wall, yet, in turn, it is folded by thrust related folds beneath it (Fig. 14). The geometrical relationship is borne out by the stratigraphic evidence that the last 7 km of displacement on the Digne thrust was across a post-Messinian land surface but earlier movements on the structure were pre-Eocene. The complexity is probably another manifestation of the buttress (buffer stop) effect of the sub-thrust structure. There are numerous other instances of the phenomena; the Sibillini Mountains of Italy (Leveccia 1978) are a particularly good example.

Arcs and out of sequence thrusts should make us suspect the presence of sub-thrust extensional structures (with or without associ-

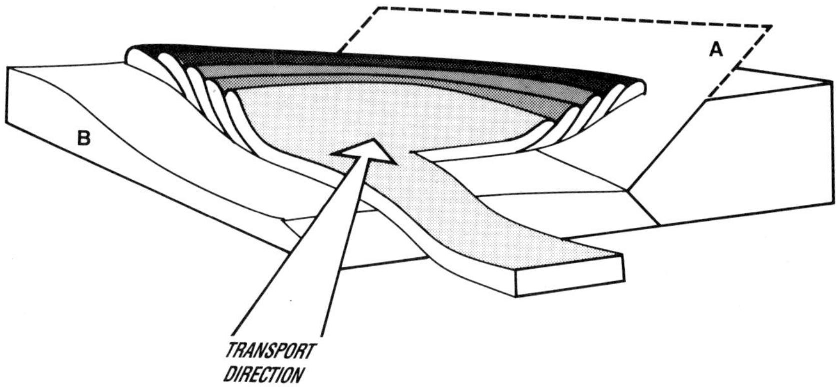

FIG. 18. Exploded block diagram showing oblique ramps and arc formation of thrust related folds associated with oblique inversion. A, a fault, B, a rollover.

ated inversion) with important implications for regional structural interpretation.

Footwall shortcuts

Footwall shortcuts like those suggested in the Belledone section of the Western Alps have been produced in the experiments done by McClay, *this volume* where pre-existing extensional faults are rotated too steeply to be reactivated themselves as reverse faults. In nature footwall shortcut thrusts produce isolated wedges of footwall rocks which may be translated in a thrust hanging wall (Fig. 19). The mechanisms allow isolated lenses of crystalline basement to be emplaced high within an apparently thin skinned sequence. This often presents interpretational difficulties. We know of two relatively large scale possible examples of this phenomenon.

The first example is the Johnstone Thrust/Benton fault system, (Brooks *et al.* 1983; Hancock *et al.* 1981) in the Hercynian fold belt of Southwest Wales (Figs 20 and 21), where an isolated lens of crystalline basement lies 5 km above regional basement level along the line of a major Caledonian basin controlling fault (Brooks *et al.* 1983). It sits on a shallow thrust (the Johnstone thrust) and is emplaced onto Upper Carboniferous rocks which directly and unconformably overlie Lower Palaeozoic slates. A major extensional fault, the Benton fault, forms the S edge of the crystalline wedge, bounding it from a thick sequence of Devonian rocks (Old Red Sandstone). The fault probably represents the N border of the Old Red Sandstone basin (Hancock *et al.* 1981) shortcut by the thrust during the Hercynian orogeny.

A second example is from the Dome de Romollen in the external Alpine zones of Southeastern France, (Figs 12 and 22) where an isolated exposure of crystalline basement occurs in the core of an anticline which folds the Digne Thrust. We have already noted that the hanging wall of the thrust contains basinal Jurassic shales thrusted some 20 km over thinner platform carbonates (Siddans 1978; Gigot *et al.* 1974). The thrust is detached in Triassic evaporites. A 20 km long thin blade cut through basement below this detachment is a most unlikely geometry in the context of the regional geology. A small fragment of footwall from an extensional fault within the original basin is a more reasonable explanation of the basement inlier.

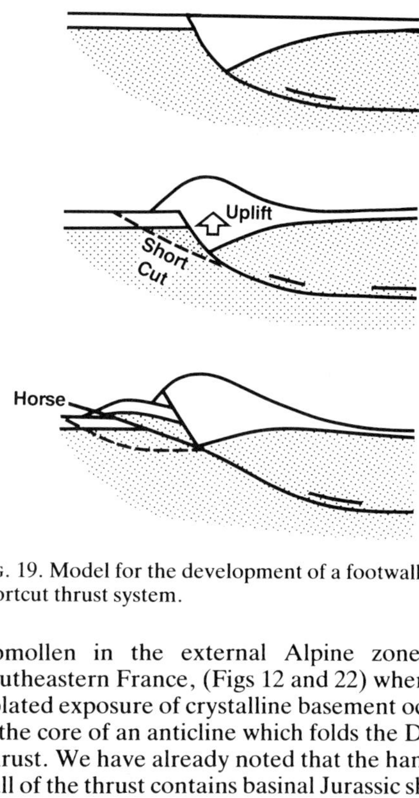

FIG. 19. Model for the development of a footwall shortcut thrust system.

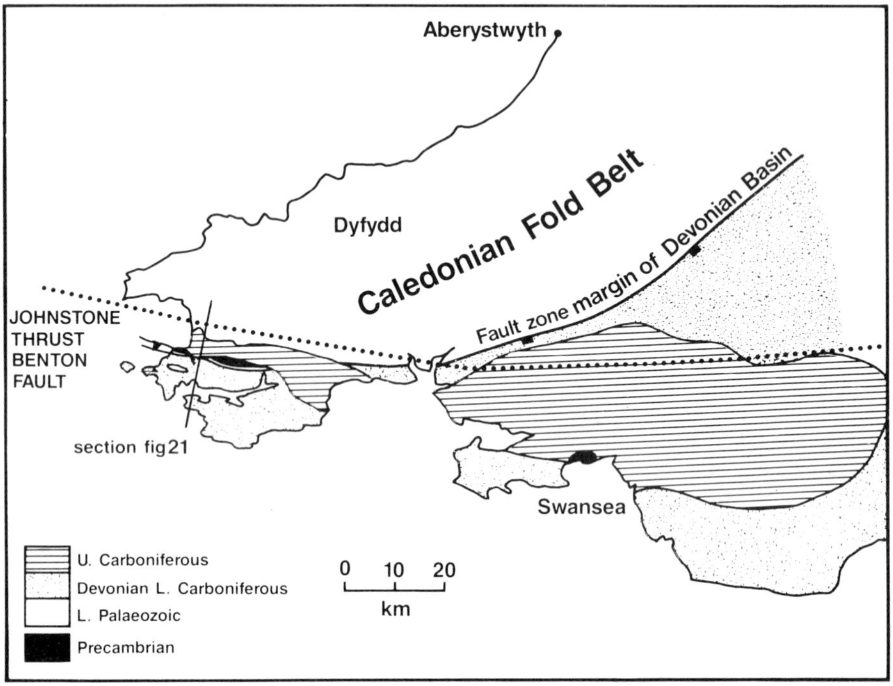

FIG. 20. Map showing the geology of Southwest Wales. Dotted line is limit of Hercynian deformation.

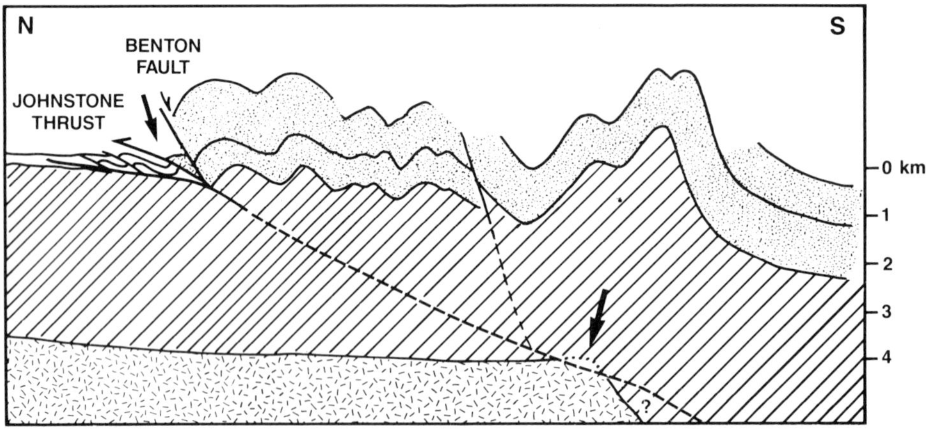

FIG. 21. Sketch section showing the Johnstone/Benton block as a footwall shortcut. Depth to basement is from Mechie 1981. Hackle shading is basement, lined shading is Lower Palaeozoic, dotted is Devonian and Lower Carboniferous. Upper Carboniferous is blank.

Inversion post-dating nappe emplacement

Commonly the floors of orogenic foredeep basins are broken by extensional faults. Some of these faults date from passive margin phases before orogenic collision. Others may have been formed by the foredeep–forebulge deformation wave and show growth sequences of foredeep basin fill. Commonly such structures are inverted late in the deformation history of the mountain belt.

In the foredeep basins of the Apennines and

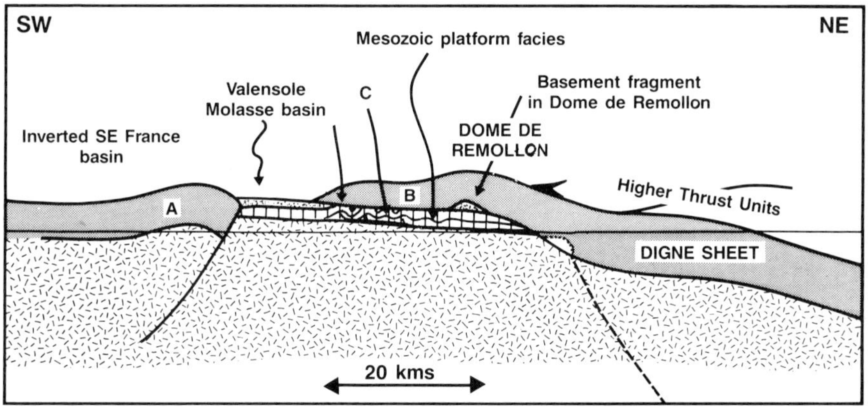

FIG. 22. Conceptual section showing the Dome de Romollon basement outcrop interpreted as a footwall shortcut in the extruded basinal rocks of the Digne thrust sheet. Letters refer to stratigraphy in Fig. 13. For location of section see Fig. 12.

in Oman, inversion of this kind developed after the emplacement of far travelled thrust sheets into the foredeep basins. The thin skinned thrusts have been deformed by 'thick skin' inversion-related anticlines whose scale is greater than earlier thin skinned structures and whose orientation may also be distinct.

Jebel Akdar and Hamrat Duru in Oman and the structure which elevates the main Lagonegro outcrop in the Southern Apennines are possible examples of large scale late inversion structures.

The interplay of thin skinning and inversion — concluding remarks

We have seen that pre-existing extensional faults may trigger ramp generation and complicate thrust sequence and trend. Where inversion occurs later than far travelled thrusting, complex interference of the two types of structure may occur.

The existence of extensional faults also complicates considerably the construction of balanced cross-sections. Although some of the geometric principles still apply, clearly the absolute rigour of the method is lost where there is no table-top basal detachment. Some inversions (e.g. in Southern England, see Simpson *et al., this volume*) 'heal' with the basement once more continuous across the old fault. This creates a potential interpretative pitfall of which we need to beware. It is nevertheless important that we attempt to restore sections. Even though there may be no definitive answer, the technique still provides valuable insights into regional geometry.

The potentially complex inter-relationships between thin skinned thrusting and basin inversion are a prime problem to be tackled by the structural geologist working in orogenic terrains. Inversion is one of a number of possible ways in which pre-existing extensional fault geometry influences later thrusting to produce a more complex geometry than that which we see in areas of pure thin skinned thrusting. If stratigraphic and seismic data are good, interpretative problems are small. However seismic data are commonly equivocal in orogenic regions. Examples like those we have cited may have some empirical interpretative value. Clearly, the application of thin skinned tectonic models without sufficient regard to detailed stratigraphy can be misleading and may result in erroneous structural prediction and shortening estimates.

It is our contention that two broad structural provinces can be recognized in a stretched continental margin which later suffers orogenic contraction. The area of fault-controlled stretching is likely to be the site of basin extrusion, fault block restacking, footwall shortcutting, buttressing and complex structure. It is an area where we should expect bedlength mismatches of cover and basement. Further up the passive margin we may pass into a region which suffered only thermal subsidence with the development of 'steers head' shaped tapering sedimentary wedges. Here, thin skin tectonics on 'table top' basement such as we see in the Rocky Mountain foothills is likely to be dominant. Localized inversions on isolated passive margin faults like those described earlier in this paper spread far into the foreland beyond the limit of any thin skinned deformation. In the external parts of a mountain chain we may see interference of these two sorts of structures, we may also see inversion of extensional faults which formed coevally with foredeep basin development.

ACKNOWLEDGMENTS: We are grateful to BP PLC for permission to publish this paper. Many of the ideas on the offshore Holland area were developed in conjunction with our colleagues in BP Holland. Professors M. Lemoine and P. C. de Graciansky opened our eyes to the geology of the Western Alps. Chris Banks made many useful comments on an earlier version of this manuscript.

References and further reading

BALLY, A. W. 1981. Musings over sedimentary basin evolution *Philosophical Transactions of the Royal Society of London*, **A305**, 325–338.

—— 1984. Tectonogenese et sismique reflection: *Bulletin de la Sociéte Géologique de France*, **26**, 279–286.

——, GORDY, P. L. & STEWARD, G. A. 1966. Structure, seismic data and orogenic evolution of southern Canadian Rockies. *Bulletin of Canadian Petroleum Geology*, **14**, 337–381.

BODENHOUSE, J. W. A. & OTT, W. F. 1981. Habitat of the Rijswijk Oil Province Onshore Netherlands. *In*: ILLING, L. V. & HOBSON, G. D. (eds) *Petroleum Geology of the Continental Shelf of Northwestern Europe*. Institute of Petroleum, London and Heyden & Sons Ltd, pp. 289–300.

BOILLOT, G. 1984. Le Golfe de Gascogne et les Pyrenees. *In*: BOILLOT, G., MONTADERT, L., LEMOINE, M. & BIJU-DUVAL, B. (eds) *Les Marges Continentales Actuelles et Fossiles Autour de la France*. Masson, Paris.

BOYER, S. E. & ELLIOT, D. 1982. Thrust systems. *Bulletin of the American Association of Petroleum Geologists* **66**, 1196–1230.

BROOKS, M. LLEWELYN, D. J. & MECHIE, J. 1983 Geophysical investigations in the Variscides of southwest Britain. *In*: Hancock, P. L. (ed.) *The Variscan Fold Belt in the British Isles*. Adam Hilger, Bristol, 186–197.

BUTLER, R. W. H. 1983. Balanced cross-sections and their implications for the deep structure of the NW Alps. *Journal of Structural Geology*, **5**, 125–137.

—— 1985. The restoration of thrust systems and displacement continuity around the Mont Blanc massif, NW external Alpine thrust belt. *Journal of Structural Geology*, **7**, 569–582.

—— The influence of pre-existing basin structure on thrust system evolution in the western Alps, *This volume*.

——, MATHEWS, S. J. & PARISH, M. 1986. The NW external Alpine thrust belt and its implications for the geometry of the western Alpine orogen. *In*: COWARD, M. P. & RIES, A. C. (eds) & *Collision Tectonics Special Publication of the Geological Society of London*, **19**, 245–960.

CHADWICK, R. A. 1986. Extension tectonics in the Wessex basin, southern England: *Journal of the Geological Society of London*, **143**, 465–486.

CHOUKROUNE, P. & MATTAUER, M. 1978. Tectonique des et Pyrenees: Sur le fonctionnement de la faille transformante nord Pyreneene; comparison avec des modeles actuels. *Bulletin de la Sociéte Geólogique de France*, **7**, 689–700.

DAHLSTROM, C. D. A. 1969. Balanced cross-sections. *Canadian Journal of Earth Sciences*, **6**, 743–57.

DAVIES, V. M. 1982. Interaction of thrust and basement faults in the French external Alps. *Tectonophysics*, **86**, 325–331.

DEBELMAS, J. 1974. *Geologie de France, II, La Chaine Recente*.

DE GRACIANSKY, P., BOURBON, M., CHENET, P. Y. & LEMOINE, M. 1979. Genese et evolution comparees de deux marges continentales passives marge iberique de l'ocean Atlantique et marge Europeane de la Tethys dana les Alpes occidentales *Bulletin de la Société Géologique de France*, **21**, 663–674.

——, DE DARDEAU, G., LEMOINE, M. and TRICART, P. The inverted margin of the French Alps. *This volume*.

DEWEY, J. F. 1982. Plate tectonics and the evolution of the British Isles. *Journal of the Geological Society of London*, **139**, 371–412.

ETHERIDGE, M. A. 1986. On the reactivation of extensional fault systems. *Philosophical Transactions of the Royal Society of London*, **A 317**, 179–194.

GIBBS, A. D. 1983. Balanced cross-section construction from seismic sections in areas of extensional tectonics. *Journal of Structural Geology*, **5**, 153–160.

—— 1984. Structural evolution of extensional basin margins: *Journal of the Geological Society of London*, **141**, 609–619.

GIGOT, P., GRANDJAQUET, C. & HACCARD, D. 1974. Evolution tectono-sedimentaire de la septentrionale du bassin tertaine de Digne depuis l'Eocene. *Bulletin de la Société Géologique de France*, **16**, 128–139.

GILLCRIST, R., COWARD, M. & MUGNIER, J. L. Structural inversion examples from the Alpine foreland and the French Alps. *Geodynamica Acta (Paris)* **1**, 5–34.

——, COWARD, M. P., TRUDGILL, B., PECHER, A. MUNGIER, J. L., Structural inversion in the external French Alps (Abstract). *This volume*.

HANCOCK, P., DUNNE, W. M. & TRINGHAM, M. E. 1981. Variscan structures in South-West Wales. *In*: ZWART, H. J. & DORNIEPEN, V. F. (eds) *The Variscan Orogeny in Europe. Geologie en*

Mijnbouw, **60**, 81–88.

HARDING, T. P. 1985. Seismic characteristics and identification of negative flower structures; positive flower structures, and positive structural inversion: *Bulletin of the American Association of Petroleum Geologists*, **69**, 582–600.

JACKSON, J. A. 1980. Reactivation of basement faults and crustal shortening in orogenic belts. *Nature*, **283**, 343–346.

LEMOINE, M. 1987. Pre oceanic rifting in the Alps. *Tectonophysics*, **133**, 305–320.

——, BAS, T., ARNAUD-VANNEAU, A. *et al.* 1986. The continental margin of the Mesozoic Tethys in the Western Alps. *Marine and Petroleum Geology*, **3**, 179–199.

——, GIDON, M. & BARFETY, J. C. 1981. Les massifs crystalline externes des Alpes occidentaless d'anciens blocs bascules age du Lias Lors du rifting tethysin. *Compte Rendu de l'Academie des Sciences, Paris*. **292**, 917–920.

LEVECCIA, G. 1978. Confronto fra due sistemi a pieghe L'Appennino Umbro-Marchigiano ed il Giura Franco-Svizzero. *Bolletin Société Geologique* **H** **101** 69–76.

MCCLAY, K. R. Analogue models of inversion tectonics, *This volume*.

MECHIE, J. 1981. Seismic studies of deep structure in the British Channel area. Unpublished PhD thesis, University of Wales.

MENARD, 1976. Relations entre structures profondes et structures superficielles dans le Sud-Est de la France. Essai d'utilisation des données geophysiques. Thése 3iéme cycle Grenoble I, 169 pp.

OELE, J. A., HOL, A. C. P. J. & TIEMANS, J. 1981. Some Rotliengend gas fields of the K. and L. blocks, Netherlands offshore (1968–1978) — A case history. *In*: L. V. Illing & G. D. Hobson (eds) *The Petroleum Geology of the Continental Shelf of Northwest Europe*. Institute of Petroleum, London & Heyden & Sons Ltd, 289–300.

ROSENDAHL, B. R., REYNOLDS, D. J., LONBEN, P. M., BURGESS, C. F., MCGILL, J., SCOTT, D., LAMBIASE, J. J. & DENTSEN, S. J. 1986. Structural expressions of rifting: lessons from Lake Tanganyika Africa. *In*: Frostick, L. E. *et al.* (eds) *Sedimentation in African Rifts. Special Publication of the Geological Society of London*, **25**, 19–28.

SIDDANS, A. W. B. 1979. Arcuate fold and thrust patterns in the subalpine chains of south-east France. *Journal of Structural Geology*, **2**, 117–126.

SIMPSON, I. R., GRAVESTOCK, M., HAM, D. LEACH, H. & THOMSON, S. D. 1988, Inversion tectonics of the Wessex Basin, *This volume*.

TRÜMPY, R. 1980. An outline of the geology of Switzerland. *Proceedings of the 26th International Geological Congress Section, G10*, A, 5–105.

WILLIAMS, G. & FISCHER, M. 1985. A balanced section across the Pyrenean orogenic belt. *Tectonics*, **3**, 773–780.

ZIEGLER, P. A. 1983. Inverted basins in the Alpine foreland. *In*: Bally, A. W. (ed.) *Seismic Expression of Structural Styles*: *A Picture and Work Atlas American Association of Petroleum Geologists Studies in Geology 15*, **3**, 3.3–3–3.3.–12.

A. B. HAYWARD, British Petroleum Development Ltd, Britannic House, Moor Lane, London EC2Y 9BU, UK.

R. H. GRAHAM, British Petroleum Development Ltd, Britannic House, Moor Lane, London EC2Y 9BU, UK.

Analogue models of inversion tectonics

K. R. McClay

SUMMARY: Inversion of dip–slip fault systems by horizontal recompression has been modelled using sand, sand–mica and sand–clay analogues. Three extension–inversion systems have been tested: uniform basal detachment, simple listric detachment and ramp/flat detachments. Extension above a uniform basal detachment produces a domino or conjugate array of near planar faults. Upon inversion only faults at angles < 60° to the layering and suitably oriented with respect to the compression direction undergo reactivation. With increased contractional strains all faults are rotated to high angles and reactivation ceases. Shortcut faults may develop in the footwall of steeply dipping faults. Extension above a simple listric detachment fault produces a characteristic rollover anticline and associated crestal collapse graben structure. Upon inversion asymmetric uplift is associated with the major detachment. For models with no competency contrast the rollover anticline is accentuated. The crestal collapse graben faults are reactivated and the graben tightens producing a 'flower-like' fault system. Reactivation of synthetic faults ceases when they become rotated into angles too steep for contractional deformation. Extension of ramp/flat detachment systems produces characteristic rollover anticlines and crestal collapse graben systems above each concave-up segment of the listric fault system. It also produces a hanging wall syncline associated with the convex-up segment of the detachment. Inversion results in asymmetric uplift at the emergence of the reactivated detachment and above the convex-up portion of the detachment surface. The crestal collapse graben faults become reactivated and rotated into steep angles resulting in flower-like geometries. The rollover anticline is accentuated resulting in nappe-like structures. The results of these analogue studies of inversion structures indicate that considerable further research is needed to understand the geometries and kinematics of inversion fault systems.

Introduction

Analogue modelling is a powerful tool for developing an understanding of the geometries and kinematics of fault systems. It has been widely used to simulate extensional regimes (Cloos 1968; Horsfield 1977, 1980; Brun & Choukroune 1983; Mandl 1984; McClay & Ellis 1987a; 1987b; Vendeville *et al.* 1987; Ellis & McClay 1988), strike–slip regimes (Wilcox *et al.* 1973; Tchalenko 1970; Naylor *et al.* 1986) and for thrust systems (Davis *et al.* 1983). Little work apart from that of Koopman *et al.* (1987), however has been published on modelling inversion structures where previously extended models have been subjected to subsequent contractional deformation.

In this paper the results of a series of inversion models are presented in which simple extensional fault geometries (Ellis & McClay 1988; McClay & Ellis 1987a; 1987b) have been inverted by simply reversing the sense of motion in the experimental deformation rig and contracting the previously extended models by horizontal recompression. This contrasts with the sand box experiments of Koopman *et al.* (1987) where inversion was achieved by uplift or rotation of rigid basement fault blocks. The experimental results are compared with inversion examples cited in the literature.

Inversion tectonics as discussed in detail elsewhere in this volume is the turning inside out of a sedimentary basin (defined as an area of negative structural dynamics, i.e. subsidence below a pre-extensional regional datum) such that the basin is uplifted and may eventually become elevated above the pre-extensional regional. The inversion is usually accomplished by reactivation and re-utilization of pre-existing extensional fault systems. However, as will be demonstrated in this paper, bulk shortening strains and interbed slip may also be important factors in inversion structures.

The experiments summarized in this paper are the first of a planned wider study of inversion dynamics. Attention is focused on the geometries of inversion models extended above a fixed footwall detachment, the reactivation of faults and bulk strain patterns of inversion.

Experimental method

The experiments were carried out in a glass sided deformation rig shown schematically in Fig. 1. The initial model dimensions were

From COOPER, M. A. & WILLIAMS, G. D. (eds), 1989, *Inversion Tectonics*
Geological Society Special Publications No. 44, 41–59.

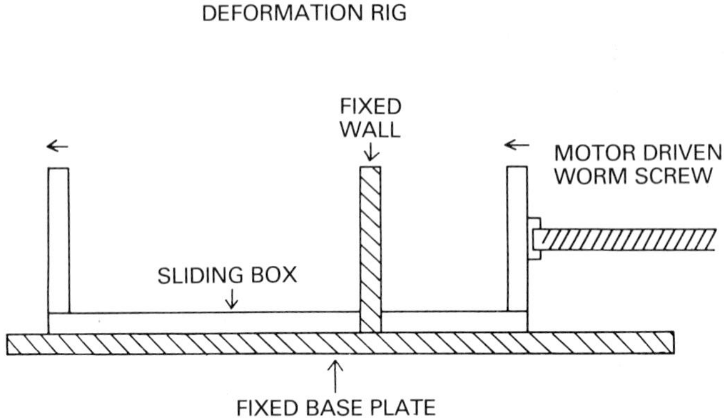

FIG. 1. Sketch of simple 'pull–push' apparatus used for modelling experiments (after McClay & Ellis 1987a).

25 cm long, 16 cm wide and 10 cm deep. Deformation was achieved by using a motor driven worm-screw at a constant displacement rate of 4.16×10^{-3} cm/s. A full description of the deformation apparatus is given in McClay & Ellis (1987a; 1987b).

Three basic detachment geometries were used in this series of experiments (Fig. 2). Uniform extension (Fig. 2A) was achieved using a rubber sheet as a basal detachment (McClay & Ellis 1987a). The simple listric and ramp/flat listric detachments (Figs 2B & C), consisted of a moulded footwall block of fixed geometry over which a thin flexible plastic base plate attached to the moving wall of the deformation rig is pulled (cf. Ellis & McClay 1988).

Data were recorded using 35 mm time-lapse photography and 16 mm cinematography through the glass walls of the deformation rig. Slight edge effects, such as blurring some of the finer detail, are present but serial sectioning of models has demonstrated the essential cylindricity of structures developed in the experiments (McClay & Ellis 1987a; 1987b; Ellis & McClay 1988).

In order to simulate upper crustal deformation of sedimentary sequences an extremely weak near cohesionless model material must be used (see Horsfield 1977; 1980; McClay & Ellis 1987a; 1987b; Ellis & McClay 1988; for details).

In these experiments the models were constructed by carefully sieving into the deformation apparatus layers of homogeneous coloured and white sand (grain size 300 μm), layers of sand–mica mixtures and thin layers of mica. Mica flakes were used to introduce anisotropy into the models and to facilitate interlayer slip. In particular, during extension the mica flakes become entrained in the fault zones and hence may be expected to promote

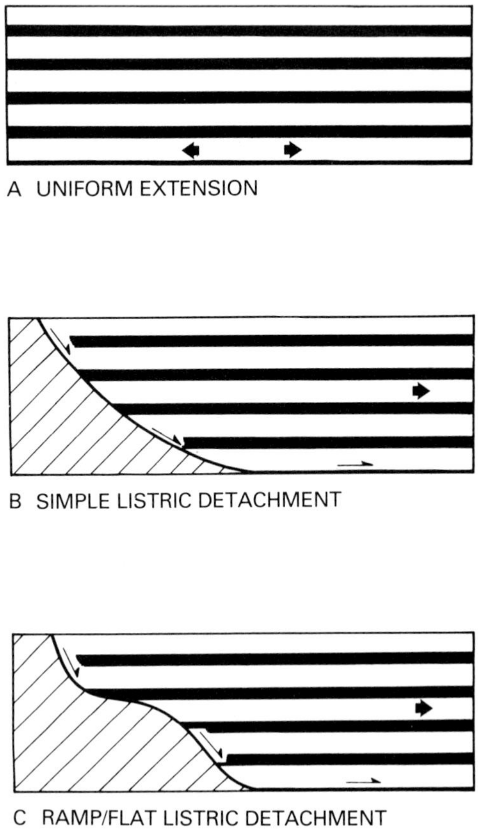

FIG. 2. Basement configurations for the extension and inversion models: A, Uniform extension above a homogeneously stretching basement (rubber sheet); B, Extension above a simple listric detachment (fixed footwall geometry); C, Extension above a ramp/flat listric detachment (fixed footwall geometry). In all cases inversion was achieved by reversing the sense of motion (indicated for extensional deformation by the arrows in A–C above).

reactivation during inversion. During the course of each experiment layers of sand and sand–mica mixtures are regularly and incrementally infilled during extension (Ellis & McClay 1987a; 1987b) to simulate synrift sedimentation. In some experiments (Table 1) postrift layers are introduced in order to monitor inversion features. In other experiments material was

TABLE 1. *Summary of extension and inversion structures*

	Extension				Inversion		
Experiment series	Starting geometry	Model materials	Total extension	Summary of structures	Total contraction (% of extended model)	Fault reactivation	Summary of structures
1	Uniform extension horizontal detachment	Homogeneous sand	50%	Domino array of listric faults	33.3%	None. Extension faults cut by thrusts	New thrusts extension faults. No reactivation
2	Uniform extension	Sand–mica mica layers	50%	Positive listric. Minimum fault angle 40° Wedge grabens	19.1%	Reactivation of thrust faults. Local reactivation on parts of faults	All rift faults rotate to high angles. Out of graben faults. Wedge structures
3	Uniform extension	Sand–mica mica layers	50%	Conjugate planar faults; horst and graben structures. Minimum fault angle 55°	25%	Reactivation of extension faults; array of thrust faults with small displacements	All extension faults rotated to high angles. Prerift shearing at high bulk strains
4	Simple listric	Homogeneous sand	48.7%	Rollover anticline. Superposed crestal collapse graben structures	31.6%	Reactivation of major graben bounding faults	Asymmetric uplift. Accentuation of rollover anticline. Out of graben movement. Footwall shortcut in fault 1
5	Simple listric fault	Sand with 5 mm clay layers	50%	Rollover anticline. Superposed crestal collapse graben structures	47%	Reactivation of graben bounding faults	Asymmetric uplift. Major detachment becomes shear zone. Change in fault motion in synthetics
6	Ramp/flat listric	Sand with mica layers	50%	Rollover anticline. Hanging wall syncline. Crestal collapse grabens	25.4%	Reactivation of graben bounding faults	Asymmetric uplift at major detachment and above ramp. Out of graben faults. Accentuation of upper rollover anticline
7	Ramp/flat listric fault	Sand and mica with mica layers	50%	Rollover anticlines. Hanging wall syncline. Crestal collapse graben structures	40%	Reactivation of crestal collapse graben faults	Asymmetric uplift. Nappe structures. Out of graben faults. Shear zone at major detachment

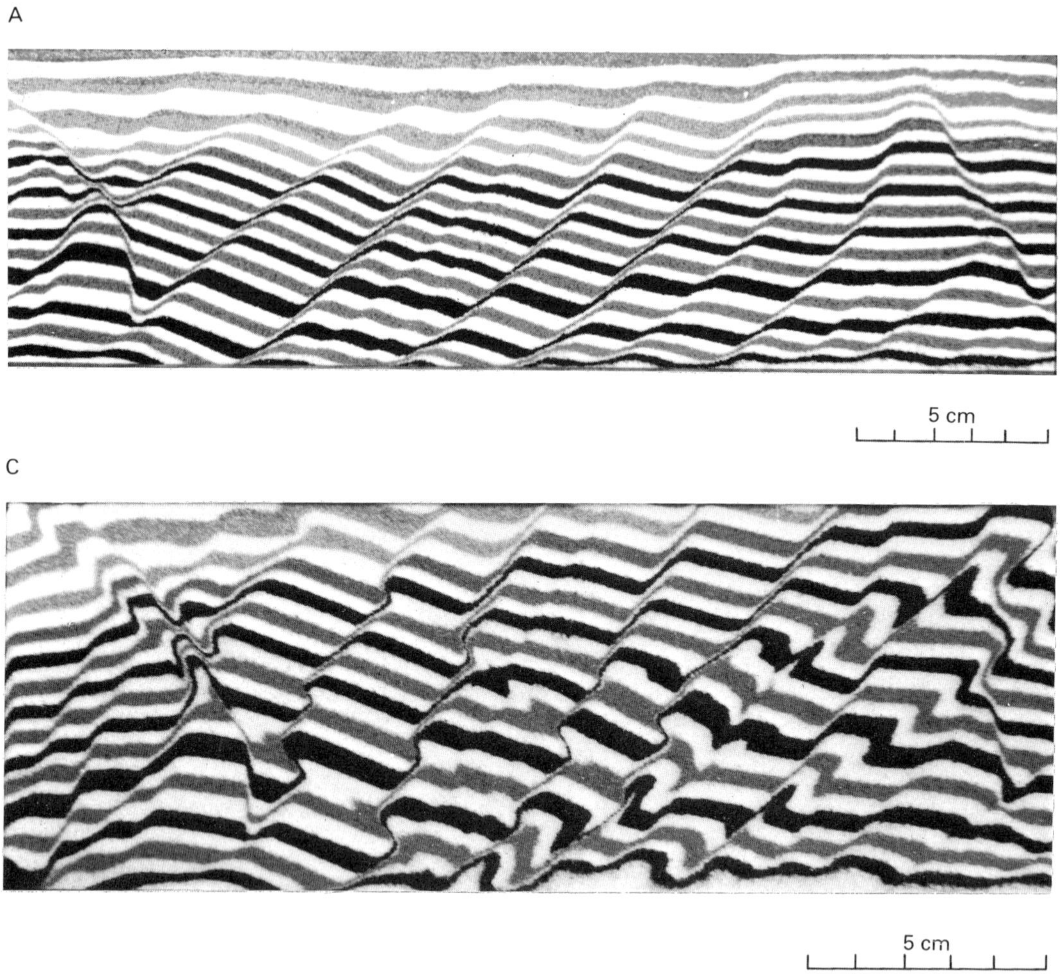

FIG. 3. Series 1 experiment. Homogeneous sand model with uniform extension above a horizontal detachment followed by contraction above the same detachment: A, End of extension (50%). The upper seven layers (light grey and white sand layers) are the synrift sediments added incrementally during extension. B, Line diagram of a series 1 experiment depicting end extension and showing fault sequences. C, End of contraction (33%) showing the complex array of deformed extensional faults (from A above) and new formed kink-like contractional faults (contractional shear zones). D, Line diagram at the end of contraction showing interpreted fault sequences.

removed from the top of the apparatus during inversion (Table 1).

Results

The extensional geometries used here as the starting architecture for the inversion models have been described in detail by McClay & Ellis (1987a; 1987b) and Ellis & McClay (1988). In this paper only the end results of extension will be presented and briefly summarized. Attention is focused upon the geometries of inversion structures produced during the contraction of three groups of inversion experiments (Table 1). The first group of experiments (series 1–3, Table 1) examine the inversion of domino fault models. The second group of experiments (series 4 and 5, Table 1) examine the inversion of simple listric fault systems. The third group of experiments involve inversion of ramp/flat listric geometries (series 6 and 7, Table 1). In all

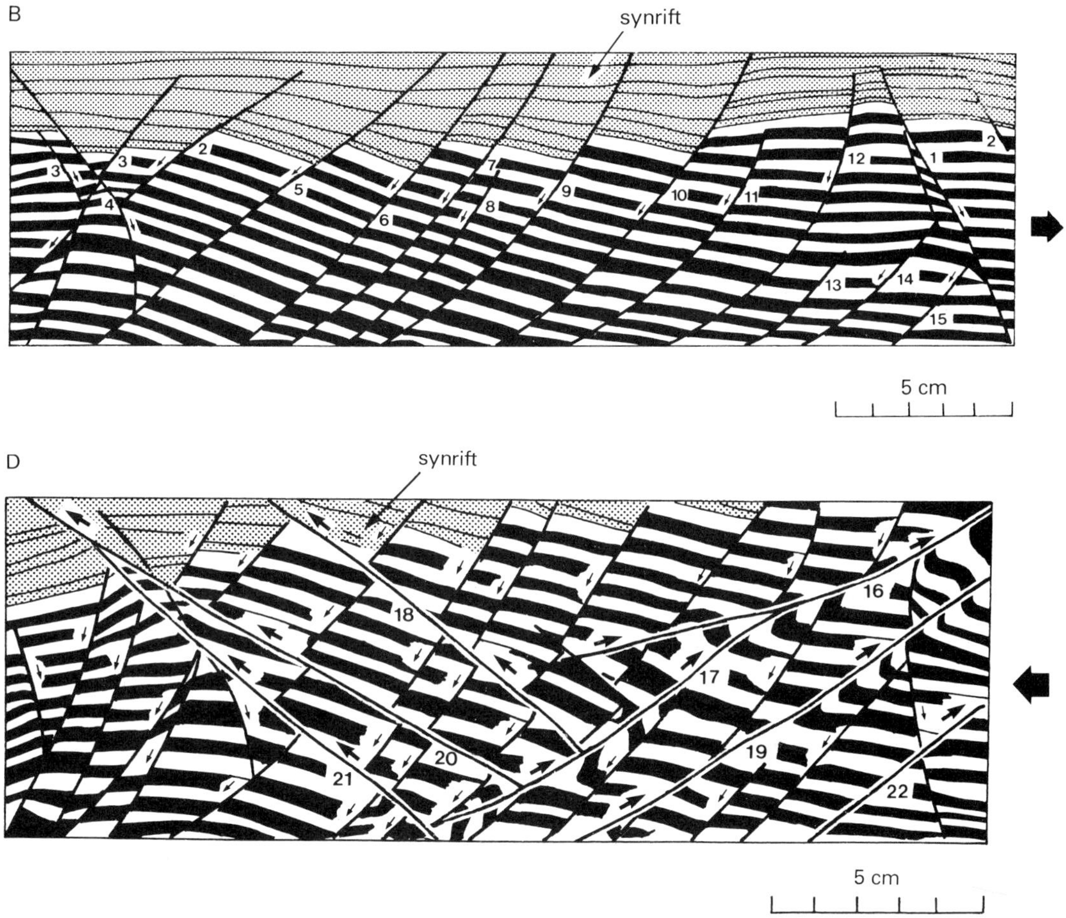

of the experiments described in this paper extension was achieved by pulling the right hand wall of the apparatus away (as indicated by the arrows in Figs 3–9) and contraction was achieved by pushing the right hand wall in (Figs 3–9).

Inversion of domino extensional fault systems (series 1–3)

Series 1 experiments using homogeneous sand as a modelling material produced an array of domino style planar to listric faults (Fig. 3A) with variable fault nucleation patterns (Fig. 3B). This kind of fault pattern has been found in numerous experiments of this type (McClay & Ellis 1987a; 1987b). Inversion of this series of experiments produces bulk shortening with formation of contractional faults that deform the pre-existing extensional faults (Figs 3C & D). The homogeneous isotropic sand models have little or no memory of the extensional deformation. The listric to planar extensional faults (Fig. 3A) are not reactivated.

Extension of series 2 experiments using alternating sand and thin mica layers (Table 1) produced an array of slightly listric to listric faults (Fig. 4A), similar to, though not as abundant, as those in series 1 models (Fig. 3A). In extension, fault nucleation does not show a regular pattern (Fig. 4B).

Inversion of this series of experiments shows a complicated pattern (Figs 4C & D). Upon initial contraction, inversion occurs by limited reactivation of low angle faults (45–60° dip) particularly in the synrift sediments. Small but marked thrust faults occur in the postrift sediments (Figs 4C & D). At higher bulk contractional strains initial low angle extensional faults become rotated into higher angles (>60°) and

A

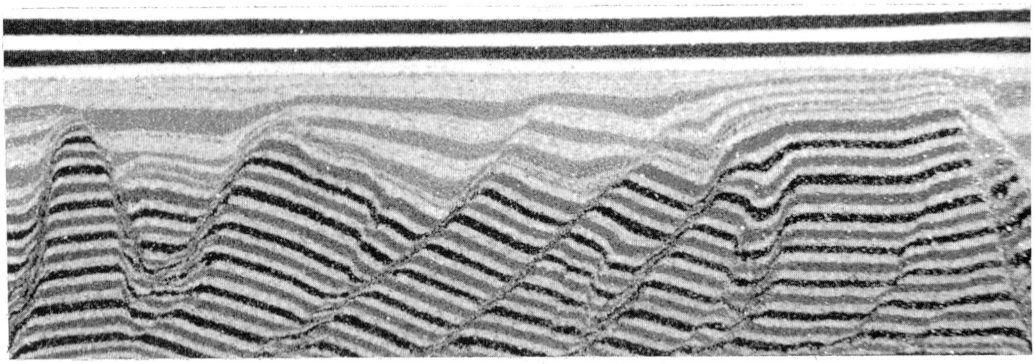

5 cm

C

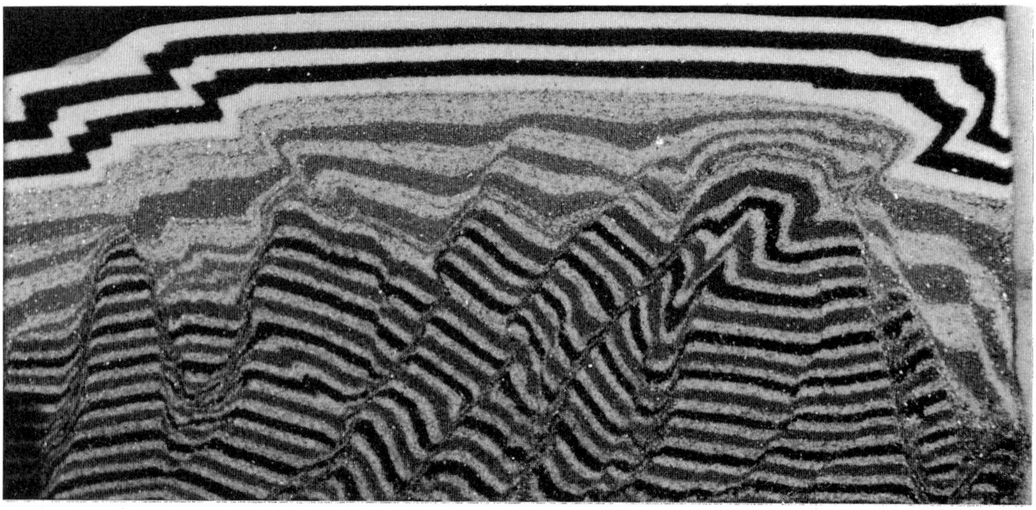

5 cm

FIG. 4. Series 2 experiment. Sand–mica and mica layered model with homogeneous extension and subsequent contraction above a horizontal detachment: A, End of extension (50%). The upper 6 light grey layers are the synrift sequence of sand–mica and mica layers added incrementally during extension. B, Line drawing of end extension showing fault sequence 1–16. C, End of contraction (19.1%) showing the upper four (black and white) layers of homogeneous sand (postrift sequence) deformed by moderately dipping contractional faults (kink-like shear zones). D, Line drawing of end contraction (19.1%). Faults 17–24 are generated in part by reactivation of pre-existing faults.

cease to be active. Synrift sediment wedges are pushed out of the half-grabens and the prerift sequence becomes deformed by contraction faults. At higher contractional strains only suitably oriented parts of the extensional faults become reactivated. In particular, the upper high angle parts of the extension faults are bypassed by shortcut contraction faults (cf. fault 19, Fig. 4D).

Series 3 experiments were sand–mica models (Table 1) that had a 5° tilt of the basal detachment (Fig. 5A) in order to attempt to produce extensional faults at a low angle to the layering. An array of conjugate faults was produced

B

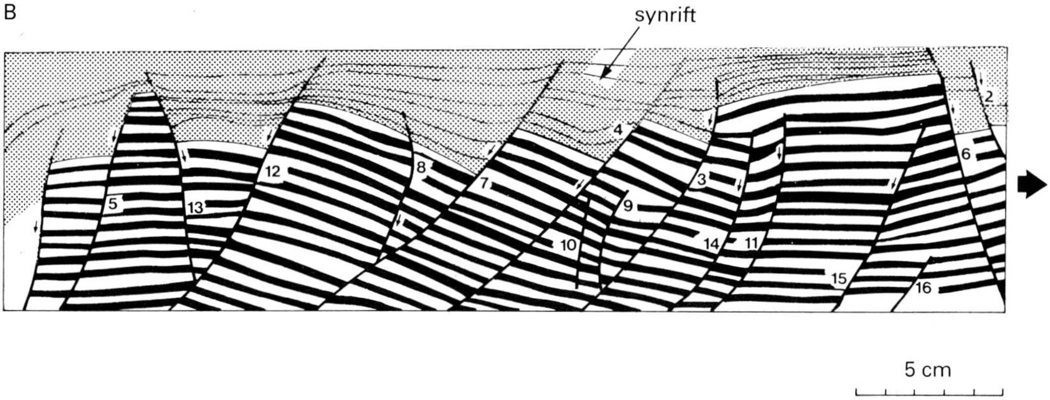

D

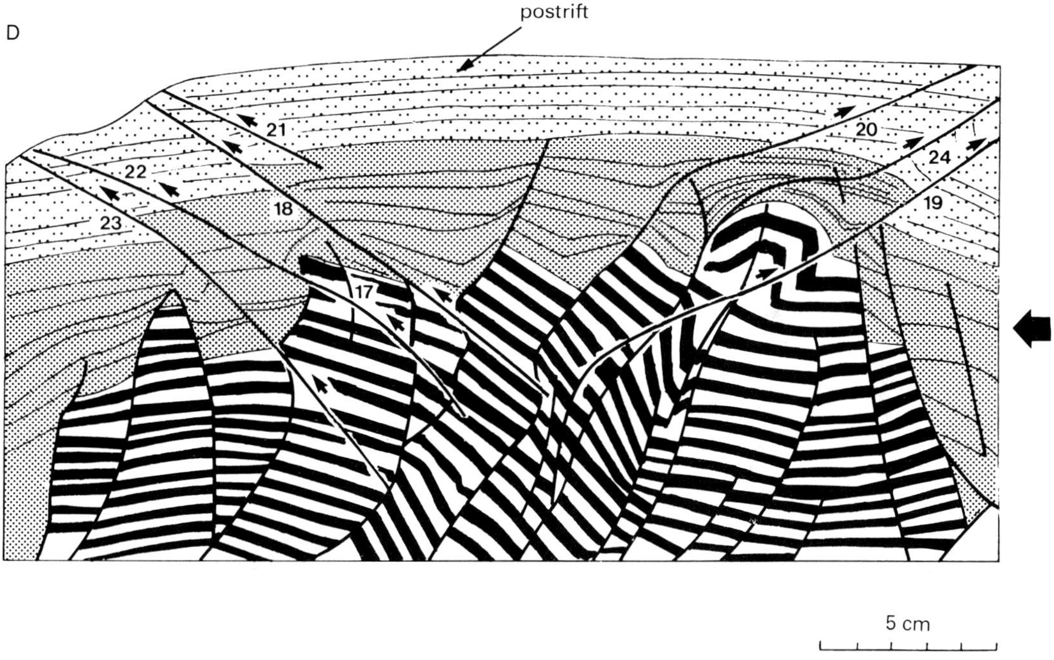

giving rise to well developed horst and graben structures (Figs 5B & C). For the contractional deformation, the basal detachment was returned to horizontal. Upon inversion with the contractional push from the right hand end of the model (Fig. 5E) those faults favourably oriented (e.g. faults 3, 6 and 13; see Fig. 5C) were reactivated. This produced minor thrust faults in the postrift sediments (Figs 5D & E) whilst retaining *net extension* at depth in the prerift sequence. The sequence of propagation of these faults is towards the foreland away from the tectonic push (Fig. 5E faults 20–24). At high bulk shortening strains, all of the extensional faults are rotated into steep angles with respect to the basal detachment and cease to be reactivated. Instead the prerift sequence begins to deform and through-going contractional faults with a back-thrust sense (e.g. fault 25 in Fig. 5E) develop. Series 3 experiments, in particular, show that only faults suitably oriented with respect to the direction of compression will become reactivated.

Inversion of simple listric fault systems

In extension, simple listric detachments produce a characteristic hanging wall geometry (Experiment series 4 and 5; Figs 6 & 7). The hanging wall extensional structures are characterized by

A

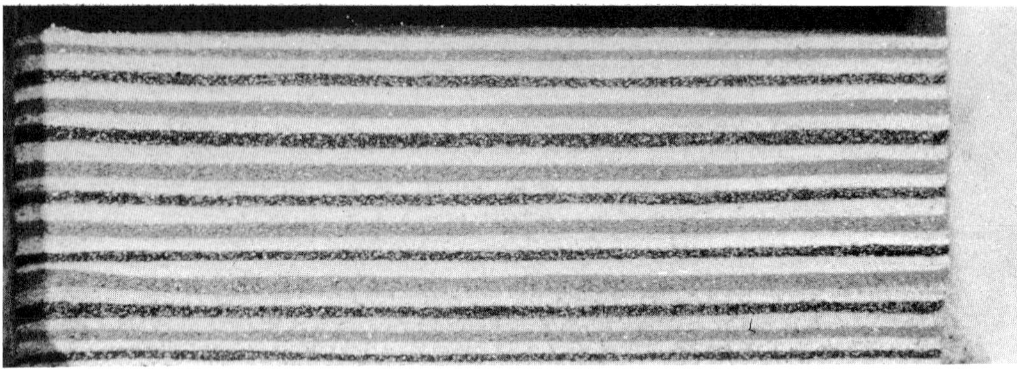

C

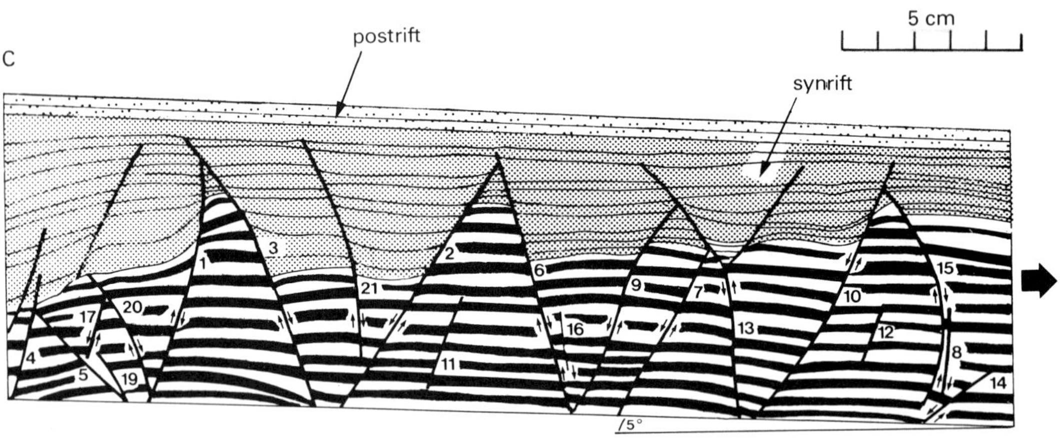
5 cm
postrift
synrift
1
2
3
4
5
6
7
8
9
10
11
12
13
14
15
16
17
19
20
21
/5°

E

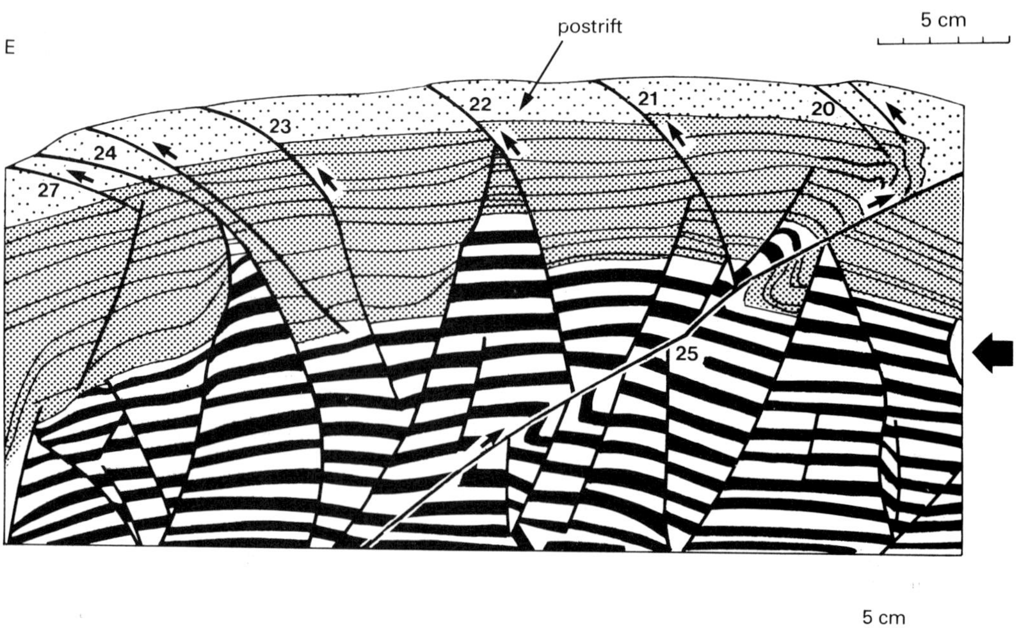
5 cm
postrift
20
21
22
23
24
25
27
5 cm

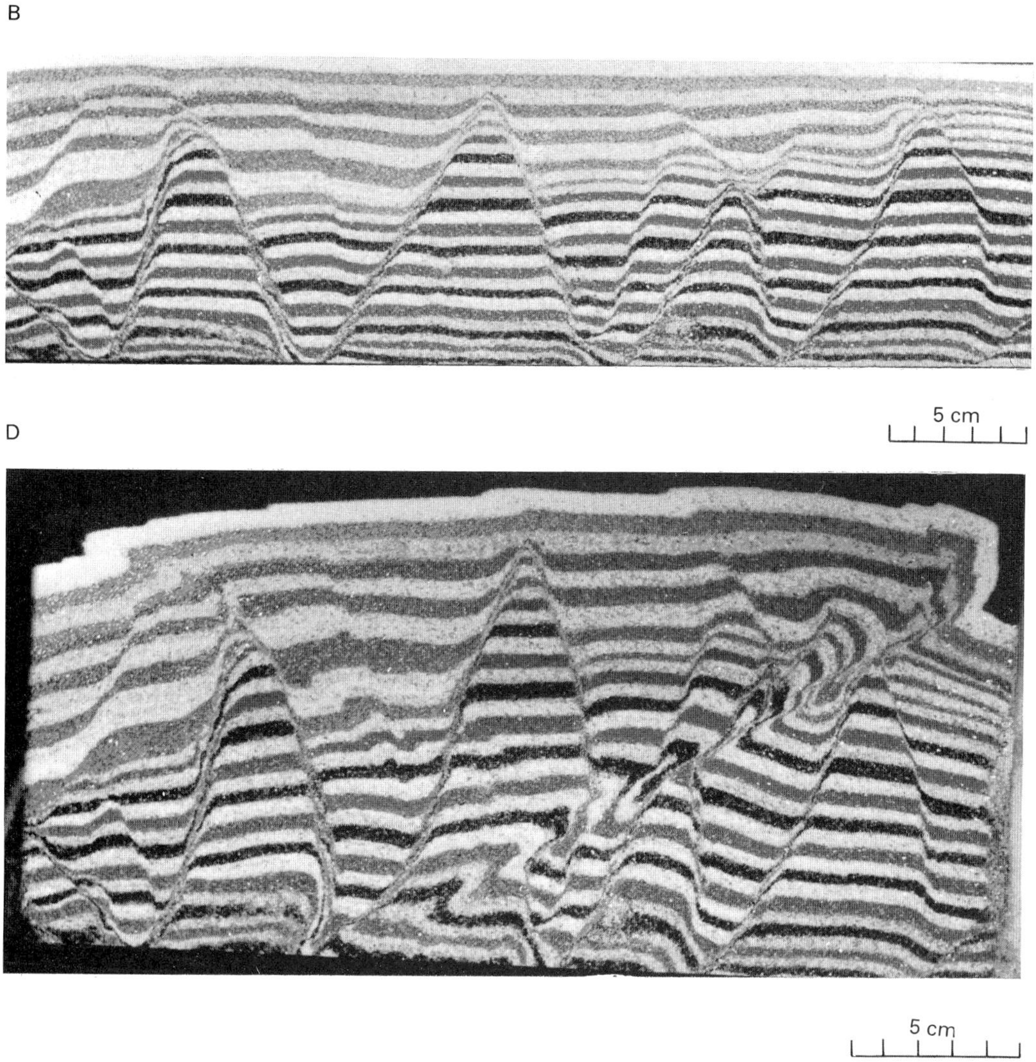

FIG. 5. Series 3 experiment. Sand–mica and mica layered model with homogeneous extension above a tilted detachment and subsequent inversion above a horizontal detachment. A, Initial starting geometry before extension. Basal detachment tilted at 5°. B, End extension 50%. The upper 10 light grey and white layers are the synrift sequence of sand–mica and mica layers added incrementally during extension. C, Line drawing of the end extension (50%) tilted down to the right by 5°. Faults 1–21 were generated in extension. Synrift sediments are stippled. D, End contraction (25%) with one white layer of postrift sediments. E, Line drawing of end contraction (25%) horizontal lower detachment. Faults 22–27 were generated during inversion.

a classical rollover anticline with a crestal collapse graben system of antithetic and synthetic faults (Figs 6A & B, 7A & B). These have been described in detail by McClay & Ellis (1987b) and Ellis & McClay (1988). The results of extension above a simple listric detachment for models using homogeneous sand and sand–clay layers (thus producing a competence contrast) are broadly similar (Figs 7A & B).

Extension of series 4 experiments with homogeneous sand produced a classic rollover anticline structure with multiple superposed crestal collapse graben faults (Figs 6A & B). Upon inversion the rollover anticline is accentuated and the major listric detachment (Figs 6C & D) becomes reactivated with maximum uplift generated above the listric segment of the detachment. The synrift sequence is pushed out of

A

C

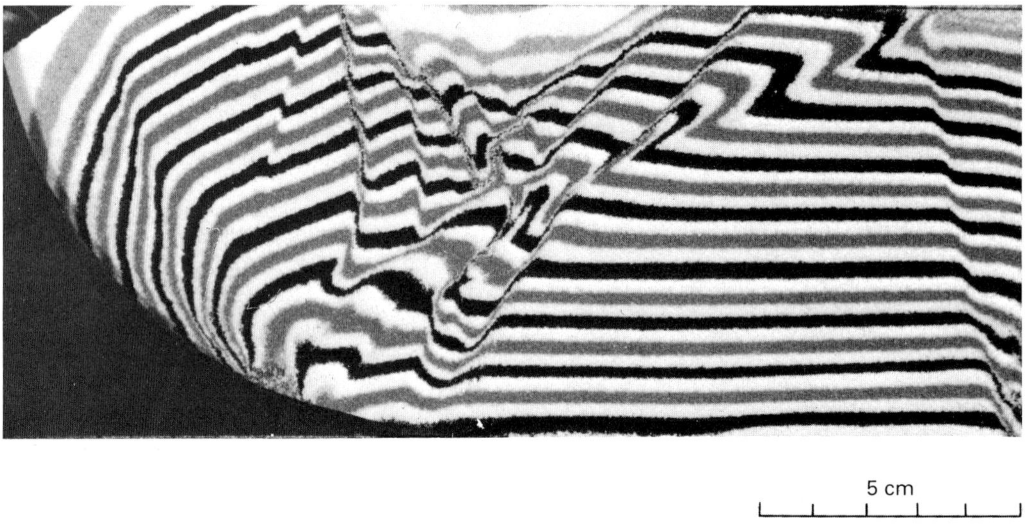

FIG. 6. Series 4 experiment. Homogeneous sand model above a simple listric detachment: A, End extension (48.7%). The upper eight layers (light grey and white) are synrift sediments deposited incrementally during extension. B, Line drawing of end extension (48.7%) showing fault sequence 1–14. C, End inversion (31.6%). D, Line drawing of end contraction (31.6%).

the crestal collapse graben, initially with both the synthetic and antithetic faults being reactivated. At higher shortening strains the synthetic faults become rotated into steep angles and cease to be reactivated. At approximately 22.5% shortening a major through going short-cut fault (back thrust) developed in the prerift sequences (Figs 6C & D).

Series 5 experiments using sand and dry clay layers to simulate competency contrasts (see Ellis & McClay 1988 for a full discussion) produced a classic rollover and crestal collapse graben structure in extension (Figs 7A & B). Upon inversion the major listric detachment was reactivated with material being pushed up and out of the apparatus and the relatively stiff and planar layers of the rollover being rotated back up the listric fault. There were only minor drag effects adjacent to the fault (Figs 7C & D). The rollover anticline is *not* accentuated as in series 4 experiments. The uplift associated with inversion is asymmetric concentrated above the listric detachment with both the antithetic and synthetic crestal collapse faults (faults 6–8,

B

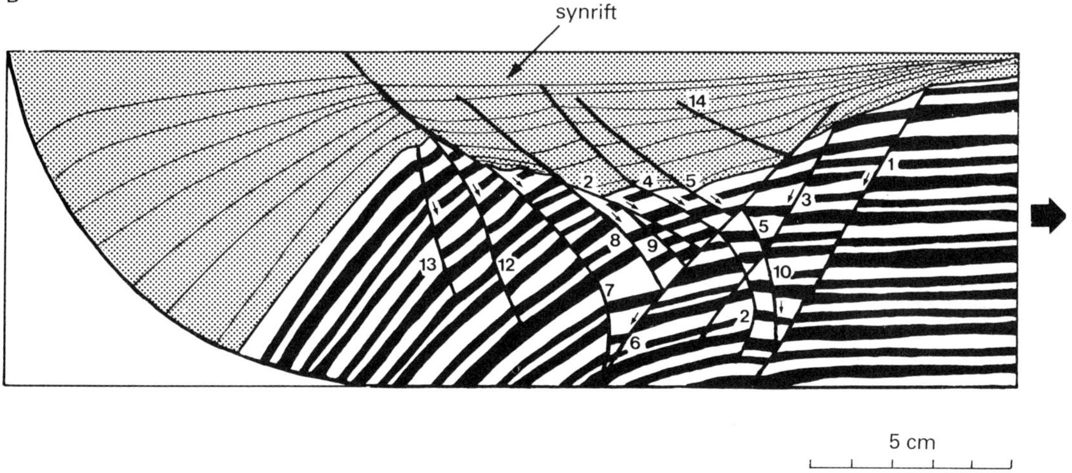

D

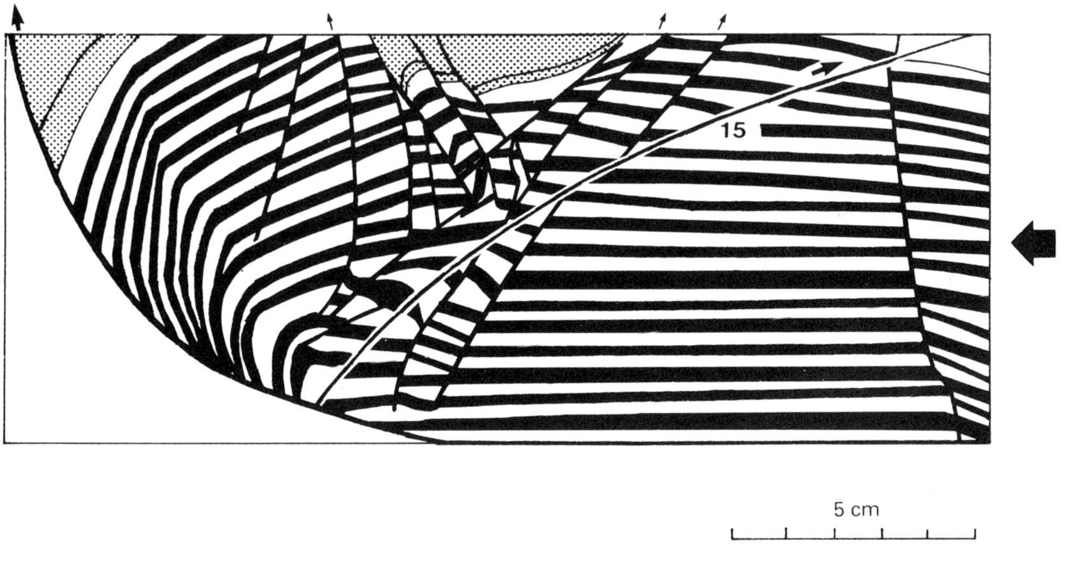

Fig. 7B) being reactivated. The synthetic crestal collapse faults are initially reactivated with hanging wall up. However, with increased inversion they become progressively rotated until at 22% contraction they pass through the vertical and the relative motion switches so that the original footwall becomes the hanging wall and the faults show true reverse displacement. At approximately 30% shortening reactivated fault 2 (Fig. 7B) takes a footwall shortcut and becomes a major back thrust (Figs 7C & D). At high values of inversion the crestal collapse structure has the appearance of a 'flower' or 'tulip structure' (Fig. 7D).

Inversion of ramp/flat listric fault geometry

Extension above a ramp/flat listric detachment (experiment series 6 and 7) shows typical rollover anticlines associated with each steepening upward segment of the detachment geometry, separated by a hanging wall syncline (Figs 8 & 9). Crestal collapse graben structures are associated with each rollover anticline.

Inversion of series 6 experiments (sand and mica layers) produced asymmetric uplift with reactivation of the major detachment and major uplift above the ramp. Synrift sediments were pushed out of the crestal collapse graben with

A

C

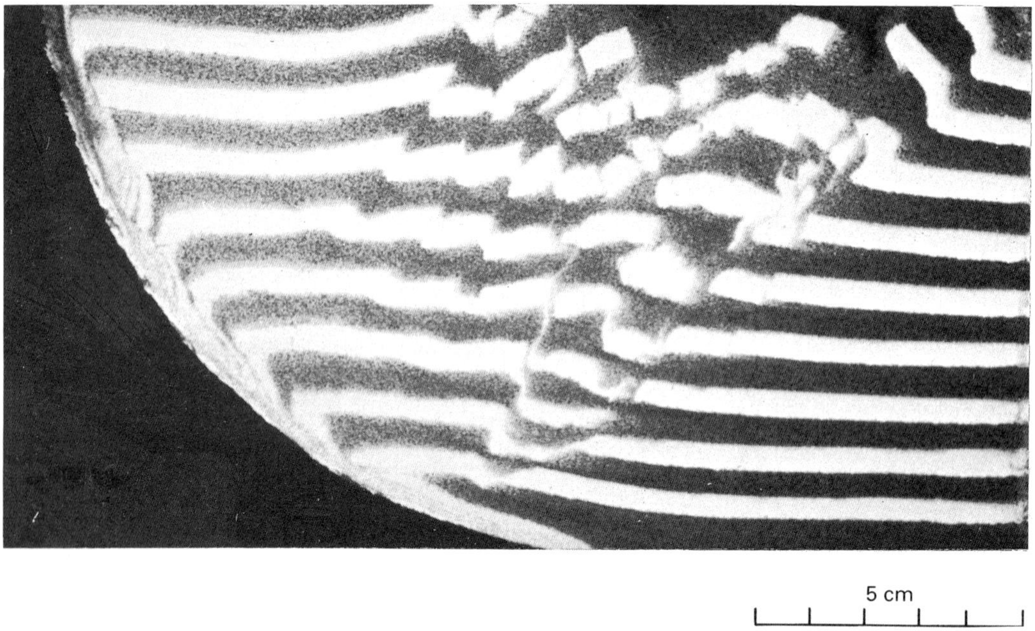

FIG. 7. Series 5 experiment. Sand and clay layers above as simple listric detachment: A, End extension (50%). The upper ten layers (black and white) are homogeneous sand synrift sediments deposited incrementally during extension. B, Line drawing of end extension (50%). Faults 1–8 were generated (note some faults appear to have nucleated simultaneously). C, End contraction (47%). D, End contraction (47%). Most extensional faults reactivated. The only new fault is the shortcut fault (9) on the extreme right.

reactivation of both the synthetic and antithetic graben boundary faults (faults 2–5, Fig. 8B). All of the synrift faults are rotated into higher angles upon contraction and a shortcut fault (fault 20 in Fig. 8D) develops in the footwall to the major graben bounding fault (fault 3 in Fig. 8B). The rollover anticline associated with the upper listric segment of the major detachment becomes accentuated and a remnant of synrift sediment is left in the footwall to the overturned nappe-like structure (Fig. 8B). The asymmetric character of the inversion leaves remnants of the synrift basins in the inverted structure.

Extension of series 7 experiments (sand–mica and mica layers) produced almost identical

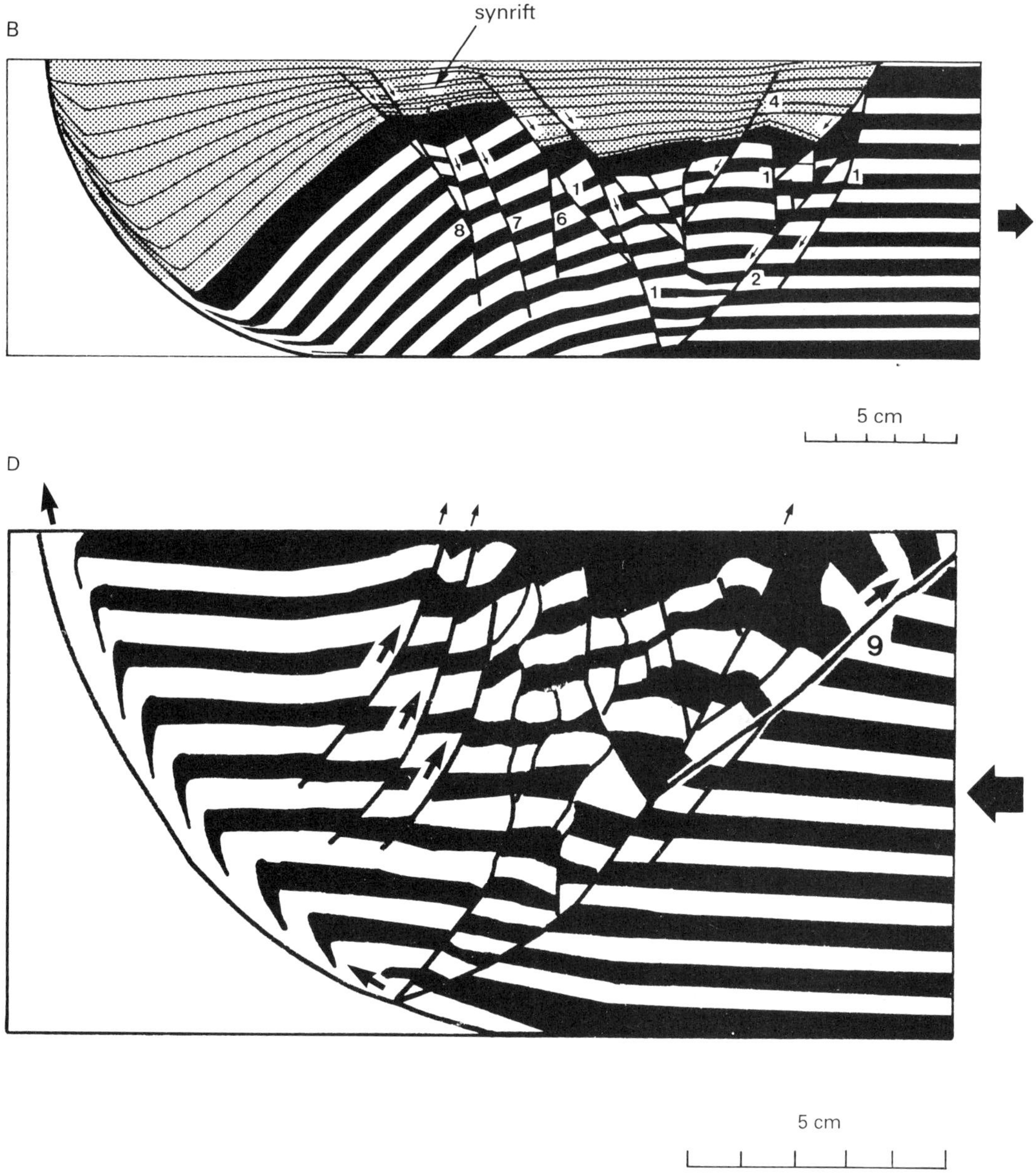

structures to those formed in series 6 experiments (Figs 8A & 9A). Inversion also produced similar results with a large nappe structure developing at high contractional strains. Out-of-the-graben faulting develops and maximum uplift occurs at the emergence of the major (reactivated) detachment and above the ramp. The greater shortening in series 7 experiments highlights the accentuation of the rollover anticline structure and the formation of a nappe-like structure (Figs 9C & D).

Discussion and conclusions

The results of inversion experiments presented in this paper are the first in an on-going programme of inversion modelling of a wide variety

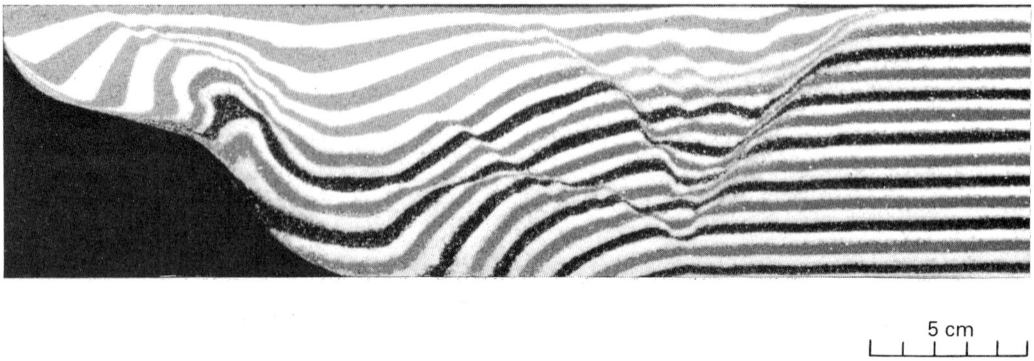

C

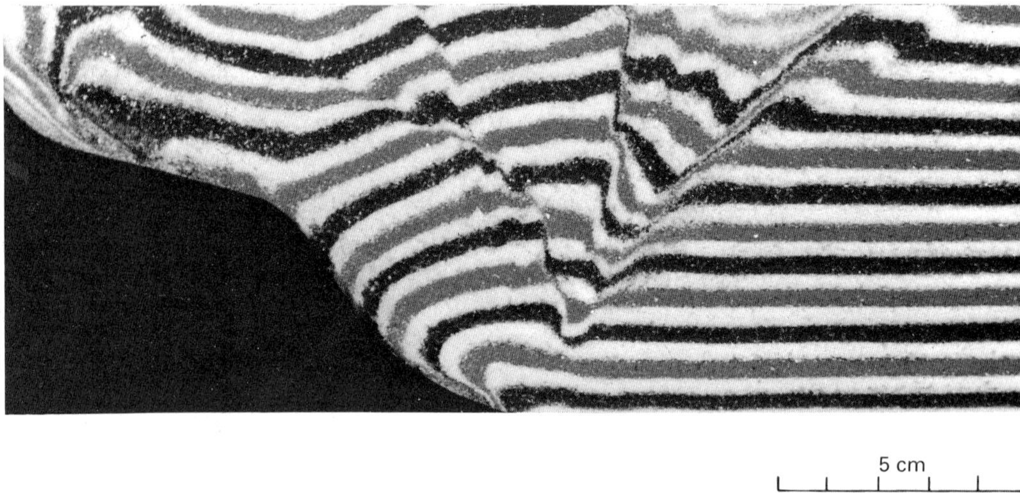

FIG. 8. Series 6 experiment. Sand and mica layers above a ramp/flat listric detachment: A, End extension (50%). Upper seven layers (light-grey and white) are homogeneous sand synrift sediments deposited incrementally during extension. B, Line drawing of end extension (50%). Fault sequence 1–15. C, End contraction (25.4%). D, Line drawing of end contraction (25%). Faults 2, 3, 4, 12, 15 were reactivated (B) whereas fault 20 is a footwall shortcut to fault 3.

of basin geometries. Attention has been focused on the inversion of dip–slip fault systems by horizontal compression above a flat basal detachment. Although problems were encountered in obtaining reactivation in some fault systems due to a lack of 'memory' in the modelling material, the basic features of inversion of dip–slip systems can be illustrated.

Inversion above a uniform basal detachment – domino fault arrays

All of the inversion experiments of this type (Table 1) produced similar results. Only suitably oriented faults (i.e. those at angles less than 55–60° to the compression direction) become reactivated. The amount of displacement during reactivation was generally small and reactivation ceased when the faults became rotated to steeper angles. For half grabens where bounding faults were not suitably oriented for inversion, the synrift sediments became pushed out of the half graben along out-of-the-half-graben thrust faults, parallel to the tilted layering. At higher values of shortening (>15%), shortcut faults develop in the footwalls to steep fault segments.

The structures developed during the inversion of planar or conjugate fault arrays are similar to those found in the S North Sea, e.g. Sole Pit inversion (Glennie & Boegner, 1980) or Quadrant 53 (Badley & Price, 1988). In these cases

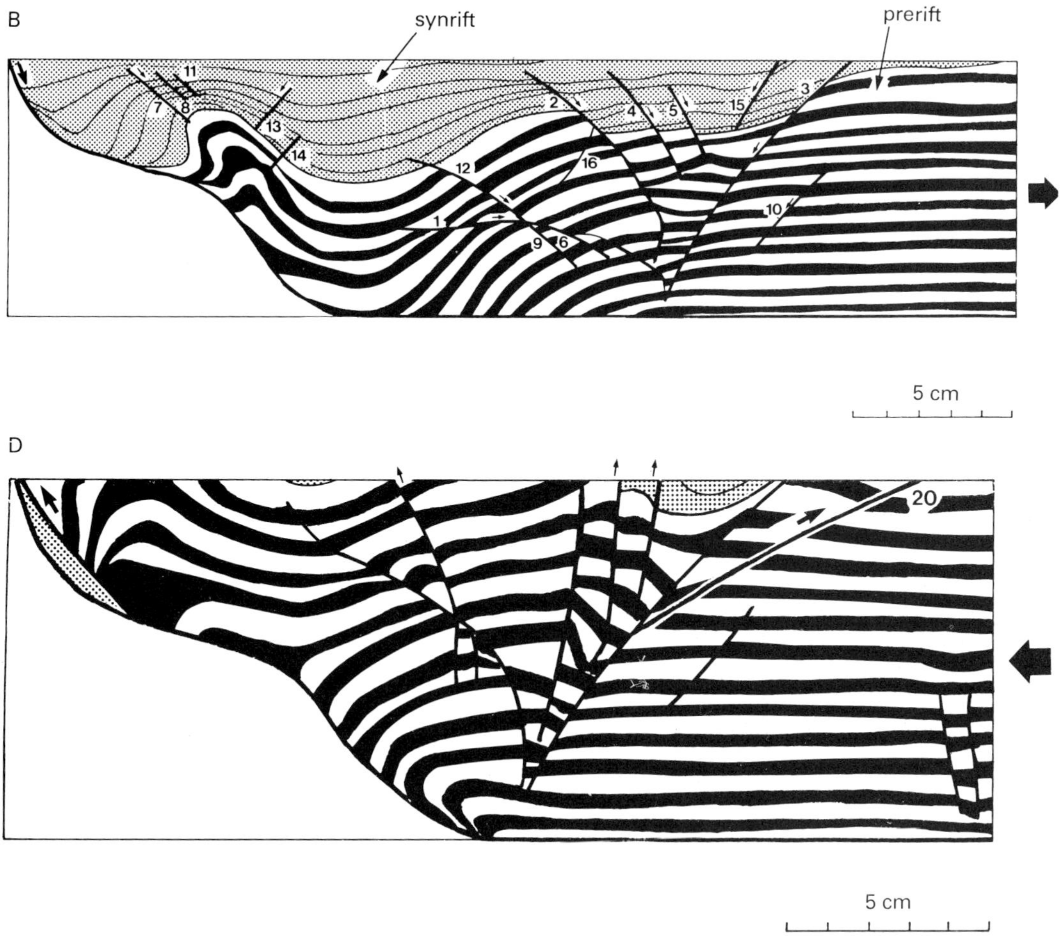

the planar faults have limited reactivation and retain net extension at depth.

Inversion above a simple listric detachment

Inversion of simple listric extensional fault systems results in reactivation of the main detachment surface with marked asymmetric uplift centred at the emergence of the major detachment. In homogeneous sand and sand–mica models the geometry of the rollover anticline is accentuated upon inversion. In sand–clay models (the clay layers behave as competent rigid beds) this effect is not found. The crestal collapse graben fault systems become reactivated. The synthetic faults undergo major *rotation* until they become too steep (>65°) for inversion. However, in the case of the sand–clay models (series 5 experiments) the synthetic convex up extension faults rotate through the vertical and become *reutilized* as thrust faults (Fig. 7D). At high contractional strains the crestal collapse graben structures narrow and develop an apparent 'flower' form. The graben bounding antithetic faults develop significant penetrative shortcut faults that allow contraction to continue. The solid footwall geometry of this experimental configuration places a significant constraint on the inversion geometry in that footwall shortcut faults cannot develop.

Inversion of a ramp/flat listric detachment

Inversion of ramp/flat listric extensional fault systems result in reactivation of the major basal detachment. Uplift is greatest at the emergence of the detachment and above the ramp segment of the detachment. As in the case of simple listric geometries, the crestal collapse graben fault systems become reactivated and rotated so

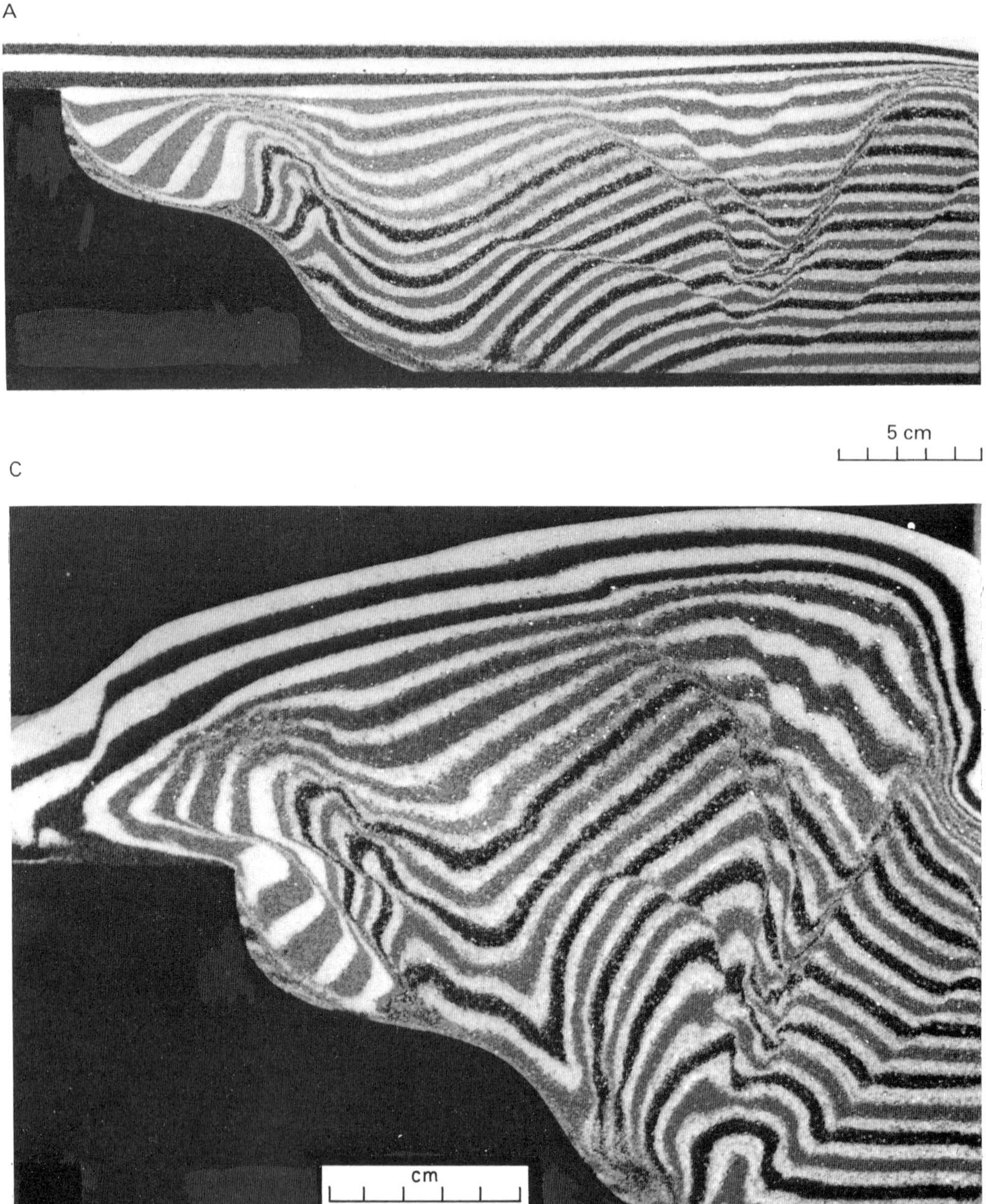

FIG. 9. Series 7 experiment. Sand–mica and mica layers above a ramp/flat listric detachment: A, End extension (50%). The top four layers (undeformed) are post-rift homogeneous sand layers. In the deformed model the upper ten layers (alternating light-grey and white) are sand/mica and mica synrift sediments deposited incrementally during extension. B, Line drawing of end extension (50%). Fault sequence 1–10. Synrift sediments are stippled. C, End contraction (40%). The model is now shorter than the original pre-extension geometry. D, Line drawing of end contraction (40%). The model is shorter than the original pre-extension geometry. New faults 14 to 24 are highlighted. Synrift sediments are stippled.

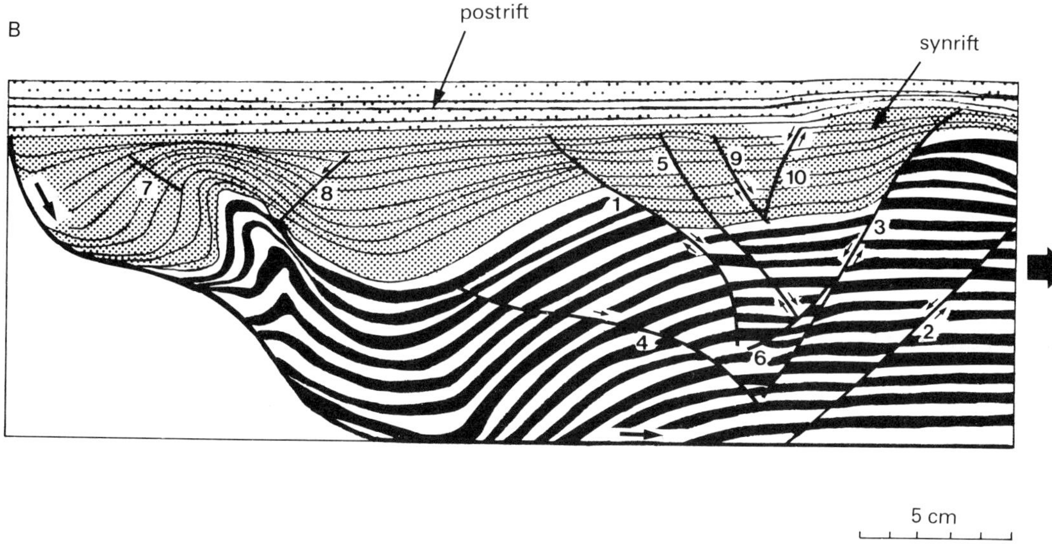

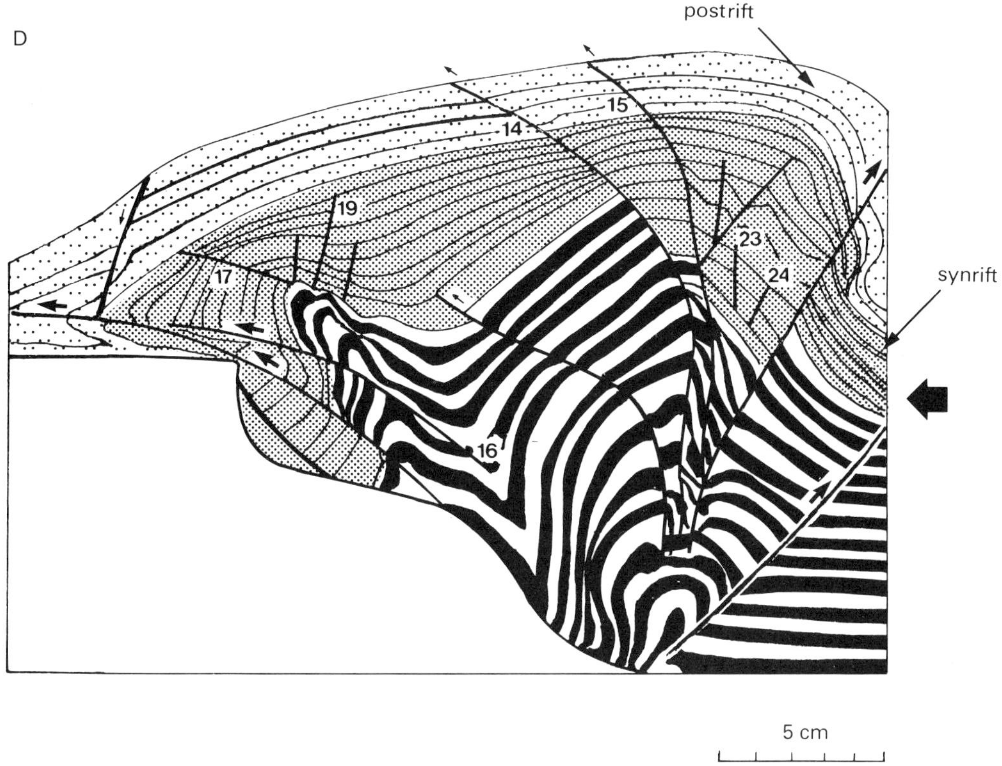

that the graben narrows. The resultant fault geometry of the crestal collapse graben has a 'flower-like' appearance which resembles structures often cited as evidence of strike–slip or oblique–slip tectonics (Harding 1985; Gibbs 1986; and others). The geometry of the upper rollover anticline becomes strongly accentuated and the formation of an overturned limb leads to the formation of nappe-like geometries.

In the experiments described in this paper there are a number of common elements in the contraction and inversion process. High angle extension faults which may initially undergo reactivation, subsequently rotate and lock, ceasing to be active. Once locked up these faults are bypassed by footwall shortcuts that develop significant displacement. In all of the experiments contraction is accompanied or accomplished by significant folding and bulk strain that thickens layers. In the cases of contraction above listric fault geometries maximum uplift occurs in the region above the position where the fault flattens out into the basal detachment. Significant back thrusts are nucleated in this region. For contraction of ramp/flat listric geometries significant uplift occurs above the convex up ramp segment of the basal detachment surface.

These preliminary inversion experiments of dip–slip fault systems have provided some models similar to those found in the southern North Sea. In particular, new fault patterns and complexities for listric systems have been found and their applicability needs to be tested. Flower structures can be developed by simple inversion of dip–slip fault systems. Strike–slip movement need not be involved. The formation of nappe structures from the compression of listric ramp/flat geometries may be of great importance in explaining the formation of large recumbent nappe structures in orogenic belts.

ACKNOWLEDGMENTS: This research was financed in part by Royal Holloway and Bedford New College and in part by British Petroleum PLC who are thanked for permission to publish. John Simmonds gave great assistance with the experiments and Craig Hildrew ably drafted the diagrams. Sarah Viggers is thanked for typing the manuscript.

References and further reading

BADLEY, M. & PRICE, J. D. & BACKSHALL, L. C. Inversion, reactivated faults and related structures; seismic examples from the southern North Sea. *This volume*.

BRUN, J. P. & CHOUKROUNE, P. 1983. Normal faulting and décollement in a stretched crust. *Tectonics*, **2**, 345–356.

CLOOS, E. 1986. Experimental analysis of Gulf Coast fracture patterns. *Bulletin of the American Association of Petroleum Geologists*, **52**, 420–444.

DAVIS, G. A., SUPPE, J. & DAHLEN, F. A. 1983. Mechanics of fold and thrust belts and accretionary wedges. *Journal of Geophysical Research*, **88**, 1153–1172.

ELLIS, P. G. & MCCLAY, K. R. 1988. Listric extensional fault systems — results of analogue model experiments. *Basin Research* **1**, 55–70.

GIBBS, A. D. 1986. Strike–slip basins and inversion: a possible model for the Southern North Sea Gas Areas. *In*: BROOKS, J., GOFF, J. C. & VAN, HOORN, B. (eds) *Habitat of Palaeozoic Gas in NW Europe. Special Publication of the Geological Society of London*, **23**, 23–35.

GLENNIE, K. W. & BOEGNER, P. L. F., 1980. Sole Pit Inversion Tectonics. *In*: ILLING, L. V. & HOBSON, G. D. (eds). *Petroleum Geology of the Continental Shelf of Northwest Europe*. Heyden, London, 110–120.

HARDING, T. P. 1985. Seismic characteristics and identification of negative flower structures, positive flower structures and positive structural inversion. *Bulletin of the American Association of Petroleum Geologists*, **69**, 582–600.

HORSFIELD, W. T. 1977. An experimental approach to basement controlled faulting. *Geologie en Mijnbouw*, **56(4)**, 363–370.

—— 1980. Contemporaneous movement along crossing conjugate normal faults. *Journal of Structural Geology*, **2(3)**, 305–310.

KOOPMAN, A., SPECKSNIJDER, A. & HORSFIELD, W. T. 1987. Sand. box studies of inversion tectonics. *Tectonophysics*, **137**, 379–388.

MANDL, G. 1984. Rotating parallel faults — 'bookshelf' Mechanism. Abstract in *Bulletin of the American Association of Petroleum Geologists*, **68**, 502–503.

MCCLAY, K. R. & ELLIS, P. G. 1987a. Analogue models of extensional fault geometries. *In*: COWARD, M. P., DEWEY, J. F. & HANCOCK, P. L. (eds) *Continental Extensional Tectonics, Special Publication of the Geological Society of London*, **28**, 109–125.

—— & —— 1987b. Geometries of extensional fault systems developed in model experiments. *Geology*, **15**, 341–344.

NAYLOR, M. A., MANDL, G. AND SIJPESTEIN, C. H. K. 1986. Fault geometries in basement-induced wrench faulting under different initial stress states. *Journal of Structural Geology*, **8**, 737–752.

TCHALENKO, J. S. 1970. Similarities between shear zones of different magnitudes. *Geological Society of America Bulletin*, **81**, 1625–1640.

VENDEVILLE, B., COBBOLD, P. DAVY, P., BRUN,

J. P. & Choukroune, P. 1987. Physical Models of extensional tectonics at various scales. *In*: Coward, M. P., Dewey, J. F., & Hancock, P. L. (eds) *Continental Extensional Tectonics*; *Special Publication Geological Society of London*, **28**, 95–107.

Wilcox, R. E., Harding, T. P. & Seely, D. R. 1973. Basic wrench tectonics. *Bulletin of the American Association of Petroleum Geologists*, **57(1)**, 74–96.

K. R. McClay, Royal Holloway and Bedford New College, University of London, Egham Hill, Egham, Surrey, TW20 0EX, UK.

Inversion in the Alps and Alpine Foreland

Geodynamic model for Alpine intra-plate compressional deformation in Western and Central Europe

P. A. Ziegler

SUMMARY: The Mesozoic grabens and wrench induced troughs of Western and Central Europe developed in response to Triassic to Early Cretaceous intra-plate tensional stresses which affected the Arctic–North Atlantic and the Tethys borderlands during the rifting phases preceding the opening of the respective oceans. During the Alpine orogeny, collision-related compressive stresses exerted on the foreland induced the reactiviation of pre-existing fracture systems, and caused the inversion of Mesozoic grabens and the uplifting of major basement blocks at distances up to 1300 km to the north of the present Alpine deformation front. The structural style of these inversion structures is indicative of compression and transpression. Total crustal shortening is unlikely to exceed a few tens of kilometres. These displacements require a coupling between the foreland and the orogen at the Alpine A-subduction zones and, within the foreland, a decoupling at intra-crustal levels between the crust and mantle and/or at deeper lithospheric levels.

The post-Hercynian tectonic framework of Western and Central Europe is characterized by a complex array of highs composed of Palaeozoic and older metamorphics and sediments, and intervening Mesozoic and Cenozoic sedimentary basins. At the top basement level many areas are characterized by a high structural relief that often involves intense faulting, with a throw on individual faults in the order of several thousand metres (Fig. 1).

Western and Central Europe owes its intense post-Hercynian structuring to multiple deformation phases ranging in age from Permo–Carboniferous to Cenozoic. During this time span regional stress regimes changed repeatedly so that pre-existing crustal discontinuities became reactivated time and again. This applies particularly to Mesozoic basins which developed under a tensional to transtensional setting and which became inverted during the Late Cretaceous and/or Cenozoic in response to compressional/transpressional stresses. Some of these tectonic features subsequently became overprinted by the Eocene and younger Rhône–Rhine rift system or were buried under the thick Cenozoic sediments of the North Sea Basin or those of the Alpine and Carpathian foredeeps.

The objective of this paper is to set out a geodynamic rationale for the Late Cretaceous and Cenozoic compressional foreland deformation evident in Western and Central Europe, the timing and distribution of which appears to be related to Alpine plate collision. However, the hypotheses advanced in this paper require further qualitative and quantitative evaluation, both on a local and a regional scale. They should, therefore, be regarded as working hypotheses.

Main phases of post-Variscan intra-plate tectonics

The observed post-Hercynian structuring of the Alpine foreland areas of Western and Central Europe, as summarized in Fig. 1, is the result of a sequence of different intra-plate processes which reflect repeated changes in the mega-tectonic setting of these areas. Basically, four deformation cycles can be recognized since the latest Carboniferous consolidation of the Variscan fold belt.

Permo–Carboniferous wrench faulting

The first phase of structuring, post-dating the Late Westphalian termination of crustal shortening in the Variscan fold belt, occurred during the Stephanian and Early Permian. This deformation phase coincides with the terminal phases of the Hercynian orogeny during which continued crustal shortening in the Urals and the Appalachians was accompanied by important dextral translations between Europe and Africa. In Europe these plate movements induced the development of a complex set of conjugate shear faults (Ziegler 1982; 1984; 1986). Activity along these faults was associated with local transpressional deformations, the subsidence of half-grabens in response to diverging wrench movements (e.g. Saar–Nahe and Jura Mountain troughs, basins in the Bohemian Massif and

From Cooper, M. A. & Williams, G. D. (eds), 1989. *Inversion Tectonics*
Geological Society Special Publications No. 44, 63–85.

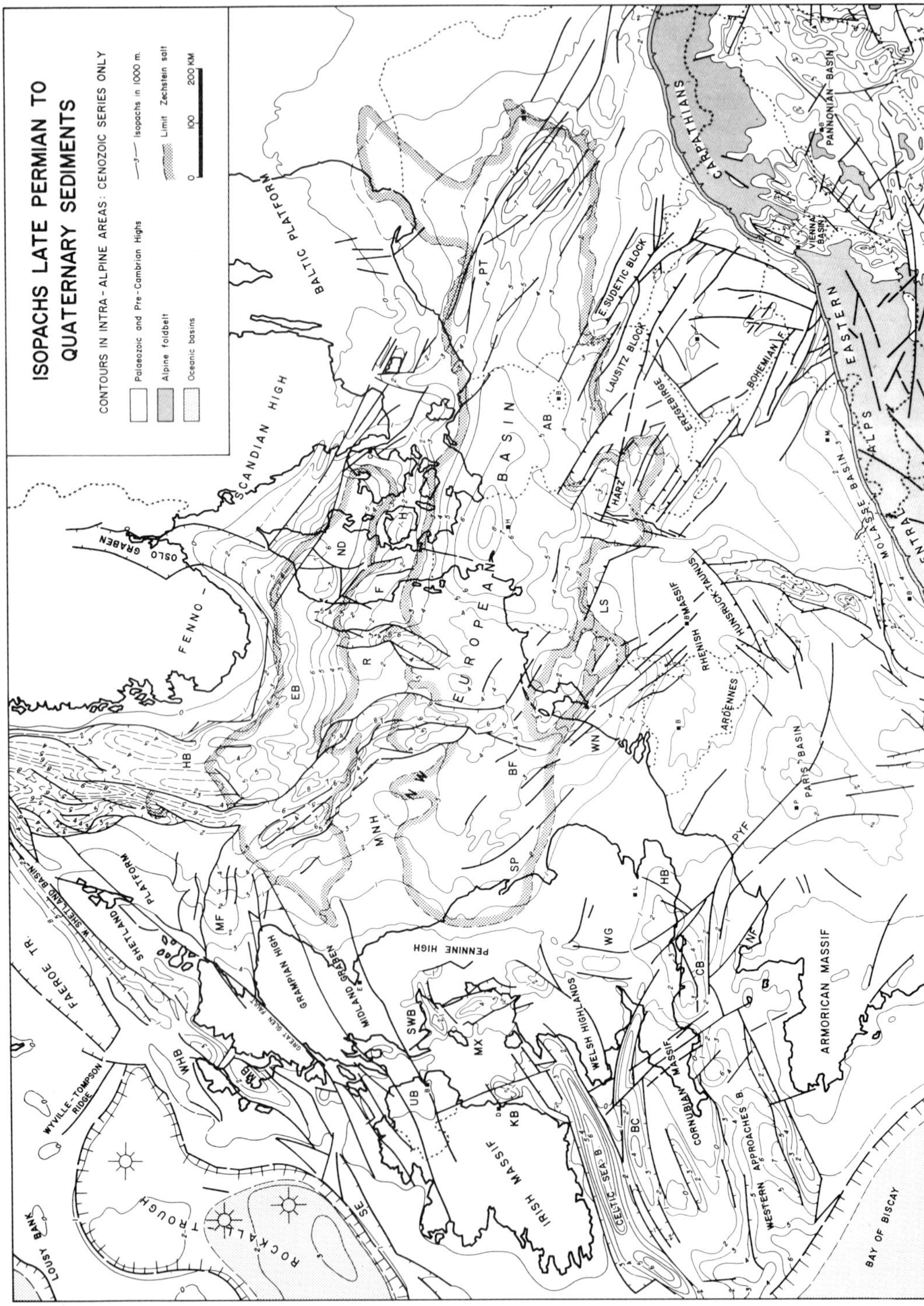

FIG. 1. Isopachs of Late Permian to Quaternary sediments.

Massif Central) and of pull-apart basins at the termination of major wrench faults (e.g. Oslo Graben, Schneverdingen Graben; Drong *et al.* 1982) as well as widespread volcanism. The development of this extensive Permo–Carboniferous fracture system, which dissected the Variscan fold belt and its foreland areas, reflects a diffuse plate boundary between Africa and Europe.

After the late Early Permian Alleghanian consolidation of the Appalachian fold belt this fracture zone became largely inactive. Decay of the thermal anomalies induced by the Permo–Carboniferous tectonism in the area of the North Sea, northern Germany and Poland governed the subsidence of the Northern and Southern Permian basins of Northwestern and Central Europe (Ziegler 1982; 1988). However, certain elements of the Permo–Carboniferous fracture system became repeatedly reactivated during the Mesozoic and Cenozoic evolution of the Alpine foreland (e.g. Tornquist–Teisseyre Line, Bohemian Massif, Sole Pit and West Netherlands Basin). This demonstrates that the Late Hercynian fracture systems had significantly weakened the cohesion of the crust of Western and Central Europe.

Mesozoic rifting and wrench faulting

The second phase of structuring of Western and Central Europe is related to the break-up of Pangea in the Mesozic. During this episode intra-plate tensional stresses affected the Arctic–North Atlantic and the Tethys borderlands until crustal separation was achieved and oceanic basins began to open in the Atlantic and Tethys realms. These stresses partly reactivated the Permo–Carboniferous fracture system. This localized some of the Triassic–Early Cretaceous rift and wrench systems of Western and Central Europe (see Ziegler 1982; 1988, Fig. 2).

Rifts associated with the northern Tethys margin remained active until the mid-Jurassic opening of the south Penninic–Piedmont–Ligurian and the Alboran oceans (Trümpy 1982; Debelmas *et al.* 1983; Lemoine 1985; Ziegler 1988). Troughs branching off the Aquitaine–Bay of Biscay rift/wrench zone, such as the Celtic Sea and Western Approaches troughs remained active until the mid-Aptian crustal separation between the Grand Banks area, Europe and Iberia and the onset of sea-floor spreading in the Bay of Biscay (Montadert *et al.* 1977; Olivet *et al.* 1984).

The North Sea rift system and the wrench induced troughs associated with it remained active until the Aptian–Albian. Last rifting movements in the Central and Northern North Sea preceded the Late Palaeocene–Early Eocene crustal separation in the Norwegian Greenland Sea (Eldholm & Talwani 1981; Ziegler 1982; Bukowicz & Ziegler 1985).

The subsidence of the Mesozoic rifts and wrench-induced basins of Western and Central Europe was accompanied by substantial crustal thinning. This is seen, for instance, in the Central Graben of the North Sea (Barton & Wood 1984). This, in combination with wrench faulting, resulted in a further fragmentation and weakening of the crust of Western and Central Europe. Some of the wrench deformation was transpressional and caused local basin inversion. This is evident, for example, in the Late Jurassic–Early Cretaceous development of the West Netherlands Basin (Bodenhausen & Ott, 1981) and in the Early Cretaceous evolution of the Celtic Sea and Western Approaches area as well as along the SW Bohemian border zone (see Ziegler 1987).

Cretaceous and Cenozoic foreland compression

The third phase of intra-plate deformation, which spanned the latest Cretaceous and Cenozoic, involved the compressional–transpressional uplift of the Variscan Massifs and the inversion of major Mesozoic, and in part also Permo–Carboniferous, grabens and troughs at distances up to 1300 km to the N and NW of the present day Alpine deformation front (Fig. 2). This deformation was contemporaneous with the Alpine orogeny (Trümpy 1980). A causal relationship is therefore suggested between the Alpine plate collision and the observed intra-plate compressional and transpressional deformation of the Alpine foreland.

Cenozoic rifts

The latest Cretaceous and Cenozoic compressional–transpressional foreland deformation was contemporaneous with the last rifting phases in the Norwegian–Greenland Sea and with the Late Eocene and younger development of the Rhône–Bresse–Rhine rift, the Eger Graben and the Vienna–Pannonian system of intra-Alpine grabens (Ziegler 1982; 1988).

The evolution of the Rhône–Rhine rift systems and the Eger Graben represents the fourth phase of intra-plate tectonics in the Alpine foreland. This graben system transects the northwestern and western foreland of the Alps and extends southward into the Western Mediterranean. Subsidence of this graben system was coupled with extensive volcanism and

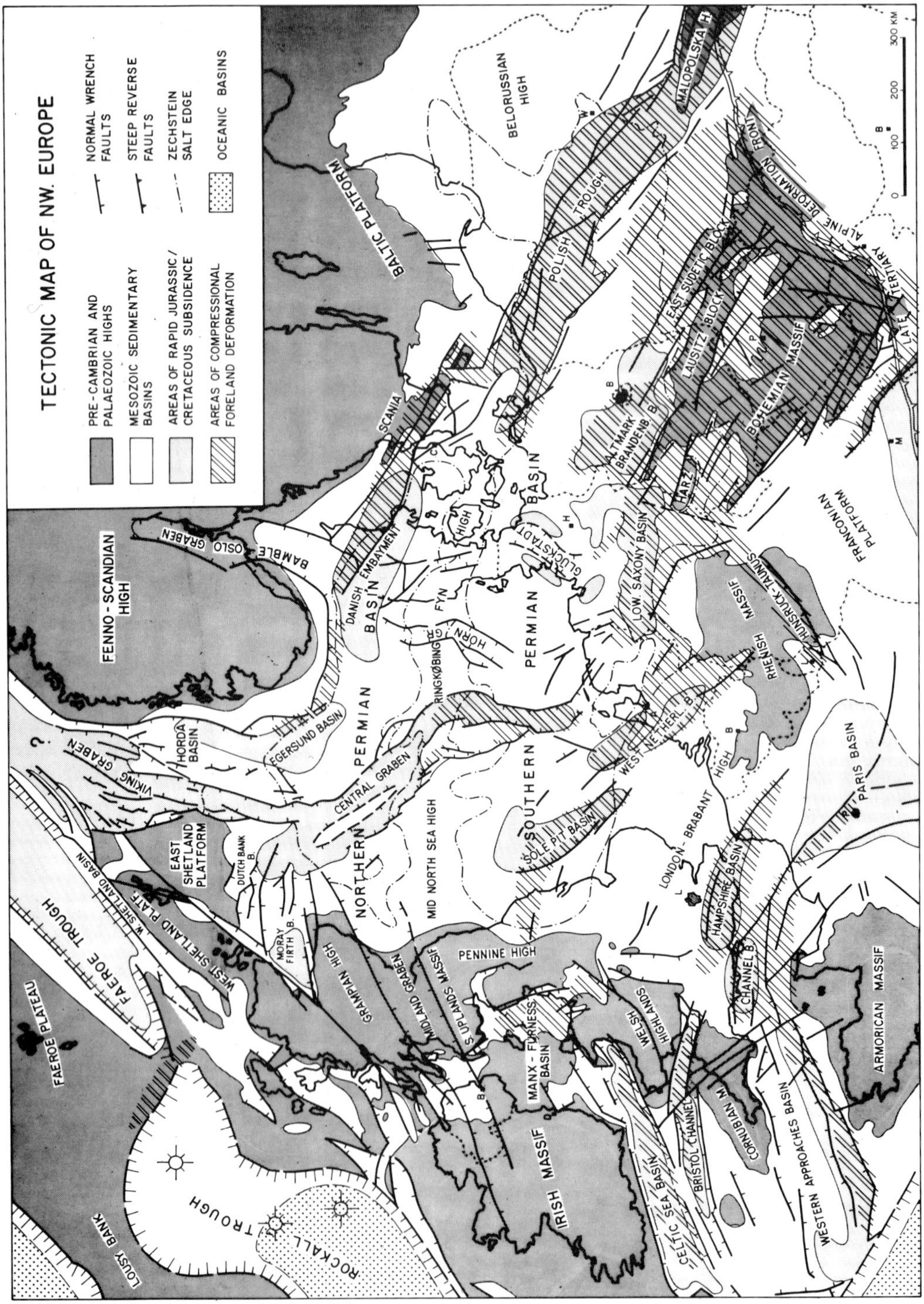

FIG. 2. Mesozoic tectonic map of NW Europe.

the uplift of broad thermal domes inducing deep truncation of the Mesozoic cover of the affected areas. Development of the Rhône–Rhine Rift is broadly synchronous with the Eocene and younger inversion of Mesozoic basins in the Celtic Sea–Western Approaches, Channel and southern North Sea areas. Periods of active rifting appear to alternate with the main inversion phases of these basins.

Compressional foreland deformation

The latest Cretaceous and Cenozoic intra-plate compressional deformations affected most of the major Mesozoic rifts and wrench-induced troughs in the Alpine foreland. The most distal deformation is located in the Central North Sea, 1300 km NW of the present day Alpine thrust front (Fig. 2). Overall, a decrease in the intensity of deformation can be observed with increasing distance from the Alpine front. The only non-inverted rifts which fall within the radius of these foreland deformations are the essentially Triassic Horn Graben in the Danish part of the North Sea and the Glückstadt Graben in Schleswig–Holstein in Germany and the Permian Oslo Graben.

The stratigraphic record of the several inverted basins of Central Europe shows that their deformation was not synchronous. A systematic shift of tectonic activity from the N to the NW Alpine foreland can be observed through time (for details, see Ziegler 1987).

Sub-Hercynian inversion phase

In the Northern Alpine foreland the first phase of basin inversion, which can be related to the Alpine plate collision, occurred during the Late Turonian to Early Senonian. This deformation, colloquially referred to as the 'sub-Hercynian' phase, culminated variably during the Santonian to Campanian. Deformations of this age are restricted to areas between the Polish Trough in the E and the Sole Pit and West Netherlands–Broad Fourteens Basin to the W. They are also recognized along the Fennoscandian border zone and in the Danish and Dutch sectors of the North Sea Central Graben (Fig. 3).

The initial phase of inversion of the Sole Pit Basin, the Lower Saxony Basin and the Danish Central Graben are dated as Late Turonian to Santonian. They thus predate the Santonian–Campanian main sub-Hercynian deformation phase.

In the area of the Bohemian Massif the transpressional uplifting of major basement blocks along Permo–Carboniferous and Late Jurassic fracture zones also began during the Late Turonian–Early Senonian.

Laramide inversion phase

In the Alpine foreland tectonic activity apparently abated during the Maastrichtian and Danian but resumed again during the Middle Palaeocene. This gave rise to the so-called Laramide or main phase of inversion of the Polish Trough, the Altmark–Brandenburg, sub–Hercynian, Lower Saxony and West Netherlands–Broad Fourteens basins. It also cause inversion in the Dutch and Danish part of the Central North Sea Graben and the Fennoscandian border zone. The most northerly traces of basin inversion are reported from the Egersund Basin in the offshore of Southern Norway (Pegrum 1984) (Fig. 4). The Sole Pit Basin was apparently not affected by the Laramide tectonism. Mild mid-Palaeocene inversion movements have been recorded in the Celtic Sea Trough, the Fastnet Basin, the Channel area and in the Paris Basin. In contrast, the Western Approaches Basin was hardly affected by the Laramide movements.

The main phase of deformation of the Bohemian Massif and of the areas occupied by the eastern parts of the Molasse Basin and the Northern Carpathian foredeep is probably also of a mid-Palaeocene age although the stratigraphic evidence is not conclusive. Similarly, stratigraphic evidence to determine whether the South Hunsrück–Taunus Border zone was also compressionally reactivated during the Laramide phase of foreland deformation is also lacking. However, given the location of the Rhenish Massif between the inverted West Netherlands and Lower Saxony basins and the Alpine collision front, such deformations would appear plausible.

In the Central Alps, similar foreland deformations are considered to be associated with the latest Cretaceous–Early Tertiary unconformity recognized in the area of the Aare Massif, the Sub-Helvetic and Helvetic Nappes (Herb 1965). Results of deep exploration wells drilled in the Vienna Basin and in the externides (external, non-metamorphic parts of a fold belt) of the Carpathian mountains show that the intra-Palaeocene deformations observed in the Eastern Molasse Basin and the Carpathian foredeep extend a considerable distance southward into the authochthonous substratum of the Alpine and Carpathian nappes (Brix *et al.*

FIG. 3. Sub-Hercynian foreland deformation.

Fig. 4. Laramide foreland deformation.

1977; Kolarova & Roth 1977; Wachtel & Wessely 1981; Heller & Moryc 1984).

Late Eocene–Early Oligocene inversion phase

Mild Late Eocene to Early Oligocene inversion movements are evident in the Celtic Sea–Western Approaches, Channel area and the West Netherlands–Broad Fourteens basins. Post-Palaeocene inversion plays a very subordinate role in the Lower Saxony Basin and is essentially lacking in areas located further E (Fig. 5). The main phase of inversion of the Sole Pit Basin occurred at the transition from the Eocene to the Oligocene. Minor deformation continued into the latest Oligocene.

Minor Eocene compressional and wrench deformation is evident along the lower continental slope of the Armorican margin. These are contemporaneous with the early phases of the Pyrenean orogeny (Boillot *et al.* 1985; Sibuet *et al.* 1985).

Late Oligocene–Miocene inversion phase

The main inversion of the Channel area, the Bristol Channel and Western Approaches basins occurred during the Oligocene and Miocene. Tectonic activity continued intermittently into the Pliocene and possibly into the Pleistocene (Fig. 6). Contemporaneous wrench deformation is evident in the Paris Basin and the Armorican Massif. On the other hand, in the Celtic Sea Trough and the Fastnet Basin, Late Oligocene and Miocene inversion movements play a very subordinate role.

Oligocene wrench deformation of the Irish Sea, Cardigan Bay and Bristol Channel area can be inferred from the rapid subsidence of local pull-apart basins such as the Loch Neagh Basin in Northern Ireland. The subsidence of these Oligocene basins is related by Lake & Karner (1987) to a short-lived phase of sinistral motion on pre-existing wrench faults. Activity along these faults was sinistral during the Mesozoic subsidence of the Celtic Sea, Bristol Channel and Western Approaches basins and dextral during their inversion in the Cenozoic.

The main inversion phases of the Mesozoic troughs of the Celtic Sea–Western Approaches area appear to alternate with the main subsidence phases of the Rhône–Bresse–Rhine Graben (Ziegler 1982).

Oligocene transpressional deformations are also evident on the Mid-Norway Shelf (Bukowicz & Ziegler 1985) and at the northern termination of the Rockall Trough in the area of the Wyvill–Thompson Ridge. These wrench deformations are probably related to movements along intra-oceanic transform fault zones during the intra-Oligocene reorganization of the sea-floor spreading axes in the Norwegian–Greenland Sea (Ziegler, 1988).

Structural style and crustal configuration

The compressional/transpressional nature of the Late Cretaceous and Cenozoic stress regimes which induced the inversion of major Mesozoic graben systems and the uplift of basement blocks at considerable distances from the contemporaneous thrust front of the Alpine orogen can be deduced from the observed deformation patterns (Ziegler 1988).

As a general rule, inverted basins in the Alpine foreland developed out of grabens or wrench induced troughs whilst isolated inversion features, such as the Pays de Bray anticline in the Paris Basin (Pommerole 1978), are associated with wrench faults. In the case of the Pays de Bray anticline fault displacements were of a sinistral nature during the Mesozoic and were dextral during its Cenozoic reactivation.

The structural configuration of the different inverted basins of Northwest Europe, as derived from reflection seismic data, shows that basin inversion involved the uplift of the basin floor and the folding of the basin fill. This caused the throw on tensional basement faults, which controlled the original relief of the respective grabens or troughs, to become partly or totally reversed (see Ziegler 1987). This implies that a commensurate amount of shortening of the basin floor accompanied the buckling-up of major anticlinora. The crestal parts of the anticlinoria were subjected to erosion, whilst sedimentation continued in secondary, marginal basins. Owing to the greater pre-inversion sediment thickness in basinal areas, the structural relief created by the inversion movements is often considerably smaller at depth than at shallower levels in many basins (e.g. Sole Pit and Broad Fourteens basins, Ziegler 1983; van Hoorn *this volume*). The mechanical concept of basin inversion is borne out by the model experiments presented by Koopman *et al.* (1987).

The structural style of inverted basins is very variable and depends on their pre-inversion configuration, the lithologic composition of their sedimentary fill, the degree of inversion and the orientation of the basin axis relative to the greatest principal stress which induced its inversion. In this respect the presence of major

FIG. 5. Eocene–Oligocene foreland deformations.

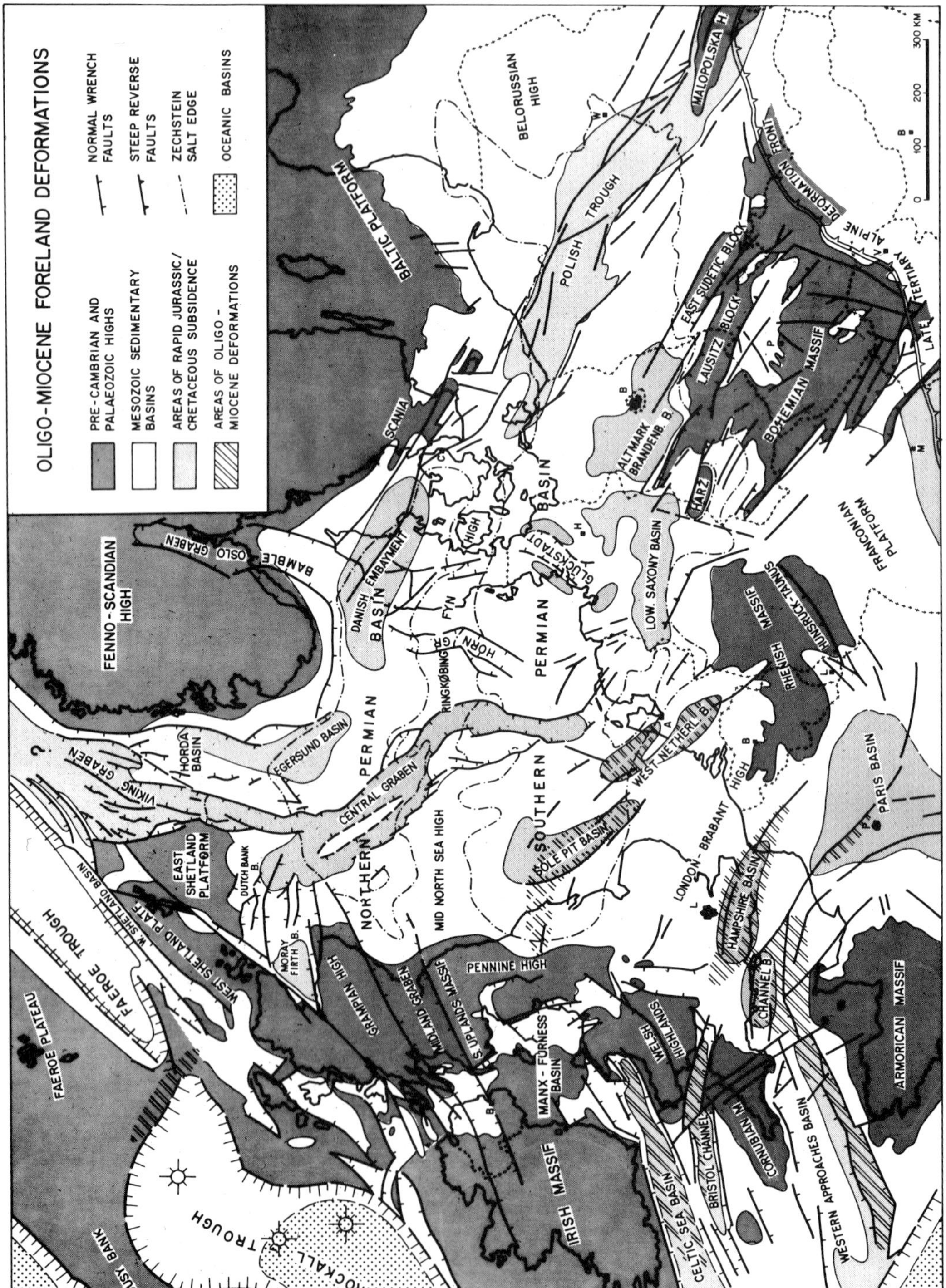

FIG. 6. Oligocene–Miocene foreland deformations.

halite and/or shale layers acting as detachment planes can play a significant role in the disharmony of the structural style observed at shallow and at deeper levels (e.g. Lower Saxony, Broad Fourteens and Sole Pit basins, see Ziegler 1987).

In general terms compression/transpression induced inversion of a basin is associated with a reduction of its volume resulting from intrabasinal crustal shortening. To the author's knowledge little effort has hitherto been devoted to quantify the amount of crustal shortening associated with the different inverted basins of Western and Central Europe. In view of the prevalence of transpressional movements during the process of inversion of many of these basins, a three dimensional approach to their palinspastic reconstruction is required.

There is still only limited data available on the deep structure and the crustal configuration of most of the inverted basins of Western and Central Europe. A notable exception is the Polish Trough. There refraction and deep reflection surveys indicate that beneath its axial inverted part the crust–mantle interface is not pulled up but is somewhat depressed and forms the so-called Guterch Graben (Guterch *et al.* 1986). Reconstructions of pre-inversion profiles across the Polish Graben indicate that in its deepest parts Mesozoic and Late Palaeozoic sediments attained thicknesses of up to 10 km (Pozaryski & Brochwicz-Lewinski 1978). In order to accommodate such a thick sedimentary sequence, isostatic considerations require that a significant amount of crustal thinning must have taken place durig the Mesozoic rifting stage of the Polish Trough. Its pre-inversion crustal configuration may have resembled that of the present day Central North Sea (Barton & Wood 1984). These features suggest that during the inversion of the Polish Trough its crust became mechanically thickened again by transpression-induced crustal shortening. As a consequence of its intense inversion the Polish Trough achieved isostatic and thermal stability during its latest Cretaceous and Palaeocene inversion.

The validity of this concept may be evaluated by a comparison of the present and pre-inversion crustal configuration of inverted basins as derived from palinspastic basin reconstructions and corresponding gravity models.

The amount of crustal shortening associated with the inversion of the different basins in the Alpine foreland and the deformation of the Variscan massifs is at present still difficult to estimate but is unlikely to exceed at best a few tens of kilometres.

Dynamics of the Alpine plate collision

In the Central and Western Tethys and in the Central Atlantic rifting and wrench tectonics culminated during the Mid-Jurassic in crustal separation and the onset of sea-floor spreading (Tollmann 1978; Trümpy 1980; Emery & Uchupi 1984; Lemoine 1985).

The Central Atlantic sea-floor spreading axis, was the more active one and already dominated over the sea-floor spreading axes in the Western and Central Tethys during the Late Jurassic (Hellenic–Dinaric, South Penninic–Piedmont–Ligurian and Alboran axes) (Ziegler 1988).

Late Jurassic to Early Cretaceous

Rapid sea-floor spreading in the Central Atlantic induced a major sinistral translation between Africa and Europe (Fig. 7). With the Italo–Dinaric promontory still firmly attached to North Africa, this caused the transtensional opening of the Alboran–Ligurian–Penninic Ocean and, at the same time, a gradual build-up of compressional stresses in the Dinaric–Hellenic Ocean. During the Late Jurassic the Central Tethys sea-floor spreading axis became inactive and a subduction system developed along the NE margin of the Hellenic–Dinaric ocean. During the latest Jurassic this oceanic basin gradually closed and the Pelagonia–Golija zone, forming the leading edge of the Italo–Dinaric block, collided during the Early Cretaceous with the Rhodope–Moesia complex on the SE margin of Europe (Vardar suture) (Dercourt *et al.* 1986; Ziegler 1988).

During the Early Cretaceous continued rapid sea-floor spreading in the Central Atlantic was accompanied by the development of the Maghrebian–South Anatolian transform fault along which the Italo–Dinaric promontory became partly decoupled from Africa (Fig. 8). This occurred in conjunction with the counter-clockwise rotation of the Italo–Dinaric block, the emplacement of major nappe systems in the internal Dinarides and Hellenides and the onset of transpressional and later, of compressional deformations in the frontal parts of the Austro–Alpine and in the Penninic realms respectively (Dercourt *et al.* 1986; Westphal *et al.* 1986). In the Central and Eastern Alps and the Carpathians high pressure metamorphics and the extensive accumulation of a synorogenic flysch series give evidence for the development of an Apto–Albian and early Late Cretaceous

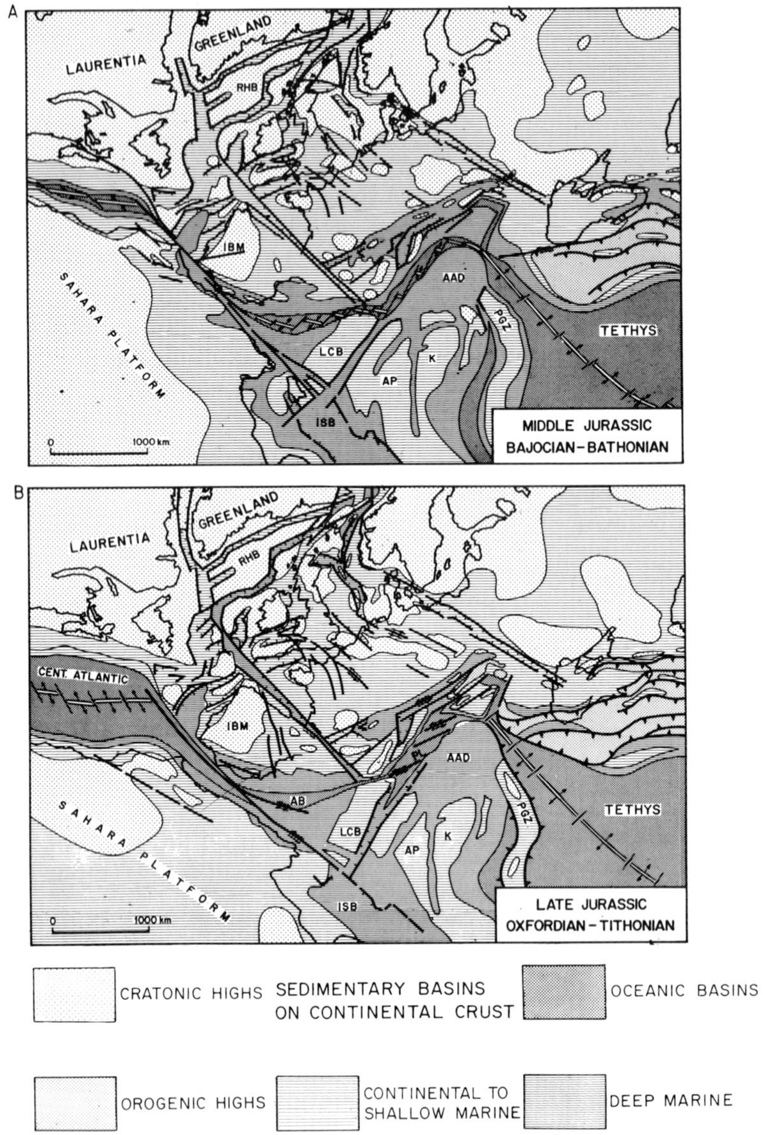

FIG. 7. Jurassic palaeotectonic reconstruction of the Atlantic–Tethys domain (by E. Haan and P. A. Ziegler). Abbreviations: AAD, Austro–Alpine domain; AB, Alboran Sea; AP, Apulia Platform; ATP, Anatolia–Taurus Platform; IBM, Iberian Meseta; ISB, Ionian Sea Basin; K, Karst Platform; LCB, Lucania–Campania Block; PGZ, Pelagonia–Golija Zone; PL, Penninic–Piedmont–Ligurian Ocean; RHB, Rockall–Hatton Bank.

southwards dipping subduction system (Tollmann 1978; Frisch 1979; Homewood *et al.* 1980; Trümpy 1980; 1982).

Continued sinistral translation between Europe and Africa, including further rotation of the Italo–Dinaric block, was accompanied by the progressive subduction of the South Penninic–Piedmont Ocean. During the Albian to Turonian the Austro–Alpine subduction system collided with the Central Penninic domain and induced the development of a new subduction zone along its northern margin in the Central (?) and Eastern Alps (Homewood *et al.* 1980; Tollmann 1980; Trümpy 1980).

During the Aptian crustal separation was achieved between the Grand Banks–Flemish

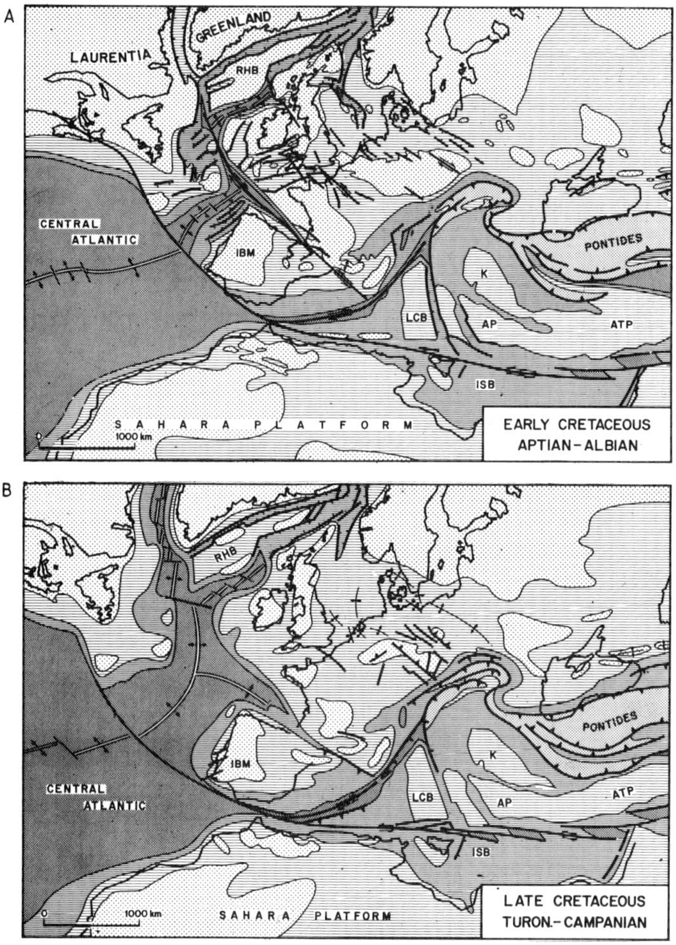

FIG. 8. Cretaceous palaeotectonic reconstructions of the Atlantic–Tethys domain (by E. Haan & P. A. Ziegler), For legend see Fig. 7 caption.

Cap and Iberia and in the Bay of Biscay. With this Iberia assumed the role of a micro-continent and rotated in a sinistral mode away from the Armorican margin. During the Albian, the North Atlantic sea-floor spreading axis propagated into the southern parts of the Rockall Trough (Roberts *et al.* 1981; Olivet *et al.* 1984).

Late Cretaceous

During the Late Cretaceous Africa–Arabia began to rotate in a counter-clockwise direction and to converge with the southern margin of Eurasia in response to the gradual opening of the Atlantic–Indian Ocean (Olivet *et al.* 1984; Livermore & Smith 1985; Savostin *et al.* 1986). This caused the rapid propagation of the Alpine subduction system into the western Mediterranean area and the gradual closure of the Piedmont–Ligurian–Alboran Ocean. At the same time the collision front between the Italo–Dinarid promontory and the Hellenic–Pontided subduction zone propagated eastward (Fig. 8).

In the Eastern Alpine–North Carpathian domain the orogenic front reached the passive margin of the Helvetic Shelf during the Late Turonian–Early Senonian after the imbrication and partial subduction of the North Penninic Trough. The latter was presumably characterized by drastically attenuated continental crust (Frisch 1979; Tollmann 1980; Trümpy 1980; Debelmas *et al.* 1983). This was probably accompanied by the development of a true A-subduction zone at the Penninic–Helvetic boundary across which compressive stresses could be transmitted onto the fractured and weakened craton of Western and Central Europe. These stresses induced the transpressional 'sub-Hercynian' reactivation of pre-existing intra-plate discontinuities in Central Europe. In particular, the Mesozoic rifts and wrench induced troughs with their thinned crust acted like shear-pins in an otherwise rigid craton. They started to collapse and were inverted to varying degrees. At the same time the Permo–Carboniferous fractures transsecting the Bohemian Massif became reactivated and governed the uplift of major basement blocks along wrench and steep reverse faults.

Intra-Senonian compressional foreland stresses induced deformations at distances of up to 1200 km to the N and NW of the Alpine collision front (Fig. 2) which at that time was presumably located some 100 km to the south of the present Alpine deformation front.

The distribution of the sub-Hercynian foreland deformations suggests that the Penninic subduction system had collided only with the East Alpine–Carpathian part of the Helvetic Shelf during the Senonian. In this context it ought to be borne in mind that in the Eastern Alps and in the northern Carpathians tectono-stratigraphic units corresponding to the distal parts of the Helvetic Shelf are deeply buried beneath the Austro–Alpine and Penninic nappes and that control on their deformation history is limited to a few deep wells in the Vienna Basin and in the Carpathian externides (Kolarova & Roth 1977; Wachtel & Wessely 1981; Heller & Moryc 1984).

In the Central Alps, on the other hand, where Neogene uplift of the external massifs caused deep truncation of the Austro–Alpine and Penninic nappes, the stratigraphic record of the exposed Helvetic nappes shows that the south Helvetic and ultra-Helvetic domain was affected by mild Campanian compressional deformation (Trümpy 1980; Debelmas *et al.* 1983). This also indicates that compressional foreland stresses gradually built up in the Central Alpine domain during the Senonian.

In the West Central and Western Alps, the Late Cretaceous closure of the Penninic–Piedmont ocean was accompanied by the obduction of oceanic crustal material of the Piedmont domain onto the southern and eastern flanks of the Briançonnais High. However, development of an A-subduction zone in the North Penninic Trough apparently did not progress further westward than the Central Alps (Debelmas *et al.* 1983). Correspondingly, compressional stresses were not yet exerted on the foreland of the Western Alps where sub-Hercynian compressional intra-plate deformations are conspicuously absent.

In the framework of this model, the Late Turonian–Early Senonian, early sub-Hercynian deformations, as evident in the Bohemian Massif, the Lower Saxony and Sole Pit basins and the Danish Central Graben, were probably induced by compressional stresses which were transmitted through the lithosphere of the Alpine foreland immediately prior to, and during the initial phases of collision between the Austro–Alpine and Penninic nappe complexes with the Helvetic Shelf. During the Santonian and Early Campanian increasing coupling between the rising orogen and cratonic Europe was accompanied by the build up of compressional foreland stresses. This is reflected by the main phases of the sub-Hercynian intra-plate tectonism (Fig. 3).

The Senonian foreland deformations were more or less contemporaneous with the abandonment of the sea-floor spreading axis in the Bay of Biscay and the Rockall Trough and the initiation of a new axis in the Labrador Sea (Roberts *et al.* 1981; Srivastava *et al.* 1981; Olivet *et al.* 1984). This reorganization of sea-floor spreading axes in the North Atlantic coincides with the sub-Hercynian fundamental change in the regional stress regime of northwest Europe and the Mediterranean domain. This can be related to the counter-clockwise rotation and northward drift of Africa–Arabia and its collision with Eurasia in response to the gradual opening of the South Atlantic–Indian Ocean (Livermore & Smith 1985).

In this context, it may be hypothesized that regional compressional stresses, which were transmitted through the lithosphere of Western and Central Europe as a consequence of the collision of the intra-Alpine and Carpathian subduction systems with its southern passive margins, were large enough to impede the northward propagation of the Rockall sea-floor spreading axis into the Norwegian–Greenland Sea and cause its abandonment in favour of the newly developing Labrador Sea spreading axis. However, these intra-plate compressional stresses were, apparently, not large enough to cause significant strain in areas located NW of the Sole Pit and West Netherlands–Broad Fourteens basins, the Danish North Sea and the Danish Embayment (Ziegler 1987; 1988).

Palaeocene

Late Campanian to Danian compressional deformations are of little importance in the Helvetic domain (Trümpy 1980) and are only locally evident in the Alpine foreland. This may be interpreted as reflecting a decrease, or even a pause, in the convergence between the Alpine nappe systems and the European foreland. These movements accelerated or resumed again during the Mid-Palaeocene. They gave rise to the second, so-called Laramide phase of foreland compression. This phase was even more pervasive than the sub-Hercynian one (Fig. 4). Northernmost deformations are recorded from the Egersund Basin, located 1300 km to the N of the present Alpine thrust front (Pegrum 1984).

In the Northern Alpine–Carpathian foreland the Laramide tectonism corresponds to the main phase of basin inversion and uplifting of basement blocks. Results of deep wells indicate that the area affected by deformation extends a considerable distance southward under the Carpathian and East Alpine nappes (Kolarova & Roth 1977; Wachtel & Wessely 1981; Heller & Moryc 1984; Ziegler 1988). In the Central Alpine domain, similar foreland deformation is considered to be responsible for the latest Cretaceous–Early Tertiary unconformity recognized in the area of the Aare Massif, the sub-Helvetic and Helvetic nappes (Herb 1965) and the upwarping of a low relief anticlinal structure in the area of the eastern Swiss Molasse Basin (Büchi & Schlanke 1977; Trümpy 1980). These deformations induced the regional truncation of the formerly extensive Cretaceous sedimentary cover in the foreland of the Central Alps (Ziegler 1982).

As intra-Palaeocene deformations are also evident in the foreland of the Western Alps it must be assumed that the collision front between the Penninic nappe system and the Helvetic Shelf had propagated westward (Fig. 9). The stratigraphic record of the North Penninic Valais Trough of the West-Central Alps suggests, however, that this basin closed only during the Eocene (Homewood *et al.* 1980; Laubscher & Bernoulli 1982; Debelmas *et al.* 1983). From this it could be inferred that the crust and upper mantle of the Valais Trough were sufficiently rigid to permit during the Palaeocene the transmission of compressional stresses into the foreland.

The strong Laramide deformation of the Alpine and Carpathian foreland indicates that the Central and Eastern Alpine and Carpathian nappe systems were mechanically coupled with the foreland at their respective A-subduction zones.

Stratigraphic and reflection seismic data from areas affected by the Laramide tectonism indicate that the latest Palaeocene and Early Eocene corresponds to a renewed period of reduced compressional foreland deformation. It is significant that during this time the North Atlantic sea-floor spreading axis propagated rapidly northward into the Norwegian–Greenland Sea area (Eldholm & Thiede 1980; Eldholm & Talwani 1981; Roberts & Ginzburg 1984). This suggests that stresses responsible for the Late Palaeocene compressional foreland deformations had relaxed again and thus ceased to compete with the tensional forces governing the evolution of the Artic–North Atlantic domain.

Eocene–Oligocene

During the Eocene and Oligocene the Helvetic Shelf and the northward adjacent autochthonous domain of the Alps and Carpathians began to subside progressively under the load of the rapidly advancing Austro–Alpine and Penninic nappes. These nappes are cored by crystalline, upper crustal material (Milnes & Pfiffner 1980; Homewood *et al.* 1980). Therefore their emplacement involved crustal delamination and the subduction of lower crustal and upper mantle material (Müller 1982). At the same time the sedimentary sequence of the Helvetic Shelf became imbricated. By the Early Miocene the Alpine–Carpathian nappes were emplaced near their present location (Homewood *et al.* 1980; Trümpy 1980; Debelmas *et al.* 1983; Pfiffner 1986) (Fig. 9).

The Eocene and Oligocene emplacement of the Austro–Alpine and Penninic nappes on the N European craton was accompanied by the development of the Alpine foredeep, the Molasse Basin. Through time, the axis of this basin migrated northward in response to the northward shifting tectonic load exerted by the advancing nappe complexes onto the foreland crust. In the eastern Molasse Basin Eocene clastics and shelf carbonates trangressed over the deeply truncated Mesozoic series which had been deformed during the sub-Hercynian and Laramide phases of foreland compression. Eocene and Oligocene sediments are dissected by a system of basin parallel tensional faults which developed during the Oligocene rapid subsidence of the Molasse Basin (see Ziegler 1982; 1987).

The lack of Eocene and younger compressional deformation in the foreland of the Eastern Alps and the Carpathians (Figs 5 & 6) indicates that during this important phase of the Alpine orogeny stresses transmitted onto the foreland were no longer large enough to cause significant strain. Based on this observation it

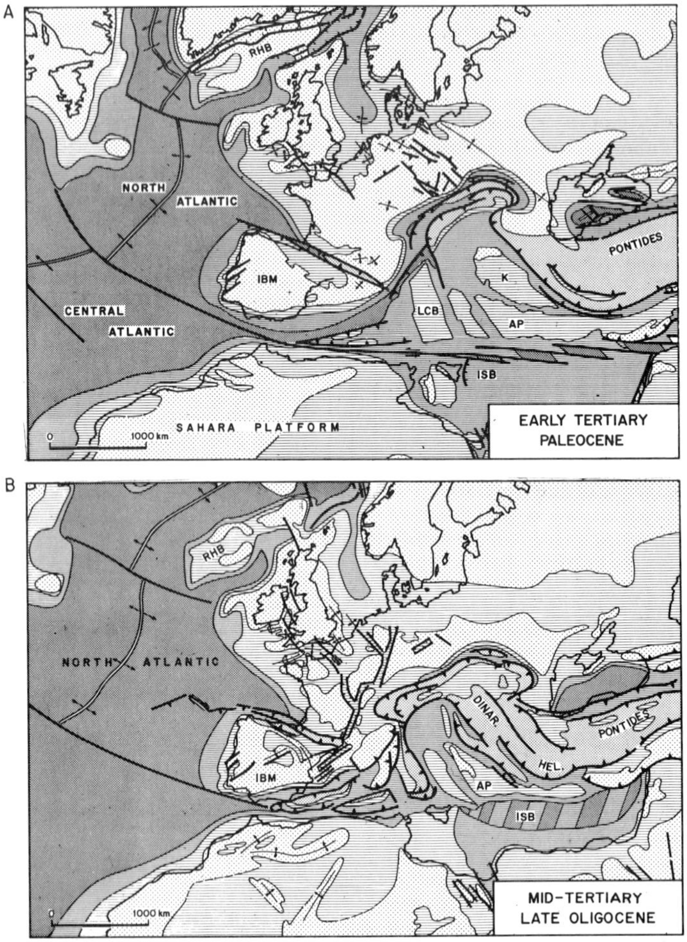

FIG. 9. Palaeogene palaeoreconstructions of the Atlantic–Tethys domain (by E. Haan & P. A. Ziegler). For legend see Fig. 7 caption.

must be assumed that the orogene and the North European craton were essentially decoupled during the emplacement of the Austro–Alpine and Penninic nappes.

On the other hand, in the Western and West Central Alps, the Eocene closure of the Valais and sub-Briançonnais troughs was accompanied by the development of prograding A-subduction zones (Debelmas *et al.* 1983) across which the orogen and the foreland became apparently increasingly coupled. This is reflected by the Late Eocene and Oligocene compressional deformations in the English Channel–Western Approaches–Celtic Sea and the Sole Pit–West Netherlands areas (Fig. 5; see Ziegler 1987).

The Eocene–Oligocene phase of sinistral wrench faulting caused the subsidence of local transtensional basins in Northern Ireland, Caernarvon and Cardigan Bay and Cornwall (Lake & Karner, 1987). The limited igneous activity in the Porupine Trough (Seemann 1984) and possibly in the W parts of the Celtic Sea Trough as well as the intrusion of the Lundy Granite may be related to changes in sea-floor spreading rates in the different segments of the Northern North Atlantic and the reorganization of sea-floor spreading axes in the Norwegian–Greenland and Labrador Sea. Contemporaneous inversion movements in the mid-Norway Basin (Bukowicz & Ziegler 1985) and wrench deformation of the Wyville–Thompson Ridge are thought to reflect tectonic activity along the intra-oceanic Jan Mayen and Iceland fracture zones which apparently projected beneath the adjacent shelf areas.

On the other hand, Eocene compressional

and wrench deformation of the Armorican margin (Boillot *et al.* 1985; Sibuet *et al.* 1985) can be related to stresses transmitted through the continental and oceanic foreland of the Pyrenees. The lack of post-Eocene deformation along the Armorican margin suggests that during the main phases of the Pyrenean orogeny most of the compressive stress was absorbed at the North Iberian subduction zone.

Miocene–Pliocene

Late Oligocene and Miocene intensified deformation of the Channel–Western Approaches area coincides with further stacking of the Helvetic nappes and the uplift of the external basement massifs of the Swiss and French Alps. These massifs represent imbrications of the European foreland crust (Trümpy 1980; Debelmas *et al.* 1983; Pfiffner 1986). Their development clearly indicates that during the Late Alpine orogenic phases, which mainly affected the Central and Western Alps, the foreland and the orogen were strongly coupled at the respective A-subduction zones which prograded stepwise into the foreland crust (Fig. 6). During the Late Miocene and Pliocene compressional stresses were apparently intermittently exerted on the NW Alpine foreland. This is reflected by the folding of the Jura Mountains, which are at least in part superimposed on a Permo–Carboniferous trough (Laubscher 1986), by the deformation of the external chains of the Western Alps (Dauphine zone, Debelmas *et al.* 1983) and by wrench faulting in the Armorican Massif and the Paris Basin (Congé 1974; Pommerol 1978).

This sequence of events is the expression of the continued convergence of Africa and Europe and of the Early Miocene reorientation of their convergence direction from one that was essentially SE–NW orientated to a more E–W orientation (Letouzey 1986; Ziegler 1988). Significantly, Miocene–Pliocene compressional deformation is lacking in the northern foreland of the Alps and Carpathians but plays an important role in the Paris Basin, the Channel area and particularly the Western Approaches Basin.

Crustal shortening accompanying the Late Palaeogene to Neogene inversion of the Mesozoic troughs in the Channel–Western Approaches area was compensated by dextral wrench deformations along pre-existing fracture systems of the Paris Basin, Channel area and southwestern England, and a net northwestward displacement of the Armorican–Massif Central block relative to the London–Brabant and Irish massifs (Fig. 6). However, the magnitude of these displacements has not yet been quantified. These movements are broadly contemporaneous with the evolution of the Rhône–Bresse–Rhine graben system. The main phases of subsidence of this rift system appear to alternate with the main phases of basin inversion in the western shelf areas (Ziegler 1982). This suggests that during the principal periods of basin inversion compressional forces exerted on the Alpine foreland were able to override the tensional stresses which governed the subsidence of the Rhône–Bresse–Rhine rift system.

During the Pliocene the southern margin of the Rhine Graben and the eastern boundary of the Bresse Graben were overridden by the frontal thrust elements of the Jura Mountains. In this process the eastern boundary fault of the Bresse Graben became locally modified by upthrust movements. This is taken as evidence that the development of the Rhône–Bresse–Rhine Graben was not directly linked genetically with the Alpine plate collision as suggested by Tapponier (1977), Sengör *et al.* (1978) and Bergerat & Geyssant (1980). It is more likely that the evolution of this rift system was governed by geodynamic processes related to the early phases of a plate reorganization which may ultimately lead to the break-up of the Late Alpine continent assembly (Illies 1974; Neugebauer 1981; Ziegler 1982; 1988).

Remnant stress measurements indicate that the Upper Rhine Graben is at present being deformed by sinistral strike–slip movements and is in an incipient state of inversion whilst the Ruhr Graben subsides actively in response to tensional stresses (Ahorner 1975; Illies & Greiner 1978; Baumann 1981).

Discussion

The indicated synchroneity between the Alpine orogeny and the latest Cretaceous and Cenozoic compressional/transpressional intra-plate deformations in Western and Central Europe suggests that tangential stresses can be intermittently exerted on and transmitted through the colliding continental plates during the collision between two continents. Depending on the intensity of the pre-existing fracturing of the respective cratons such stresses are apparently able to induce significant intra-plate compressional deformations at distances up to 1300 km from the collisional margin of a foreland plate by reactivation of crustal discontinuities (Fig. 10). Moreover, circumstantial evi-

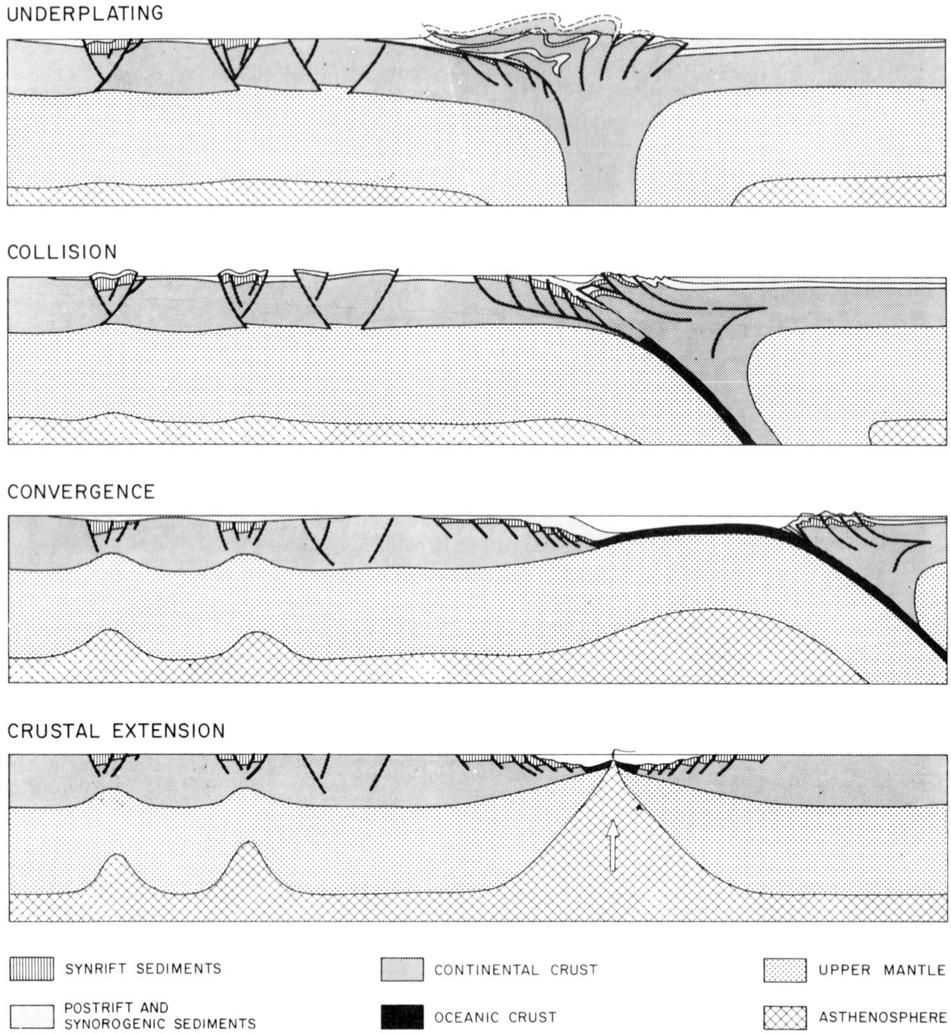

FIG. 10. Intra-plate tectonic model.

dence points to the possibility that compressional stresses can be transmitted through the lithosphere over even greater distances than the most distal observed deformation (e.g. impeded northward propagation of the North Atlantic sea-floor spreading axis during Late Senonian and Palaeocene, Ziegler 1988). However, these concepts require further testing on the basis of scaled models, such as three dimensional finite stress element models.

Amongst different types of intra-plate discontinuities, rifts with a strongly thinned crust appear to be prone to early inversion in response to collision related intra-plate tangential stresses. Major wrench faults, which penetrate much of the crust and possibly extend into the upper mantle, are also prone to compressional reactivation. On the other hand, thrust faults and even ancient A-subduction zones, such as occur in the Variscan fold-belt, appear to be less prone to compressional reactivation. This is in contrast to their tensional reactivation as demonstrated by Brewer & Smythe (1984). (See also Lake & Karner 1987).

Depending on the pattern of pre-existing intra-plate discontinuities the boundary between the collision related orogen and the compressional intra-plate deformations can become diffuse, as, for instance, in the foreland of the Western Alps. They can render futile a distinction between so-called Alpino-type and Germano-type deformations (Voigt 1962).

Collision related compressional intra-plate deformations require a certain amount of coupling between the foreland plate and the associated orogen at the A-subduction zone marking their common boundary to facilitate stress transmission. On the other hand, within the foreland plate a decoupling is required at intra-crustal levels (low velocity channel?), the crust–mantle boundary and/or possibly within the upper mantle or even at the base of the lithosphere to accommodate crustal shortening associated with the inversion of basins and the uptilting of basement blocks along steep reverse faults. It is hoped that new regional deep seismic reflection profiles will help to establish at which levels such detachment horizons occur.

The example of the Polish Trough suggests that during inversion the thinned crust of a rift can be mechanically thickened as a result of crustal shortening. Depending on the degree of inversion, an abandoned rift with a still destabilized upper mantle can attain isostatic and thermal stability by this process. This is borne out by the lack of post-inversion subsidence of, for instance, the Polish Trough and the Lower Saxony Basin. Both of these basins had apparently continued to subside differentially until shortly before the beginning of their inversion (Ziegler 1982).

The evolution of the Rhône–Bresse–Rhine graben and the North Atlantic–Norwegian–Greenland Sea rift illustrate that compressional intra-plate stresses can impede the evolution of active rift systems by counteracting the tensional stress regime which governs their development (Ziegler 1988).

Although essentially synchronous inversion movements affected wide areas, sometimes at considerable distances from the Alpine collision front, differences are evident in the timing of the main inversion phases of, for example, the West Netherlands–Broad Fourteens Basin and the Sole Pit Basin. These differences suggest that during the Laramide phase of foreland compression the bulk of the intra-plate stresses were absorbed by the deformation of the West Netherlands–Broad Fourteens Basin. This shielded the Sole Pit Basin from early inversion. However, the crust of the West Netherlands–Broad Fourteens Basin apparently became stabilized to such a degree during the Laramide deformation phase that only minor additional deformations were possible during the subsequent Late Eocene–Early Oligocene phase of foreland compression (locked system). Consequently, stresses could then be transmitted into areas located NW of the West Netherlands–Broad Fourteens Basin, where they caused the main inversion of the Sole Pit Basin. This type of 'inversion progradation' presumably requires that basins located more proximal to the collision front be characterized by a pre-inversion strongly attenuated crust which does not permit the transmission of significant stresses into more distal basins. Similarly, it appears logical that basins with the thinnest crust are prone to early inversion, even if they are located at a considerable distance from the collision front (e.g. the Laramide inversion of the Fastnet–Celtic Sea Basin and virtual lack of contemporaneous deformations in the Western Approaches Basin). However, under continued convergence of the orogenic system and the foreland plate, and as a result of the early deformation of the 'weakest' basins, stresses may build up within the foreland to the degree that they can cause the inversion of more proximally located basins, characterized by an apparently less drastically thinned crust. This type of 'inversion retrogradation' can be inferred from the timing of inversion movements in the different basins of the Western Shelf area where, for instance, the Oligocene–Miocene main inversion of the Western Approaches Basin clearly postdates the Laramide main inversion of the Fastnet–Celtic Sea Basin.

In the overall pattern of inverted Mesozoic grabens in the Alpine foreland, the non-inverted Horn and Glückstadt grabens (Best *et al.* 1983) form an exception. These grabens, which contain 7–8 km of Permo–Triassic sediments, became inactive during the Early to Middle Jurassic, or some 90–100 million years before the onset of compressional deformation of the Alpine foreland. They are virtually surrounded by the inverted Fenno-scandian border zone, the Danish–Dutch part of the Central Graben and the Lower Saxony Basin. The crustal configuration of the Horn and Glückstadt grabens is unknown (Figs 3 & 4).

It is uncertain whether the yield strength of a rift zone, which is subjected to tangential compressional stresses, is purely controlled by its crustal thickness (degree of crustal attenuation) or whether its thermal regime and lithospheric configuration also play an important role. Empirically it appears plausible that, under the assumption of similar crustal attenuation, the thermally destabilized crust and lithosphere of an active rift (e.g. Rhine Graben) has a lower yield strength than that of a thermally stabilized palaeorift.

The above considerations suggest that the Late Cretaceous–Early Tertiary compressional foreland stresses were not large enough to overcome the yield strength of the crust and lithosphere of the Horn and Glückstadt palaeorifts. On the other hand, these stresses did exceed the yield strengths of the crust and lithosphere of the North Danish Basin, the Southern North Sea Central Graben and the Lower Saxony Basin, all of which had remained tectonically active until shortly before their inversion. Quantitative models may be able to evaluate the validity of this geodynamic concept which is based on empirical relationships only.

If the above developed concepts are essentially valid, then it must follow that the stratigraphic record of collision related intra-plate compressional deformations can provide valuable information about the dating of major orogenic events which affected the collisional margin of the respective plate. They can thus complement the stratigraphic record preserved in the corresponding orogen. In this context it should be kept in mind that within orogenic systems the stratigraphic record of earlier orogenic events is often canibalized during subsequent deformation phases and that the emplacement of major nappes masks the stratigraphic record perserved in their substratum (e.g. Austro–Alpine nappes of Austria and Germany).

Although compressional intra-plate deformations appear to be most commonly associated with Himalayan-type orogens resulting from the collision of two continents (e.g. Variscan, Uralian, Himalayan and Alpine foreland deformation) they can also be associated with Pacific or accretion-type orogens that formed in response to the high convergence rate between a continental and an oceanic plate (e.g. Argentinian Andean foreland, Rocky Mountains).

The lack of compressional intra-plate deformations in the foreland of an orogen (e.g. Canadian Cordillera of Alberta and British Columbia) can be related either to the absence of major intra-plate discontinuities in the respective foreland and/or to the lack of mechanical coupling between it and the orogen during the evolution of the latter.

The mechanisms which govern the coupling, or the lack thereof, between a foreland plate and its associated orogen at an A-subduction zone are not yet understood. They could possibly be a function of the crustal configuration of the collisionally involved passive margin (abrupt crustal thinning versus gentle basement ramp), the rate of convergence and whether or not crustal delamination of the overriding plate facilitates the emplacement of major basement cored nappes.

For example, the main phases of compressional deformations in the foreland of the Eastern Alps and Carpathians appear to correlate with the initial phases of collision of the Austro–Alpine and Penninic nappe complexes with the North European foreland. During the subsequent phase of emplacement of these basement cored nappes, the orogen and the foreland were apparently decoupled (Fig. 10).

On the other hand, in the Western and Central Alps, where corresponding basement cored nappes play a far smaller role (Debelmas *et al.* 1983), coupling between the orogen and the foreland seems to have gradually increased during the Oligocene and Miocene. This is illustrated by the uplift of the external Alpine massifs and the inversion of Mesozoic basins in the Western Shelf area.

Finally, it should be noted that compressional/transpressional intra-plate deformations, which resulted in the inversion of sedimentary basins, can also occur under non-orogenic settings. For instance, the Neogene partial inversion of the mid-Norway shelf is thought to be related to wrench movements along the landward extension of the Jan Mayen fracture zone during the Oligocene reorganization of sea-floor spreading axes in the Norwegian–Greenland Sea (Bukowics & Ziegler 1985). A similar mechanism may have to be invoked for the mild Late Eocene wrench deformation in the Northern North Sea described by Pegrum & Ljones (1984) and the Oligocene wrench deformations along the Wyville–Thompson Ridge.

On the other hand, the wrench-induced partial inversion of the West Netherlands–Broad Fourteen Basin during the Late Jurassic and of the Irish Sea, Bristol Channel, Western Approaches and Channel areas during the Early Cretaceous as well as the contemporaneous wrench deformations along the SW Bohemain Border Zone (see Ziegler 1987), reflect fundamental changes in the overall stress pattern affecting the NW European rift systems. This change in the regional stress pattern, which can be related to crustal separation in the Tethys domain during the mid-Jurassic, the Bay of Biscay and the Rockall Trough during the mid-Cretaceous and to continued crustal extension in the Arctic–North Atlantic rift system, caused the progressive 'polarization' of the NW European rift systems toward the Arctic–North Atlantic rift system (Ziegler 1982). The associated transpressional and transtensional deformations clearly predate the Alpine plate collision and thus are undoubtedly anorogenic.

Overall, circumstancial evidence suggests that the most important intra-plate compressional/transpressional deformations involving large scale basin inversion and the upthrusting of major basement blocks is related to orogenic events affecting the margin of the respective foreland plate.

In this context, the megatectonic setting and the scope of compressional/transpressional deformations in the Alpine foreland can be compared with the Laramide tectonism of the Rocky Mountains of Colorado and Wyoming (Brewer *et al.* 1980), the Permian deformations in the Timan–Pechora area forming the northwestern foreland of the Urals (Ulmicheck 1982; Matviyevskaya *et al.* 1986) and the Late Carboniferous and Cenozoic deformations of the Sahara Platform involving the inversion of the Ougarta Trough in the foreland of the Variscan and Alpine chains of northwest Africa (Ziegler 1988). The most distal compressional/transpressional foreland structures occur in the Timan–Pechora area, 500 km to the W of the Uralian thrust front, the Rocky Mountains, about 600 km to the E of the Cordilleran thrust front, NW Africa, 800 km to the S of the Maghrebian thrust front and Europe, about 1300 km NW of the Alpine thrust front.

These examples illustrate that during orogenic events large stresses can be transmitted into the foreland where they can induce important intraplate deformations often at long distances from the collisional plate margin.

References and further reading

AHORNER, L. 1975. Present-day stress field and seismotectonic block movements along major fault zones in Central Europe. *Tectonophysics*, **29**, 233–249.

BARTON, P. & WOOD, R. 1984. Tectonic evolution of the North Sea Basin: crustal stretching and subsidence. *Geophysical Journal of the Royal Astronomical Society*, **79**, 987–1022.

BAUMANN A. 1981. Regional stress field and rifting in Western Europe. *Tectonophysics*, **73**, 105–111.

BERGERAT F. & GEYSSANT J. 1980. La fracturation tertiary de l'Europe de Nord: résultat de la collision Afrique-Europe. *Compte Rendue de l'Academie des Sciences, Paris, Série D*, **390**, 1521–1254.

BEST, G., KOCKEL, F. & SCHÖNLAUB, H. 1983. Geological history of the southern Horn Graben. *In*: KAASSCHIETER, J. P. H. & REIJERS, T. J. A. (eds) *Petroleum Geology of the south eastern North Sea and adjacent onshore areas. Geologie en Mijnbouw*, Special issue, 47–50.

BODENHAUSEN, J. W. A. & OTT, W. H. 1981. Habitat of the Rijswijk Oil Province, Onshore The Netherlands. *In*: ILLING, L. V. & HOBSON, G. D. (eds) *Petroleum Geology of the Continental Shelf of North-West Europe*. Heyden and Son, London, 301–309.

BOILLOT, G., TÉMIME D. & MALOD, J. A. *et al.* 1985. Exploration par submersible de la marge continentals au Nord-Ouest de l'Espagne. *Bulletin de la Société géologique de France*, **(8)I(1)**, 89–102.

BREWER, J. A. & SMYTHE, D. K. 1984. MOIST and the continuity of crustal reflector geometry along the Caledonian–Appalachian Orogen. *Journal of the Geological Society, London*, **41(1)**, 105–120.

——, SMITHSON, S. B., OLIVER, J. E., KAUFMAN, S. & BROWN, L. D. 1980. The Laramide orogeny: evidence from COCORP deep crustal seismic profiles in the Windriver Mountains, Wyoming. *Tectonophysics*, **62**, 165–189.

BRIX, F., KRÖLL, A. & WESSELY, G. 1987. Die Molassezone and deren Untergrund in Niederösterreich. *Erdöl–Erdgas Zeitschrift*, Special Issue, **93**, 12–35.

BÜCHI, U. P. & SCHLANKE, S. 1977. Zur Paläogeographie der schweizerischen Molasse. *Erdöl-Erdgas Zeitschrift*, Special Issue **93**, 57–69.

BUKOWICZ, C. & ZIEGLER, P. A. 1985. Tectonic development of the Mid-Norway continental margin. *Marine and Petroleum Geology*, **2(1)**, 12–22.

COGNÉ, J. 1974. Le Massif Armoricain. *In*: DEBELMAS, J. (ed) *Géologie de France, Vol. I Vieux Massifs et grands Bassins Sédimentaires*, Doin Ed., Paris, 105–161.

DEBELMAS, J., ESCHER, A. & TRÜMPY, R. 1983. Profiles through the Western Alps *In*: RAST, N. & DELANY, F. M. (eds) *Profiles of Orogenic Belts*. Geodynamic Series, Vol. 10, American Geophysical Union, Washington, 83–86.

DERCOURT, J. ZONENSHAIN, L. P., LE PICHON, X. 1986. Geological evolution of the Tethys Belt from the Atlantic to the Pamirs since the Lias. *Tectonophysics*, **123**, 241–315.

DRONG, H. J., PLEIN, E., SANNEMAN, D., SCHUEBACH, M. A. & ZIMDARS, J. 1982. Der Schneverdingen – Sandstein des Rotliegend — eine äolische Sedimentfüllung alter Graben-strukturen. *Zeitschrift der deutschen geologischen Gesellschaft*, **133**, 699–725.

ELDHOLM, O. & TALWANI, M. 1981. The passive margins of northern Europe and East Greenland. *In*: *Dynamics of Passive Margins*, Geodynamics Series, Vol. 6, American Geophysical Union, Washington, 30–44.

—— & THIEDE, J. 1980. Cenozoic continental separation between Europe and Greenland. *Palaeogeography, Palaeoclimatology, Palaeoecology*, **30**, 243–259.

EMERY, H. O. & UCHUPI, E. 1984. *The geology of the Atlantic Ocean*. Springer Verlag, New York 1050–1000.

FRISCH, W. 1979. Tectonic progradation and plate-tectonic evolution of the Alps. *Tectonophysics*, **60**, 121–139.

GUTERCH, A., GRAD, M., MATERZOK, R. & PERCHUC, E. 1986. Deep structure of the Earth's Crust in the contact zone of the Palaeozoic and Precambrian platform of Poland (Tornquist–Teisseyre zone). *Tectonophysics*, **128**, 251–279.

HELLER, I. & MORYC, W. 1984. Stratigraphy of the Upper Cretaceous deposits in the Carpathian foreland. *Biuletyn Instytut Geologicznego, Warszawa*, **346(XXIV)**, 63–116

HERB, R. 1965. Das Tertiär der helvetischen Decken der Ostschweiz. *Bulletin der Vereinigung schweizerischen Petroleum Geologen und Ingenieure*, **31**, 135–151.

HOMEWOOD, P., GOSSO, E., ESTHER, A. & MILNES, A. 1980. Cretaceous and Tertiary evolution along the Besançon-Biella Traverse (Western Alps). *Eclogae Geologicae Helvetiae*, **73(2)**, 635–649.

ILLIES, J. H. 1974. Intra-Plattentektonik in Mitteleuropa und der Rheingraben. *Oberrheinische geologische Abhandlungen*, **23**, 1–24.

—— 1979. Holocene movements and state of stress in the Rhine Graben rift system. *Tectonophysics*, **52**, 349–359.

—— & GREINER, G. 1978. Rhine Graben and the Alpine system. *Geological Society of America Bulletin*, **89**, 770–782.

KOLAROVA, M. & ROTH., Z. 1977. Geology of the platform entering below the Carpathian Foredeep and its Carpathian nappes in S.E. Moravia. *Bulletin of the Geological Survey, Prague*, **52(5)**, 257–265.

KOOPMAN, A., SPEKSNIJDER, A. & HORSEFIELD, W. T. 1987. Sandbox model studies of Inversion Tectonics. *Tectonophysics*, **137**, 379–388.

LAKE, S. D. & KARNER, G. D. 1987. The structure and evolution of the Wessex Basin, southern England, an example of inversion tectonics. *Tectonophysics*, **137**, 347–378.

LAUBSCHER, H. P. 1986. The eastern Jura: Relations between thin-skinned and basement tectonics, local and regional. *Geologische Rundschau*, **75(3)**, 535–553.

—— & BERNOULLI, D. 1982. History of deformation of the Alps. *In*: HSÜ, K. J. (ed) *Mountain Building Processes*. Academic Press, London, 169–180.

LEMOINE, M. 1985. Structuration des Alpes occidentales et palinspastique de la Tethys ligure. *Bulletin de la Société géologique de France*, **(8)I(1)**, 126–137.

LETOUZEY, J. 1986. Cenozoic palaeo-stress pattern in the Alpine Foreland and structural interpretation in a platform basin. *Tectonophysics*, **132**, 215–231.

LIVERMORE, R. A. & SMITH, A. G. 1985. Some boundary conditions for the evolution of the Mediterranean Region *In*: STANLEY, D. J. & WEZEL, F. C. (eds) *Geological Evolution of the Mediterranean Basin. Raimono Selli Commemorative Volume*, Springer Verlag, Berlin, 83–100.

MATVIYEVSKAYA, N. O., BOGATSKY, V. I., VASSERMAN, B. YA., GOBANOV, L. A. & SHAFRAN, YE. B. 1986. Geological history of oil-gas potential of the Pechora–Kolvin anlagocen. *Petroleum Geology*, **21(7)**, 301–305.

MILNES, A. G. & PFIFFNER, O. A. 1980. Tectonic evolution of the Central Alps in the cross-section St Gallen–Como. *Eclogae geologicae Helvetiae*, **73(2)**, 619–633.

MONTADERT L. ROBERTS, D. G., AUFFRET, G. A., BOCK, W., PEUBLE, P. A. du, HAILWOOD, E. A., HARRISON, W, KAGANI, H. LUMSDEN, D. N., MÜLLER, C., SCHNITKE, D., THOMPSON, R. W., THOMPSON, T. L., & TIMOTEEV, P. P. 1977. Rifting and subsidence on passive continental margins in the North-East Atlantic. *Nature*, **268**, 305–309.

MÜLLER, S. 1982. Deep structure and recent dynamics in the Alps. *In*: HSÜ, K. J. (ed) *Mountain Building Processes*. Academic Press, London, 181–199.

NEUGEBAUER, H. J. 1981. Crustal uplift and the propagation of failure zones. *In*: ILLIES, J. H. (ed) *Mechanisms of Graben Formation, Developments in Geotectonics* Vol. 17, 33–52.

OLIVET, J. C., BONNIN, J., BEUZART, P. & AUZENDE, J. M. 1984. Cinematique de l'Atlantique Nord et Central. *Publication Centre National de l'Exploration des Oceans, Rapport Scientific et Technique*, No. 54, 107 pp.

PEGRUM, R. M. 1984. The extension of the Tornquist Zone in the Norwegian North Sea. *Norsk Geologiske Tidsskrift*, **64**, 39–68.

—— & LJONES, T. E. 1984. 15/9 Gamma Gas filed offshore Norway, new trap type for North Sea Basin with regional structional implications. *Bulletin of the American Association of Petroleum Geologists*, **68(7)**, 874–902.

PFIFFNER, O. A. 1986. Evolution of the north Alpine foreland basin in the Central Alps. *Special Publication of the International Association of Sedimentologists*, **8**, 219–228.

POMMEROL, C. 1978. Evolution paléogéographique et structurale du Bassin de Paris du Précambrien à l'actuel, en relation avec les régons avoisinantes. *Geologie en Mijnbouw*, **57**, 533–543.

POZARYSKI W. & BROCHWICZ-LEWINSKI, W. 1978. On the Polish Trough. *Geologie en Mijnbouw*, **57**, 545–558.

ROBERTS, D. G. & GINTZBURG, A. 1984. Deep crustal structure of SW Rockall Plateau. *Nature*, **308**, 435–439.

——, MASSON, D. G. & MILES, P. R. 1981. Age and structure of the Southern Rockall Trough: new evidence. *Earth Planetary Science Letters*, **52**, 115–128.

SAVOSTIN, L. A., SIBUET, J.C., ZONENSHAIN, L. P., LE PICHON, X. & SIBUET, M. J. 1986. Kinematic evolution of the Tethys Belt from the Atlantic Ocean to the Pamirs since the Triassic. *Tectonophysics*, **123**, 1–35.

SEEMANN, U., 1984. Tertiary intrusives on the Atlantic continental margin of southwest Ireland. *Irish Journal of Earth Sciences*, **6**, 229–236.

SENGÖR, D. M. C., BURKE, K. & DEWEY, J. F. 1978. Rifts at high angles to orogenic belts: test for their origin and the upper Rhine graben as an example. *American Journal of Science*, **278**, 24–40.

SIBUET, J. -C., MATHIS, B., PASTOURET, L. *et al.* 1985. Morphology and basement structure of the Goban Spur continental margin (northeast Atlantic) and the role of the Pyrenean orogeny. *In*: DE GRACIANSKY, P. C., POAG, W. C. *et al.* (eds) *Initial Reports DSPD* Vol. 80. US Government Printing Office, Washington, DC, 1153–1165.

SRIVASTAVA, S. P., FALCONER, R. K. H. & MACLEAN, B. 1981 Labrador Sea, Davies Strait, Baffin Bay: Geology and Geophysics — a review. *In*: KERR, J. W., FERGUSSON, A. J. & MACHAN, L. C. (eds). *Geology of the North Atlantic Borderlands. Canadian Society of Petroleum Geologists, Memoir*, **7**, 333–398.

TAPPONIER, P. 1977. Evolution tectonique du système Alpin en Méditerranée: poinçonnement rigide-plastique. *Bulletin de la Société Géologique de France*, **19(3)**, 437–460.

TOLLMANN, A. 1978. Plattentektonische Fragen in den Ostalpen und der plattentektonische Mechanismus des Mediterranen Orogens. *Mitteilungen der Osterreichischen Geologischen Gesellschaft*, **69**, 291–351.

—— 1980. Grosstektonische Ergebnisse aus den Ostalpen im Sinne der Platten-tektonik. *Mitteilungen der Osterreichischen Geologischen Gesellschaft*, **71/72**, 32–44.

TRÜMPY, R. 1980. *Geology of Switzerland — a guide book. Part A. Outline of the Geology of Switzerland*. Wepf & Co., 102 pp.

—— 1982. Alpine Palaeogeography: a reappraisal. *In*: HSÜ, K. J., (ed) *Mountain Building Processes*. Academic Press, London, 149–180.

ULMISHEK, G. 1982. *Petroleum Geology and Resource Assessment of the Timan-Pechora Basin, USSR and the adjacent Barents–Northern Kara Shelf.* US Department of Commerce National Technical Information Service, No. DE83–004421. 197 pp.

VAN HOORN, B. Structural evolution, timing and tectonic style of the Sole Pit inversion (Abstract). *This volume*.

VOIGT, W. A. 1962. Uber Randtröge und Schollenränder und ihre Bedeutung im Gebiet der Mitteleuropäischen Senke und angrenzender Gebiete. *Zeitschrift der deutschen geologischen Gesellschaft*, **114**, 378–418.

WACHTEL, G. & WESSELY, G. 1981. Die Tiefbohrung Berndorf in den östlichen Kalkalpen und ihr geologischer Rahmen. *Mitteilungen der österreichischen geologischen Gesellschaft*, **74**, 137–165.

WESTPHAL, M., BAZHENOV, M. L., LAUER, J. P., PECHERSKY, D. M. & SIBUET, J. C. 1986. Palaeomagnetic implications on the evolution of the Tethys Belt from the Atlantic Ocean to Pamir since Trias. *Tectonophysics*, **123**, 37–82.

ZIEGLER, P. A. 1982. Geological Atlas of Western and Central Europe. Elsevier Scientific Publishing Co., Amsterdam.

—— 1983. Inverted Basins in the Alpine Foreland. *In*: BALLY, A. W. (ed.) *Seismic Expression of Structural Styles — A Picture and Work atlas. American Association of Petroleum Geologists, Studies in Geology, Series No. 15*, Vol. 3, 3.3.3–3.3.12.

—— 1984. Caledonian and Hercynian crustal consolidation of Western and Central Europe — a working hypothesis. *Geologie en Mijnbouw*, **63(1)**, 93–108.

—— 1986. Geodynamic model for the Palaeozoic crustal consolidation of Western and Central Europe. *Tectonophysics*, **126**, 303–328.

—— 1987. Compressional intra-plate deformations in the Alpine foreland. *Tectonophysics*, **137**, 420 pp.

—— 1988. Evolution of the Arctic–North Atlantic and the Western Tethys. *American Association of Petroleum Geologists, Memoir*, **43**, 198 pp.

P. A. ZIEGLER, Shell Internationale Petroleum Mij. B. V., PO Box 162, 2501 AN The Hague, The Netherlands. Present address: Geological–Palaeontological Institute, University of Basel, Bernoullistrasse 32, CH-4058 Basel, Switzerland.

The inverted margin of the French Alps and foreland basin inversion

P. C. de Graciansky, G. Dardeau, M. Lemoine & P. Tricart

SUMMARY: The whole of the Western Alpine realm can be considered to result from the inversion of the European margin of Ligurian Tethys as a consequence of the collision between Apulia and Europe. The margin was developed due to Tethyan rifting during the Liassic and Middle Jurassic and regional thermal subsidence as the Ligurian ocean was spreading in the Late Jurassic and Early Cretaceous. Coevally, the margin underwent additional stretching in response to North Atlantic rifting. Inversion was initiated by Late Cretaceous times when the Ligurian oceanic crust began to be subducted, with localized obduction, beneath the previously passive margin. The most recent inversion of the Western Alps occurred in the external Molasse basin of Manosque–Digne–Valensole during the Late Miocene and Pliocene. Regions of mild inversion allow useful comparisons to be made with more strongly inverted basins. The geometric characteristics of the inverted basins depend on the lithology, geometry of the extensional structures, orientation of extensional faults with respect to the compressive Alpine stresses and the amount of compression. Comparisons of the external and internal zones suggest that the ductilely deformed thrust sheets may have originated from inverted half-grabens.

Introduction

The concept of inversion tectonics can be applied at various scales to mountain belts such as the Western Alps. At the scale of hundreds of kilometres the whole Alpine belt can be considered to be the inverted margin of the Tethyan–Ligurian ocean in response to the collision of Apulia against Europe. On a smaller scale of a few tens of kilometres, the various sub-basins of the Ligurian Tethys margin have been inverted to produce the main Alpine thrust sheets or folded zones. This occurred successively from Alpine inner (eastern) zones to external (western) zones.

Besides the obduction of Ligurian ophiolites during the Cretaceous, which is poorly documented, the earliest thrusting, which is that of the Nappe du Flysch à Helminthoides de l'Autapie, occurred in a deep-sea environment by Late Eocene times. The youngest thrust is the Nappe de Digne which was emplaced onto a subaerially eroded surface during the Early Quaternary in the external Molasse foreland basin of Valensole. The tectonic units of internal origin suffered more intense compressional deformation as shown by metamorphic assemblages. This has obscured the field evidence for inversion. In the external Alpine zones, especially in the Molasse basins, mild tectonism allows comparison with the N European intracratonic inverted basins.

In many of the intracratonic basins inversion occurred only once, for example, in the Aptian on the Bray and Rouen faults. In orogenic belts both extensional and compressional developments are essentially multiphasic. Hence a consistent analysis of the inversion has to account for variations in both time and space. In this chapter we aim to describe some geometrical aspects of Alpine basin inversion, including the chronological and geodynamic implications.

Main stages of Tethyan and Alpine development

Extensional tectonics during the development of Ligurian Tethys

Extensional tectonics, as recorded in Alpine stratigraphy, is related to the 'disintegration' (Bernoulli & Jenkyns, 1974) of Pangea during the Mesozoic, in two main phases. The first phase (Figs 1A & 2) is dated as Early and Middle Jurassic. It corresponds to the early stages of the rifting process. This rifting phase was essentially discontinuous. There were several episodes of faulting, each of which was followed by subsidence of the whole rift zone. The exception is the Briançonnais (Figs 3A & 4), a major positive block in the rift zone, where tilting induced emergence with subaerial erosion of hundreds of metres of Triassic rocks

From Cooper, M. A. & Williams, G. D. (eds), 1989, *Inversion Tectonics*
Geological Society Special Publications No. 44, 87–104.

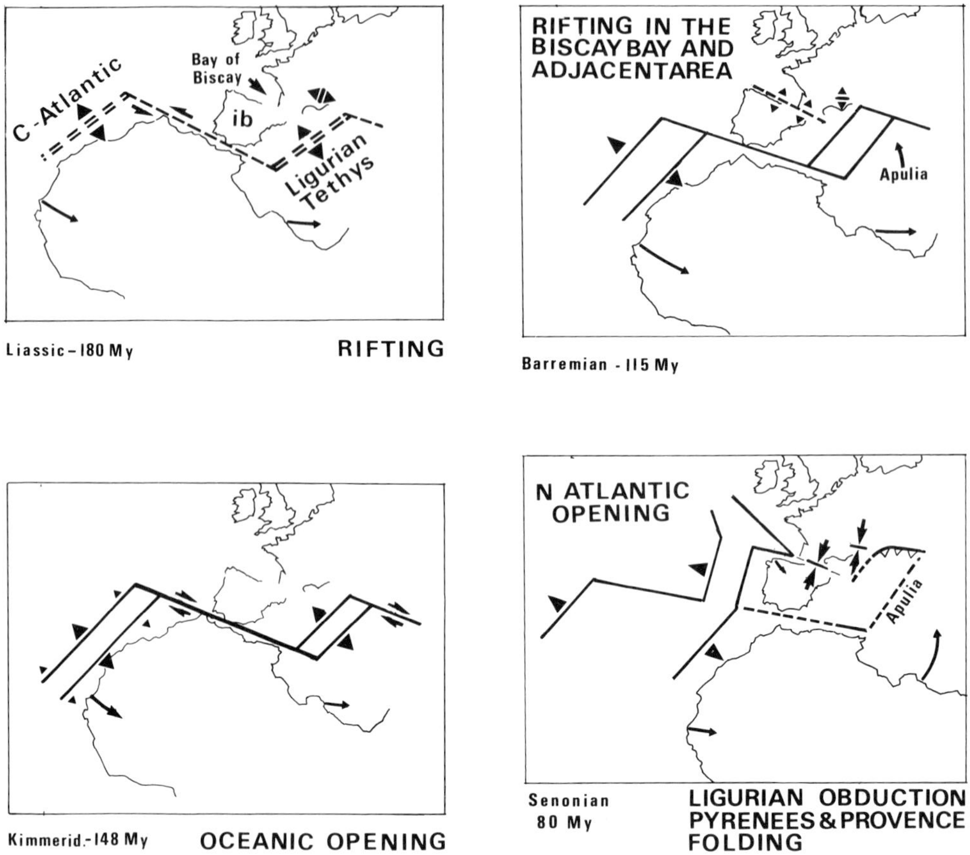

FIG 1. Schematic plate movements in the Tethyan–Alpine domain.

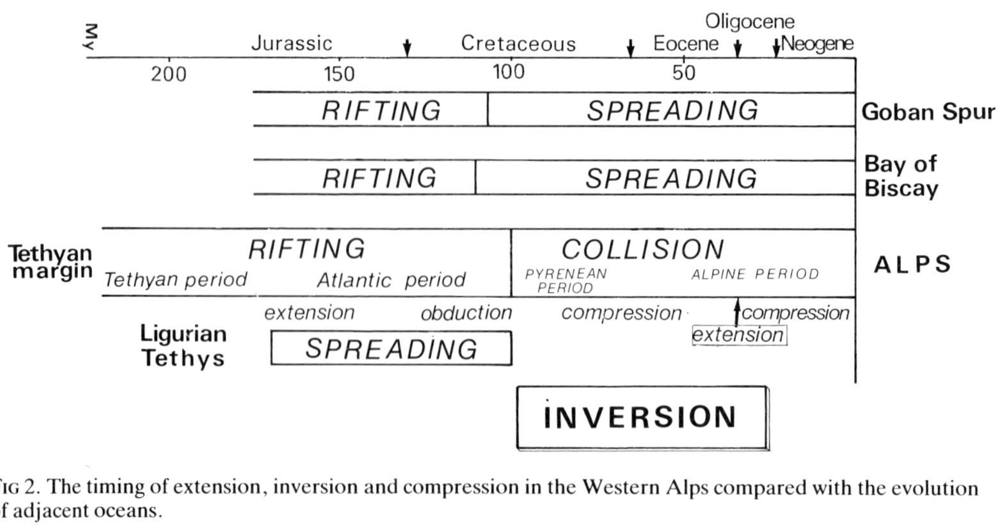

FIG 2. The timing of extension, inversion and compression in the Western Alps compared with the evolution of adjacent oceans.

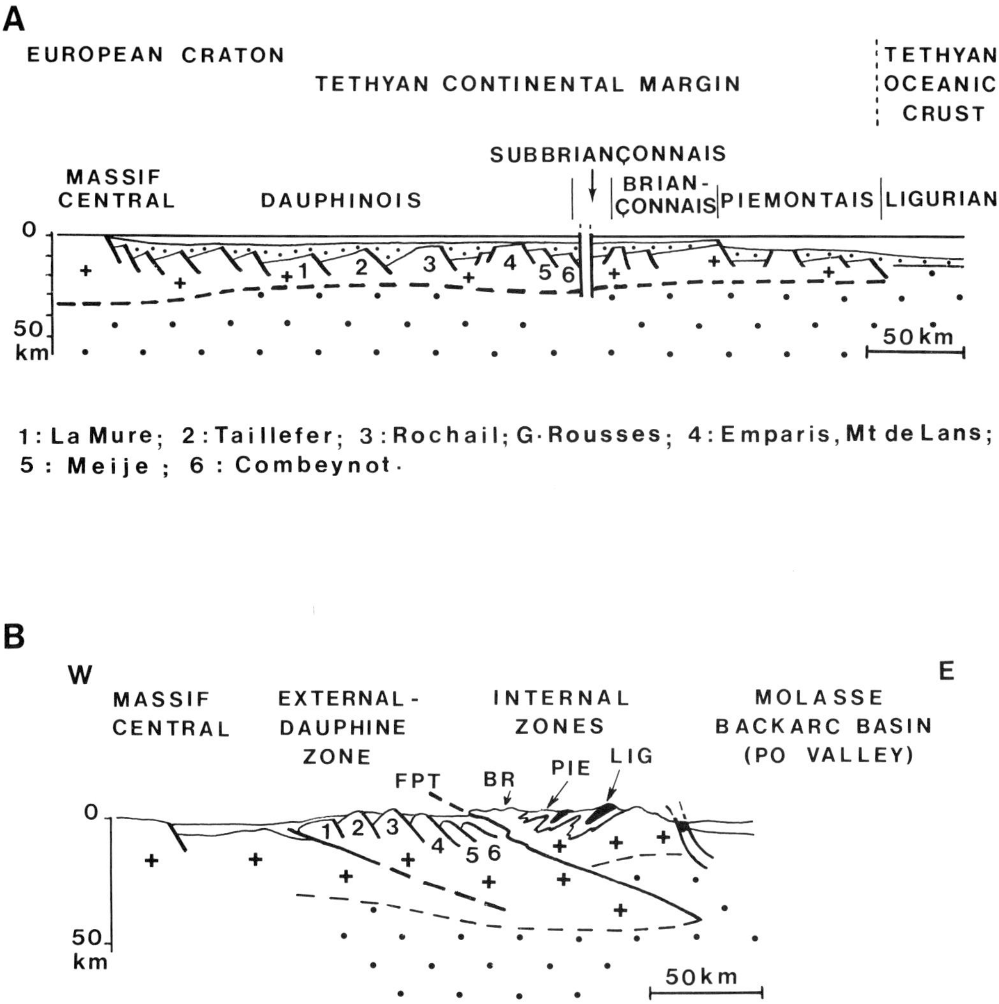

FIG 3. Schematic pre-inversion (A) and post-inversion (B) structural sections across the Alps. BR, Briançonnais; PIE, Piedmontese; Lig, Ligurian. FPT, Frontal Penninic thrust. The locality of the section is shown on Fig 4. The Dauphiné zone is external. Internal nappes and palaeogeographical zones from the west (external) to the east (internal) are the sub-Briançonnais, the Briançonnais, the Piedmontese, and the Ligurian zones. The Externides and Internides are separated by a major thrust (previously named the *chevauchement pennique frontal* or Frontal Penninic thrust). Its importance was probably overestimated in the past when the presence of Jurassic listric faults and tilted blocks were not appreciated by alpine geologists. The sub-Briançonnais, Briançonnais, and Piedmontese together constitute the Penninic, an old-fashioned, but convenient, term for the units which all derived from the European continental margin. Ligurian units were derived from the Tethyan ocean and were tectonically mixed with the Piedmontese units as a consequence of the collision. On this transect the Austro–Alpine units, which are of Apulian origin, are still buried below the Molasse back-arc basin of the Pô valley.

during Middle Jurassic times. This is analogous to the scenario on the western Atlantic margin across the Long Island platform (USGS Line 36; Hutchinson, Klitgord & Detrick 1986). The main phases of extensional faulting are dated Late Triassic, Sinemurian, Late Toarcian–Early Aalenian and Callovian. The Callovian phase coincides with the onset of Ligurian oceanic spreading (Lemoine *et al.* 1986). Directions of extension have been estimated from

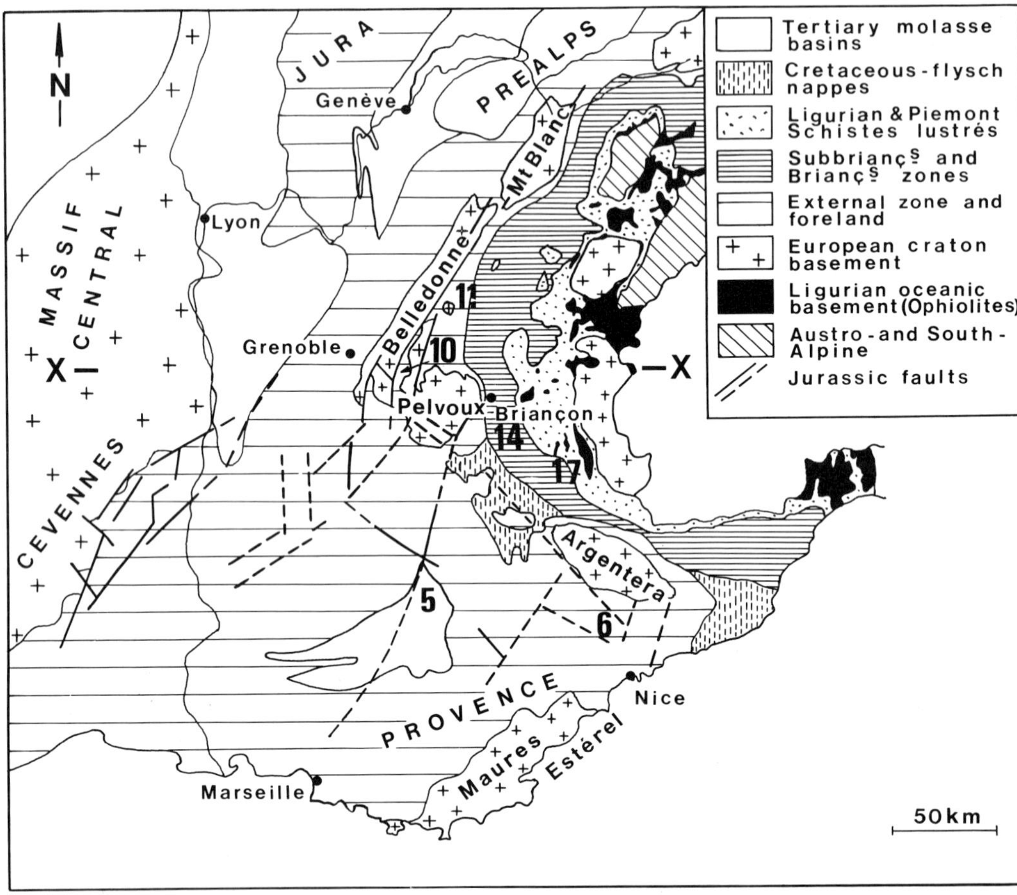

FIG 4. Structural sketch map of the Western Alps. The numbers indicate the location of other figures in this paper. X–X is location of the section in Fig 3.

microtectonic data in the Bourg d'Oisans area (Grand, 1988). Extension occurred successively from N–S (Late Triassic), NE–SW (Early Liassic), and WNW–ESE (Late Liassic).

Extensional tectonics linked with the development of the Atlantic

The second phase of extension is dated Late Jurassic to Albian in the sub-Alpine ranges (Figs 1C & 3). It created a complex set of extension faults and tilted fault blocks which intersected and/or reactivated the older faults. The Cretaceous extension directions are poorly constrained but are dominantly N–S in the Baronnies (Gerlier & Mascle, 1987) and in the Alpes Maritimes (Dardeau 1987).

Alpine compression

The compressional history involves three successive phases: Ligurian oceanic closure during the Cretaceous (Fig. 1D); initial Apulian collision against Europe during Late Cretaceous or Palaeogene; collision in the Neogene which corresponds to the maximum convergence between the two cratons (Fig. 2).

The following stages can be recognized in the compressional history:

1. Closure of the Ligurian Tethys ocean, inferred to have been of Middle Cretaceous age, (Fig. 1D) caused obduction and subduction of oceanic lithosphere onto the European craton. This is documented by the existence of sub-greenschist facies ophiolite

bodies on top of the nappe pile and very high pressure metamorphic assemblages in the inner Alpine zones (Cortesogno *et al.* 1979; Chopin 1987).

This initiated tectonic inversion in the Alpine inner zones. In the sub-Alpine ranges the picture was complicated by the stresses in the European Craton. The eastward movement of Iberia began at the end of the Early Cretaceous (Fig. 1C) and initiated the opening of the Bay of Biscay. This could have initiated the closure of Ligurian Tethys. It probably induced Late Cretaceous folding in the Pyrénées, Provence and Devoluy (sub-Alpine Ranges).

2. The Apulian collision is interpreted by assuming that the 'Apulian craton' (Fig. 1C; Argand 1924) was a part of Africa during the northward motion of Africa since mid-Cretaceous times. Two main phases of folding are dated as Eocene and Oligocene.

The Eocene phase deformed the Penninic zone in the lawsonite–albite to lawsonite–glaucophane metamorphic facies and corresponded with the emplacement of the first thrust sheets onto the margin.

The history of Alpine compression was interrupted by a short episode of extension dated as latest Eocene to early Oligocene. This induced the development of deep marine flysch troughs such as Flysch Noir in Briançonnais. In the internal Dauphinois, Grès d'Annot, Grès du Champsaur and flysch des Aiguilles d'Arves were deposited on an erosional surface which intersected tilted blocks of the Hercynian basement and/or Mesozoic cover (Ravenne *et al.* 1987). Coevally, the early Molasse troughs appeared in continental environments at the west in the Rhône valley and in the district of Manosque (Gigot *et al.* 1975), i.e. in the foreland basins.

The Oligocene phase of compression developed in the pumpellyite–actinolite metamorphic facies and corresponded to renewed thrusting. This terminated flysch deposition and initiated inversion in the external Alpine zones.

3. The final collision is Neogene and produced pumpellyite–actinolite facies assemblages in the Inner Alpine zones. It caused the tightening of earlier structures and conspicuous eastward backthrusting from the eastern Briançonnais (Figs 2 & 3B; Tricart & Lemoine 1986).

Both the extensional and compressional history of the Alpine chain record the evolution of the short-lived Ligurian ocean which existed for only 50 MY and closed around the time that North Atlantic oceanic spreading commenced. (Fig. 1C & D)

Inversion on the scale of the whole chain

Pre-inversion geometry

The European margin of the Ligurian Tethys developed from the Late Jurassic to the Early Cretaceous (Fig. 3A). It extends from the Massif Central in the W to the Ligurian ocean in the E. The Massif Central remained part of the European craton and probably had a typical crustal thickness of 30–35 km. Stretching of the margin occurred on a system of extensional faults. This produced tilted blocks which are a common characteristic of passive continental margins. They can be reconstructed despite Alpine tectonism.

Reconstructions of the Briançonnais, which emerged during Late Liassic and Middle Jurassic times, suggest a locally thinned continental crust with respect to adjacent areas (J. L. Rudkiewicz, personal communication, 1987; Fig. 3A).

The Ligurian oceanic crust is composed predominantly of gabbros and serpentinite. Pillow basalts are relatively scarce. Extensional fault blocks are still preserved even in tectonic units of oceanic origin within the ophiolitic bodies of the Queyras and Viso areas (Tricart & Lemoine 1986).

Present day geometry

The present day geometry results from the overriding of the European margin by fragments of the Ligurian oceanic crust together with elements of the Apulian continental crust. The consequence of the collision is the inversion of a stretched and thinned continental crust which was thickened to a minimum of 60 km thickness to the west of the Ivrea zone (Hirn 1980).

The visible structures are mainly thrust sheets which have been emplaced from the distal part of the margin towards the European craton. In addition to the dominant westward tectonic transport some eastward transport of superficial internal units was caused by a late back-thrusting phase (Fig. 3B).

A comparison of tectonic styles on a traverse through the chain shows that the amount of tightening and shortening increases from foreland zones to internal units. Therefore, the early Alpine deformation structures, which mark the onset of inversion, can be recognized much more easily in the sub-Alpine ranges than in the Schistes Lustrés, where several phases of isoclinal folding developed.

Inversion in the foreland 'molasse' basin of Digne–Valensole–Manosque

The molasse basin of Valensole and Manosque shows inverted structures that compare well with those observed from seismic data in Jurassic and Cretaceous basins in the North Sea. However, the structural evolution is complicated by the progressive migration of Alpine deformation with time.

The evolution of a composite section running through Barrême, Digne, Manosque, and Banon illustrates the inversion (Fig. 5). Three regions can be distinguished structurally: the domain of Alpine folding *sensu stricto* in the E, which is bounded by a line of wrench and thrust faults such as the Digne thrust; the Digne–Valensole trough with a sequence of Miocene and Pliocene age several kms thick; the Manosque trough with an Oligocene sequence which is 1–2 km thick.

During Oligocene times the Manosque area underwent subsidence due to block faulting under an extensional regime. The fault scarp of the Durance produced huge olistoliths which were carried westward and progressively buried under coarse fluviatile conglomerates. In the Early Miocene (Burdigalian) the whole molasse basin was flooded and shallow marine deposits concealed the earlier extensional faults.

The inversion of the Manosque trough occurred during the Late Miocene and Pliocene when uplift and E–W folding caused emergence of the area. The Durance fault was also inverted and acted subsequently as a sinistral strike–slip fault.

The Digne–Valensole molasse basin was in a

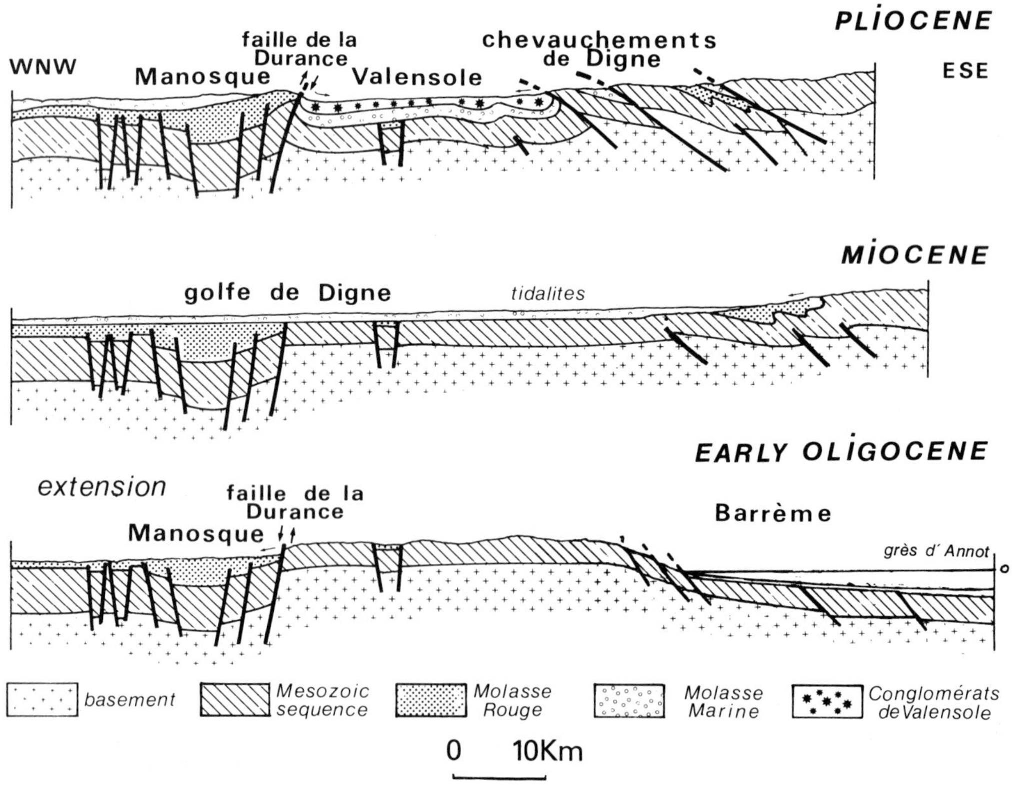

FIG 5. Complex inversion in the Digne–Valensole–Manosque molassic troughs. Note the thickening of the Molasse and the Mesozoic to the W across the Durance fault. Modified from Gigot *et al.* (1975).

relatively elevated position during the Oligocene due to the Pyrenees–Provence phase of folding. However, as a consequence of the Late Miocene to Pliocene inversion it has been overridden by both the Durance fault to the W and by the sub-Alpine thrust sheets to the east. It simultaneously underwent rapid subsidence due to the thrust load. This consists of several thousand metres of continental sediments which have been syndepositionally folded.

Inversion in the Alpes Maritimes

The external sub-Alpine ranges in the Alpes Maritimes extend from the Argentera crystalline massif in the north southwards to the Maures–Esterel massif on the Mediterranean coast. Alpine folds trend E–W to ESE–WNW. They developed during the Late Cretaceous to Palaeocene and Late Cenozoic deformation phases. The structures are combined with three main NE–SW trending wrench fault zones (Rouaine, Vesubie, and Roya) and with a network of thrust faults. The NNW–SSE Esteron and Gourdan–Vial thrust faults connect the Rouaine and the Vesubie faults (Figs 6 & 7). The thrust faults are reactivated extensional palaeofaults (Dardeau 1987) that can be identified by the stratigraphic thickening of Jurassic and Cretaceous sections on the NE sides of the faults.

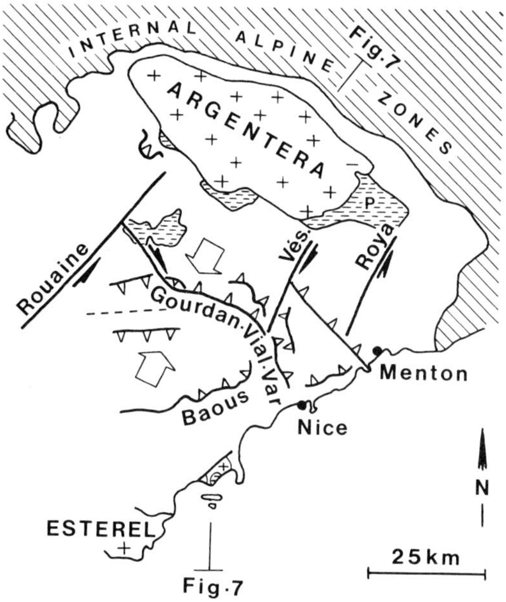

FIG 6. Structural map of the Alpes Maritimes.

Pre-inversion geometry

The pre-inversion geometry (Fig. 7) can be documented along a reconstructed section running south from the Argentera Massif. The Estéron and the Gourdan–Vial faults separated two major tilted extensional fault blocks at the northern edge of the subsiding carbonate platform of Provence during the Jurassic and Early Cretaceous. To the N, the Var valley half-graben was filled with a thick sequence (2000 m) of alternating marl and limestone. These uniform lithologies were deposited in a pelagic environment dated by ammonites to be of Hettangian to Late Cretaceous ages. The Var half-graben was the deepest part of this transect during the Jurassic. It was connected with the Dauphinois and Vocontian palaeogeographic domains of the sub-Alpine ranges.

The Estéron tilted block and half-graben was located at an intermediate depth between the Provence platform and the Var half-graben. It was bounded by the Estéron fault to the S and the Gourdan–Vial fault to the N. Sediment thicknesses are less than in the Var half-graben but a similar ammonite-bearing limestone and marl sequence was deposited in pelagic to hemipelagic environments. However, the raised edge of the Estéron block was uplifted sufficiently during the Jurassic to allow the deposition of shallow water carbonates which are very similar to the Calcaires de Provence.

Post-inversion geometry

As a consequence of Alpine inversion the N–S trending faults (Rouaine, Vesubie, Roya) which were roughly parallel to the transport direction have been reactivated into strike–slip faults. Those trending at a high angle to the maximum stress were strongly inverted to become thrusts.

The Calcaires de Provence, which had been deposited upon an elevated platform since the Early Liassic, were pushed upward again during Alpine compression. This is due to the combined effect of the stiffness of the thick carbonate sequence and to the dip of the bounding faults back into the orogen.

The Gourdan–Vial master extension fault, which is now a thrust fault, acted as a buttress during Alpine tectonism. The Var valley half-graben was transformed into an asymetric syncline whose gently dipping northern limb on the Argentera massif contrasts with its steeply northward inclined to overturned southern limb. The original thickness of the Mesozoic sequence

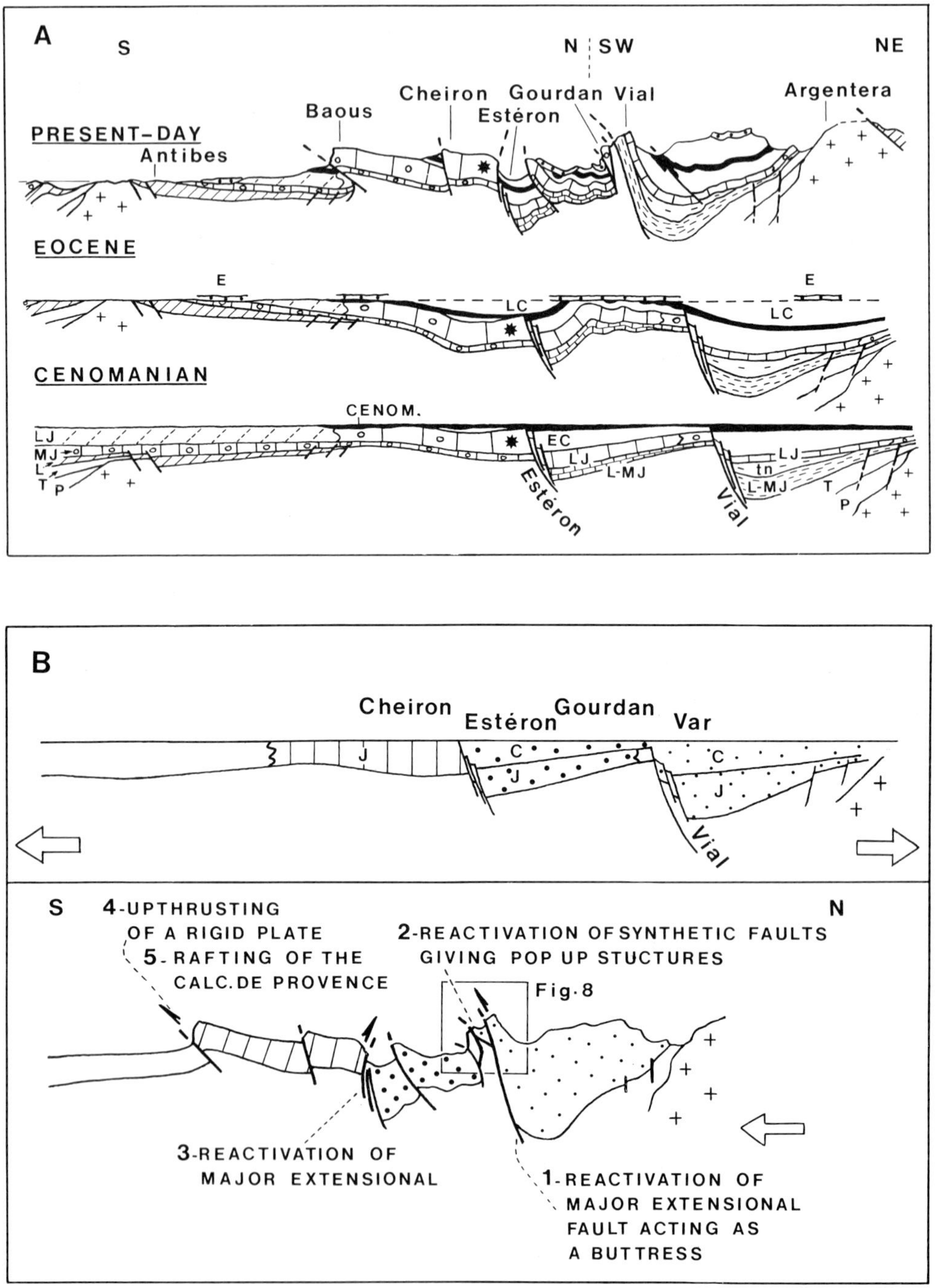

FIG 7. Pre-inversion (Cenomanian) and post-inversion geometry in the Alpes Maritimes. A, Actual profile and restorations. B, Schematic model.

was progressively increased by mesoscopic folding against the buttress. A pop-up structure is associated with the master fault in the Rourebel region where a slab of an attenuated sequence of Jurassic and Early Cretaceous age is bound to the N and S by thrust faults of opposed dip which originated as minor extensional faults closely associated with a major fault (Fig. 8).

This transect across the Alpes Maritimes illustrates typical mild inversion. Compression has reactivated earlier extensional structures, accentuating variations in elevation such that platform areas remain high and half-grabens are still structural lows. A similar history of evolution can be found in other parts of the external sub-Alpine ranges such as the Baronnies.

Inversion in the inner sub-Alpine zones

An appropriate area for studying inversion is a transect across the Massifs Cristallins Externes which extend from la Mure through the Taillefer to the Combeynot, which is the most internal satellite of the Pelvoux crystalline massif (Fig. 9).

Pre-inversion geometry

The pre-inversion geometry results from the phase of extension dated mainly as Liassic. This classic section (Fig. 9A) shows that the external crystalline cores were already elevated during Liassic times and corresponded with the crests of major tilted blocks (Lemoine *et al.* 1986). Eastward facing master faults which delineated the tilted blocks are still preserved at the eastern boundaries of the La Mure (Bas 1988), Taillefer (Barfety & Gidon 1984), and Grandes Rousses (Gidon *et al.* 1983) fault blocks. The corresponding half-grabens comprise fault scarps with local olistoliths, e.g. Ornon (Barfety & Gidon 1984) or Grandes Rousses. In contrast, the Liassic beds onlap the relatively gently sloping side of the half-graben opposite the fault. The original geometry is distorted by deformation and the rollover geometry is assumed by analogy with structures on seismic sections from extensional regimes (Gibbs 1984).

Post-inversion geometry

Alpine tectonism and inversion resulted in the development of megamullions as a consequence of the high contrast in ductility between the sedimentary cover and the basement during compression (Tricart & Lemoine 1986). This implies an upward bending of the basement surface between narrow pinches of the overlying pile of highly ductile sediments, an idea initially proposed by Ramsay (1967). However, the existence of pre-compression faults was not perceived at that time.

The increasing importance of strain across strike can be shown from the external to the internal part of the transect in Fig. 9B. It is documented by increasing steepness of the master faults from the W to the E.

Other regional features include a shear zone at the base of the sedimentary cover and the eastward dipping cleavage within the Jurassic sequence. In the well studied Bourg d'Oisans 'syncline' (Gratier & Vialon 1980; Gilcrist *et al.*

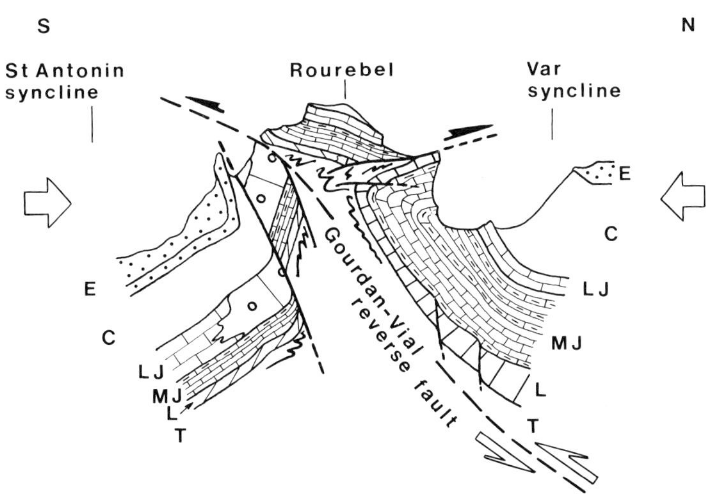

FIG 8. A pop-up structure along the Gourdan–Vial reverse fault. Location shown on Fig 7; see text for discussion.

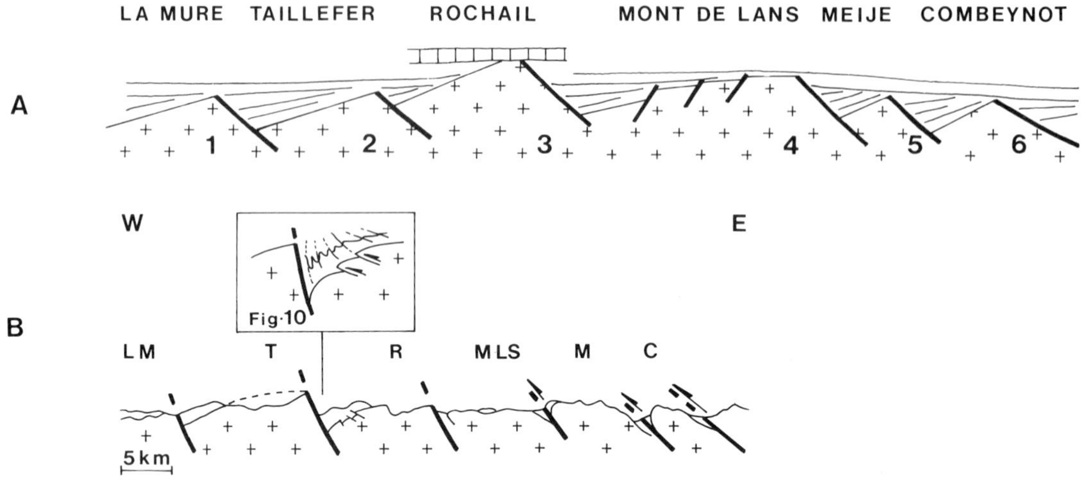

Fig 9. A, Pre-inversion and B, post-inversion structural sections across the Massif Cristallins Externes.

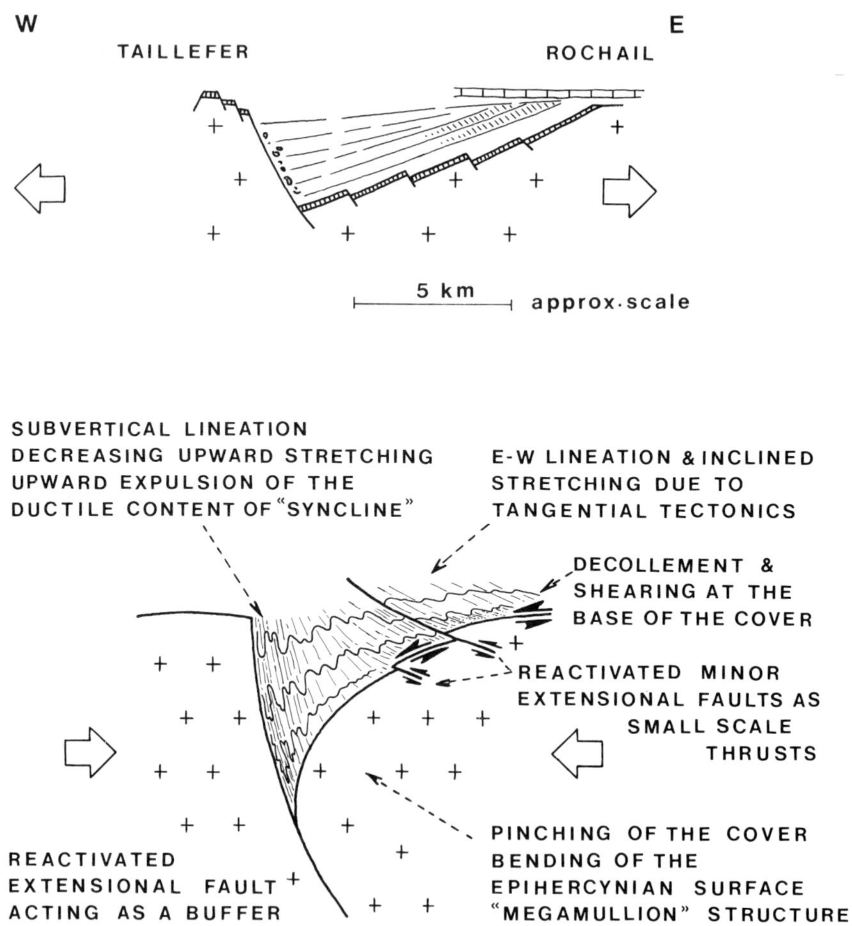

Fig 10. Pre-inversion and post-inversion geometry in the Bourg d'Oisans syncline. See Fig 4 for location.

1987; Grand, 1988), the characteristics and orientation of these structures (Fig. 10) are modified close to major extensional faults as a consequence of the buttressing effect of the faults. The folds generally face upwards and the associated steep cleavage contains a subvertical lineation indicating stretching perpendicular to the sub-horizontal fold axes which trend N–S.

These structures suggest that the horizontal E–W regional contraction resulted in the upward expulsion and shortening of the 'ductile' contents of the half-grabens to produce pseudosynclines. Therefore, inversion resulted here in small scale structures and thrust faults that cross-cut the older structures.

Inversion in the external Penninic zones

The sub-Briançonnais nappes constitute the most external of the inner Alpine zones (Fig. 3). They are well developed in the Maurienne valley, where the Nappe du Pas du Roc with its two subunits of the Perron de Encombres and of the Grande Moendaz were defined by R. Barbier (1948).

Pre-inversion geometry

At the Perron des Encombres a Jurassic half-graben can be reconstructed (Fig. 11A) with a westward palaeoslope on the supposed rollover. This geometry is similar to those of more external zones. Sponge spicule bearing marls and limestones constitute a relatively thick sequence in the deeper part of the half-graben. Crinoidal limestone of Liassic to Middle Jurassic age caps the raised edge of the footwall of the tilted block at the summit of the Perron des Encombres. Pre-inversion salt domes, plugs and diapirs are still preserved in the footwall and are dated as Early and Middle Jurassic (de Graciansky *et al.* 1986; Samec *et al.*, 1988). The Oxfordian Terres Noires (i.e. black shales) is the oldest clearly postrift formation in this area.

Post-inversion geometry

The sub-Briançonnais nappes (Fig. 11B) are now a series of complex folds with westward vergence between two main thrust surfaces. The western thrust separates internal from external Alpine units (*chevauchement pennique frontal*). The sub-Briançonnais nappe overrides

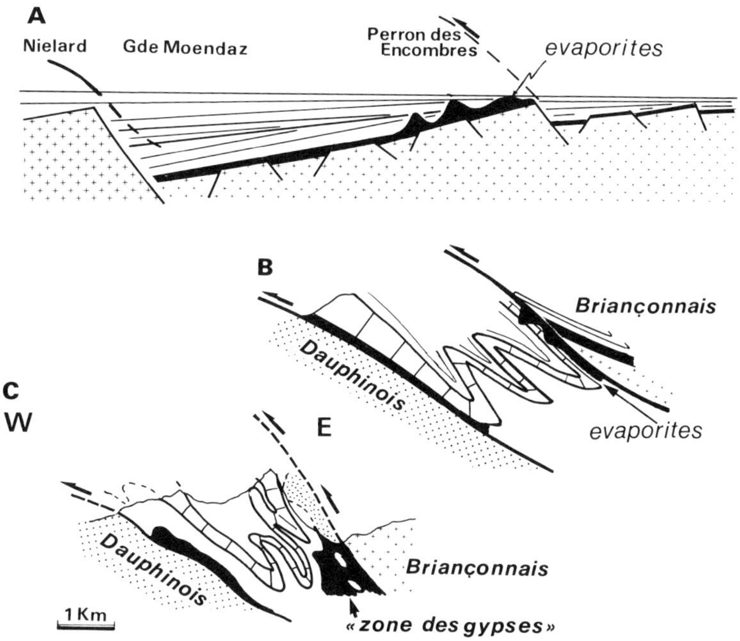

FIG 11. Reconstruction of structures of the sub-Briançonnais nappe du Pas du Roc at the Perron des Encombres. A, Reconstructed pre-inversion geometry. Note the salt domes near the raised edge of the tilted block which developed during Jurassic times. B, Schematic reconstruction during thrusting. Evaporites at the décollement surface in the Briançonnais were emplaced onto autochtonous sub-Briançonnais evaporites. C, Present day geometry.

the Flysch des Aiguilles d'Arves of the internal Dauphinois zone. The other thrust has a strike–slip component and separates the sub–Briançonnais from the Briançonnais. Both tectonic boundaries coincide with sharp lithological changes and strong variations in stratigraphic thicknesses. This strongly suggests that the old palaeogeographical boundaries, which generally coincide with extensional faults, have been reused during Alpine inversion. The contents of the half-graben were expelled as complexly folded, foliated thrust-sheets during Alpine deformation. The sedimentary cover has been completely detached from the basement which cannot be seen on this transect.

As a consequence of the emplacement of the Briançonnais above the overturned limb of the eastern sub-Briançonnais, the Briançonnais evaporites have been transported onto sub-Briançonnais evaporites creating a melange. The well-known *zone des gypses* with its exotic blocks of calpionellid-bearing limestones at the boundary between the two main palaeogeographic and structural units probably developed in this way. This created an exceptionally weak décollement horizon which was utilized as Alpine deformation continued.

Inversion in the Briançonnais nappes: the reverse approach

The Briançonnais nappes are situated more internally in the Penninic zones and thus originally lay closer to the Ligurian ocean with respect to the sub-Briançonnais (Fig. 3). Alpine tectonism resulted in a pile of thrust sheets detached from the basement. The increased intensity of deformation compared to the external zones is documented by the appearance of the lawsonite–albite metamorphic zone. The interpretation of the inversion structures is thus more difficult than in the external zones and requires the careful reconstruction of the pre-inversion geometry.

The existence of Jurassic synsedimentary extensional faults can be documented from three observations.

1. Small scale faulted blocks and extensional faults with displacements of up to 10 m are still preserved within thrust sheets. The late Middle and Late Jurassic postrift sequence largely obscures the structures (Figs 12, 13 & 14).
2. The basal unconformity of the postrift sequence has not been activated as a detachment.
3. Abrupt changes of stratigraphy and lithology correlate with major tectonic boundaries (Figs 15 & 16) suggesting that old extensional faults could have been reactivated during Alpine orogenesis. It is therefore considered reasonable to assume by comparison with the external zones that major extensional faults (1 km displacement at least) also exist in the Briançonnais.

If this initial reconstruction is accepted, the implication is that every thrust sheet in the Briançonnais was derived from an individual half-graben. The thrusts reactivated the major extensional faults but minor faults have been preserved and transported with the thrust sheets (Fig. 12).

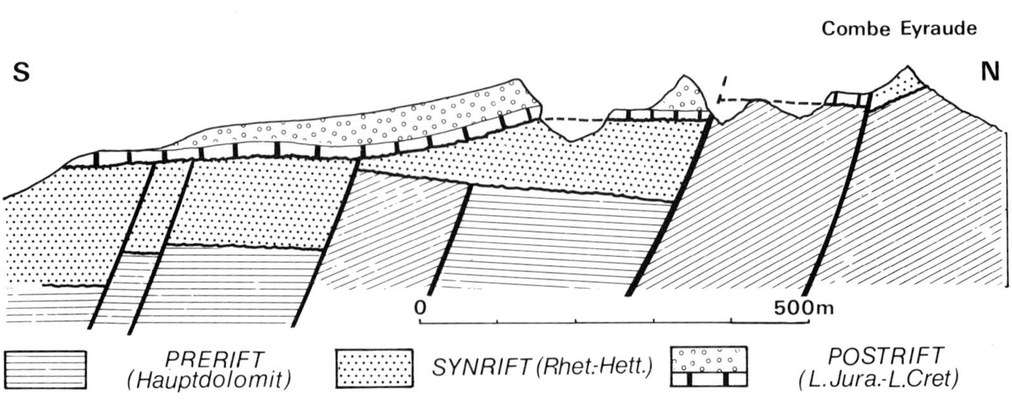

FIG 12. Evidence for extensional block faulting of Jurassic age in the Nappe de Peyre Haute (Briançonnais). Jurassic extensional faults have been transported passively with the thrust sheet.

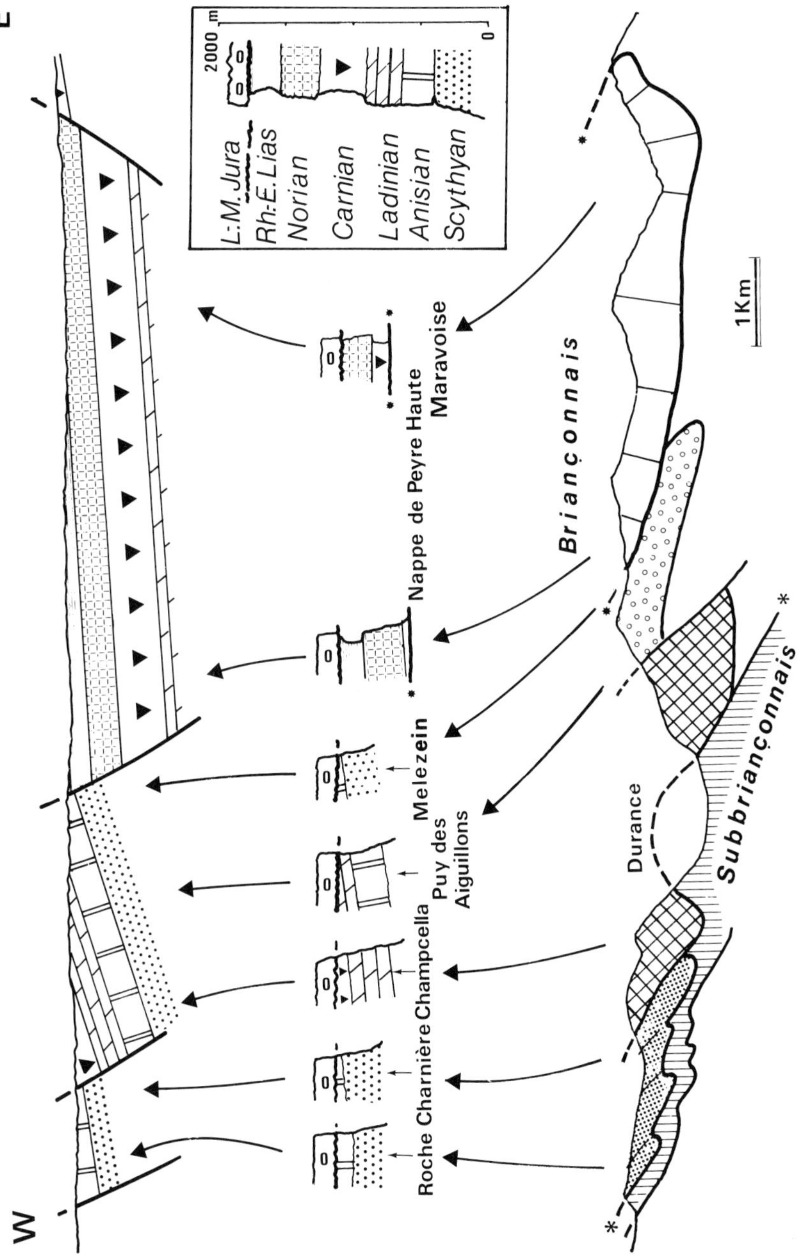

FIG. 13. Reconstructed tilted blocks of the three main nappes in the external Briançonnais.

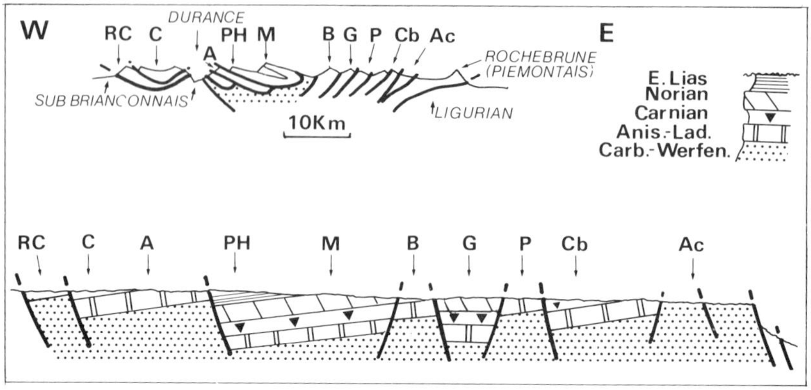

FIG 14. Reconstructed pre-inversion and post-inversion geometry across the Briançonnais and external sub-Briançonnais. Each Briançonnais nappe is considered to have been expelled from an individual half-graben. A, Aiguillons; Ac, Acceglio; B, Baudouis; C, Champcella; Cb, Clôt la Cîme and Côte Belle; G, Crête des Granges; M, Maravoise; P, Peygus; PH, Peyre-Haute; RC, Roche Charnière.

Inversion in the Schistes Lustrés west of the Viso area

The Schistes Lustrés comprise distinct units which were tectonically juxtaposed during obduction of the Ligurian oceanic crust onto the European margin in Cretaceous times (Fig. 3). One type of stratigraphical sequence comprises ophiolites overlain by a condensed sedimentary sequence dated as Late Jurassic to Cretaceous. The other involves the Hercynian basement overlain by Triassic dolomites and then by a classical Jurassic and Cretaceous sequence (Fig. 17). All units were metamorphosed and deformed during several Tertiary episodes, producing lawsonite–glaucophane–jadeite and zoisite–glaucophane–jadeite facies.

Ophiolite bodies range from less than one metre to several kilometres in size. The ophiolites are found either as slices along tectonic contacts, as olistoliths within Jurassic and Cretaceous calcschists or as larger cohesive remnants of the Ligurian oceanic basement, up to several kilometres in size which carry their original sedimentary cover (Tricart & Lemoine 1986).

Within the ophiolites the mafic bodies are stratified slabs, a few hundred metres thick. They comprise slightly flattened and stretched pillow lavas and metagabbros which still display subsequent granular textures. Certain gabbro bodies are intensively foliated with penetrative deformation associated with high temperature metamorphism. However, they are cut by virtually undeformed basaltic dykes. This suggests that both foliation and metamorphism of the gabbros must predate the emplacement of basalts on the Tethyan sea floor and also predates Alpine deformation. Remnants of oceanic crust were rigidly deformed during Alpine contraction, as were isolated slabs of Norian Hauptdolomit.

The isolated Triassic dolomite slabs (remnants) and the ophiolitic bodies (Fig. 17), appear to result from boudinage on a regional scale during the Eocene (Tricart & Lemoine 1986). However, field observation suggest that Alpine fragmentation of Triassic dolomites was controlled by synsedimentary palaeofaults linked with the Tethyan rifting (cf. the Briançonnais). Similarly, the faults which bound the ophiolitic blocks and are injected by serpentinite can be regarded as extensional palaeofaults that cut the Tethyan–Ligurian oceanic crust (Tricart & Lemoine 1986). The megaboudinage of both Piedmont Triassic dolomites and Ligurian ophiolites resulted from the strong contrast in ductility with the adjacent, highly deformable, calcschists under Alpine metamorphic conditions. It was facilitated by segmentation due to previous Tethyan extensional and strike–slip faults.

The concept of inversion in Alpine geology

The generally accepted usage of tectonic inversion is for areas which have been successively subsided and then uplifted or vice versa. Such

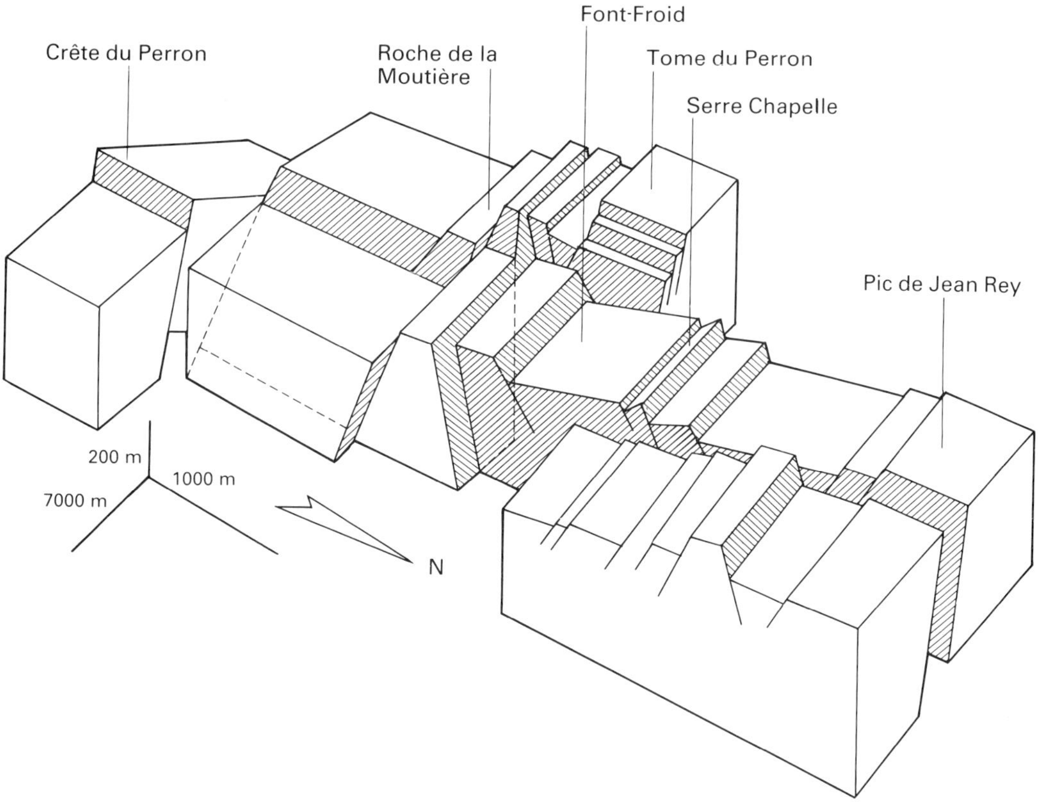

FIG 15. Reconstruction of the extensional fault pattern in a part of the nappe de Peyre-Haute (Briançonnais).

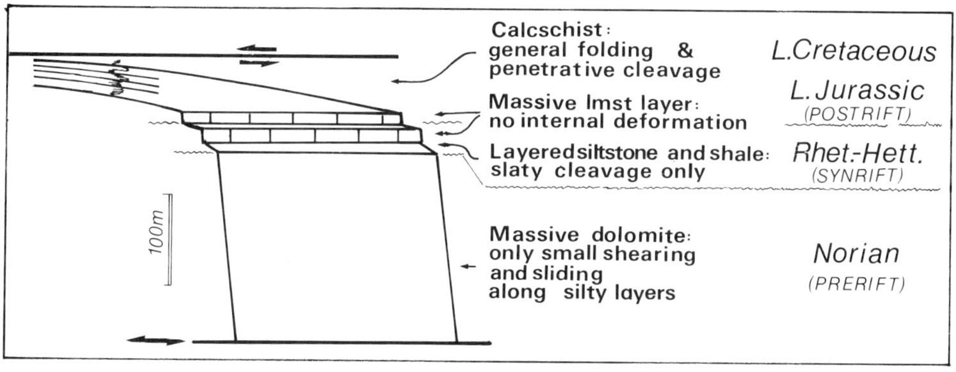

FIG 16. Vertical zonation of the style of deformation in the stratigraphic units of the Briançonnais nappes. (from Tricart 1980, p. 75).

an informal concept is convenient for explaining the structural evolution of sedimentary basins. It arose from seismic data in areas which are still below sea level or areas of flat topography such as the North Sea basins (Ziegler 1983) which can be considered as part of the Alpine foreland. The successive phases of extension and compression can be correlated in time with the main phases of Alpine tectonic evolution. The boundary between 'typical' inversion and an orogenic belt remains somewhat ambiguous.

Some of the most demonstrative examples of inversion can be found in molassic basins and in the external parts of the sub-Alpine ranges

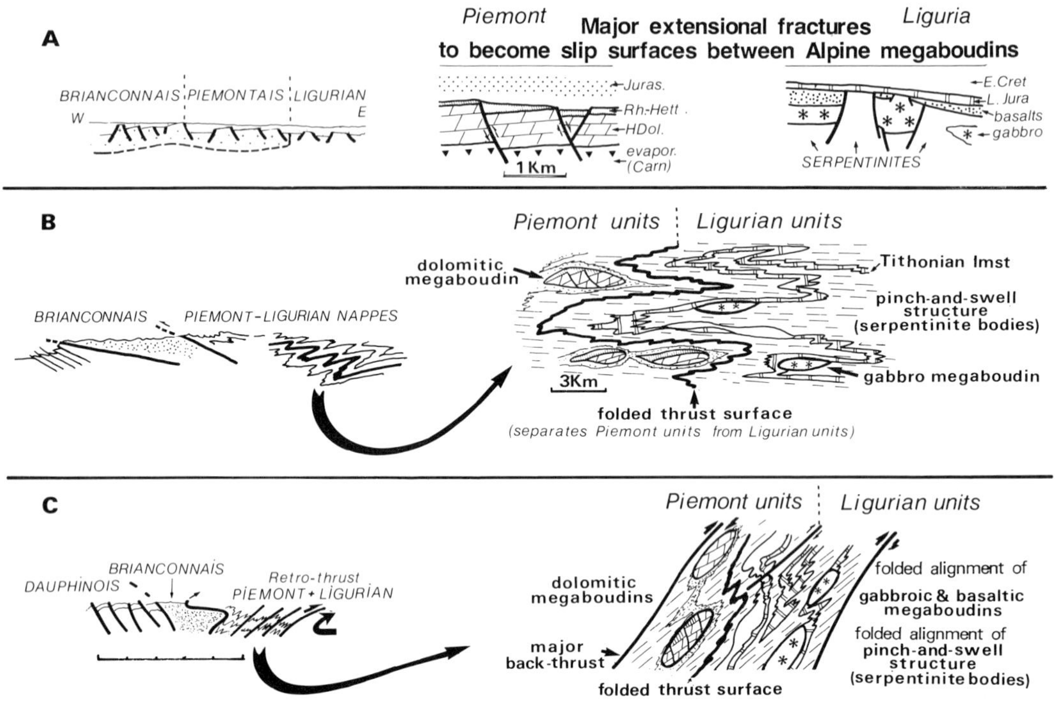

FIG 17. Schematic reconstruction of pre-inversion, post-inversion and post back-thrusting geometry in Piedmont and Ligurian units. (simplified after Tricart & Lemoine 1986).

which suffered relatively mild tectonism. The concepts of inversion may help in explaining the geometry of even the more severely tectonized areas. This can be documented by the gradient in the amount of strain from external to internal zones; the comparison of structural styles at any scale between external and internal zones; and the possible reconstruction of the geometry of the early stages of compression in each zone.

Where the original extension and subsequent compression are of the same order, the concept of inversion is directly useful for structural reconstruction whatever the scale of the region and whatever the geodynamic environment. This includes cratonic and oceanic basins, epeiric basins, marginal environments and mountain belts.

References and further reading

AMAUDRIC DU CHAFFAUT, S. & FUDRAL, S. 1986. De la marge océanique à la chaîne de collision dans les Alpes du Dauphiné. *Bulletin de la Société Géologique de France*, **8 (II)2**, 197–231.

ARGAND, E. 1924. La tectonique de l'Asie, *Comptes Rendus du XIIIeme Congrés Géologique International, Bruxelles*, 171–372.

BARBIER, R. 1948. Les zones ultradauphinoise et subBriançonnaise entre l'Arc et l'Isère, *Mémoire Expl. Carte Géologique détaillée de la France*, 291 pp.

BARFETY, J. C. & GIDON, M. 1984. Un exemple de sédimentation sur un abrupt de faille fossile: le Lias du versant Est du massif du Taillefer (zone dauphinoise, Alpes Occidentales). *Revue Géologie Dynamique et Géographie Physique*, **25**, 267–276.

BAS, T. 1988. Rifting liasique dans la marge passive téthysienne : le haut-fond de la Mure et le bassin de Beaumont (Alpes Occidentales). *Bulletin de la Société Géologique de France*, **8(IV)5**, 717–723.

BERNOULLI, D. & JENKYNS, H. C. 1974. Alpine,

Mediterranean, and Central Atlantic Mesozoic facies in relation to the Early evolution of the Tethys. *American Association of Petroleum Geologists Special Publication*, **19**, 119–160.

CHOPIN, C., 1987. Very high pressure metamorphism in the Western Alps: implications for subduction of continental crust. *Philosophical Transactions of the Royal Society of London*, **A 231**, 183–197.

CORTESOGNO L., GRANDJACQUET C. & HACCARD D. 1979. Contribution à l'étude de la liaison Alpes-Apennins: évolution tectono-métamorphique des principaux ensembles ophiolitiques de Ligurie (Apennins du Nord). *Ofioliti*, **4(2)** 157–172.

DARDEAU, G. 1987. Inversion du style structural et permanence des unités structurales dans l'histoire mésozoique et alpine du bassin des Alpes Maritimes, partie de l'ancienne marge passive de la Tethys. *Comptes Rendus Académie des Sciences, Paris*, **305(II)** 483–486.

—— 1988. Tethyan evolution and Alpine reactivation of Jurassic extensional structures in the French Alpes maritimes, *Bulletin de la Société Géologique de France, Paris*, **8(IV)4**, 651–657.

DE GRACIANSKY, P. C., RUDKIEWICZ, J. L. & SAMEC, P. 1986. Tectonique salifère d'âge Jurassique dans la zone sub Briançonnaise (Alpes de Savoie, France). Rôle dans le découpage en nappes de charriage et leur progression, *Comptes Rendus Académie des Sciences, Paris*, Série II, **302** 891–896.

GERLIER, A. & MASCLE, G. 1987. Distensions d'âge Crétacé inférieur dans les Baronnies, *Compte Rendus 11ème réunion des Sciences de la Terre, Clermont-Ferrand*, Société Geólogique de France, Paris, p. 77.

GIBBS, A. D. 1984. Structural evolution of extensional basin margins, *Journal of the Geological Society of London*, **141**, 609–620.

GIDON M., BARFETY J. C., MOUTERDE R., LEMOINE M., BRAVARD C. & PAIRIS J. L. 1983. Accidents stratigraphiques liés à une activité tectonique synsédimentaire dans la couverture des Massifs Cristallins Externes des Alpes Françaises au SE de Grenoble. 108ème Congrès national des Sociétés Savantes, Grenoble. *Sciences*, **1**, 153–163.

GIGOT, P. GUBLER, Y. & HACCARD, D. 1975. Relations entre sédimentation et tectonique (en compression et en extension): exemples pris dans des bassins Tertiaires des Alpes du Sud et de Haute Provence. *Comptes Rendus IXème Congrès International de Sédimentologie*, **IV**, 157–162.

GILLCRIST, R., COWARD, M. & MUGNIER, J. L. 1987. Structural inversion examples from the Alpine foreland and the French Alps. *Geodinamica Acta*, **1**, 5–34.

GOFFE, B. & CHOPIN, C. 1986. High pressure metamorphism in the Western Alps: zoneography of metapelites, chronology and consequences. *Schweizerische Mineralogische Petrografische Mitteilungen*, **66**, 41–52.

GRAND, T. 1988. Jurassic faulting in the Dauphiné palaeo-tilted blocks between Grenoble and La Grave. *Bulletin de la Société Géologique de France*, **8(IV)4**, 613–621.

GRATIER, P. J. & VIALON, P. 1980. Deformation pattern in a heterogenous material: folded and cleaved sedimentary cover immediately overlying a crystalline basement (Oisans, French Alps), *Tectonophysics*, **65**, 151–180.

HIRN, A. 1980. Le cadre structural profond d'après les profils sismiques. *In*: AUTRAN, A. & DERCOURT, J. (eds), *Image Géophysique de la France, 26th International Geological Congress, Mémoire Bureau Recherches Géologique et Minières*, **107**, 34–42.

HUTCHINSON, D. R., KLITGORD, K. D., & DETRICK, R. S. (1986). Rift basins of the Long Island Platform, *Bulletin of the Geological Society of America*, **97**, 688–702.

LEMOINE, M. 1984. La marge occidentale de la Téthys ligure. *In*: BOILLOT, G. (ed.) *Les Marges Continentales Actuelles et Fossiles Autour de la France*, Part 3, Masson Editeur Paris, 157–248.

——, MARTHALER, M., CARON, M., SARTORI, M., AMAUDRIC DU CHAFFAUT, S., DUMONT, T., ESCHER, A., MASSON, H., POLINO, R., & TRICART, P. 1984. Découverte de foraminifères planctoniques du Crétacé supérieur dans les schistes lustrés du Queyras (Alpes Occidentales). Conséquences paléogéographiques et tectoniques. *Comptes Rendus Académie des Sciences, Paris*, Série II, **299**, 727–732.

——, Bas, T., Arnaud-Vanneau, A., Arnaud, H., Dumont, T., Gidon, M., Bourbon, M., De Gracianshy, P. C., Rudhiewicz, J. L., Megard-Gailli, Tricard, P. 1986. The continental margin of the Mesozoic Tethys in the Western Alps, *Marine and Petroleum Geology*, **3**, 179–199.

RAMSAY, J. G. 1967. *Folding and Fracturing of Rocks*, McGraw Hill, New York.

RAVENNE C., VIALLY R., RICHE P., & TREMOLIERES P. (1987). Sédimentation et tectonique dans le bassin marin Eocène supérieur–Oligocène des Alpes du Sud, *Revue Institut français du Pétrole*, **42(5)**, 529–553.

SAMEC, P, DE GRACIANSKY, P. C. & RUDKIEWICZ, J. L. 1988. Tectonique synsédimentaire jurassique dans la zone sub Briançonnaise au Nord de l'Arc (Massif du Perron des Encombres), *Bulletin de la Société Géologique de France, Paris*, **8(IV)4**, 659–667.

TRICART, P. 1980. *Tectoniques superposées dans les Alpes occidentales au Sud du Pelvoux. Evolution structurale d'une chaîne de collision*, Thesis Université Louis Pasteur, Strasbourg, 407 pp.

—— 1984. From passive margin to continental collision: a tectonic scenario for the Western Alps, *American Journal of Science*, **284**, 97–120.

—— & LEMOINE, M. 1986. From faulted blocks to megamullions and megaboudins: Tethyan heritage in the structure of the Western Alps. *Tectonics*, **5(1)**, 95–110.

——, BOURBON, M., CHENET P. -Y., CROS P., DELORME, M., DUMONT T., DE GRACIANSKY P. C., LEMOINE M., MEGARD-GALLI J. RICHEZ M., 1988. Tectonique synsédimentaire triasicojurassique et rifting téthysien dans la nappe Briançonnaise de Peyre-Haute (Alpes Occidentales). *Bulletin de la Société Géologique de France* **8(IV)4**, 669–680.

ZIEGLER, P. A. 1983. Inverted basins in the Alpine foreland, *In*: BALLY, A. W. *Seismic expression of structural styles—a picture and work atlas* AAPG Studies in Geology, Series No. 15, Vol. 3, 3.3–3–3.3–12.

P. C. DE GRACIANSKY, Ecole Nationale Supérieure des Mines, 60 Blvd St Michel, 75272 Paris Cédex 06, France.

G. DARDEAU, Université des Sciences et Techniques, 28 Parc Valrose, 06034 Nice, France.

M. LEMOINE, Laboratoire de Géodynamique Sous-marine, BP 48, 06230 Villefranche-sur-Mer, France.

P. TRICART, Ecole Nationale d'Ingenieurs, Département des Sciences de la Terre, BP W, 3038 Sfax, Tunisia.

The influence of pre-existing basin structure on thrust system evolution in the Western Alps

R. W. H. Butler

SUMMARY: Collision mountain belts have recently been interpreted as being large scale stacked sedimentary basins, where the wholesale reactivation of pre-existing normal faults exerts the primary control on orogenic style. This model is explored here using examples from the Western Alps. The use of stratigraphic separation across faults, in the absence of corroborative sedimentological data, provides an ambiguous criterion to determinate fault reactivation. Variations between 'compressional' and 'extensional' stratigraphic separations across the Frontal Pennine thrust are more likely to be due to regional unconformity patterns developed in lower Tertiary times rather than normal fault reactivation along the eastern margin of the external zones. Mesozoic–Tertiary basin structure has been important in controlling the sites of major thrust ramps where stratigraphic horizons suitable for activation as thrust detachment zones have been offset. Examples include the thrust front in the Jura (Triassic detachment dropped to W by faults bounding the Bresse basin) and Vercors (basement uplifted to W at Faille de l'Isere, bounding the Mesozoic sub-Alpine basin to W). Buttressing structures, where thrust displacement is converted into vertical pure shear, may indicate buried normal faults but can occur at either hinterland or foreland-directed basin faults. The overall structural style in the sub-Alpine thrust belt indicates detachment between basement and cover with basement involvement along the Sole thrust which carries the external basement massifs. There is no indication from surface geology that this thrust originates from reactivation of a Mesozoic basin fault although this process may have occurred at depth. Similar styles of detachment-dominated thrusting with large horizontal relative to vertical displacements also characterize internal Alpine structure. Preliminary results from the ECORS-CROP deep seismic reflection profile across the Western Alps support the detachment thrusting model although they do not rule out minor normal fault reactivation.

The reactivation of faults on a crustal scale has been debated in structural geology for many years (e.g. Watson 1980). In cratonic settings, such as western and central Australia, long-lived lineaments appear to have controlled sedimentation and mineralization throughout late Proterozoic and Phanerozoic times (Hills 1956; Obee & White 1986). In the Alpine orogenic belt of Switzerland Argand (1911) proposed that basement highs and lows inherited from Mesozoic basin development pre-destined the sites of Tertiary compressional deformation and hence directly evolved into the patterns of nappes observed today. Jackson (1980) interpreted moderate to steeply dipping compressional fault plane solutions from the Zagros as thrust reactivation of normal faults and also proposed that this mechanism was of major importance in restacking continental margins and basins into mountain belts. Bally (1984) suggested that the evolution of mountain belts such as the Pyrenees is controlled by the pre-existing basin geometry, particularly the position of normal faults. Since then it has been widely assumed that thrusts usually result from the 'rejuvenation of old listric normal faults' (Dewey *et al.* 1986). So what is the geometry of these structures?

The large scale geometry of normal faults which accommodate crustal extension in sedimentary basins and on continental margins, as distinct from collapsing orogens such as the Basin and Range (Coney 1987) or Norwegian Caledonides (Hossack 1984), is indicated by several lines of evidence. These include: fault plane solutions from large magnitude earthquakes and associated aftershocks from actively extending regions such as the Gulf of Corinth (Jackson 1987); seismic reflection data across the North Sea (Badley *et al.* 1984) and elsewhere in Northwest Europe (McGeary *et al.* 1987; Pinet *et al.* 1987); analogue model experiments (McClay & Ellis 1987; Venderville *et al.* 1987); direct observations in areas such as the Gulf of Suez (Chenét *et al.* 1987) and western Turkey (Hancock & Barka 1987); and predicted from simple geometric constructions (e.g. Verral 1981). All show that the faults tend to dip at moderate to steep angles (>45°) in the upper crust with local listricity into mid-lower crustal detachments or continuing with moderate dips into the upper mantle. Activation of

From COOPER, M. A. & WILLIAMS, G. D. (eds), 1989, *Inversion Tectonics*
Geological Society Special Publications No. 44, 105–122.

upper crustal detachments is relatively rare and decoupling within the overlying sediments is weak although moderate ramp/flat extensional fault profiles can develop when facilitated by evaporite-rich or shale layers (Walker & Cooper 1987). Spacing between the major normal faults at the basin floor generally exceeds 20 km (e.g. Pinet *et al.* 1987) so that even quite broad basins contain only relatively few normal faults. Earlier faults within such arrays can be rotated to lower absolute angles by displacements elsewhere in the system although they will still retain high cut-off angles with respect to the stratigraphy within the fault blocks. Seismic reflection profiles across parts of the Northwest Atlantic continental margin show arrays of normal faults dipping oceanward, their listric forms merging into lower crustal shear zones (Montadert *et al.* 1979). Structures such as these, which control the extension of continental crust, are now considered to be preferentially suited for thrust-sense reactivation during the collision between continental margins.

Consider the range of structures which might occur if an Atlantic-type continental margin with its array of down-to-the-ocean normal faults became compressed, with thrusts preferentially located along the old normal faults (Fig. 1). Prior to shortening the extended continental crust is likely to have been overlain by a wedge of sediments which, provided there was abundant source material, will thicken stratigraphically towards the ocean. On sediment-starved margins the deposits would become progressively of a deeper water facies towards the ocean and may actually be thinner than on the fault blocks near the craton. Upon shortening, displacements can localize on some of the old normal faults, leaving others with their extensional geometry preserved. Discrete displacements near surface may pass down into syn-metamorphic straining zones of pure shear. This could explain the steep root zones found in deeply eroded interiors of mountain belts (Argand 1911) together with the steep fault plane solutions on seismogenic faults in the upper crust (Jackson 1980). Substantial shortening within the sedimentary cover can be balanced with only marginally overthickened crust at depth because prior to shortening the crust was attenuated (Helwig 1976). The overall geometry is one of *in situ* shortening without major translation of basement sheets. There is little activation of detachment horizons above the old lower crust and hence no significant juxtaposition of originally different lithospheric segments. The original passive margin palaeogeographic template, although somewhat telescoped, will not have been mixed up. The sediments with greatest oceanic affinites should lie along the suture.

A number of features of mountain belt geology suggest that, despite its apparent attraction, the reactivation model outlined above may not have universal application. Seismicity along

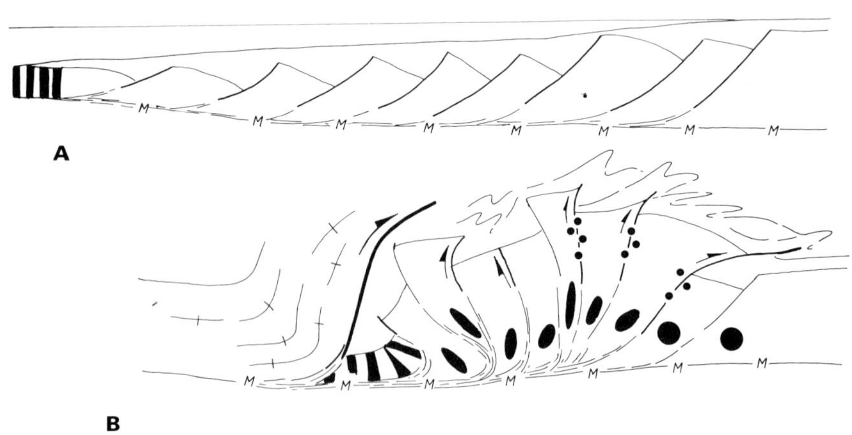

FIG. 1. Idealized geometry of a continental collision zone B formed by the reactivation of normal faults (see Bally 1984; Dewey *et al.* 1986) which existed in the precursor continental margin A, Substantial shortening in cover rocks can occur above a crust of little more than normal, cratonic thickness B, Faults and shear zones within the mountain belt should steepen at depth or pass into zones of dominantly vertical pure shear strains (represented by ellipses). Actively deforming mountain belts would show steep fault plane solutions (represented by dots; see Jackson 1980). Little of the internal structure of such mountain belts above the Moho should be visible on deep seismic reflection profiles.

active thrusts in the Himalayas (e.g. Ni & Barazangi 1984) depicts a low-angle (<10°) detachment at depths of 10–15 km beneath the frontal ranges. Seismic reflection profiling in the Appalachians (Cook *et al.* 1979) shows detachment for several hundred kilometres in the mid-upper crust beneath crystalline thrust sheets. Low-angle thrusts in basement terrains have long been recognized from the Norwegian (Tornebohm 1888) and Scottish Caledonides (Peach *et al.* 1888). This style of deformation is considered to dominate central Alpine tectonics (Boyer & Elliott 1982; Laubscher & Bernoulli 1982) where basement sheets only a few kilometres thick have been piled up along shear zones. Such large scale detachment is not predicted by the simple reactivation model. This chapter further examines the model and also the more general role of pre-existing basin faults in controlling structural style and thrust geometry in parts of the Western Alps (Fig. 2).

The stratigraphic framework essential for the understanding of larger scale aspects of structural geology in the western Alps has been the subject of several excellent reviews (e.g. Lemoine *et al.* 1986). Extensional basin development occured at various times throughout the Mesozoic, superimposed on a deeply eroded and peneplaned Hercynian (late Palaeozoic) orogenic edifice. The main phase of continental extension developed during mid-late Jurassic times with regional subsidence throughout the Cretaceous. Only relatively thin stratigraphic sequences were built up (<5 km) due either to relatively slight extension or sediment starvation. The main periods of orogenic shortening in the areas to be considered here were in late Eocene to latest Miocene times. More com-

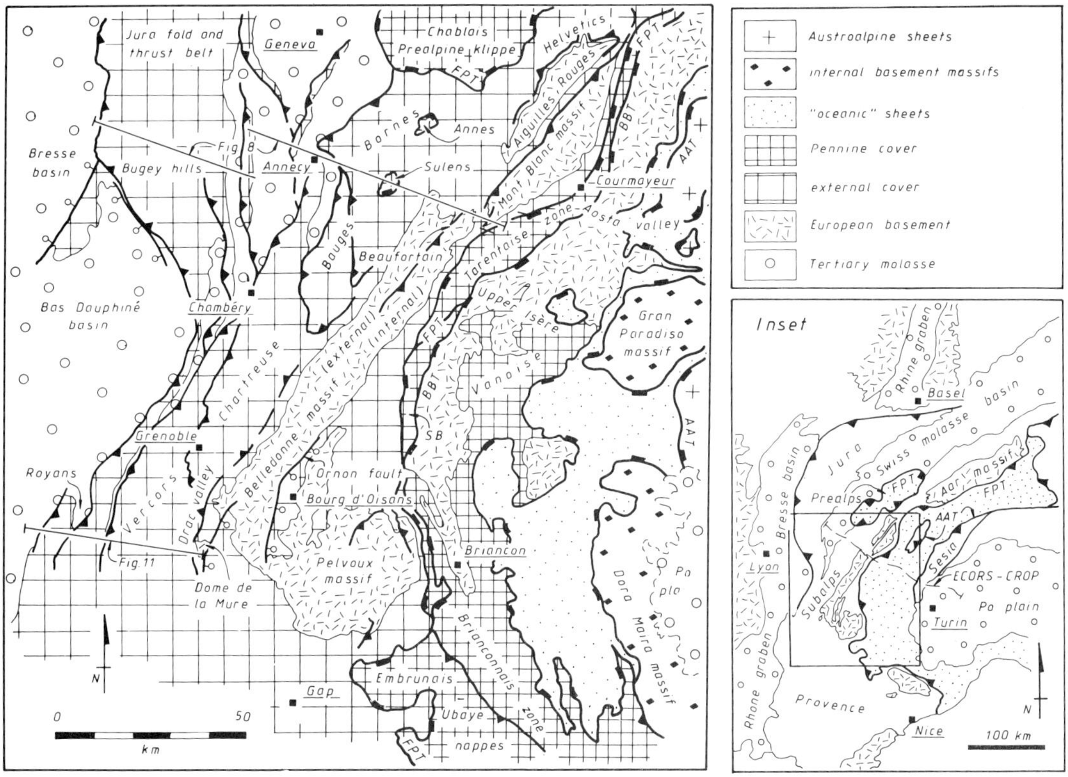

FIG. 2. Simplified tectonic map of the NW Alps illustrating the main structural units and locations of detailed cross-sections (Figs 8 & 11). FPT, Frontal Pennine thrust; BBT, Basal Briançonnais thrust; SB, sub-Briançonnais zone; AAT, Austro-alpine thrust. Location given by inset map of the Alps and adjacent European basin areas (open circles). The line of the ECORS-CROP seismic experiment (Bayer *et al.* 1987; see Fig. 14) is indicated. The basement of the European foreland is indicated by the random fleck ornament; the internal Alps are stippled.

plete structural appraisals are presented elsewhere (Butler 1986; 1988).

Stratigraphic separation as criteria for fault reactivation

The emplacement of stratigraphically younger rocks onto older rocks with the apparent omission of rock sequences on faults which otherwise show kinematic evidence for compressional displacements has been taken as good evidence for the thrust sense reactivation of pre-existing extensional faults (Soper & Anderton 1984; Powell 1987). A normal fault which has been reactivated by compressional displacements which are less than those associated with extension can show variable stratigraphic separations on maps and cross-sections (Fig. 3). The higher structural levels, where thrusting of late to post-rift sediments occurs, may show conventional, older on younger, stratigraphic separations. However, the early to prerift sediments may maintain extensional offsets with younger rocks carried on top of older and the net loss of stratigraphy across the fault zone. The strength of these relationships as criteria for fault reactivation can be examined by considering the Frontal Pennine thrust in the western Alps. This thrust is a major tectonic boundary which separates distinct Mesozoic stratigraphies and divides the external and internal tectonic domains (Debelmas & Keckhove 1980).

The Frontal Pennine thrust can be mapped throughout the western Alps from Mont Blanc to Pelvoux (Fig. 4). It is characterized by eastward (SE–E) plunging stretching lineations

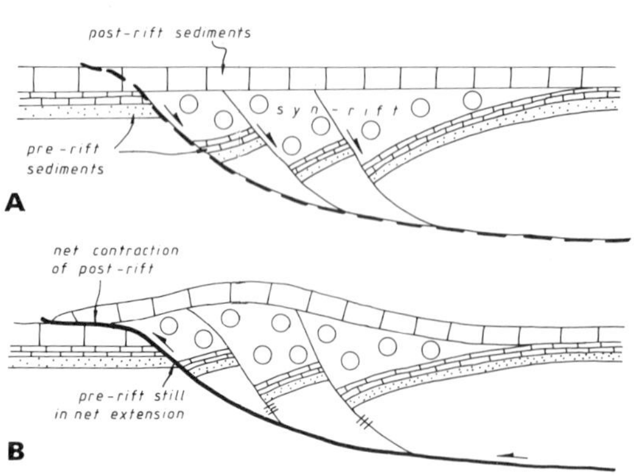

FIG. 3. Idealized view of normal fault reactivation (A–B in time), illustrating the variable stratigraphic separations at depth in the section. The pecked line (A) denotes the route for thrust reactivation (B).

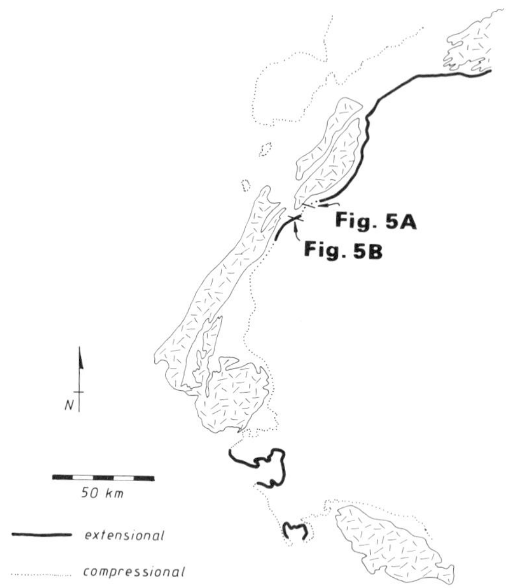

FIG. 4. Sketch map of the variable stratigraphic separation geometries along the Frontal Pennine thrust and its relationship to the external basement massifs (random flecks). The locations of detailed sections (Fig. 5) are indicated. Where there is no net stratigraphic separation the thrust is shown in compression.

with shear criteria indicating westward-directed movement (see Butler *et al.* 1986 for discussion). In the East Pelvoux sector, Cretaceous carbonates and marls, Liassic limestone and Triassic evaporites of the sub-Briançonnais zone lie on upper Eocene–lower Oligocene flysch sandstones. As such, the thrust places older rocks on top of younger. In contrast, the Mont Blanc sector displays younger on older relationships with upper Cretaceous–Eocene Tarentaise flysch sandstones immediately above the thrust plane lying directly on top of Liassic slates of the external zones (Antoine 1971; Butler 1988; see Fig. 5A). Elsewhere, as in the Courmayeur district (Fig. 5B) the hanging wall lies in the older Mesozoic substratum of the Tarentaise (frontal Pennine) zones while the footwall remains in the Liassic slates. The Tarentaise flysch outcrops about 50 m above the thrust plane. Hence in the Mont Blanc area the Frontal Pennine thrust has variable stratigraphic separation. Perhaps it originated from compressional reactivation from a major down-to-the-east normal fault which originally separated the internal and external palaeogeographic domains?

Unfortunately, the same stratigraphic separations can be produced without any fault reactivation. Critically important is the absence of

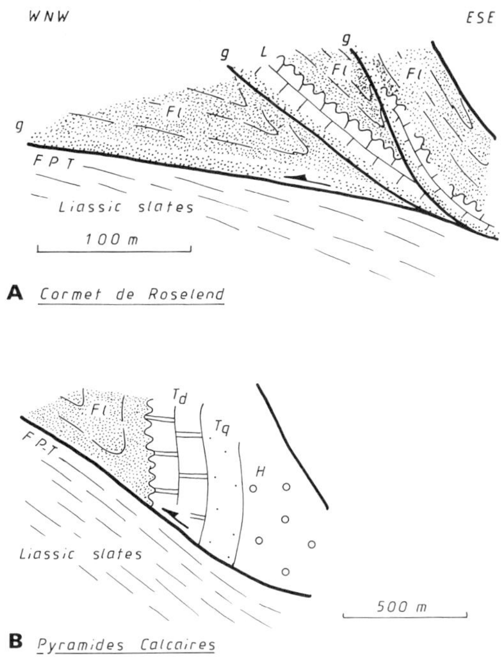

FIG. 5. Detailed profiles through the Frontal Pennine thrust in the SW Mont Blanc area (see Fig. 4 for locations). Stratigraphy in the hanging wall is from the Tarentaise (Valais) zone: H, Carboniferous basement; Tq, Triassic quartzites; Ta, Triassic dolomites; L, Liassic limestones; Fl, Upper Cretaceous flysch sandstones; g, exotic gypsum intruded and flowing along thrust planes.

Cretaceous rocks in eastern parts of the external Alps around SW Mont Blanc (Landry 1978). Eocene limestones uncomformably overlie Jurassic limestone suggesting either a period of non-deposition or erosion in early Tertiary times. To define relative fault offsets a datum surface must first be established. An upper Eocene time line has been chosen from which to hang the hanging wall and footwall stratigraphies (Fig. 6). Since the Tarentaise (pre-upper Eocene, Antoine 1971) flysch sandstones are over 500 m thick, the older Jurassic units of the frontal Pennine zones lie a greater depth below the datum than do those of the external zones to the W. Indeed, the Tarentaise basin could well have been deeper than the entire Mesozoic succession preserved beneath the datum surface in the external zones. Thus, with respect to this datum a thrust could cut up towards WNW but carry Tarentaise Upper Cretaceous flysch across external zone Liassic shales. Any more vigorous westward upcutting would carry flysch over Eocene rocks, as occurs SE of Pelvoux.

Compressional stratigraphic separations could also occur if the Frontal Pennine thrust lay within lower Jurassic and Triassic substrata of the Tarentaise zone. Thus there is evidence for different elevations of the lower Mesozoic units prior to thrusting at the Pennine Front. Fig. 7 shows how this type of variable stratigraphic separation can occur by trunction of a basin-bounding normal fault by a low-angle thrust. Using stratigraphic separations alone it is not possible to distinguish between this

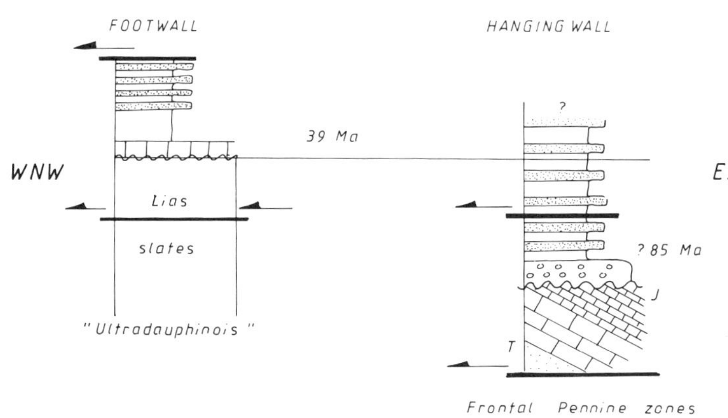

FIG. 6. Stratigraphic separation diagrams for the Frontal Pennine thrust in the SW Mont Blanc sector (see Fig. 4). The stratigraphic columns (after Landry 1978 in footwall, Antoine 1971 in hanging wall) are hung from a late Eocene datum (39 Ma) corresponding to the age of the Nummulitic limestone in the Beaufortain area (Priabonian, Landry 1978). A horizontal thrust segment running just beneath this datum would carry Tarentaise flysch (late Cretaceous and younger) onto Liassic slates. The net stratigraphic omission results from the nature of the pre-Nummulitic unconformity rather than any fault geometry. Thrusts cutting up to higher levels in the footwall would show 'conventional' stratigraphic separations.

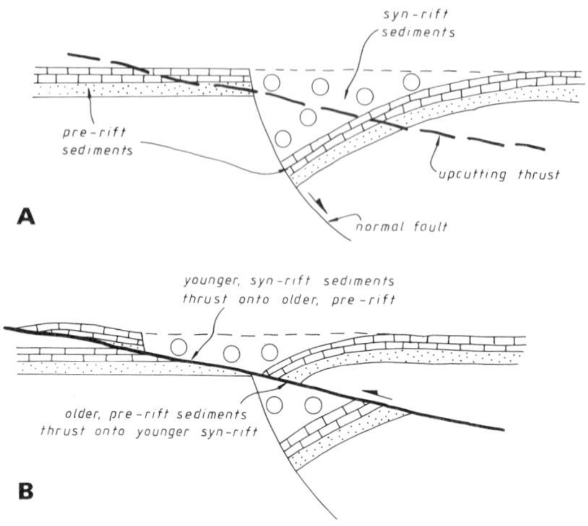

FIG. 7. Idealized geometry of stratigraphy and faults resulting from a thrust surface truncating a pre-existing basin structure (A–B in time). The same range of stratigraphic relationships can develop here as in the reactivation model (Fig. 3). The two models can be differentiated between only with additional stratigraphic and sedimentological information (distribution of fault proximal facies etc.).

geometry (Fig. 7) and reactivation (Fig. 3). Work is underway to use detailed intra-basin stratigraphy to establish the positions of growth faults in the Tarentaise zone which can then be compared with the geometry of the Tertiary thrust structures.

Basement cover detachment and buttressing

The generation of regional ramp/flat geometries in foreland thrust belts is dependent on the initial deposition of sediments as simple, sub-horizontal layers (Elliott & Johnson 1980). Any pre-existing normal faults which disrupt the bedding anisotropy to offset any easy-slip horizons might be predicted to impede flat thrusting and so generate folds and ramps (Schedl & Wiltschko 1987). So too would some intra-basin unconformities and lateral facies variations. In Alpine geology the concept of basement cover decoupling is well established. Classic examples include the Jura (Fig. 8) where detachment of the Mesozoic cover along Triassic evaporites has allowed folding and minor imbrication above an essentially undeformed and rigid basement. Clearly in this setting there can be few normal faults which disrupt the decoupling surface. Minor faulting nucleating local folds and ramps could occur provided the throw on each fault was less than the thickness of the layer of evaporites in order to allow continuity of the detachment zone.

The frontal termination of the Jura detachment lies on the edge of the Bresse basin where Mesozoic cover is carried out onto the Tertiary basin fill. Enay (1982) proposed that this emergent thrust front overstepped and truncated the Bresse basin faults (Fig. 8) which

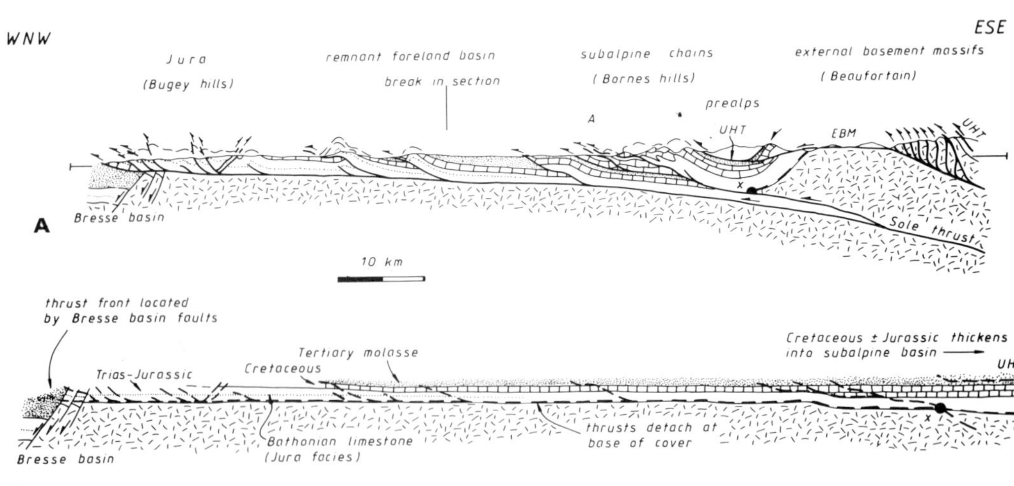

FIG. 8. Simplified balanced (A) and restored (B) cross-section through the Jura and sub-Alpine chains along the Bornes to Bugey transect (see Fig. 2 for location). Section originally constructed at 1 : 25 000. The section implies in excess of 30 km horizontal shortening and displacement on the Sole thrust. UHT, Ultrahelvetic thrust (just below the Frontal Pennine thrust); EBM, External Belledonne massif, A, Annecy town. Structure of the eastern Beaufortain basement thrusts is idealized.

originally dipped NW. Clearly in this orientation the earlier normal faults were unfavourable for reactivation during compression except as gravitational collapse structures. However, their presence prevents the activation of the detachment in the Triassic evaporites within the Bresse basin and so they control the site of the emergent Alpine thrust front.

The effect of pre-existing normal faults which dip towards oncoming thrust sheets has been well documented by Gidon and co-workers (Barfety & Gidon 1984) working in the Bourg d'Oisans area of the Dauphine Alps (Fig. 2). They describe a Liassic age synsedimentary normal fault which forms the present eastern margin of the Belledonne basement massif. The sediments in its hanging wall (i.e. to the W) are highly folded and cleaved, indicating substantial vertical pure shear stretching (see also Gillcrist *et al.* 1987). These structures decouple on the underlying Triassic evaporites. This type of process is referred to as 'buttressing' (e.g. Welbon 1988) although the term 'buffer-stop' has been used colloquially by Lemoine (R. H. Graham, personal communication, 1987). Folding develops because the thrust detachment at the basement cover contact is blocked by running into rigid basement in the footwall to the pre-existing normal fault (Fig. 9). Thrust translation is converted into *in situ* shortening. It is tempting then to consider narrow belts of folding and intense deformation as diagnostic of buttressing at early normal faults which dip back to the orogenic hinterland (Gillcrist *et al.* 1987).

Welbon (1988) and Butler (*in press*) describe local high strain zones around some thrust ramps in the sub-Alpine chains. In both cases stratigraphic data and regional mapping show these areas to be sites of pre-existing normal faults which control primary sedimentary thickness variation. However, the faults so defined dipped away from the orogen (i.e. eastwards) yet they still generated buttressing strains and folding during subsequent compression. These fault zones lie at high levels within the sedimentary pile where massive middle Cretaceous limestones (the Urgonian) are brought down against older (Hauterivian) thin limestones and shales. Thrusts within the shales ran up against the downthrown limestones, locked and then generated distributed strain zones (Fig. 10). Thrust wedging and foreland-vergent folding rotated the pre-existing normal fault into a hinterland-dipping orientation suitable for partial reactivation. In neither case is there any evidence for deeper parts of the normal fault being used for thrust displacement.

Buttressing of thrust displacements against pre-existing faults may explain local sites of

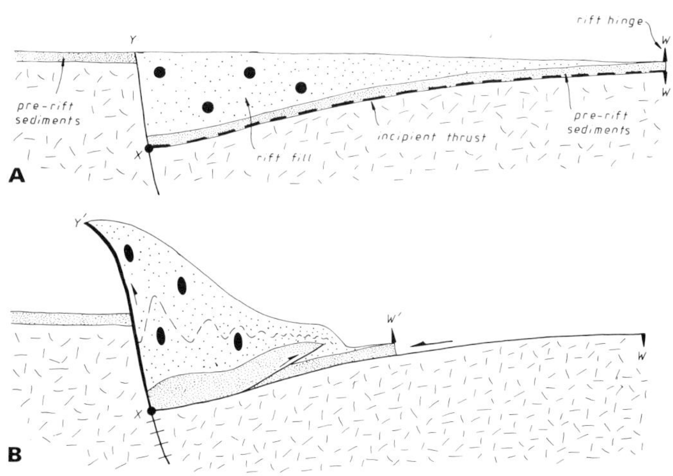

FIG. 9. Idealized development of buttressing structure (A–B in time) produced by the change in deformation mechanisms from thrusting at the base of the cover units (stippled) to vertical pure shear. This is caused by the reduced propensity for thrust migration into the basement (random flecks) in the footwall to an old normal fault. The basin stratigraphy must be restored (W–W′) before it can be used to predict the deep structure of the normal fault. In this model the normal fault has not been reactivated from depth but minor activation is required at higher levels to maintain strain compatibility between the shortened basin sediments and those on the normal fault footwall block (Y–Y′).

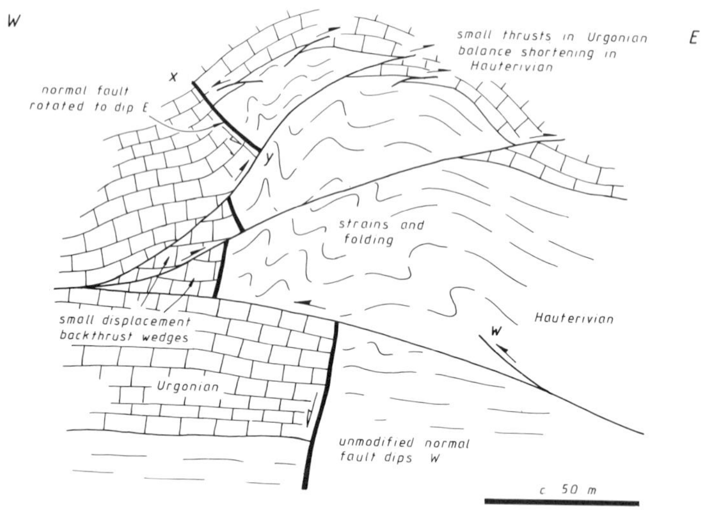

FIG. 10. Schematic section through buttressing structure generated by a thrust attempting to migrate across from relatively weak Hauterivian limestones and shales into massive Urgonian limestones. These units have been juxtaposed originally by a down-to-the-west normal fault which becomes folded into an E-dipping orientation during compression (X–Y). Continued deformation may bypass such structures (W–Y–X) to reactivate a short segment of the normal fault. Based on the Col de la Bataille area of the Vercors (B on Fig. 11; see Butler 1988, Welbon 1988).

upright folds, strain and cleavage intensification in thrust belts (Fig. 9). However, the switch from localized thrusting into this type of distributed deformation is not of itself diagnostic of pre-existing basin faults. Similar patterns could result from detachment zones terminating for stratigraphic reasons or from break-back regional thrust sequences.

Patterns of thrust detachment activation

Detachment between basement and cover, such as occurs beneath the Jura (Fig. 8) also occurs beneath the adjacent sub-Alpine thrust systems (Fig. 2, see Butler 1988 for discussion and sections). Geophysical studies (Menard 1979; Bayer *et al.* 1987) show the basement top beneath these thrust systems to be dipping gently eastwards until it reaches the buried leading edge of the the external Belledonne basement massif (Fig. 2). This massif is the surface expression of a hanging wall anticline lying on the frontal Alpine basement thrust (Menard & Thouvenot 1984; Butler, 1988 referred to henceforth as the Sole thrust (Fig. 8). Throughout its strike length of 100 km the external Belledonne massif retains a remarkably consistent crestal uplift, suggesting a similar lateral continuity of displacement on the underlying Sole thrust. These displacements pass up dip onto the sub-Alpine and Jura thrust systems before emerging into the molasse basins. Stratigraphic information from these basins bracket the age of thrusting as being between middle Miocene and Pliocene times (approximately 12–6 Ma).

Although the uniform crestal uplift on the external Belledonne massif suggests lateral continuity of displacements on the Sole thrust and hence across the sub-Alpine and Jura thrust belts there are radical changes in structural style and thrust geometry along the strike length. In the N areas, along the Annecy sector (Fig. 2), the thrust belt is wide. The frontal fold belt of the Jura lies ahead (W) of the main sub-Alpine (Bornes) thrust systems. The two are separated by a weakly uplifted portion of the molasse basin (Fig. 8). Further south, in the Chambery area (Fig. 2), the Jura and sub-Alpine structures converge so that in the southern part of the region (Vercors) the two belts are continuous and rather narrow. Perhaps this pattern of thrusting represents lateral variations in the activation of basement cover decoupling, in turn controlled by the presence or absence of normal faults.

In the Jura the thrust front appears to be

coincident with the eastern margin of the Bresse basin. At depth a normal fault is seen to downthrow to the W. Drill hole and sparse seismic data (BRGM 1968; 1975) in the W Vercors district suggest that the thrust front is also coincident with a buried basin margin fault, the Faille de l'Isere (Fig. 11). However, these data together with stratigraphic information indicate that the Faille de l'Isere downthrows eastwards. Gillcrist *et al.* (1987) proposed that this structure caused ramp climb on the sub-Vercors thrusts, hence localizing the thrust front. As in the Jura, the emergent frontal thrust shows little displacement (less than 2–3 km) since it cannot be mapped as a continous structure along strike to the south (Butler *in press*). Structures within the Vercors are simple without any major thrusts. Shortening is likely to be only about 10 km in total (Butler *in press*) so there is a clear discrepancy between shortening in the cover and that predicted to occur at depth to uplift the external Belledonne massif.

Thrust evolution in the Vercors

Vann *et al.* (1986, following Jones 1982) proposed a solution for reconciling the absence of emergent frontal thrusts ahead of well developed thrust belts: the presence of a back thrust in the mountain front area. Fig. 11 is a cross-section through the Vercors which illustrates this solution, however the back thrust must be located between the Vercors and the external Belledonne massif (Dome de la Mure). Gidon (1981) reported easterly vergent structures in just this locality, and a major back thrust repetition of Jurassic stratigraphy is evident on published maps (e.g. Gidon 1979). The hanging wall of the back thrust contains a 2 km thickness of Liassic limestones and shales which form the base of the Vercors stratigraphic section. In contrast, a condensed sequence of 200 m of crinoidal limestones represents the Liassic succession on the external Belledonne massif (Amaudriac de Chaffaut *et al.* 1986). Similar ranges of Liassic stratigraphic thicknesses have been reported from areas to the east (Barfety & Gidon 1984) and related to eastward-directed synsedimentary extensional faults (see also Lemoine *et al.* 1986). Thus it appears that the Vercors thrust structures are spatially coincident with pre-existing basin faults and associated stratigraphic variations. Let us now examine these relationships as interpreted on the cross-section (Fig. 11).

The Vercors section (Fig. 11) is underpinned by Menard's (1979) depth to basement compilation. The surface geology exposes stratigraphic units down to upper Jurassic (Tithonian) limestones; deeper levels are exposed in the ground between the Vercors plateau and the external Belledonne massif. In cross-section the subsurface structure of the Vercors corresponds to a westward tapering wedge of Mesozoic rocks, now carried by the back thrust to the east. When this wedge is restored (Fig. 11B) the

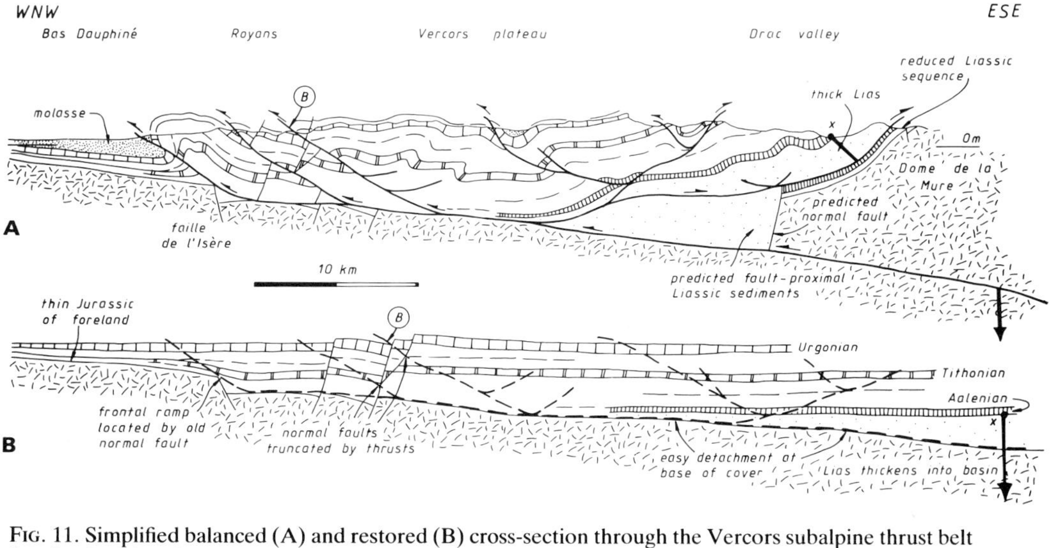

FIG. 11. Simplified balanced (A) and restored (B) cross-section through the Vercors subalpine thrust belt (section location shown on Fig. 2). Originally constructed at 1 : 25 000. B, Col de la Bataille area (actually 5 km S of section line, see Fig. 10 and Butler *in press*).

basin geometry which existed prior to thrusting becomes clearer. The external Belledonne basement must have been a high while the Vercors region was an eastward deepening low. This arrangement is best explained by siting a down-to-the-west normal fault near the present W margin of the external Belledonne massif (Dome de la Mure). This massif lies at a change in normal fault polarity, down-to-the-east in the E and down-to-the-west in the W (Fig. 12).

Large scale backthrusts require relatively easy thrust detachment so that the underthrusting wedge can move forwards (Butler 1987). Therefore the Vercors plateau must have been underlain by a relatively simple basement cover contact (Fig. 11B). A sequential model suggests itself whereby the Sole thrust activated this 1basement cover contact from which was spawned
the major backthrust now found in the Eastern Vercors area. Displacements on this backthrust 1probably exceed 15 km, generated as the external
Belledonne massif underthrust the Vercors plateau as a wedge. This wedge will have contained the basin sediments proximal to the postulated down-to-the-west normal fault, and will have continued to drive beneath the Vercors until simple basement cover detachment became hindered. This could occur either by a facies change in the Liassic sediments, perhaps representing ramp carbonate production on the Vercors basin hinge (Fig. 12), or by the presence of faults associated with the Faille de l'Isere. At this stage the structural style will have changed to conventional foreland-directed thrusting and folding, possibly nucleating on basement steps (Schedl & Wiltschko 1987).

The Western Vercors area contains several important normal faults which disrupt the Mesozoic succession and therefore presumably generate basement steps at depth (Fig. 11).

Stratigraphic data (BRGM 1978) indicate these faults to have been active in early Tertiary times and show that they are cut by thrust structures generated in the mid-late Miocene. The normal faults form part of the eastern limit of the Rhone graben, a basin system which connects up through the Bresse and into the Rhine graben. Similar orientations of steep normal faults have been found elsewhere in the sub-Alpine chains (Welbon 1988).

In summary, the outer parts of the Western Alps are dominated by classic foreland thrusting generated by the regional Sole thrust climbing into the Mesozoic cover ahead of the present external Belledonne massif. There is considerable evidence of normal faulting in the Mesozoic rocks, generated during sedimentation in Jurassic and early Tertiary times. These faults nucleate folds which can themselves be truncated, folded and locally reactivated (Welbon 1988) by compressional movements. It is probable that the presence of normal faults controls aspects of the structural style and the spatial area of the foreland thrust belt. However, there is no evidence for the normal faults themselves being reactivated from depth. This would have

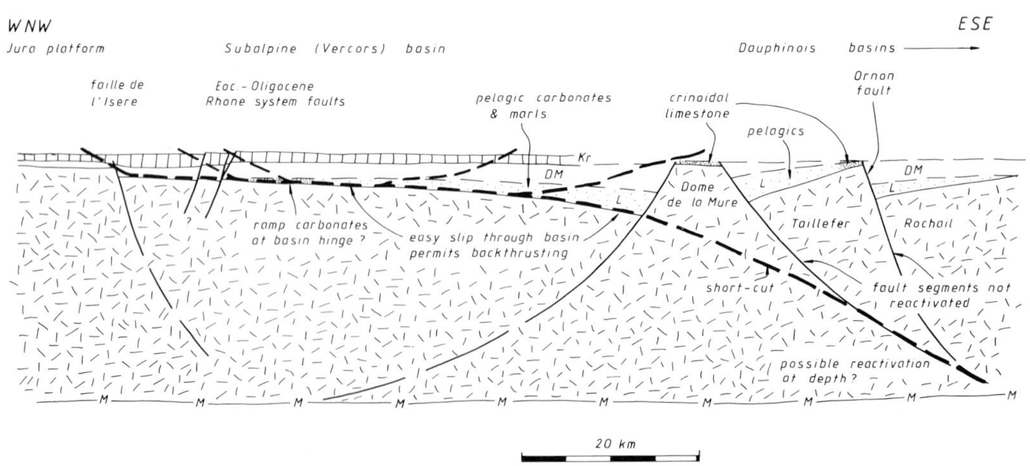

FIG. 12. Speculative true scale restored crustal section through the Vercors and neighbouring districts showing the distribution of Mesozoic basins and the predicted geometry of their bounding normal faults. Pecked line denotes the future position of the Alpine Sole thrust system which may splay from depth along a buried portion of a normal fault. Stratigraphic information from Lemoine, *et al.* (1986). L, Liassic; DM, Dogger-Malm; Kr, Cretaceous (eastern extent uncertain).

generated spaced regional uplifts because the normal faults are steep and probably penetrate deeply into the crust before acquiring a more inclined orientation. Structural styles in the sub-Alpine chains are indicative of detachment tectonics rather than the basement-rooting reverse faulting. The Sole thrust may represent deep reactivation of basin faults. It certainly coincides with the eastern margin of the sub-Alpine (Vercors) Jurassic basin. However, the bounding fault to this basin dips west while the Sole thrust must dip east to pass displacements onto the foreland structures. It is possible that the thrust splays from one of the down-to-the-east normal faults (see Beach 1987) which outcrop in the basement massifs hindward of the Belledonne (e.g. Barfety & Gidon's (1984) Ornon fault). However, the surface geology cannot clearly demonstrate this (see Butler, *in press*, for discussion).

Thrust detachment versus normal fault reactivation in the Western Alps

The foreland thrust structures discussed so far are clearly influenced by the presence of normal faults as well as other factors such as the relative proportions of limestones and shales within the stratigraphic column and the lateral extent of this multi-layer sequence. However, the compressional tectonics has generated entirely new fault systems within the sedimentary cover rather than reactivate pre-existing faults from depth. This behaviour need not typify the whole mountain belt where perhaps originally more attenuated crust has been restacked. Gillcrist *et al.* (1987) suggest that the external basement massifs around Pelvoux (Fig. 2), east of the Vercors section, represent the restacked portions of originally highly extended crust. They consider that major Alpine thrusts have inherited their geometry from the normal faults which can be defined from patterns of Mesozoic sedimentation. The relatively large (approximately 120 km) amounts of shortening across the Pelvoux district implied by their sections (Gillcrist *et al.* 1987), generates only the moderate crustal thickness of 45 km (Menard 1979). Their section implies about 60 km of horizontal extension during Liassic sedimentation in the same region. This figure which would generate an average crustal thickness of just 20 km by the end of the Jurassic.

The relatively large thrust displacements contained within the external western Alps implied by Gillcrist *et al.* (1987) are also implicit in accounts of Helvetic thrust structures further north in western Switzerland (Fig. 2). Well over 100 km of crustal shortening can be traced throughout the external zones (Butler *et al.* 1986). However, in the northern sectors there is little evidence for large scale reactivation of normal faults (cf. Gillcrist *et al.* 1987). In their type area, where stratigraphic thicknesses are greatest, the Helvetics represent decoupled cover sheets (Heim 1921; Ramsay 1981). Basement is involved at the lateral margins around the Aar and Mont Blanc massifs. Strains within the Helvetic sheets are, for the most part, rather low suggesting easy decoupling. High simple shear strains occur at the base of the Helvetic pile (Ramsay & Huber 1985) and more complex flattening patterns occur along their internal margin. So it appears likely that the Helvetic basin floor had a simple geometry permitting the detachment and emplacement of large thrust sheets. Strain are patterns not consistent with buttressing against basin faults (cf. Gillcrist *et al.* 1987). Stratigraphic variations within the Helvetics are systematic and gradual (e.g. Ramsay 1981) rather than abrupt as in the Pelvoux district. So the northern parts of the external western Alps probably experienced more regional subsidence, presumably controlled by a few, relatively large-displacement normal faults. As in the outlying portions of the Alps (sub-Alps, Jura), the original basin geometry changes along strike and the structural style generated during compressional tectonics reflects this.

As indicated above, the Helvetic thrust structures pass laterally into basement thrust systems in the Beaufortain (Butler *et al.* 1986) and Aar (Masson *et al.* 1980) districts. Both regions are characterized by condensed Mesozoic successions (Landry 1978, Masson *et al.* 1980) and clearly represented highs during Jurassic basin formation. The thrust systems pass from old basins into old basin margins and there is little evidence for any major structural inheritance or 'tramlining' (cf. Gillcrist *et al.* 1987; Fry, *in press*) as has been suggested for more SW Alpine regions.

Stacking geometries in the internal Alps

The preservation of stratigraphic relationships and the relatively low displacements in the external Alps allow a reasonable insight into the role of basin structures in controlling compressional tectonics. This becomes progressively more complex as we move into the Alps. At the start of this contribution the ambiguities of recognizing normal fault reactivation on large Alpine faults such as the Frontal Pennine thrust were outlined. Simple correlations between stratigraphy, pre-existing faults and present offsets on thrusts are rarely possible. Thus we are left with investigating the types of structural styles in the mountain belt and discussing whether these are consistent with simple basin fault reactivation. (e.g. Fig. 1).

Basement cover detachment is widespread throughout most of the Alps. The concept of cover substitution, the detachment of one sedimentary sequence from its basement, emplacement onto more external basement and its replacement by more internal cover is entrenched in Alpine literature (see discussion by Vearncombe 1985). Classic examples of this style of tectonics are the pre-Alpine nappes (Masson 1976) which lie as klippen on the sub-Alpine and Helvetic cover, far ahead of the internal Alpine basement on which they were originally deposited (Fig. 13A & B). These basement rocks are now overlain by sediments of an oceanic nature, the schistes lustres (Argand 1911). The pre-Alpine sheets contain Triassic and younger sediments. Presumably the basement cover contact from whence they came had a simple form prior to thrusting, without any significant normal faults. Certainly the 100–150 km displacement on the pre-Alpine sheets (Badoux 1963) necessary to carry them out onto the external zones is incompatible with reactivation of steeply rooting basement thrusts.

The domination of detachment tectonics in the central Alps (Laubscher & Bernoulli 1982) has long been recognized (Argand 1911; Staub 1924). This style of tectonics is now recognized as being widespread in the western sector too (Butler 1988). Steep rooting basement thrusts with large displacements are absent so that the type of crustal structure indicated on Fig. 1 does not now seem appropriate. (cf. De Jong 1973; see also the discussion by Butler 1986).

There are parts of the internal Alps, for example around the Briançon district (Fig. 2), where Mesozoic cover retains an unslipped stratigraphic contact on basement rocks. In this region the Briançonnais facies belts form a platform sequence broken up by several relatively minor, normal faults (Tricart 1984). Pioneer mapping by Debelmas (1955) defined several major thrust sheets within the Briançonnais. Tricart (1984; Tricart *et al.* 1985) documents numerous examples of the major thrusts truncating pre-existing normal faults. At this particular erosion level there appears, yet again, to have been no significant reactivation of normal faults. However, as in the the sub-Alpine areas, the normal faults may have lacked potential basement cover detachment surfaces in the Briançonnais type area so that the Mesozoic cover remains preserved in the internal Alps rather than carried out onto outlying regions.

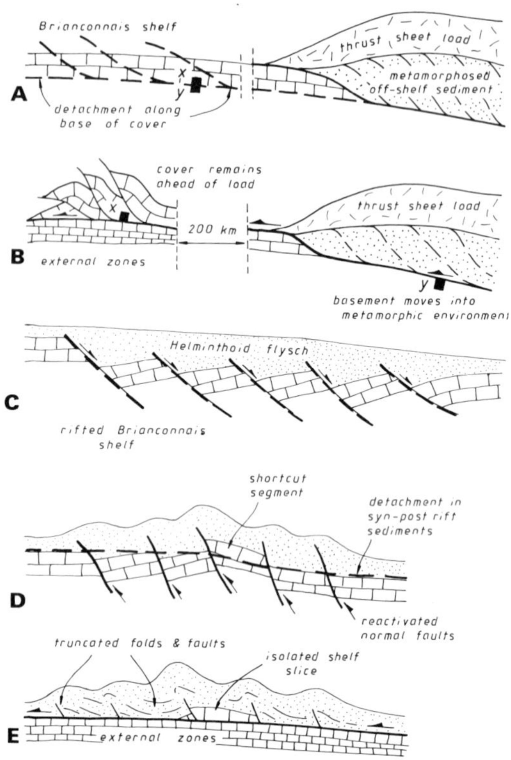

FIG. 13. Thrust detachment models and cover substitution in the internal Western Alps. A and B illustrate the emplacement of the Prealpine klippen (X) onto the external zones while their basement (Y) moves into a metamorphic environment beneath the obducted oceanic and continental thrust sheets. This large scale activation of detachment horizons occurred relatively early in the Eocene–Pliocene thrust evolution in the NW Alps and was followed by complex deformations in the internal zones and thrusting in the external zones. C, D and E illustrate in stylized form the evolution of the Embrunais–Ubaye thrust sheets. The Frontal Pennine thrust here is decorated by isolated blocks of Briançonnais units but the Helminthoid flysch has been detached from its basement and has experienced large lateral translations. The Briançonnais blocks may be shortcut pods developed when detachment occured during emplacement onto the external zones. Minor normal fault reactivation (D) may have occured prior to the detachment tectonics (E).

Detachment within the Briançonnais has occured at higher parts of the succession. The platform carbonates and their associated normal faults are blanketed by a thick sequence of postrift deep water flysch sandstones (Fig. 13C, D & E). These upper Cretaceous Helminthoid flysch sandstones have been detached to form far-travelled nappes, the Embrunais-Ubaye sheets (e.g. Tricart 1984; Fig. 2). Detachment of postrift sediments is clearly not diagnostic of areas with only weakly developed basin structures (Fig. 13) because even large normal faults can remain buried below the detachment surface.

Discussion

From the preceeding discussions it appears that thrusts at the present exposure level in the western Alps generally do not result from the simple rejuvenation of old listric normal faults (cf. Dewey *et al.* 1986). It is possible that substantial portions of normal faults are reactivated at deep levels in the crust (e.g. Fig. 12) but there is no requirement for this geometric coincidence from the exposed geology. The upper crust appears to have been stacked up by detachment thrusting. The geometry of these thrusts was controlled locally by pre-existing structures. Thrust ramps, folds and distributed straining zones may be located along some normal faults, elsewhere the thrusts can truncate pre-existing structures. At the very least normal fault reactivation 'should not be assumed because it is conceptually appealing' (Etheridge 1986).

In 1987 the joint Franco–Italian deep seismic reflection profiling consortia (ECORS-CROP) acquired a section through the Western Alps along the Annecy–Aosta valley transect, recorded to a two-way time (TWT) in excess of 20 s (Bayer *et al.* 1987; see Fig. 14). An additional wide-angle reflection experiment in 1985 located other prominent events although their precise site on the ECORS-CROP line is poorly constrained. Collectively these two

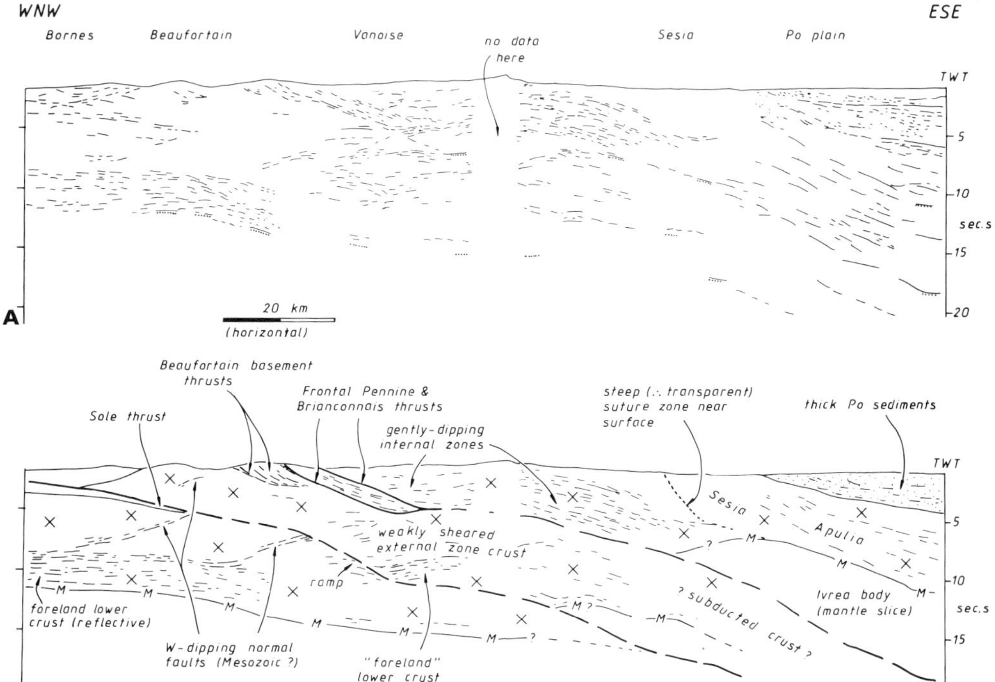

FIG. 14. A, Line drawing of the ECORS-CROP deep seismic reflection profile (time section) through the western Alps (location shown on Fig. 2), incorporating the prominent events (dotted lines) from the wide angle reflection experiment (after Bayer *et al.* 1987). B, Interpretation of the line incorporating surface geology using section balancing concepts (see Butler 1986, 1988). Only the major thrusts are illustrated. There are substantial thrusts within the gently dipping reflective zone beneath the Vanoise.

experiments provide a unique data base to combine with surface geology to interpret deep Alpine structure. The lower crust of the European foreland is characterized by a 3–4 s (TWT) band of reflectors, as beneath other basin areas of Europe (Bayer *et al.* 1987). This is lost beneath the internal Alps although the prominent Moho reflector detected from the 1985 data set can be traced at 15–18 s (TWT). At shallower depths the Sole thrust can be located running beneath the external basement massifs of Beaufortain as a gently E-dipping structure. The ECORS-CROP data suggest that this thrust is not a steeply penetrating structure (cf. Gillcrist *et al.* 1987) but rather it only cuts down to lower crustal levels beneath the Vanoise. This Sole thrust may also cut a pair of W dipping reflective zones which link down into the foreland lower crust. These may represent old Mesozoic normal fault zones, as proposed for the Vercors area (Fig. 12).

The subsurface structure of the E external and W-central internal zones is characterized by gently-dipping events (Fig. 14). Particularly prominent are the Frontal Pennine and Basal Briançonnais thrust zones which apparently dip at 20° to 5–6 s (TWT). Bayer *et al.* (1987) consider these thrust zones to pass continuously to 15–20 s (TWT). However, balanced cross-sections based on external zone geology (Butler *et al.* 1986) favour a shallower level trajectory (Fig. 14B), merging into the zone of prominent reflectors beneath the eastern Vanoise (5–6 s TWT). The dominant shallowly dipping structure of the internal Alps suggests detachment tectonics. The transparent regions beneath the Sesia zone, including the Alpine suture, coincide with a region of steep foliation (Bayer *et al.* 1987) but do not pass to great depths (>6 s TWT). The dominant gentle E dipping grain is preserved across the Alps and under the Po plain.

Although the ECORS-CROP line (Fig. 14A) incorporates limited seismic velocity information, the seismic refraction experiments reported by Menard (1979) and interpretation of gravity data (Menard & Thouvenot 1984) indicate that mantle slices (e.g. the Ivrea body) are uplifted to high crustal levels under the internal Alps. So the deep reflectors on the eastern part of the ECORS-CROP line correspond both to thrusts and to the imbricated Moho. This tectonic pattern does not preclude minor amounts of normal fault reactivation, particularly if it occurred early in local tectonic histories (e.g. Fig. 13D), but it does strongly suggest detachment thrusting (Fig. 15). The wholesale reactivation of basin faults (e.g. Fig. 1) is unlikely to generate the dominant low-angle reflectors. Some minor normal fault reactivation may occur out on the European foreland, not covered by the ECORS-CROP profile at the time of writing.

The processes of collision orogenesis are not merely the scaled up versions of those which operate during 'inversion' of sedimentary basins where fault reactivation appears dominant (see other papers in this volume). Possibly this difference in behaviour reflects a rate control on fault activation. Knipe (1985) ¨escribes how fault segments in unfavourable orientations are trimmed off by new short-cut structures. These new fault strands develop when strain rates in the wall-rocks adjacent to faults exceed their strength, and these imposed strain rates will be a function of fault geometry and the displacement rate on the fault. Orogenic studies in the Himalayas and Alps have gained revised rates of shortening of the continental crust during collision. These cluster between 1–1.5 cm yr^{-1} (Coward & Butler 1985; Lyon-Caen & Molnar 1985; Butler, 1988), based on various time-averaged calculations. Incidentally, these figures are greater than many others (e.g. approximately 4 mm yr^{-1} in the Scandinavian Caledonides; Hossack & Cooper 1986) due to historical underestimation of displacements in mountain belts (see Gayer *et al.* 1987) but are more in accord with plate convergence values during collision (e.g. Patriat & Achache 1984). For such high rates of horizontal shortening, steep faults will show very fast actual slip rates (e.g. a 70° fault will show an actual slip rate three times that of the rate of horizontal displacement). Thus greater amounts of fault reactivation will be expected for slowly shortening basins while detachment tectonics will typify most collision zones (Fig. 15).

One remaining attraction of the normal fault reactivation model of orogenesis is its solution to the crustal space problem beneath mountain belts; namely that upper crustal shortening greatly exceeds the available crustal cross-sectional area. Helwig's (1976) solution of pre-existing thinned crust being re-stacked can still be applicable in the absence of fault reactivation. The only requirement is that the magnitude of crustal shortening does not greatly exceed that of the preceding extension. However, Gillcrist *et al.* (1987; see also Butler *et al.* 1986) imply that there is a great imbalance between shortening and extension, even in an area of enhanced basin development like Pelvoux. This pattern is repeated throughout the Alps. There is far more horizontal displacement associated with compressional

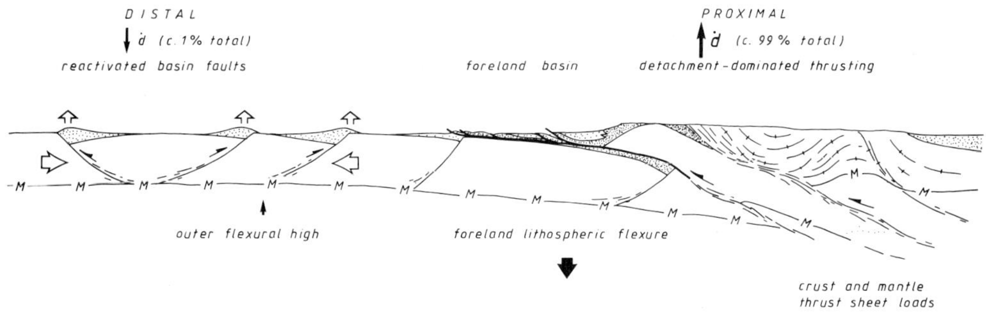

FIG. 15. Idealized structure of a mountain belt developed by detachment-dominated thrust tectonics, based on the Western Alps. Uplift within the foreland may be in response to lithospheric loading (minor) together with actual fault reactivation. Horizontal displacement rates on reactivated normal faults are considered to be greatly exceeded by those on detachment thrust systems so that most of the contaction between two continents is accommodated on new thrust surfaces rather than by reactivated normal faults. Compare final structure with ECORS-CROP line (Fig. 14) and the idealized reactivation model (Fig. 1).

Alpine tectonics than can be accommodated by extension during Mesozoic basin formation (Butler 1986). Despite its simplicity, Helwig's (1976) solution cannot solve the Alpine space problem. Clearly to gain further understanding we must consider the problem of restacking Alpine continental crust not solely in terms of remoulding the earlier passive continental margin but must also include subduction processes.

ACKNOWLEDGMENTS: For discussions on Alpine geology I am grateful to Donny Hutton, Maurice Tucker, Martin Casey, Norman Fry, Tim Needham, Alastair Welbon and Steve Matthews who also provided useful comments on an early draft. My ideas were greatly clarified by reviews, as usual, from Rod Graham and John Platt. This study forms part of a research programme, funded through a Royal Society (Jaffe Donation) Research Fellowship held initially in Durham and latterly at the Open University, on the processes of convergent continental tectonics. Field work in the internal Alps was funded by the Royal Society. Work in the sub-Alpine chains of France was funded by NERC (GR3/6172).

References

AMAUDRIAC DE CHAFFAUT, S. FUDRAL, S., BARFETY, J. C., DUMONT, T., GIDON, M., TRICART, P. & VIALON, P. 1986. De la marge oceanique a la chaine de collision dans les Alpes de Dauphine. *Bulletin de la Société Géologique de France*, **8**, 197–231.

ANTOINE, P. 1971. La zone des breches de Tarentaise entre Bourg-St.-Maurice (Vallée de l'Isere) et la frontiere Italo–Suisse. *Geologie Alpine*, **9**, 367 pp.

ARGAND, E. 1911. Les nappes de recouvrement des Alpes pennines et leurs prolongements structuraux. *Materiali per la Carta Geologica della Svizzera*.

BADLEY, M., EGEBERG, T. & NIPEN, O. 1984. Development of rift basins illustrated by the structural evolution of the Oseberg feature, Block 30/6, offshore Norway. *Journal of the Geological Society of London*, **141**, 639–649.

BADOUX, H. 1963. Les unites ultrahelvetiques de la zone des cols. *Ecologae Geologicae Helvetiae*, **56**, 1–13.

BALLY, A. W. 1984. Tectonogenese et sismique reflexion. *Bulletin de la Société Géologique de France*, **26**, 279–285.

BARFETY, C. & GIDON, M. 1984. Un exemple de sedimentation sur un abrupt de faille fossile: le Lias du versant est du massif du Taillefer (Zone Dauphinois, Alpes Occidentales). *Revue de Geologie Dynamique et Geographie Physique*, **25**, 267–276.

BAYER, R. CAZES, M., DAL PIAZ, G. V., DAMOTTE, B., ELTER, G., GOSSO, G. HIRN, A., LANZA, R., LOMBARDO, B., MUGNIER, J-L., NICOLAS, A., THOUVENOT, F., TORREILLES, G., & VILLIEN, A. 1987. Premiers resultats de la traversee des Alpes occidentales par sismique reflexion verticale (Programme ECORS-CROP). *Comptes Rendus de l'Académie des Sciences, Paris*, **305**, 1461–1470.

BEACH, A. 1987. A regional model for linked tectonics in north-west Europe. *In*: BROOKS, J. & GLENNIE, K. (eds) *Petroleum Geology of NW Europe*. Graham & Trotman, London, 43–48.

BOYER, S. & ELLIOT, D. 1982. Thrust systems. *Bulletin of the American Association of Petroleum Geologists*, **66**, 1196–1230.

Butler, R. W. H. 1986. Thrust tectonics, deep structure and continental subduction in the Alps and Himalayas. *Journal of the Geological Society*, **143**, 857–873.
—— 1987. Thrust sequences. *Journal of the Geological Society of London*, **144**, 619–634.
—— 1988. The geometry of crustal shortening in the western Alps. *In*: Sengor, A. M. C. (ed.) Tectonic evolution of the Tethyan Regions. *Proceeding of NATO Advanced Studies Institute*.
——. Thrust evolution within previously rifted regions: an example from the Vercors, French subalpine chains. *Memoria Societa Geologica Italiana* **389**, (in press)
—— Matthews, S. J. & Parish, M. 1986. The NW external Alpine thrust belt and its implications for the geometry of the Western Alpine Orogen. *In*: Coward, M. P. & Ries, A. C. (eds) *Collision Tectonics. Special Publication of the Geological Society of London*, **19**, 245–260.
BRGM (Bureau des Recherches Géologiques et Miniéres) 1968. Carte geologique de la France a 1:50 000, feuille Charpey.
—— 1975. Carte geologique de la France a 1:50 000, feuille Romans-sur-Isere.
—— 1978. Carte geologique de la France a 1:50 000, feuille Grenoble.
Chenét, P. Y., Colletta, B., Letouzey, J., Desforges, G., Ousset, E & Zaghloul, E. A. 1987. Structures associated with extensional tectonics in the Suez rift. *In*: Coward, M. P., Dewey, J. F. & Hancock, P. L. (eds) *Continental Extensional Tectonics. Special Publication of the Geological Society of London*, **28**, 551–558.
Coney, P. J. 1987. The regional tectonic setting and possible causes of Cenozoic extension in the North American Cordillera. *In*: Coward, M. P., Dewey, J. F. & Hancock, P. L. (eds) *Continental Extensional Tectonics. Special Publication of the Geological Society of London*, **28**, 177–186.
Cook, F. A., Albaugh, D. S., Brown, L. D., Kaufman, S., Oliver, J. E. & Hatcher, R. D. 1979. Thin-skinned tectonics of the crystalline Appalachians: COCORP seismic reflection profiling of the Blue Ridge and Piedmont. *Geology*, **7**, 563–567.
Coward, M. P. & Butler, R. W. H., 1985. Thrust tectonics and deep structure of the Pakistan Himalaya. *Geology*, **13**, 417–420.
Debelmas, J. 1955. Les zones sub-Briançonnaise et Briançonnaise occidentales entre Vallouise et Guillestre (Hautes Alpes), *Memoire Explicative de La Carte Geologique de La France* 171 pp.
—— & Kerckhove, C. 1980. Les Alpes franco-italiennes. *Geologie Alpine*, **56**, 21–58.
De Jong, K. A. 1973. Mountain building in the Mediterranean region. *In*: De Jong, K. A. & Scholten, R. (eds) *Gravity and Tectonics*. Wiley, New York 125–139.
Dewey, J. F., Hempton, M. R., Kidd, W. S. F., Saroglu, F. & Sengor, A. M. C. 1986. Shortening of continental lithosphere: the neotectonics of Eastern Anatolia — a young collision zone. *In*: Coward, M. P. & Ries, A. C. (eds) Collision Tectonics. *Special Publication of the Geological Society London*, **19**, 3–36.
Elliott, D. & Johnson, M. R. W. 1980. The structural evolution of the northern part of the Moine thrust zone, NW Scotland *Transactions of the Royal Society of Edinburgh*, **71**, 69–96.
Enay, R. 1982. Notice explicative de la feuille Saint Rambert en Bugey a 1:50 000. Bureau des Recherches Géologiques et Miniéres, Orleans, 84 pp.
Etheridge, M. A. 1986. On the reactivation of extensional fault systems. *Philosophical Transactions of the Royal Society of London*, **A317**, 179–194.
Fry, N. Thrusting to the SW in the French–Italian Alps. *In*: Coward, M. P., Dietrich, D. & Park, G. (eds) *Alpine Tectonics. Special Publication of the Geological Society of London* (in press).
Gayer, R. A., Rice, A. H. N., Roberts, D., Townsend, C. & Welbon, A. 1987. Restoration of the Caledonian Baltoscandian margin from balanced cross-sections: the problem of excess continental crust. *Transactions of the Royal Society of Edinburgh*, **78**, 197–217.
Gidon, M. 1979. Carte geologique simplifiee des Alpes occidentales (Leman a Digne) a 1:250 000. Bureau des Recherches Géologiques et Miniéres, Orleans.
—— 1981. Les deformations de la couverture des Alpes occidentales externes dans la region de Grenoble: leurs rapports avec celles du socle. *Comptes Rendus de l'Académie des Sciences, Paris*, **292**, 1057–1060.
Gillcrist, R., Coward, M. P. & Mugnier, J. L. 1987. Structural inversion: examples from the Alpine foreland and French Alps. *Geodynamica Acta*, **1**, 5–34.
Hancock, P. L. & Barka, A. A. 1987. Kinematic indicators on active normal faults in western Turkey. *Journal of Structural Geology*, **9**, 573–584.
Heim, A. 1921. *Geologie der Schweiz 2/1*, Tauchnitz, Leipzig.
Helwig, J. 1976. Shortening of continental crust in orogenic belts and plate tectonics. *Nature*, **266**, 768–770.
Hills, E. S. 1956. The tectonic style of Australia. Geoteckt. Sympos. Hans Stille, Deutsch. *Geol. Ges.*, Stuttgart, 336–346.
Hossack, J. R. 1984. The geometry of listric growth faults in the Devonian basins of Sunnfjord, W Norway. *Journal of the Geological Society of London*, **141**, 629–637.
—— & Cooper, M. A. 1986. Collision tectonics in the Scandinavian Caledonides. *In*: Coward, M. P. & Ries, A. C. (eds) *Collision Tectonics. Special Publication of the Geological Society of London*. **19**, 287–304.
Jackson, J. A. 1980. Reactivation of basement faults and crustal shortening in orogenic belts. *Nature*, **283**, 343–346.
—— 1987. Active normal faulting and crustal extension. *In*: Coward, M. P., Dewey, J. F. & Hancock, P. L. (eds) *Continental Extensional*

Tectonics. Special Publication of the Geological Society of London, **28**, 3–17.

JONES, P. B. 1982. Oil and gas beneath east-dipping underthrust faults in the Alberta foothills. *In*: POWERS, R. B. (ed.) *Geologic Studies of the Cordilleran Thrust Belt*. Rocky Mountain Association of Geologists, Calgary, 61–74.

KNIPE, R. J. 1985. Footwall geometry and the rheology of thrust sheets. *Journal of Structural Geology*, **7**, 1–10.

LANDRY, P. 1978. Donnees nouvelles sur la couverture sedimentaire des massifs externes au sud du Mont-Blanc. *Geologie Alpine*, **54**, 83–112.

LAUBSCHER, H. & BERNUOLLI, D. 1982. History and deformation of the Alps. *In*: HSÜ, K. J. (ed.) *Mountain Building Processes*. Academic Press, London, 169–180.

LEMOINE, M., BAS, T., ARNAUD-VANNEAU, A., ARNAUD, H., DUMONT, T., GIDON, M., DE GRACIANSKY, P. C., RUDKIEWICZ, J. L., MEGARD-GALLI, J. & TRICART, P. 1986. The continental margin of the Mesozoic Tethys in the western Alps. *Marine Petroleum Geology*, **3**, 179–199.

LYON-CAEN, H. & MOLNAR, P. 1985. Gravity anomalies, flexure of the Indian plate, and the structure, support and evolution of the Himalaya and Ganga basin. *Tectonics*, **4**, 513–538.

MASSON H. 1976. Un siècle de geologie des Prealpes: de la decouverte des nappes a la recherche de leur dynamique. *Eclogae Geologicae Helvetiae*, **69**, 527–575.

——, HERB, R. & STECK, A. 1980. Helvetic Alps of Western Switzerland, Excursion no. 1. *In*: *Geology of Switzerland, a guide book*. Swiss Geological Commission, Wepf & Co., Basel, 109–153.

MCCLAY, K. R. & ELLIS, P. G. 1987. Analogue models of extensional fault geometries. *In*: COWARD, M. P., DEWEY, J. F. & HANCOCK, P. L. (eds) *Continental Extensional Tectonics. Special Publication of the Geological Society of London*, **28**, 109–125.

MCGEARY, S., CHEADLE, M. J., WARNER, M. R. & BLUNDELL, D. J. 1987. Crustal structure of the continental shelf around Britain derived from BIRPS deep seismic profiling. *In*: BROOKS, J. & GLENNIE, K. (eds) *Petroleum Geology of NW Europe*. Graham & Trotman, London, 33–41.

MENARD, G. 1979. *Relations entre structures profondes et structures superficielles dans le sudest de la France: essai d'utilisation de données geophysiques*. Thesis, University of Grenoble, France.

—— & THOUVENOT, F. 1984. Ecaillage de la lithosphere europeene sous les Alpes occidentales: arguments gravimetriques et sismiques lies a l'anomalie d'Ivrea. *Bulletin de la Société Géologique de France*, **26**, 875–884.

MONTADERT, L., ROBERTS, D. G., DE CHARPAL, O. & GUENNOC, P. 1979. Rifting and subsidence of the northern continental margin of the Bay of Biscay. *Initial Report of the Deep Sea Drilling Project*, **48**, 1025–1060.

NI, J. & BARAZANGI, M. 1984. Seismotectonics of the Himalayan collision zone: geometry of the underthrusting Indian plate beneath the Himalaya. *Journal of Geophysical Research*, **89**, 1147–1163.

OBEE, H. K. & WHITE, S. H. 1986. Microstructural and fabric heterogeneities in fault rocks associated with fundamental faults. *Philosophical Transactions of the Royal Society of London*, **A 317**, 99–109.

PATRIAT, P. & ACHACHE, J. 1984. India-Eurasia collision chronology has implications for crustal shortening and driving mechanism of plates. *Nature*, **311**, 615–621.

PEACH, B., HORNE, J. GUNN, W., CLOUGH, C. T., HINXMAN, L. W. & CADELL, H. M. 1888. Report on the recent work of the Geological Survey in the NW Highlands of Scotland, based on field notes and maps. *Quarterly Journal of the Geological Society of London*, **64**, 378–441.

PINET, B., MONTADERT, L., MASCLE, A., CAZES, M. & BOIS, C. 1987. New insights on the structure and formation of sedimentary basins from deep seismic profiling in Western Europe. *In*: BROOKS, J. & GLENNIE, K. (eds) *Petroleum Geology of NW Europe*. Graham & Trotman, London, 11–31.

POWELL, C. M. 1987. Inversion tectonics in Dyfed. *Proceedings of the Geologists Association*, **98**, 193–203.

RAMSAY, J. G. 1981. Tectonics of the Helvetic nappes *In*: MCCLAY, K. R. & PRICE, N. J. (eds) *Thrust and Nappe Tectonics. Special Publication of the Geological Society of London*, **9**, 293–309.

—— & HUBER, M. I. 1985. *The Techniques of Modern Structural Geology. Volume 1: Strain Analysis*. Academic Press, London, 307 pp.

SCHEDL, A. & WILTSCHKO, D. V. 1987. Possible effects of pre-existing basement topography on thrust fault ramping. *Journal of Structural Geology*, **9**, 1029–1037.

SOPER, N. J. & ANDERTON, R. 1984. Did the Dalradian slides originate as extensional faults? *Nature*, **307**, 357–360.

STAUB, R. 1924. Der bau der Alpen. *Materiali per la Carta Geologica della Svissera*, **82**, 272 pp.

TORNEBOHM, A. E. 1888. Om fjallproblemet. *Geologiska Foren Stockholm Forhandlingar*, **10**, 328 p.

TRICART, P. 1984. From passive margin to continental collision: a tectonic scenario for the western Alps. *American Journal of Science*, **284**, 97–120.

——, DUMONT, T & LEMOINE, M. 1985. Evolution d'une portion de marge continentale: blocs bascules et charriages alpins dans la nappe prepiemontaise de Rochebrune (Alpes occidentales). *Revue de Geologie Dynamique et Geographie Physique*, **26**, 3–17.

VANN, I. R., GRAHAM, R. H. & HAYWARD, A. B. 1986. The structure of mountain fronts. *Journal of Structural Geology*, **8**, 215–227.

VEARNCOMBE, J. 1985. The structure of the Gran

Paradiso basement massif and its envelope, western Alps. *Eclogae Geologica Helvetiae*, **78**, 49–72.

VENDEVILLE, B., COBBOLD, P. R., DAVYS, P., BRUN, J. P. & CHOUKROUNE, P. 1987. Physical models of extensional tectonics at various scales. *In*: COWARD, M. P., DEWEY, J. F. & HANCOCK, P. L. (eds) *Continental Extensional Tectonics. Special Publication of the Geological Society of London*, **28**, 95–107.

VERRAL, P. 1981. Structural interpretation with application to North Sea problems. *Joint Association of Petroleum Exploration Courses* (UK), Geological Society of London, **49**.

WALKER, I. M. & COOPER, W. G. 1987. The structural and stratigraphic evolution of the northeast margin of the Sole Pit Basin. *In*: BROOKS, J. & GLENNIE, K. (eds) *Petroleum Geology of NW Europe*. Graham & Trotman, London, 263–275.

WATSON, J. V. 1980. Flaws in the continental crust. *Mercian Geologist*, **8**, 1–10.

WELBON, A. 1988. The influence of intrabasinal faults on the development of a linked thrust system. *Geologische Rundschau*, **77**, 1–11.

R. W. H. BUTLER, Department of Earth Sciences, The Open University, Walton Hall, Milton Keynes MK7 6AA, UK.

Notes and cross-sections illustrating inversion tectonics in the Wessex Basin

I. R. Simpson, M. Gravestock, D. Ham, H. Leach & S. D. Thompson

SUMMARY: Four cross-sections based on seismic and well data illustrate the Tertiary inversion tectonics in the Wessex Basin. Inversion is concentrated in Early Cretaceous depocentres with Permian and Triassic depocentres undergoing limited uplift. In the Portland–South Wight Basin inversion formed a northward facing monocline with a major reverse fault in the S. In contrast, in the Weald Basin a broad arch of 50 km wavelength was formed. Both basins show an overall uplift of 1.5 km. These variations in inversion style are probably a function of the geometry of the basin-controlling faults and underlying detachments.

The Wessex Basin (Kent 1949) extends from the London Platform to the Central Channel anticline in the S (Fig. 1) and consists of four main sub-basins bounded by E–W trending, southward-dipping listric extension faults (Dewey 1982; Stonley 1982). Significant NW–SE and WNW–ESE faults are also observed, e.g. the Bembridge–St Valery Line.

The basins are underlain by sediments of predominantly Cambrian to Carboniferous age (Smith 1985; Sellwood & Scott 1986) which were deformed by the Variscan orogeny. Reactivation of certain Variscan thrusts as extensional faults controlled Jurassic to early Cretaceous subsidence in the sub-basins (Chadwick *et al.* 1983; Chadwick 1985; Williams & Brooks 1985).

The development of inversion structures in the Wessex Basin commenced in Maastrichtian to Late Palaeocene times (Plint 1982) and culminated in the Oligocene to Miocene. The widespread, Early Cretaceous ('Kimmerian') erosion phase occurred during regional extension and did not produce compressional inversion structures.

The subsidence histories of the four sub-basins based on well and seismic data are shown in Fig. 2. No decompaction has been allowed for and the curves are only intended to illustrate the relative subsidence patterns. Significantly, it is the Weald and Portland–South Wight Basins, with the thickest sediments of Late Jurassic and Early Cretaceous age, which have undergone most inversion. The Permian depocentres have not responded significantly to the Tertiary compressional stresses.

Basin geometry

A series of cross-sections have been constructed across the Wessex Basin using seismic and well data (Figs 1 & 3). The interpretation below top Triassic must be treated with caution, particularly in basin depocentres where seismic data quality deteriorates below base Lias.

The western part of the Wessex Basin

The section (Fig. 4A) shows the Permian half-graben of the Pewsey Basin with a progressive onlap of the sequence southwards and only minor inversion on the basin bounding fault. Inversion of the Mere sub-basin is also minor (Fig. 4B). The Wessex Shelf is relatively undeformed. The section illustrates the 'Kimmerian' unconformity on the Purbeck High. The faults do not cut the Lower and Upper Cretaceous sequences. The most complex inversion structures occur along the northern margin of the Portland–South Wight Basin (Fig. 4C), e.g. the folding at Lulworth Cove (Arkell 1947).

The inversion of the Jurassic and Early Cretaceous half-graben in the Portland–South Wight Basin produced the broad anticlinal

From: Cooper, M. A. & Williams, G. D. (eds), 1989, *Inversion Tectonics*
Geological Society Special Publications No. 44, 123–129.

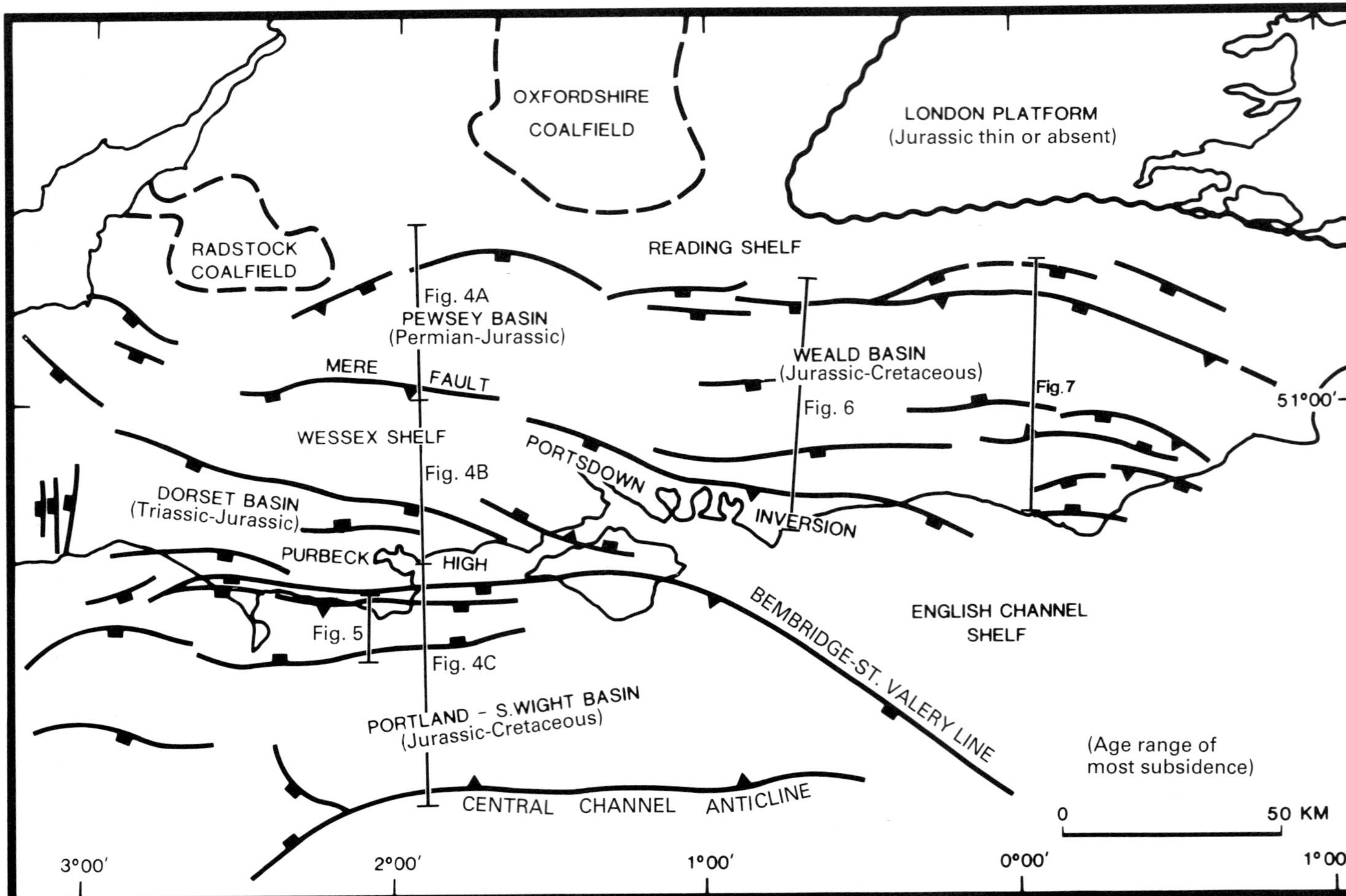

Fig. 1. Wessex Basin structural elements. The locations of cross-sections Figs 4–7 are also shown.

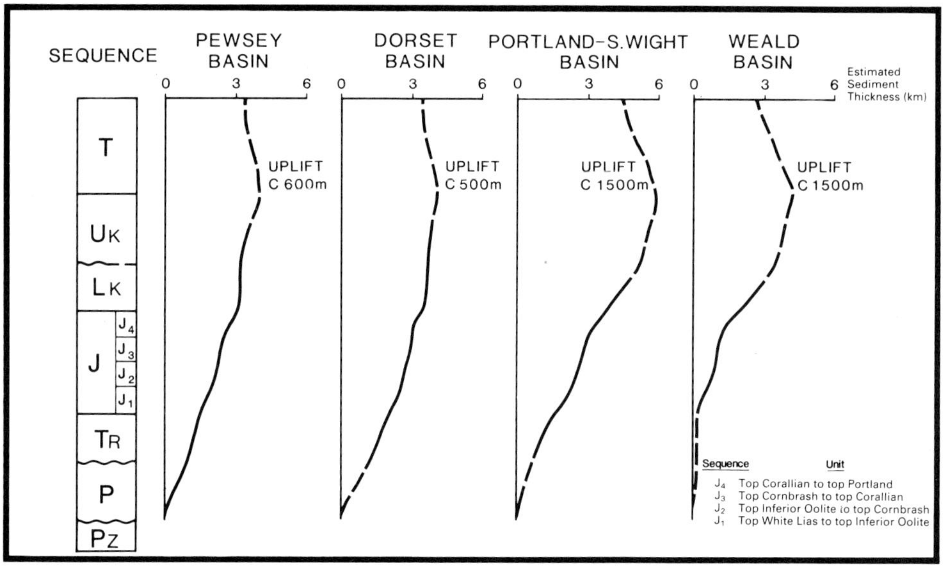

FIG 2. Comparative subsidence histories of Wessex sub-basins. The thicknesses are those observed present day and no decompaction has been allowed for.

structure and the Central Channel anticline reverse fault.

A similar, but tighter, inversion anticline occurs at Kimmeridge to the west (Fig. 5) produced by 1.5 km of uplift with a null point (Williams *et al.*, 1988) in the Middle Jurassic. The tightest inversion fold occurs where subsidence has been greatest.

North of the Kimmeridge fault inversion has been less severe. The Upper Cretaceous sequence is still present, although House (1969) and Ridd (1973) have demonstrated the presence of shortcut reverse faults.

The central and eastern parts of the Wessex Basin

In contrast, inversion of the Weald Basin produced a broad arch of about 50 km wavelength. In the Western Weald (Fig. 6), the Jurassic depocentre, inversion elevated the Hogs Back hinge and developed minor inversion anticlines in the hanging walls of major faults.

In the centre of the Weald Basin the structural style is more intense (Fig. 7). This may be related to a change in strike of the Northern controlling fault from E–W to NW–SE (Fig. 1). Although uplift is estimated to be about 1.5 km in both the Weald and the Portland–South Wight Basins, the style of inversion is different. Inversion in the Weald has generally been accommodated by limited dip–slip reversal of pre-existing extensional faults and a broad inversion anticline. In contrast, in the Portland–South Wight Basin reverse faulting occurred in the S and a northward facing monocline formed adjacent to the Purbeck High. A minor northward facing monocline may also occur in the Weald but is very poorly defined (Fig. 6). Major reverse faults, which offset Middle and Lower Jurassic sediments (Falcon & Kent 1960), are found to the E of Fig. 7.

Conclusion

Tertiary inversion controls the outcrop geology of the Wessex Basin and has produced many impressive structures. On a regional scale however, the maximum uplift was only about 1.5 km over a total 130 km of section (Fig. 4A, B & C). Upper crustal shortening is estimated to be only 2–3% (Chadwick, 1985). Thus, in comparison with Alpine inversion structures (de Graciansky *et al.*, *this volume*) the Wessex Basin has undergone only minor inversion.

The variations in inversion geometry from W to E in the Wessex Basin are probably a function of the geometry of the basin controlling faults and the underlying detachments. However, this relationship is at present unclear.

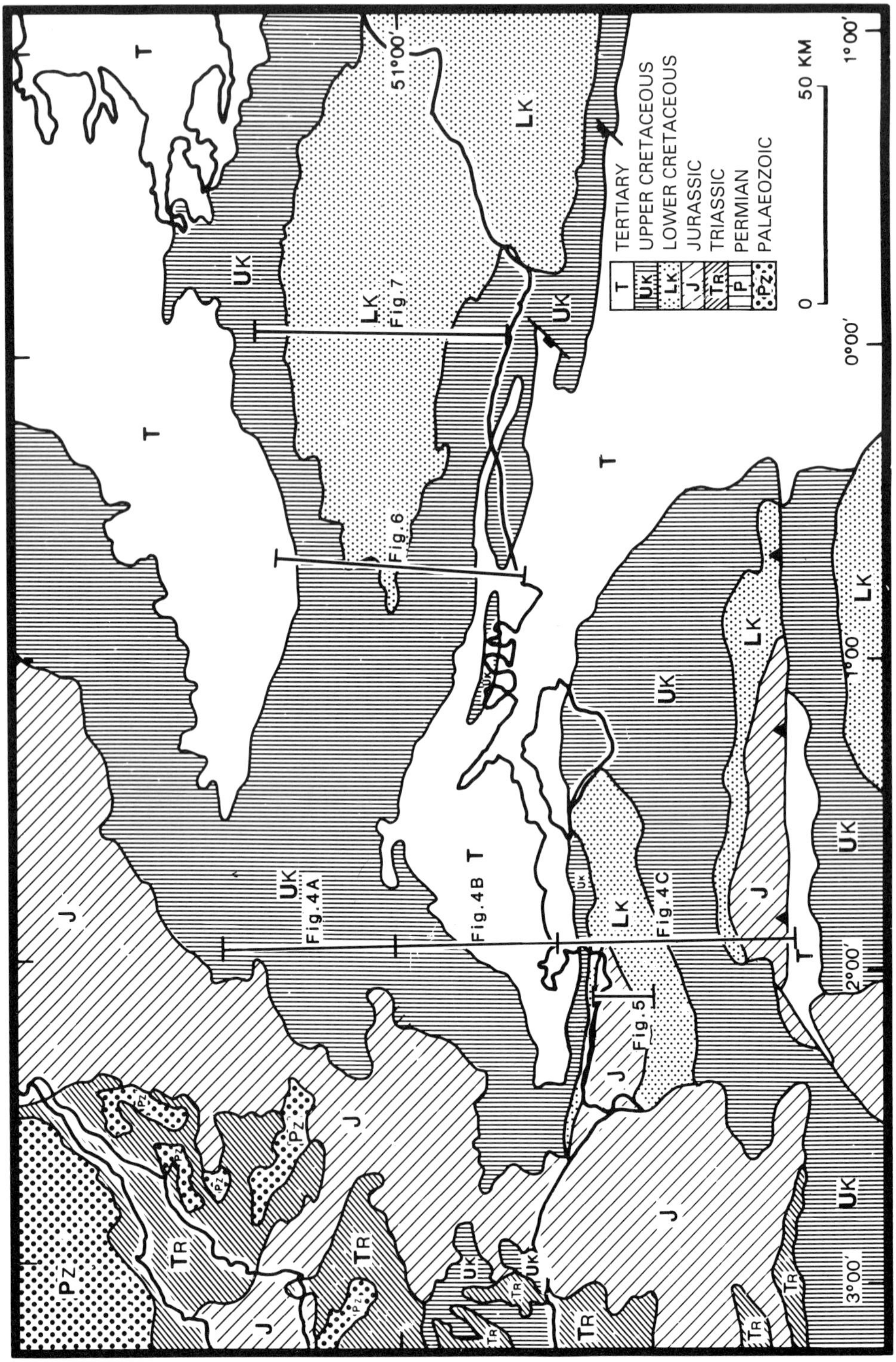

FIG 3. Wessex Basin sea bed and outcrop geology.

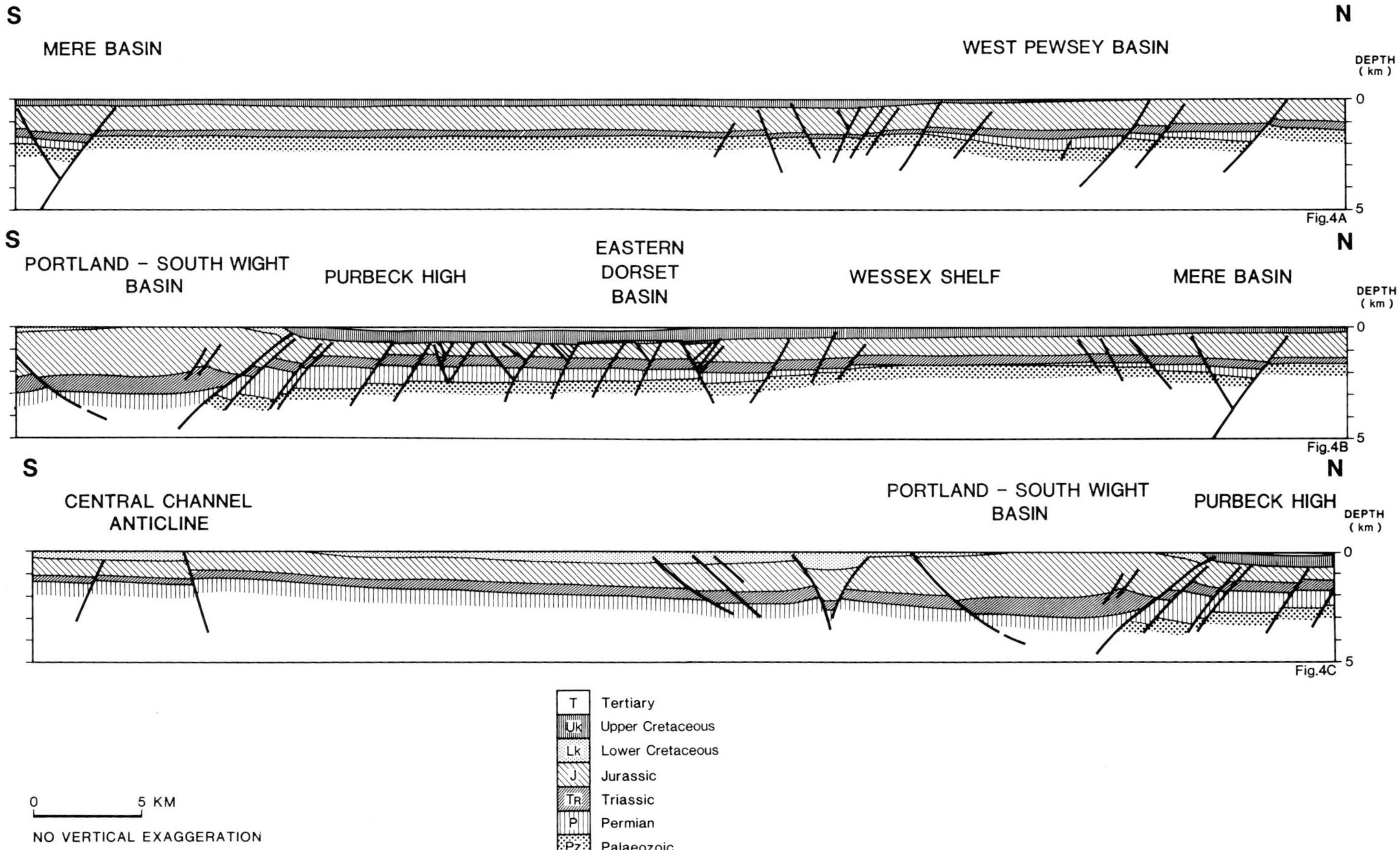

FIG. 4. Single N–S cross-section from the Pewsey Basin to the Central Channel anticline. Note the sections show a variable degree of overlap. For location see Fig. 1.

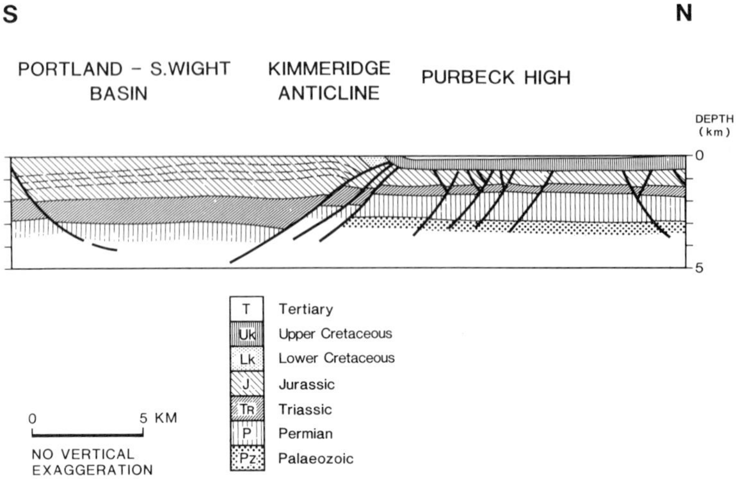

FIG 5. Section across the Kimmeridge anticline. For location see Fig 1.

References and further reading

ARKELL, W. J. 1947. The geology of the country around Weymouth, Swanage, Corfe and Lulworth. *Memoir of the Geological Survey UK*, (Nos. 341, 342, & 343) 386 pp.

CHADWICK, R. A. 1985. Cenozoic sedimentation, subsidence and tectonic inversion. *In*: WHITTAKER, A. (ed.) Atlas of *Onshore Sedimentary basis in England and Wales*: *Post-Carboniferous Tectonics and Stratigraphy*. Blackie, Glasgow, 61–63.

——, KENOLTY, N. & WHITTAKER, A. 1983. Crustal structure beneath southern England from deep seismic reflection profiles. *Journal of the Geological Society of London*, **140**, 893–911.

DEWEY, J. F. 1982. Plate tectonics and the evolution of the British Isles. *Journal of Geological Society of London*, **130**, 371–412.

FALCON, N. L. & KENT, P. E. 1960. Geological results of petroleum exploration in Britain, 1945–1947. *Memoir of the Geological Society of London*, **2**, 2–56.

HOUSE, M. R. 1969. The Dorset Coast from Poole to the Chesil Beach. *Geologists Association Guide*, No. *22* (Revised Edn).

KARNER, G. D., LAKE, S. D. & DEWEY, J. F. 1987. The thermal and mechanical development of the Wessex Basin, southern England, *In*: COWARD, M. P., DEWEY, J. F. & HANCOCK, P. L. (eds). *Continental Extensional Tectonics. Special Publications of the Geological Society of London*, **28**, 517–533.

KENT, P. E. 1949. A Structure contour map of the surface of the buried pre-Permian rocks in England and Wales. *Proceedings of the Geologists Association*, **60**, 87–104.

PLINT, A. G. 1982. Eocene sedimentation and tectonics in the Hampshire Basin. *Journal of the Geological Society of London*, **139**, 249–254.

RIDD, M. F. 1973. The Sutton Poyntz, Poxwell and Chaldon Herring Anticlines, Southern England: a reinterpretation. *Proceedings of the Geologists Association*, **83(1)**, 1–8.

SELLWOOD, B. W. & SCOTT, J. 1986. A geological map of the sub-Mesozoic floor beneath Southern England. *Proceedings of the Geologists Association*, **97(1)**, 81–85.

SMITH, N. 1985. Contours on the top of the Pre-Permian surface of the United Kingdom (south). *British Geological Survey 150th Anniversary Publication*.

STONELEY, R. 1982. The structural development of the Wessex Basin. *Journal of the Geological Society of London* , **139**, 543–554.

WILLIAMS, G. D. & BROOKS, M. 1985. A reinterpretation of the concealed Variscan structure beneath southern England by section balancing. *Journal of the Geological Society of London*, **142**, 689–695.

——, POWELL, C. M. & COOPER, M. A. Geometry and kinematics of inversion tectonics. *This volume*.

I. R. SIMPSON, British Petroleum Development Ltd, Britannic House, Moor Lane, London EC2Y 9BU, UK.

M. GRAVESTOCK, British Petroleum Development Ltd, PO Box 2749, Jakarta Selatan, Indonesia.

D. HAM, British Petroleum Development Ltd, Britannic House, Moor Lane, London EC2Y 9BU, UK.

S. D. THOMPSON, Standard Oil Production Company, 5151 San Felipe, PO Box 4587, Houston, Texas, USA.

H. Leach British Petroleum Development Ltd, 10 Quai Paul Doumer 92412, Courbevoie Cedex, Paris, France.

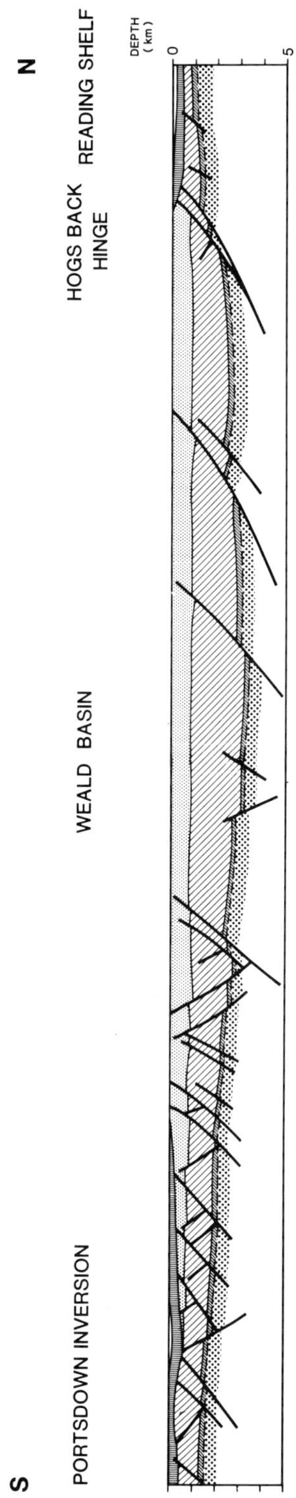

FIG. 6. Cross-section of the western Weald Basin. For location see Fig. 1.

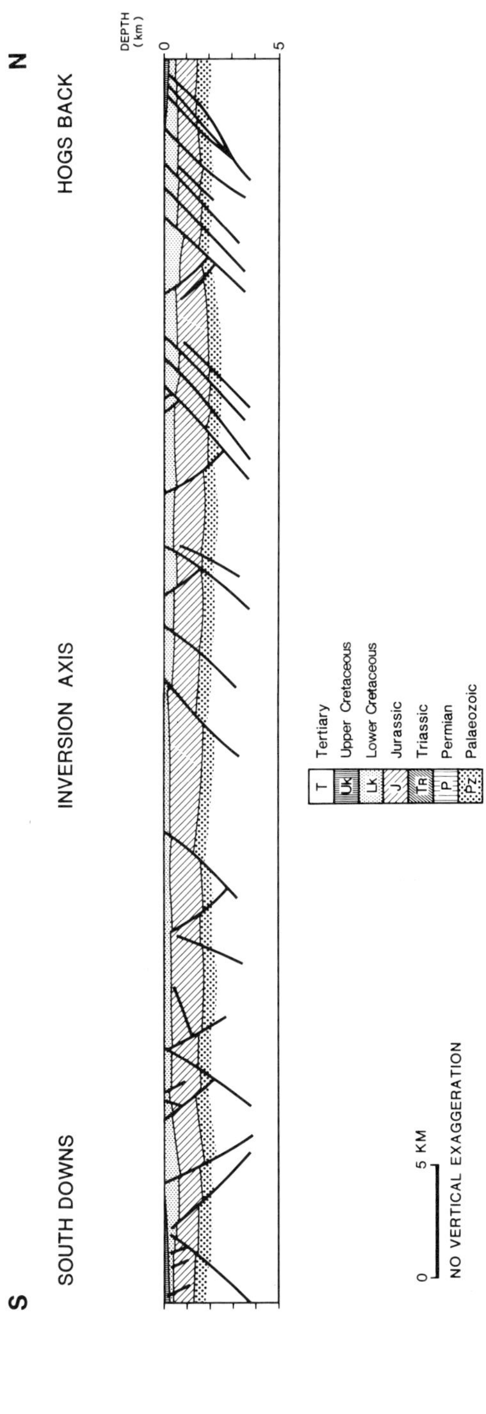

FIG. 7. Cross-section of the Weald Basin inversion axis. For location see Fig. 1.

Basin inversion in and around the British Isles

D. G. Roberts

SUMMARY: The development of basins of Mesozoic and Tertiary age both onshore and offshore the British Isles is largely a consequence of episodic rifting that led to the progressive northward opening of the North Atlantic. In these basins, extension was controlled by the structure of underlying Palaeozoic terrains and succeeded by passive thermal subsidence.

In this context, however, the British Isles represent an anomaly both in terms of their present elevation and in the widespread exposure of Pre-Cambrian to Tertiary rocks. The outcrop geology of the onshore and offshore basins is an obvious demonstration of uplift although the timing and origin and magnitude of the uplift have remained uncertain. Evidence from onshore and offshore the British Isles is used to document the influence of compressional inversion of Tertiary age in shaping the present day outcrop geology.

Introduction

The British Isles are surrounded by failed rifts and passive margins formed by the episodic rifting of the North Atlantic during Mesozoic and Tertiary time (Roberts 1975; Ziegler 1982). Given the characteristic broad regional subsidence of extensional basins, the British Isles represent an anomaly both in terms of their present elevation and in the widespread exposure of Pre-Cambrian to Tertiary rocks. The present outcrop geology obviously demonstrates uplift of the British Isles but the timing, origin and magnitude of the uplift have remained uncertain and controversial, partly because of the multiple extensional and compressional events that have affected the region during the Mesozoic and Tertiary, and also because of the fragmentary record of Mesozoic and Tertiary rocks beyond the main areas of outcrop (Walsh 1966; Moseley 1972 and Walsh *et al.* 1972).

This paper discusses the influence which compressional inversion of Tertiary age has had on shaping the present day geology of basins on and around the British Isles (Fig 1). Inversion is here defined as the reversal of throw along normal faults thereby elevating the half graben. The reversal of movement results in the elevation of syn- and postrift strata above 'regional' and ultimately the complete extrusion of the half-graben (Bally 1984; Ziegler 1983). For example, in the Wessex Basin (Simpson *this volume*) inversion has taken place along early Mesozoic extensional faults which may themselves link onto Variscan thrusts. Although the net shortening is less than 10 km, the inversion has elevated the Chalk across the 100 km width of the basin by about 1 km, and, though the inversion is large in terms of wavelength and amplitude, changes in regional dip are generally minor except on a local scale. The regional scale of this folding is quite different from the local uplift caused by strike–slip motion, elevation across transfer zones in extensional settings or by the isostatic uplift of footwall blocks. Equally, the type of inversion exemplified by the Wessex Basin must be distinguished from the broad thermal updoming on scales of several hundred kilometres which results indirectly from lithospheric extension (Steckler 1985). Examples of this phenomenon include the Gulf of Suez, where observation and modelling (Kohn and Eyal 1981; Steckler 1985) demonstrate that lateral transfer of heat by conduction from the asthenosphere has resulted in elevation of the rift shoulders thereby uplifting preexisting basins above regional. In Southern England, the progressive westward overstep of the Middle to Late Cretaceous across tilted and faulted Jurassic, Triassic and Palaeozoic strata may reflect initial thermal uplift in response to Late Jurassic–Early Cretaceous rifting in the Bay of Biscay (Roberts *et al.* 1981).

In subsequent sections of this paper, the evidence for compressional inversion of Tertiary age is discussed for onshore and offshore settings. None of the examples described is considered to be inversion due to thermal processes.

The East Midlands

The Mesozoic and Tertiary record beyond the main area of outcrop in Southern and Eastern England is fragmentary though there is evidence of a formerly more extensive Jurassic and

From COOPER, M. A. & WILLIAMS, G. D. (eds), 1989, *Inversion Tectonics*
Geological Society Special Publications No. 44, 131–150.

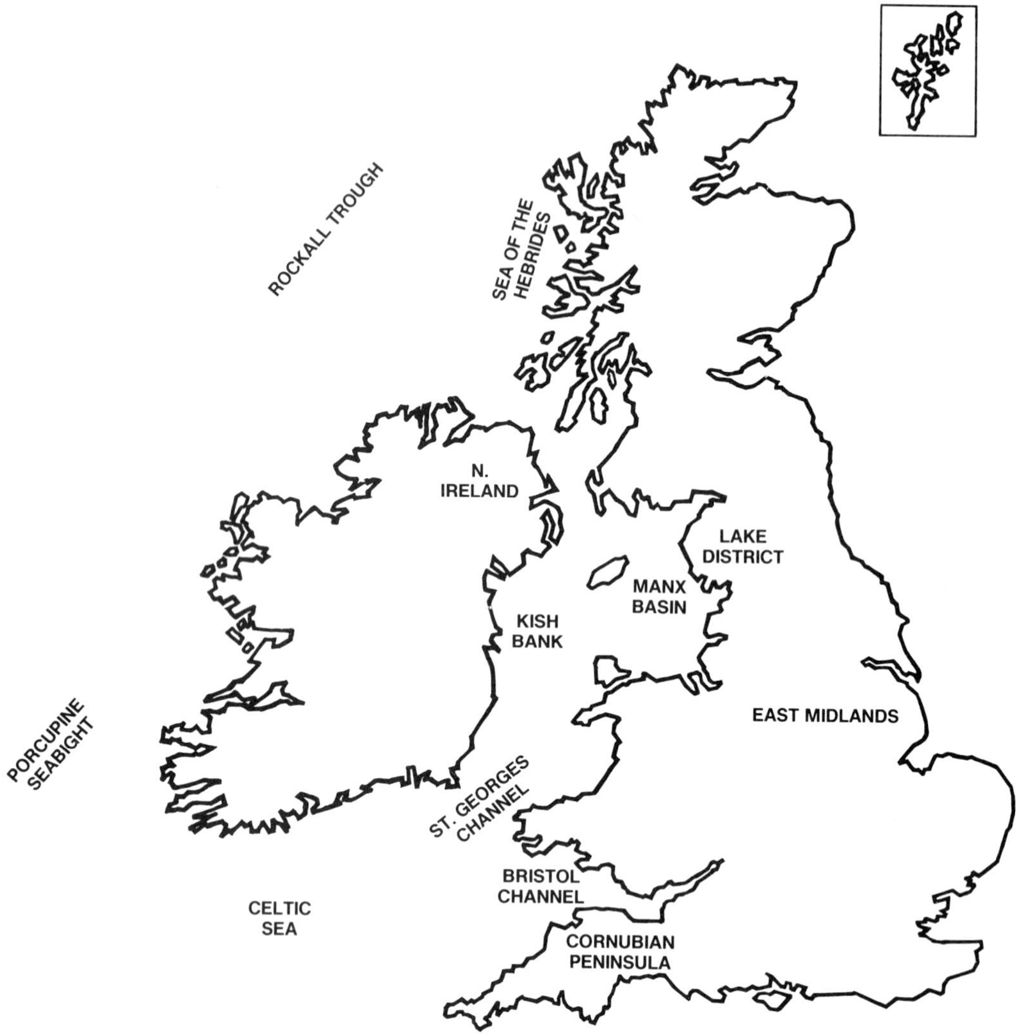

FIG. 1. Outline map of the British Isles showing the location of areas referred to in the text.

Cretaceous cover. Given the evidence of Mesozoic and Tertiary tectonism in the preserved basins, it is likely that the elevation of the Palaeozoic strata occurred during these times. However, the problem of non-uniqueness of interpretation from an outcrop of a given age as shown by comparison of three data sets (Fig. 2A, B) taken from the East Midlands area where Carboniferous, Permo–Triassic and later Mesozoic strata crop out nearby.

The seismic section (Fig. 2A) across the Eakring oil field in the Widmerpool Gulf (Kent 1968) of Nottinghamshire shows a major inversion structure of Late Carboniferous age truncated by the basal Permo–Triassic unconformity. Although the age of the inversion is indisputable in this case, a second section some 20 km to the E shows reversal of throw affecting Liassic strata. Fission track data (Fig. 2B) demonstrates, however, phases of Late Cretaceous and Early Tertiary uplift probably coeval with inversion in the Southern North Sea and Northeast Yorkshire (Kent 1980; Glennie & Boegner 1981 and van Hoorn, 1987b).

The Lake District

The Lake District provides an example of an alternative use of fission track data where determinations of the age and amount of uplift of Palaeozoic rocks at outcrop can be used as an

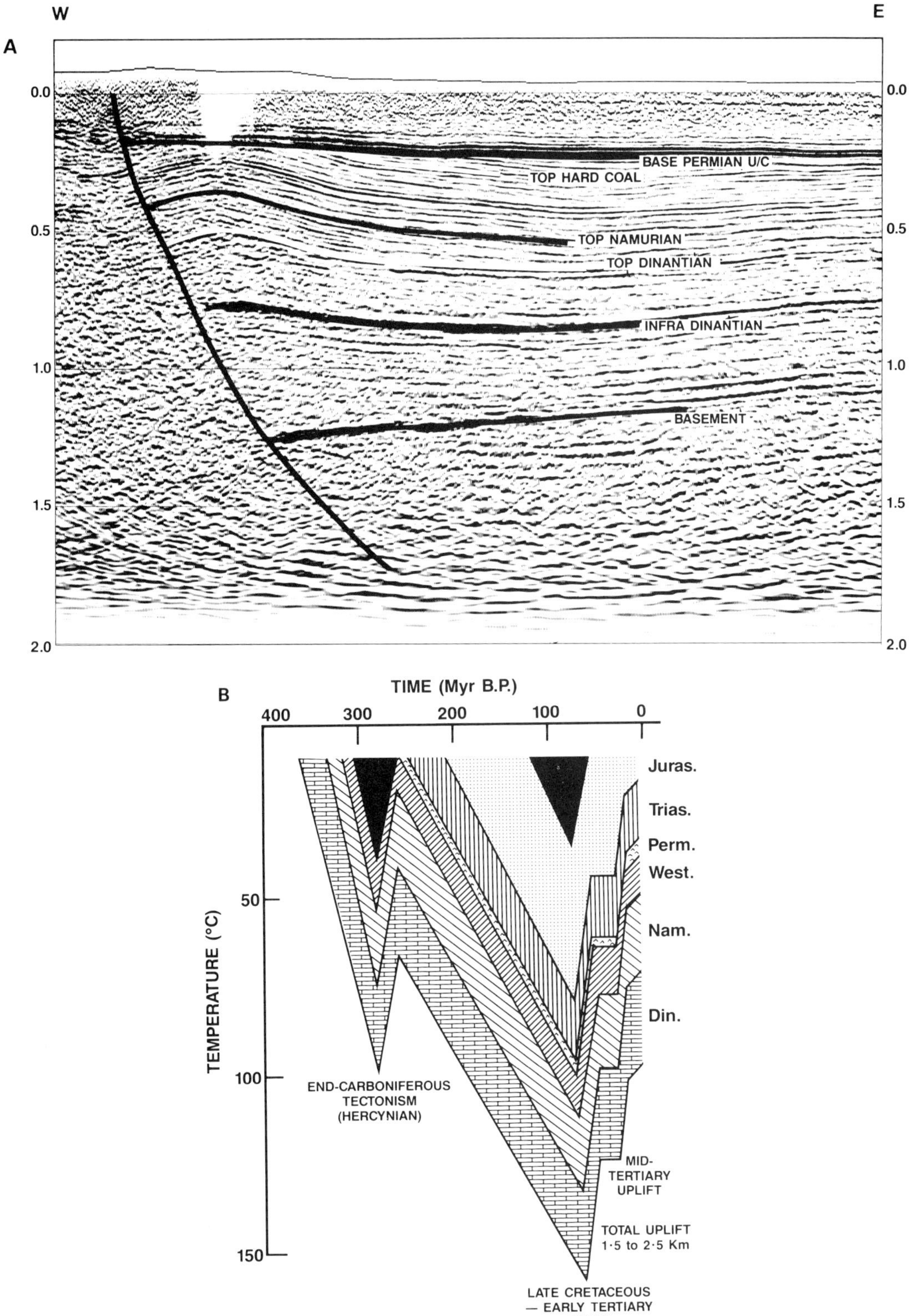

FIG. 2. A, Seismic section across the Eakring oil field, East Midlands showing pre-Permian inversion of a Dinantian extension fault. B, The uplift history of the East Midlands based on fission track data (BP Exploration, unpublished data).

effective complement to data independently gained from adjoining basins. In Northwest England, the Palaeozoic strata exposed in the Lake District represent the rollover to the Permo–Triassic (and Late Jurassic?) Vale of Eden fault and are in turn downthrown by the Lake District Boundary Fault to underlie the Permo–Triassic, and locally Jurassic, cover of the Irish Sea (Jackson *et al.* 1987). Fission track data from the Lake District document uplift of Early Tertiary age (Fig. 3) whose amount varies between 1 and 2 or 2 and 4 km depending on whether the geothermal gradient was elevated by a heat pulse or remained normal (Green 1986). Vitrinite reflectance versus depth data from wells in the adjacent Irish Sea (Fig. 3B) show uplift 2 or more km in post-Jurassic time. Exposure of Early Tertiary dykes at the seabed and onshore suggests post Early Palaeocene uplift. Although both data sets can

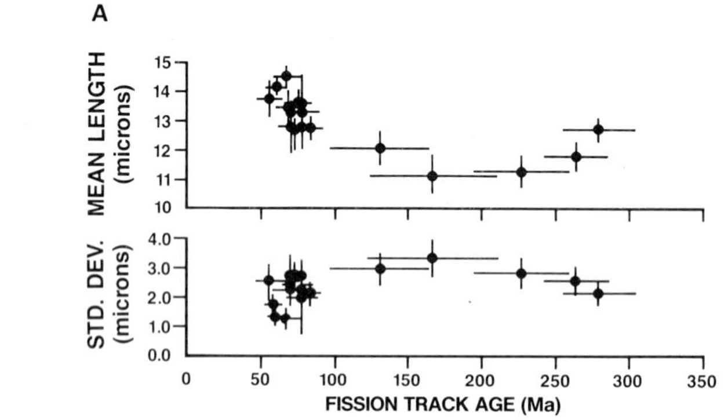

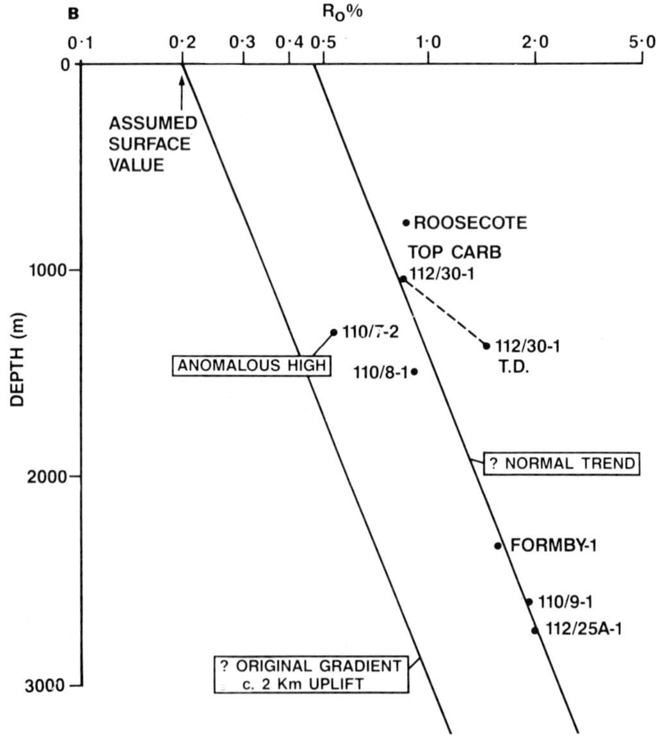

FIG. 3. A, Fission track data from the Lake District (after Green 1986). The trends indicate significant resetting of the ages at approximately 50–60 Ma, which is due to uplift. B, Vitrinite reflectance versus depth in the Manx Basin suggesting 2 km of uplift (BP Exploration, unpublished data). Well 110/7–2 is located on a local structural high and is thus anomalous.

be interpreted in terms of a broad regional uplift, three lines of evidence suggest inversion was the main factor. Porosity–depth data (R. C. Cleverley, personal communication) from the Irish Sea conform to the global curves and indicate the same order of uplift as the vitrinite reflectance data; the consistency between the two data sets is evidence of a tectonic rather than thermal origin for the uplift. Secondly, the Selker Rocks anticline closely resembles an inversion anticline and several nearby faults truncate Tertiary dykes (Figure 3 of Jackson *et al.* 1987). Onshore, Tertiary inversion is suggested by the present structure of the Lake District; Permo–Triassic strata to the N and S show an antiformal structure of c. 100 km wavelength whose age is shown by Tertiary dykes that crosscut these strata as well as the Palaeozoic basement. Nearby, Moseley (1972) has inferred Tertiary uplift of the Askrigg block.

The Bristol Channel Basin and the Cornubian Peninsula

The Bristol Channel Basin (Fig. 4) is an exhumed E–W trending half graben of Mesozoic age flanked to the N and S by Palaeozoic terrains deformed by Variscan folding and thrusting (Lloyd *et al.* 1973; Kamerling 1979; van Horn 1987a; Brooks *et al.* 1988). The major, controlling and southward dipping fault (Fig. 5A) in the N was formed in response to Late Jurassic–Early Cretaceous extension and probably links into a Variscan thrust (Brooks *et al.* 1988). The half graben is cut by NNW–SSE trending faults which may represent Variscan transfer faults rejuvenated during the Mesozoic and again during the Tertiary. Kamerling (1979) reports a Late Early Cretaceous uplift that may be associated with the thermal effects of rifting in the Bay of Biscay. Subsequently the basin passively subsided throughout the Late Cretaceous (Figs 5B & 6). Renewed deformation and uplift in Oligocene time is demonstrated by strike–slip movement along the Lustleigh–Sticklepath fault system and associated development of non-marine pull apart basins such as the Stanley Bank, Bovey Tracey and Dutson Basins (Figs 5B & 6) (British Geological Survey, 1983a; 1983b; 1985) and by angular truncation of the Late Cretaceous by the Oligocene of the Stanley Bank Basin. Onshore gentle folding has affected Cretaceous and Eocene stata in Eastern Exmoor while major

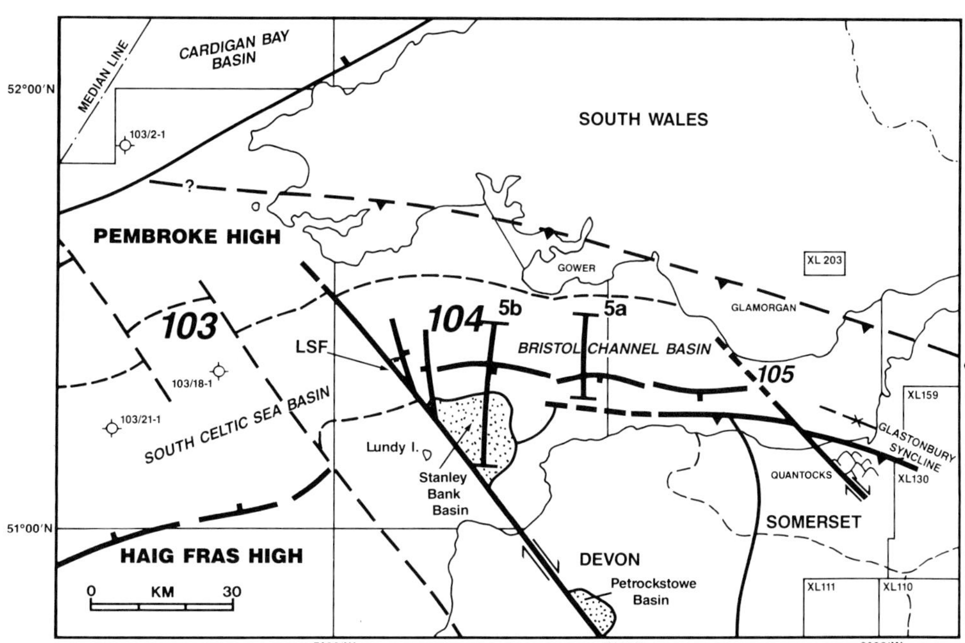

FIG. 4. Outline structure and geology of the Bristol Channel (after Lloyd *et al.* 1973; Kamerling 1979). The thin dashed line indicates the limits of Mesozoic rocks. Offshore blocks and onshore exploration licenses are numbered. Tertiary basins are stippled.

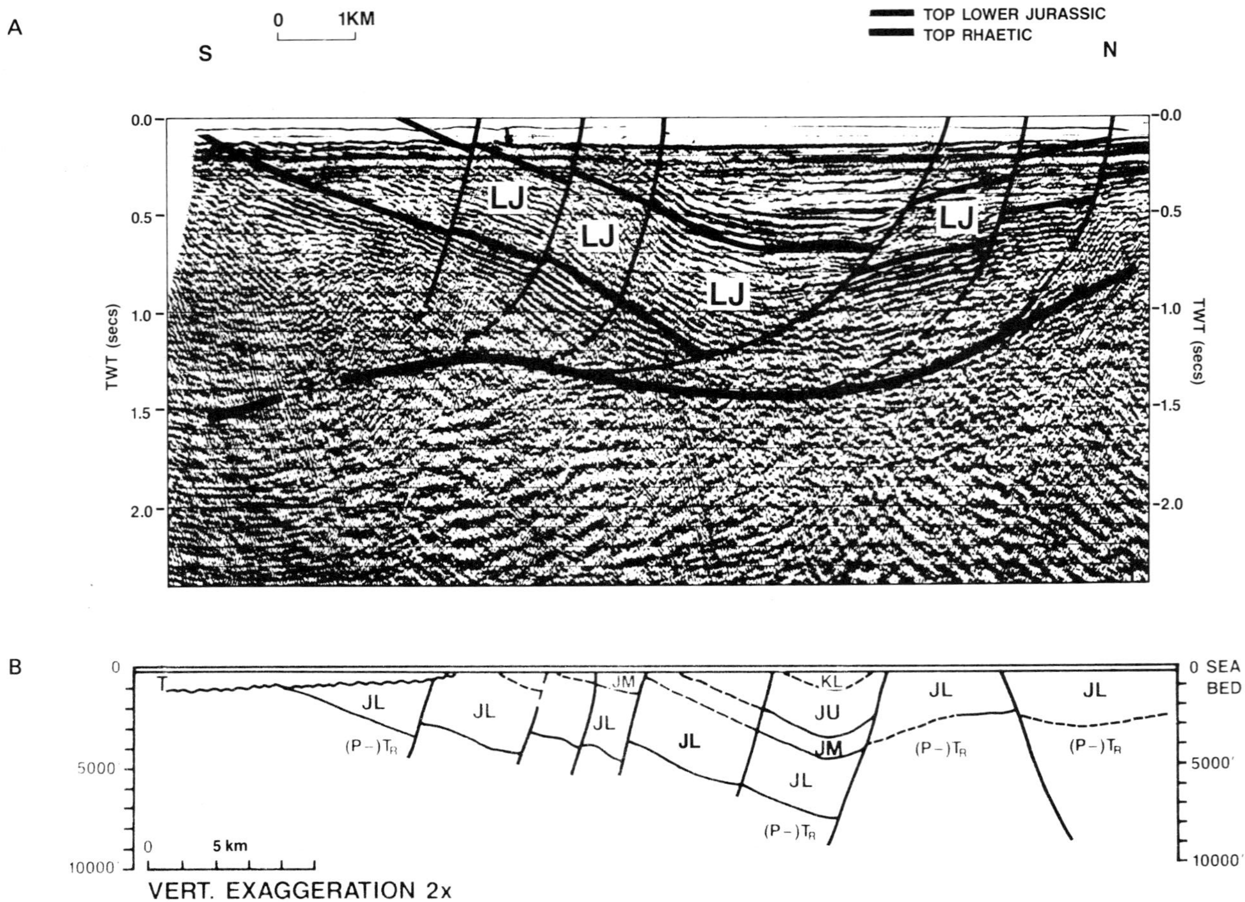

FIG. 5. A, Seismic section across the Bristol Channel. The major southward dipping and controlling normal fault is probably nucleated along a Variscan thrust (Brooks *et al*. 1988). LJ is the Lower Jurassic interval. B, Geological cross-section through the Bristol Channel and non-marine Oligocene beds of the Stanley Bank Basin (after Chappell 1986; Kamerling 1979). JL, Lower Jurassic, JM, Middle Jurassic, JU, Upper Jurassic, KL, Lower Cretaceous, T, Tertiary.

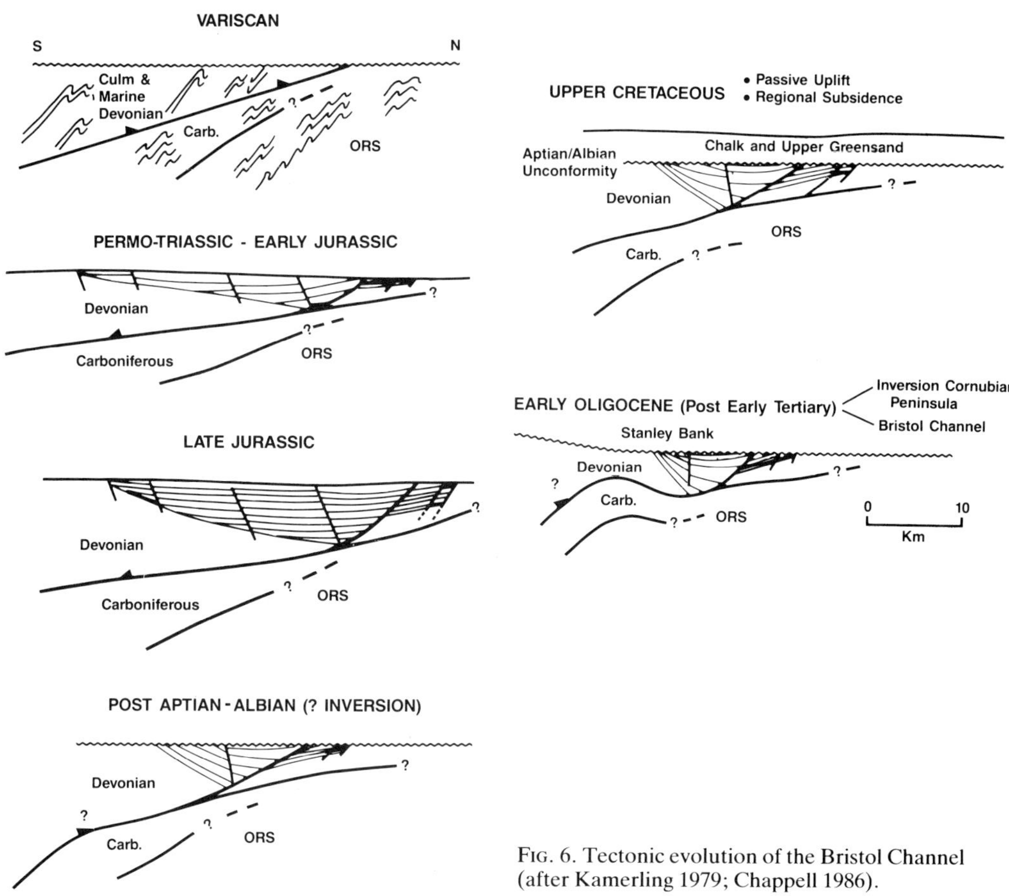

FIG. 6. Tectonic evolution of the Bristol Channel (after Kamerling 1979; Chappell 1986).

regression took place in Oligocene time in the Western English Channel (Evans & Hughes 1984).

The wider context of this uplift can be viewed by considering the regional dip of the Cretaceous N and S of the Cornubian Peninsula. Projection of the dips suggests the base Chalk surface originally extended across the peninsula, a conclusion supported by widespread occurrence of flints (K. Isaacs, personal communication), and has been elevated by approximately 1 km.

Interpretation of these data suggest that the Cornubian Peninsula is an inversion structure comparable in amplitude and wavelength to the Wessex Basin. Inversion was probably accommodated along Late Jurassic–Early Cretaceous extensional faults linked to Variscan thrusts, and by strike–slip transfer motion along the Lustleigh–Sticklepath and other parallel fault systems.

The Celtic Sea

The Celtic Sea is a NE–SW trending basin between Ireland and Cornubia (Fig. 7) formed by multiple phases of Mesozoic extension and subsequent inversion during the Tertiary (Naylor & Shannon 1982; Tucker & Arter 1987). Initial extension took place along a major eastward dipping boundary fault possibly nucleated along a Variscan thrust (Fig. 7 and 8; BIRPS and ECORS 1986).

Following the first phase of extension in Late Triassic time, the basin passively subsided until the second phase of extension in Late Jurassic–Early Cretaceous time (Naylor & Shannon 1982). The latter phase resulted in uplift and erosion of the basin margins. Renewed passive subsidence with transgression of the basin and its margins took place from Albian to Maastrichtian time. The formerly much more extensive cover of Late Cretaceous sediments

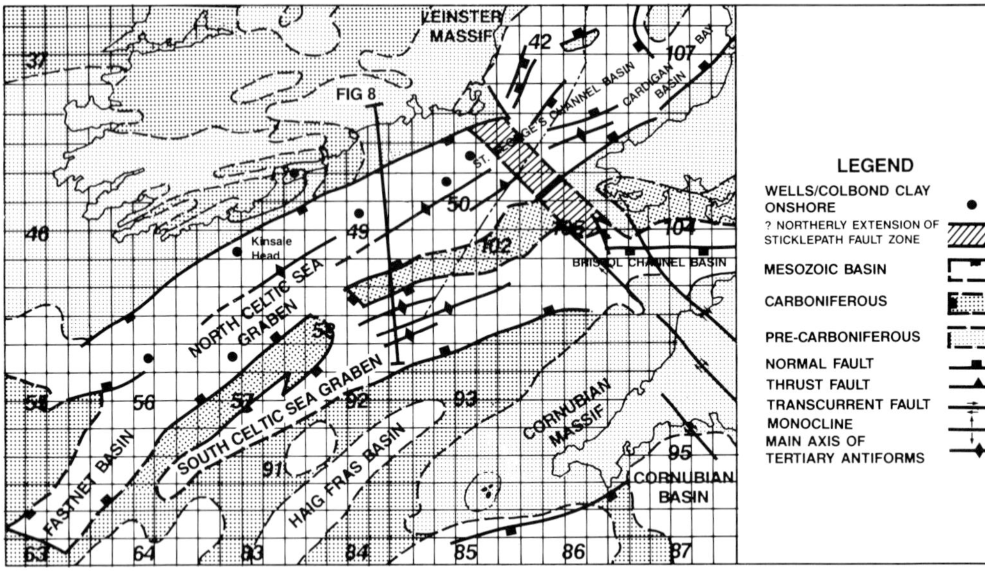

FIG. 7. Outline structural elements of the Celtic Sea. The line of the geoseismic section of Fig. 8 is indicated. The wells shown are those used in Figs 9 & 10.

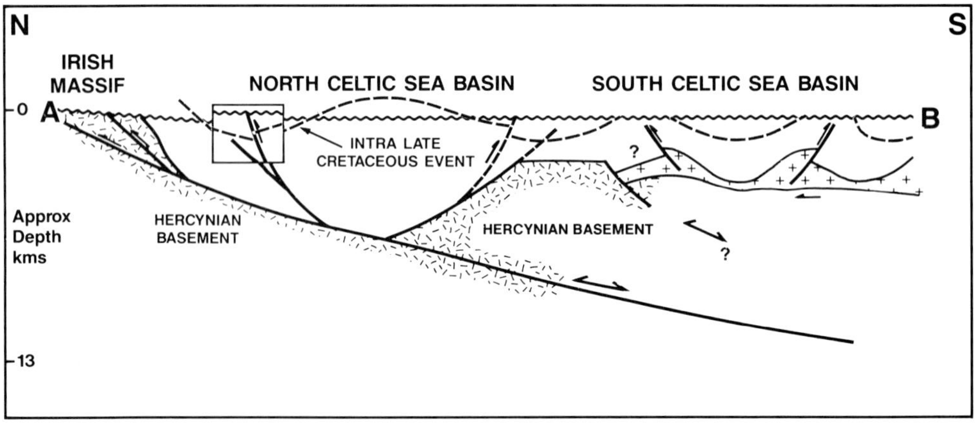

FIG. 8. Geoseismic section through the Celtic Sea based on SWAT Line 4.

deposited in this phase is shown by remnant chalk outcrops in Ireland (Walsh 1966) and implied by the chronostratigraphic diagram for the Irish Sea–Celtic Sea region (Fig. 9). The present absence of Mesozoic strata over much of Ireland and parts of the Irish Sea–Celtic Sea can be attributed to erosion of these areas following uplift caused by Tertiary inversion (Figs 9 & 10). In the Celtic Sea, vitrinite reflectance data and the configuration of the top Chalk reflector indicate uplift of 1–2 km (Fig. 10). Uplift of the adjacent onshore by perhaps as much as 2 km is shown by spore coloration data from the small outcrop of Colbond Clay of likely Middle Jurassic age near Cork (Higgs & Bees 1986) and by condont data from the Late Carboniferous (M. A. Cooper, personal communication). Although the apparent absence of Danian section and a minor angular unconformity at the 'top' of the Chalk suggests possible inversion as early as the Palaeogene, well data from the Fastnet Basin and Goban Spur (Robinson *et al.*, 1981; Cook, 1987) suggests that the major inversion is

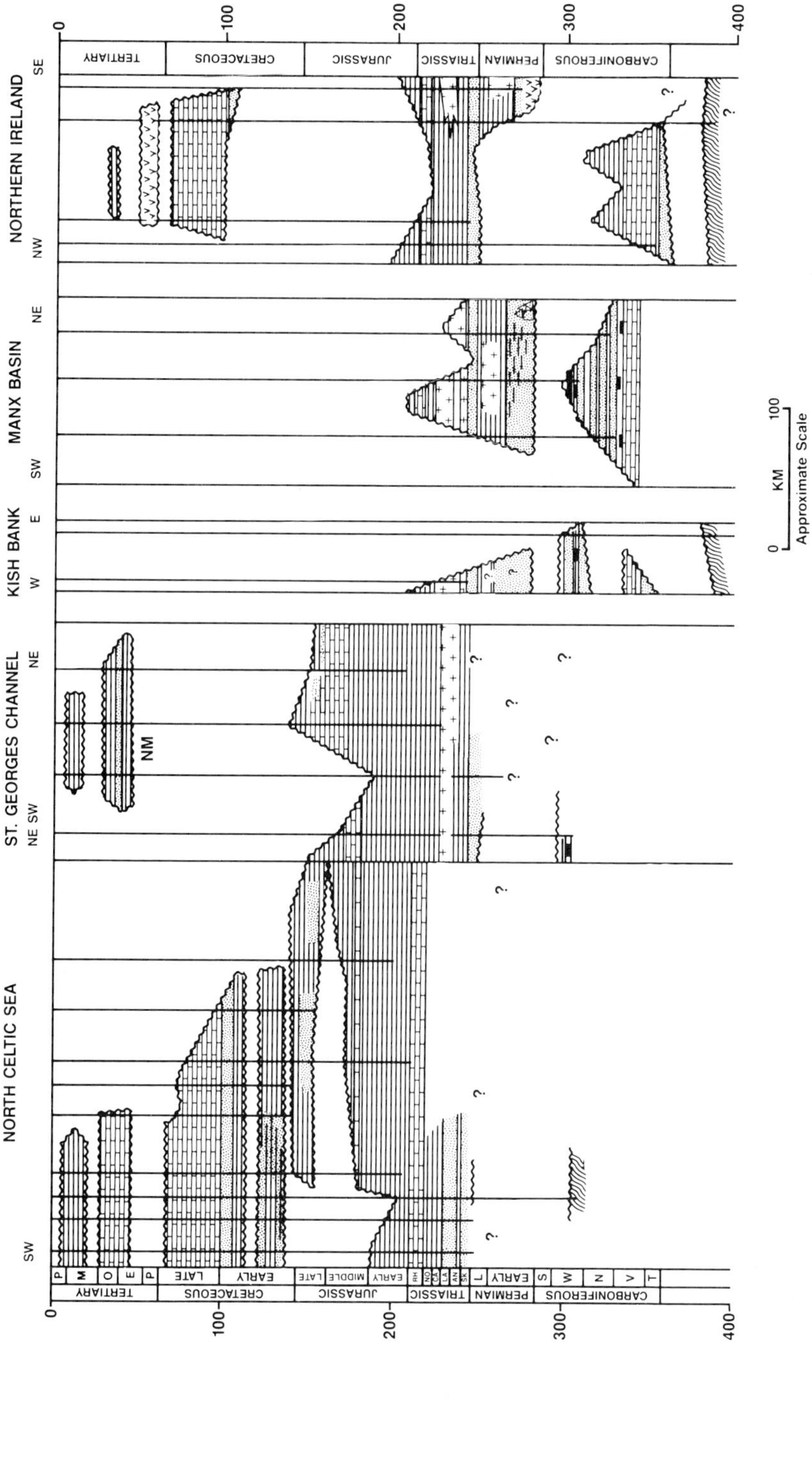

FIG. 9. Chronostratigraphic diagram for the Irish and Celtic Sea regions (after Brown 1986). Well control points shown by vertical lines; locations are approximate.

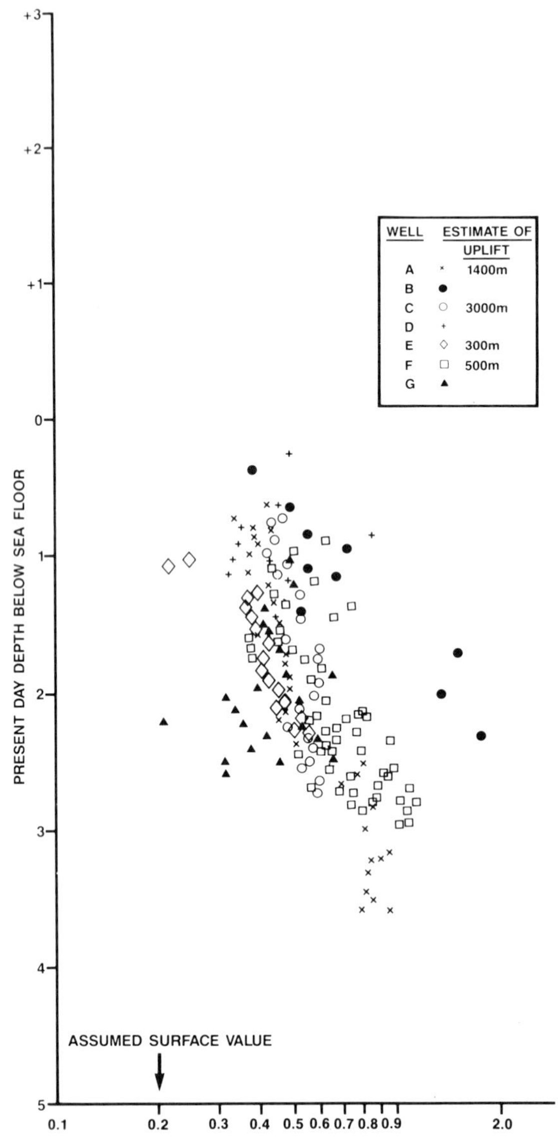

FIG. 10. Vitrinite reflectance data versus depth for wells in the Celtic Sea and the onshore outcrop of the Colbond Clay (after Higgs & Beese 1986).

Oligocene age in agreement with the western English Channel (Chapman *this volume*; Ziegler 1987).

It should be stressed that the inversion is not confined to the immediate area of onshore Ireland and rifted basins present beneath the continental shelf. West of Ireland, in the Northern Porcupine Seabight, inversion is clearly shown by the antiformal structure of Late Cretaceous and Early Tertiary strata above tilted fault blocks structured during the Late Jurassic–Early Cretaceous extension phase. In this case, the age of inversion is clearly dated by deformation of an Eocene reflector and by onlap of Miocene strata to the antiformal structure. In the S parts of the Porcupine Seabight and on the adjoining Goban Spur, well and seismic evidence (Masson & Parson 1984; Masson *et al.*

1985; Sibuet *et al.* 1985; Cook 1987) document an important phase of Oligocene uplift and compression. It should be stressed, however, that the deformation is not ubiquitously compressional and that locally extensional structures are present (Croker & Shannon 1987).

Northern Ireland

In a classic example of regional geological inference, George (1967) discussed the development of landform and structure in Ulster. In his study, he noted that the Palaeocene basalts were extruded onto an essentially horizontal Chalk surface; the thin but uniform Chalk sequence rests unconformably on tilted and faulted Mesozoic strata (Fig. 11A, B). By contrast, present day structure contours on the base of the basalts indicate elevation of several thousand feet over a wavelength of 20 km (Fig. 12A & B). Local warping of the non-marine Oligocene sequence of the Lough Neagh basin documents the age of the inversion. The interpreted regional cross section offered by George (Fig. 12B) suggests the inversion may have taken place along an older Mesozoic extensional fault. Faults, such as the Tow Valley Fault, displacing the basalts are probably related to this inversion and isoclinal folding of the basalts along splays of this fault has been reported from Rathlin Island off the North Coast of Ireland (Wilson & Robbie 1966).

Sea of the Hebrides

Non-marine basins of Oligocene age preserved beneath the Sea of the Hebrides were first reported by Smythe & Kenolty (1975). Modern seismic data (reproduced here by kind permission of JEBCO) across the example figured by them shows that the Palaeocene basalts are structured into a syncline within which rotated onlap of Oligocene strata is evident (Fig. 13). In this case, the structural and stratigraphic relationships are most readily explained by inversion along a Mesozoic (?) extensional fault or Caledonian thrust.

In the adjoining areas of Scotland beyond the main igneous centres, George (1967) has noted that exposure of Palaeocene dykes at elevations above 1000 m implies substantial post-Palaeocene uplift. His interpretation is directly substantiated by the virtual absence of marine Eocene to Miocene sediments in the Minch and Sea of the Hebrides. Indirectly, uplift is indicated by the sustained influx of clastics into the North Sea from Scotland throughout the Tertiary. However, it is quite probable that the initial post-Danian influx of clastics was caused by thermal uplift of Scotland in response to Palaeogene volcanism. Nonetheless, continued uplift is required to explain the sustained influx of clastics through Oligo–Miocene time. In the absence of fission track data, the evidence is circumstantial though there is evidence of Oligocene inversion in the Central and Northern North Sea (Ziegler 1982).

The Rockall Trough

Although the evidence from onshore Scotland is circumstantial, the stratigraphy and structure of the adjacent Rockall Trough provides strong evidence for Oligo–Miocene inversion.

The Rockall Trough is an elongate sinuous depression between the Rockall Plateau and shelf W of Scotland (Roberts 1975). Now known to be underlain by stretched continental crust (Roberts *et al.* 1988), the Trough probably formed by multiple phases of extension of which the final and most important took place in Late Jurassic–Early Cretaceous time (Roberts 1975; Roberts *et al.* 1981). Following passive subsidence in Late Cretaceous time, massive early Palaeogene volcanism took place in response to rifting between Greenland and Rockall Plateau forming the Faeroe Islands, the Wyville–Thomson Ridge and many other piles (Roberts *et al.* 1983, 1984). The lavas of the Wyville–Thomson Ridge are strongly folded and faulted (Roberts *et al.* 1983; Wood *et al.* 1987). Further S, overthrusting is evident along the continental slope N of St Kilda (Fig. 14). From regional correlation of seismic reflectors (Roberts 1975; Masson & Kidd 1987), the deformation is interpreted as Oligocene in age. S of St Kilda, the linear slope is controlled by Miocene faulting (Stride *et al.* 1969) that may reflect an underlying inversion structure. NW of Ireland, well and seismic sections show folded Mesozoic strata subcropping to an unconformity overlain by Miocene strata (Fig. 15). It appears that the present morphology of the slope reflects inversion rather than Mesozoic extension.

The inversion observed along the E margin in the Rockall Trough can be explained by recognizing the E-dipping faults bounding the major half-graben on the shelf formed part of a major extensional system developed above an E-dipping detachment implied by the structural asymmetry of the Rockall Trough (Roberts *et al.* 1988). These faults which may have nucleated on Caledonian thrusts were all inverted along the plate margin controlled by the main E-dipping detachment (Fig. 16). An apparent absence of Oligocene deformation within the Trough is suggestive of decoupling along the E margin (Roberts 1975).

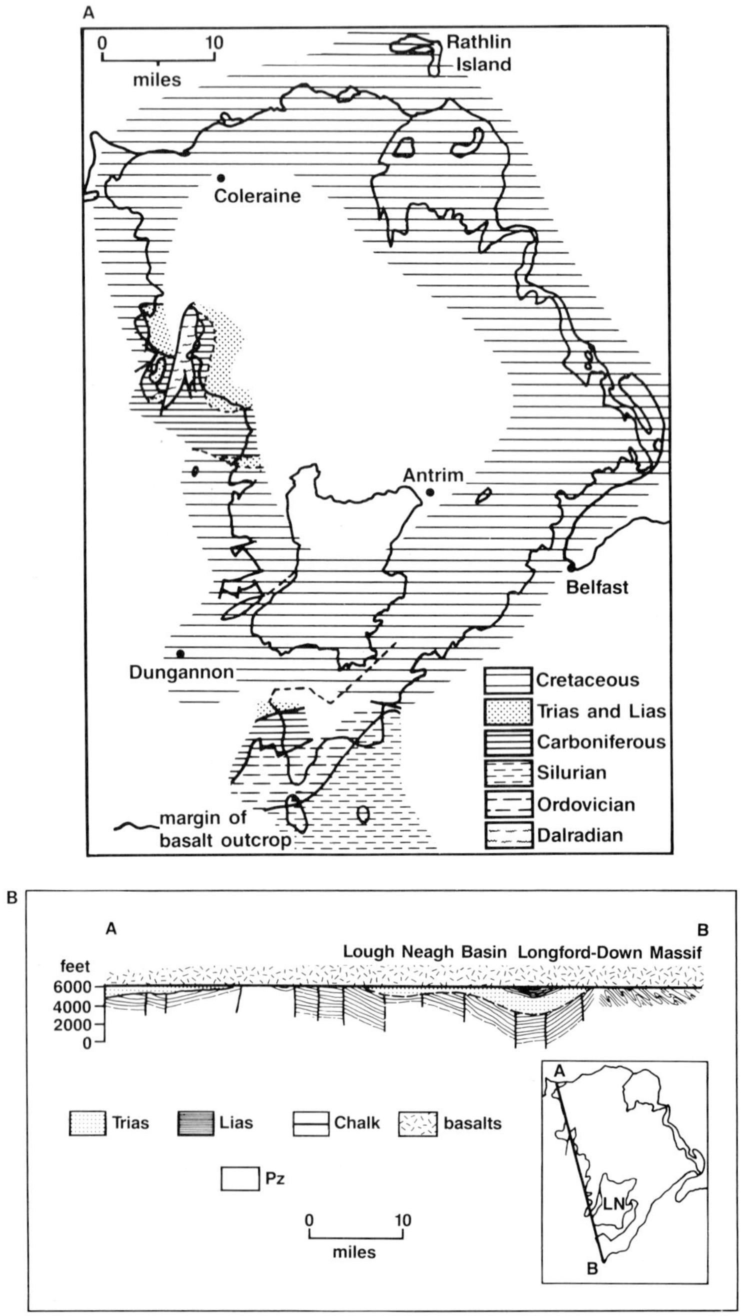

FIG. 11. A, The base Eocene subcrop in Northern Ireland (redrawn from George 1967). B, A reconstructed section of the subcrop to the base of the basalts (after George 1967). The chalk is shown as a heavy black line resting uncomformably on the Mesozoic and Palaeozoic strata.

Extent of Oligocene deformation

The preceding discussion demonstrates that compression and inversion of Oligocene age has widely affected the onshore, rifted basins on the shelf and the adjoining passive margin. It is rather less widely known, though well established, that the same deformation can be

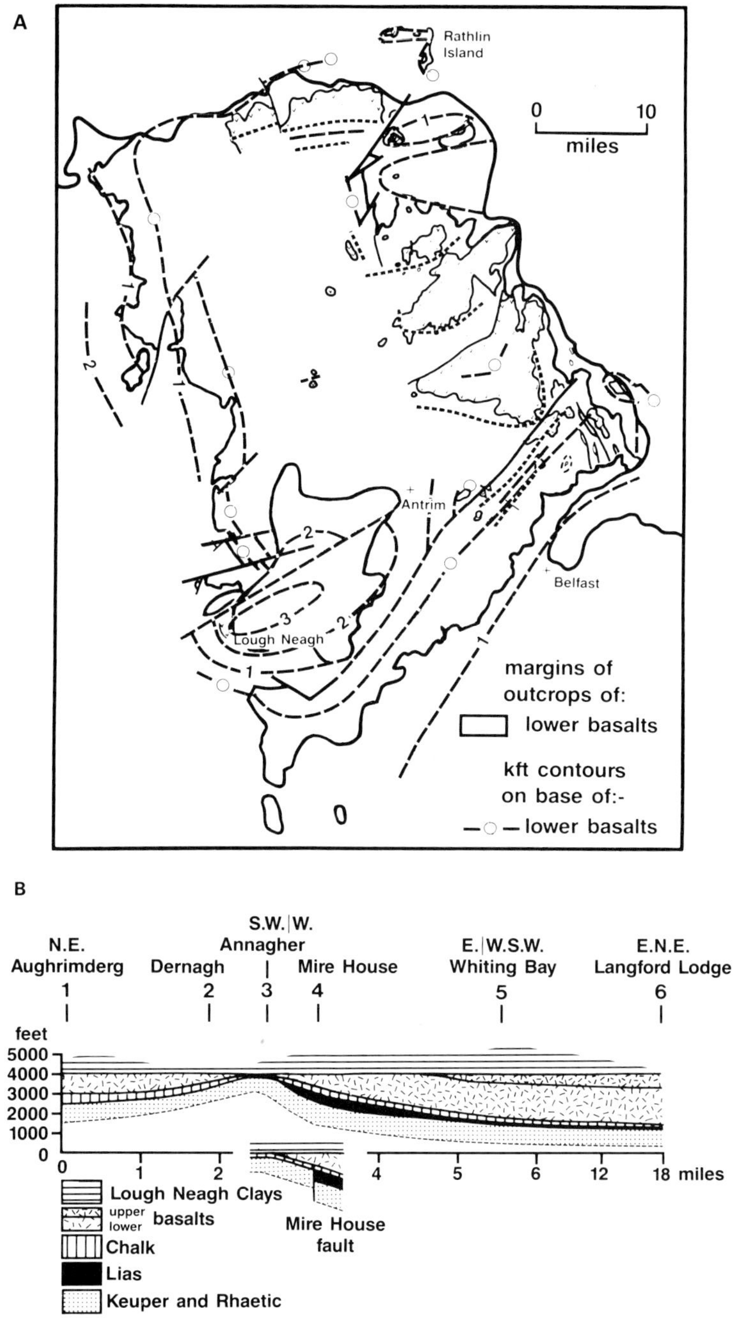

FIG. 12. A, Structural contours in kilo feet on the base of the Tertiary basalts (after George 1967). B, Reconstructed cross-section flattened on the base of the warped Oligocene Lough Neagh clays shows post-basalt deformation possibly controlled by inversion of a Mesozoic (Late Jurassic?) extensional fault.

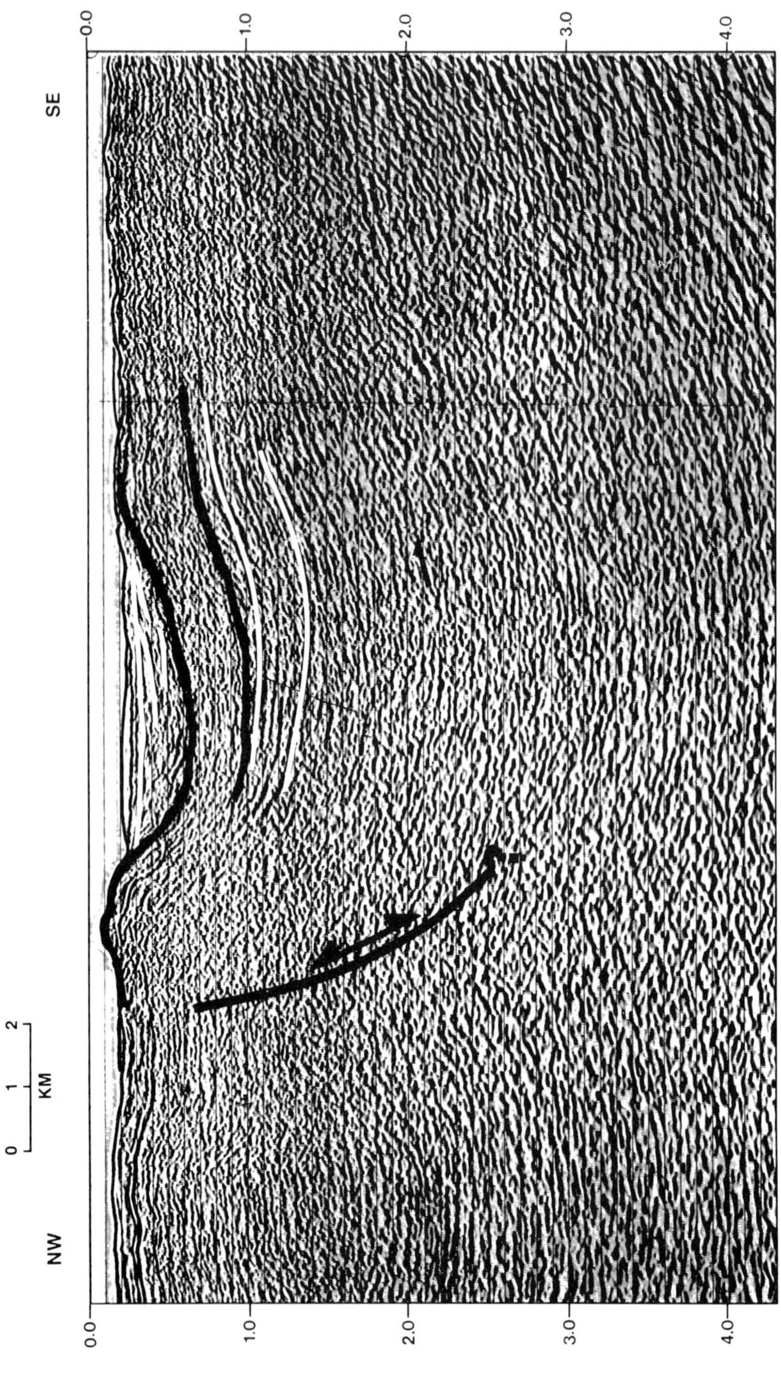

FIG. 13. Seismic section through an Oligocene non-marine basin in the Sea of the Hebrides (Smythe & Kenolty 1975). The black lines indicate the top and base of the basalts. The white lines are picks on pre-basalt horizons. Note the rotated onlap of the Oligocene sediments onto the top of the basalts, (section reproduced by kind permission of JEBCO).

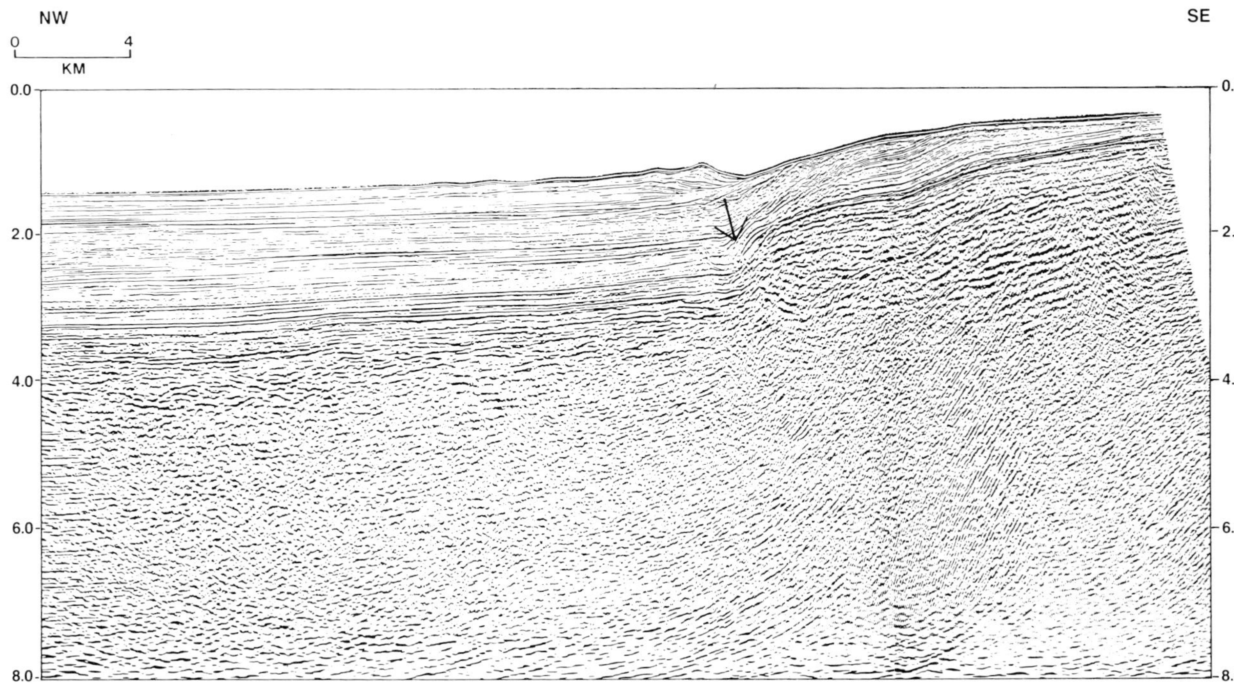

FIG. 14. Seismic section on the E margin of the Rockall Trough, N of St Kilda. The reflector shown by the arrow is the R4 reflector of Eocene–Oligocene age (Roberts 1975).

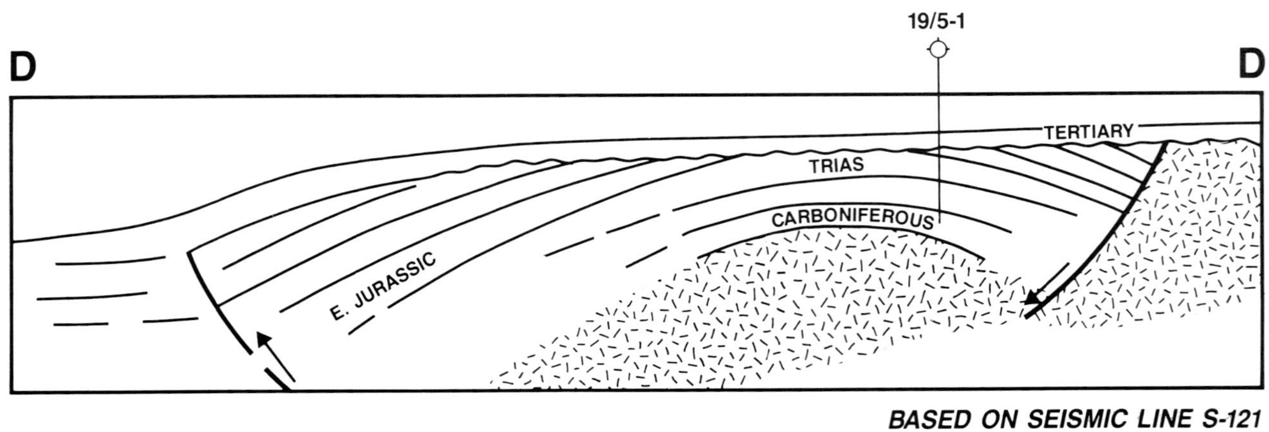

Fig. 15. Line drawing of a seismic section across an inversion structure on the continental shelf offshore Northwest Ireland. The stratigraphy is calibrated by well 19/5–1.

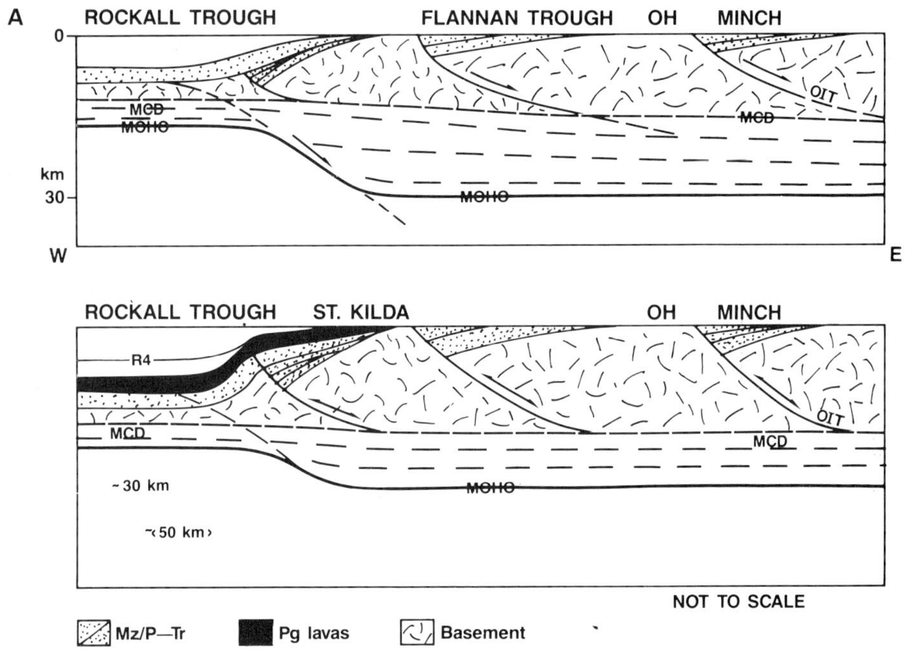

FIG. 16. Evolution of the Rockall Trough from the Late Cretaceous to the Early Miocene. OH, Outer Hebrides, MCD, Mid-Crustal Detachment. The figure illustrates the effects of Late Oligocene–Early Miocene inversion. Mz, Mesozic, P–Tr, Permo–Trias; Pg, Palaeogene.

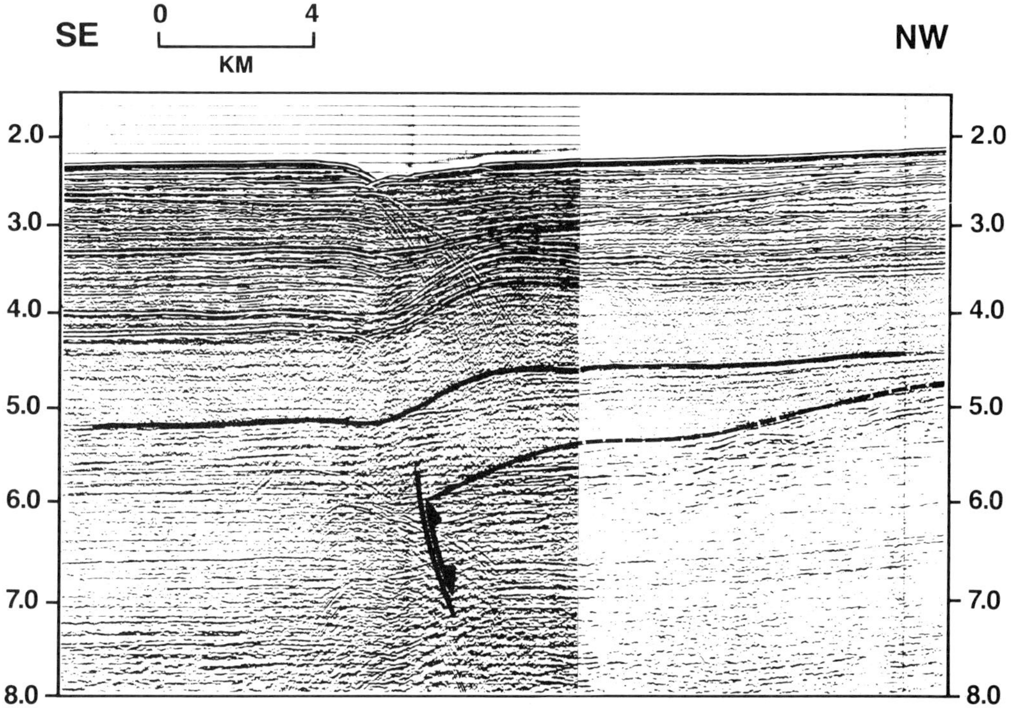

FIG. 17. Inversion structure developed on the S margin of the Porcupine Seabright–North Goban Spur (after Masson & Parson 1984; see also Sibuet *et ai.* 1985). Note the growth between the two picked reflectors. The line location is shown on Fig. 18.

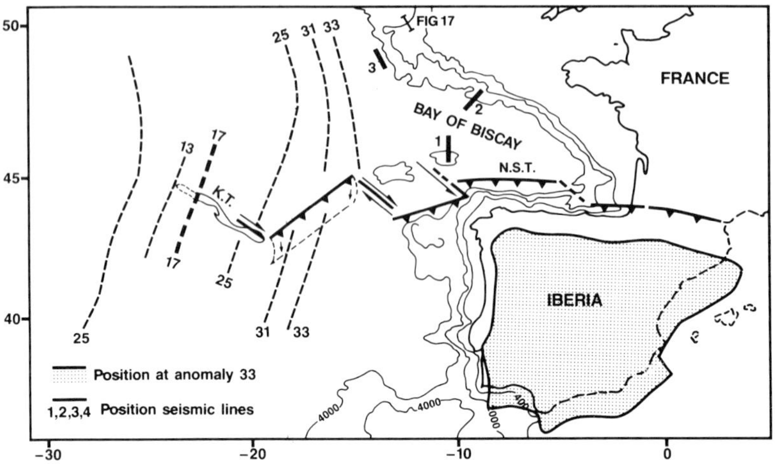

FIG. 18. Configuration of the Pyrenean Plate Boundary (after Grimaud *et al.* 1982). 1 and 2 are sections referred to in Montadert *et al.* 1975. 3 is a section referred to in Sibuet *et al.* (1975). KT, Kings Trough. NST, North Spanish Trough. The dashed lines are standard oceanic crust anomalies.

followed beyond the continent–ocean boundary to the oceanic crust. N of the Goban Spur for example (Fig. 17), overthrusting has displaced the oceanic basement and sediments no younger in age than Oligocene (Sibuet *et al.* 1985). Folding and faulting of this age can be traced across the entire width of the Bay of Biscay into the failed subduction zone (Fig. 18) of the North Spanish Trough (Montadert *et al.* 1975; Boillot *et al.* 1979).

The importance of these observations lies in the demonstration that coeval compression and inversion has affected both continental and oceanic parts of the plate. It should also be borne in mind that basin inversion in and around the British Isles has taken place at different times and has been focused in different areas (Ziegler 1987). Thus, the phase of Late Cretaceous–Early Tertiary inversion is most important in the Southern North Sea region apparently diminishing in importance westward; Oligocene inversion was of comparatively minor importance in these areas (Ziegler 1987). By contrast, the main Oligocene phase of inversion and compression was focused in basins of the Western British Isles and the Bay of Biscay.

A working hypothesis for the differences in age and geographic focus of the two phases can be developed from the histories of Alpine and Pyrenean collision in relation to extensional plate boundary (Rhine–Rhone graben) development in Western Europe. The Late Cretaceous–Early Tertiary inversion phase was driven by transmission of compressional stress through the European foreland during the early phases of Alpine collision (Ziegler 1982; 1987). This phase terminated in Eocene time probably when initiation of the extensional plate boundary along the Rhine–Rhone graben system prevented further transmission of stress. The next phase of inversion followed in response to subduction along the N margin of Biscay and collision in the Pyrenees. Initiation of a plate boundary between the European plate and Iberia plate in Late Cretaceous time led to counter clockwise rotation of Iberian plate causing collision in the Aquitaine Basin and subduction beneath the N margin of Iberia (Boillot *et al.* 1979; Grimaud *et al.* 1982). Suturing of the Iberian and European plates resulted in coupling of the plates allowing northward transmission of stress to invert the basins on the margin and beneath Western British Isles. Recent analysis of motions along this plate boundary (Fig. 18) extending from the King's Trough to the Pyrenees predict a broad diffuse region of compressive deformation known as the Porcupine plate (Srivastava & Tapscott 1986).

ACKNOWLEDGMENTS: I thank B.P. Exploration for permission to publish. JEBCO Seismic Ltd. kindly permitted publication of the section shown in Fig. 13. Discussions and input from R. Graham, M. Earle, K. Isaacs, R. Rosenthal, R. Cleverly, J. Brown, S. Petrie and I. R. Vann were invaluable. R. Graham critically reviewed an early draft of this paper.

References

BALLY, A. W. 1984. Tectogenese et sismique reflextion. *Bulletin de La Société Géologique de France*, **14(2)**, 279–285.

BIRPS & ECORS 1986. Deep seismic reflection profiling between England, France and Ireland. *Journal of the Geological Society of London*, **143**, 45–42.

BOILLOT, G., AUXIETRE, J. L., DUNAND, J. P., DUPEUBLE, P. A. and MAUFFRET, A. 1979. The North Western Iberian Margin: a Cretaceous passive margin deformed during the Eocene. *Maurice Ewing Series, American Geophysical Union*, **3**, 138–153.

British Geological Survey 1983a. Lundy Sheet 51°N–06°W, 1 : 250 000 Series Solid Geology.

—— 1983b. Lizard Sheet 49°N–06°W, 1 : 250 000 Series Solid Geology.

—— 1985. Lands End Sheet 50°N–06°W, 1 : 250 000 Series Solid Geology.

BROOKS, M., THAYNER, P. M. and TRIMBLE, T. J. 1988 Mesozoic reactivation of Variscan thrusting in the Bristol Channel area, U.K. *Journal of the Geological Society of London*, **145**, 439–444.

CHAPMAN, T. J. The Permian to Cretaceous structural evolution of the Western Approaches (Melville sub-basin) UK. *This volume*.

COOK, D. R. 1987. The Goban Spur–deep water frontier basin. *In*: BROOKS, J., & GLENNIE, K. W. (eds), *Petroleum Geology of North West Europe*, Graham and Trotman, London, 623–632.

CROKER, P. F. and SHANNON, P. 1987. Evolution and Hydrocarbon Prospectivity of the Porcupine Basin, Offshore Ireland, *In*: BROOKS, J. & GLENNIE, K. W. (eds), *Petroleum Geology of Northwest Europe*, Graham and Trotman, London, 633–642.

EVANS, C. D. R. and HUGHES, M. J. 1984. The Neogene successions of the South Western Approaches Basin, Great Britain. *Journal of the Geological Society of London*, **141**, 315–326.

GEORGE, T. N. 1967. Landform and structure in Ulster Scott. *Journal of Geology*, **3**, 413–448.

GILLCHRIST, R., COWARD, M. P., PECHER, A., MUNGIER, J. L. and TRUDGILL, B. Structural inversion in the external French Alps. *This volume*.

GLENNIE, K. W. and BOEGNER, P. L. E. 1981 Sole Pit inversion tectonics *In*: ILLING, L. V. and HOBSON G. D. (eds) *Petroleum and the Continental Shelf of North West Europe*, Institute of Petroleum, Heyden, 110–120.

GREEN, P. F. 1986. On the thermal tectonic evolution of Northern England: evidence from Fission track analysis. *Geological Magazine*, **123(5)**, 493–506.

GRIMAUD, S., BOILLOT, G., COLLETE, B. J., MAUFFRET, A., MIKS, P. R., & ROBERTS, D. G. 1982. Western extension of the Iberian–European plate boundary during the early Cenozoic (Pyrenean) convergence: a new model. *Marine Geology*, **45**, 63–77

HIGGS, K and BEESE, A. P. 1986. A Jurassic microflora from the Colbond Clay of Cloyne, County Cork, *Irish Journal of Earth Sciences*, **7**, 99–109.

VAN HOORN, B. 1987a. The South Celtic Sea – Bristol Channel Basin; origin, deformation and inversion history. *Tectonophysics*, **1376**, 309–434.

—— 1987b. Structural evolution, timing and tectonic style of Sole Pit inversion. *Tectonophysics*, **137**, 239–284.

Institute of Geological Sciences 1970. Geological map of Northern Ireland.

JACKSON, D. I. MULHOLLAND, P., JONES, S. M. & WARRINGTON, G. 1987. The geological framework of the East Irish Sea Basin. *In*: BROOKS, J. & GLENNIE, K. (eds) *Petroleum Geology of Northwest Europe*, Graham & Trotman, London, 191–204.

KAMERLING, P. 1979. Geology and hydrocarbon habitat of the Bristol Channel Basin. *Journal of Petroleum Geology*, **2**, 75–63.

KENT, P. E. 1968. The buried floor of eastern England. *In*:-SYLVESTER-BRADLEY, P. C. & FORD, T. D. (eds) *The Geology of the East Midlands*, Leicester University Press, Leicester, 138–148.

—— 1980. Subsidence and uplift in East Yorkshire and Lincolnshire: a double inversion *Proceedings of the Yorkshire Geological Society*, **42**, 505–524.

KOHN, B. P. & EYAL, M. 1981. History of uplift in the crystalline basement of Sinai and its relation to the opening of the Red Sea as revealed by Fission track dating of apatites. *Earth Planetary Sciences Letters*, **52**, 129–141.

LLOYD, A. J., SAVAGE, R. J. G., STRIDE, A. H. B. & DONOVAN, D. T. 1973. The geology of the Bristol Channel floor. *Philosophical Transactions of the Royal Society of London*, **274**, 595–626.

MACDONALD, H., ALLAN, P. M. & LOVELL J. P. B. 1987. Geology of oil accumulation in Block 26/28, Porcupine Basin, Offshore Ireland *In*: BROOKS, J. & GLENNIE, K. W. (eds) *Petroleum Geology of North West Europe*, Graham & Trotman, London, 643–651.

MASSON, D. G. & KIDD, R. B. 1987. Revised Tertiary Seismic stratigraphy of the Southern Rockall Trough. *In*: RUDDIMAN, W. B. and KIDD, R. B. *et al.* (eds) *Initial Reports DSDP, No. 94*, U.S. Government Printing Office, Washington, D.C., 1117–1128.

—— & PARSON, L. M. 1984. Eocene deformation on the continental margin southwest of the British Isles, *Journal of the Geological Society of London*, **140**, 913–920.

——, MONTADERT, L. and SCRUTTON, R. A. 1985. Regional Geology of the Goban spur continental margin *In*: DE GRACIANSKY, P. C. & POAG, C. W. *et al.* (eds) *Initial Reports DSDP*, No. 80, U.S. Government Printing Office, Washington 163–173.

MONTADERT, L., WINNOCK, E., DELTEIL, J. R. and GRAN, G. 1975. Continental margins of Galicia Portugal and the Bay of Biscay. *In*: BURK, C. A. & DRAKE, C. L. (eds) *The Geology of Continental Margins*, Springer-Verlag, New York, 323–343.

MOSELEY, F. 1972. A tectonic history of Northwest England, *Journal of the Geological Society of London*, **128**, 561–598

NAYLOR, D. and SHANNON, P. 1982. *The Geology of Offshore Ireland and West Britain*. Graham and Trotman, London, 161 pp.

ROBERTS, D. G., 1975. Marine Geology of the Rockall Plateau and Trough, *Philosophical Transactions of the Royal Society of London*, **A278**, 447–509.

——, BOTT, M. P. P. & URUSKI, C. 1983. Structure and Origin of the Wyville–Thomson Ridge. *In*: BOTT, M. H. P., SAXON, S., TALWANI, M. & THIEDE, J. (eds), *Structure and Evolution of the Greenland, Iceland, Faroe Ridge* Plenum Press, New York, 133–159.

——, GINZBERG, A., NUNN, K. T. & MCQUILLIN, R. 1988. Deep crustal structure of the Rockall Trough from seismic reflection and refraction observations, *Nature*, **332**, 632–634.

——, MASSON, D. G. & MONTADERT, L. 1981 Continental margin from the Porcupine seabight to the Amorican marginal basin. *In*: ILLING, L. V. & HOBSON, G. D. (eds) *Petroleum Geology of Northwest Europe*, Institute of Petroleum, Heyden, 455–473.

——, MORTON, A. C. & BACKMAN, J. 1984. Late Palaeocene-Eocene volcanic events in the Northern North Atlantic Ocean. *In*: ROBERTS, D. G. & SCHRUTKER, D. (eds) *Initial Reports, DSDP, No. 81*, 913–924.

ROBINSON, K. W., SHANNON, P. & YOUNG, D. G. G. 1981. The Fastnet Basin: an integrated analysis *In*: ILLING, L. V. & HOBSON, G. D. (eds) *Petroleum Geology of the Continental Shelf of Northwest Europe*, Institute of Petroleum, Heyden, 444–454.

SIBUET, J. C. *et al*. 1985. Marphology and basement structures of the Goban Spur continental margin (North Eastern Atlantic) and the role of the Pyrenean Orogeny *In*: DE GRACIANSKY, P. C. & POAG, C. W. *et al*. (eds) *Initial Reports DSDP, No. 81*, Washington, D.C. U.S. Government Printing Office, Washington D.C., 1153–1165.

SIMPSON, I. R., GRAVESTOCK, M., HAM D., LEACH, H. and THOMPSON S. D. Notes and cross-sections indicating inversion tectonics of the Wessex Basin, *This volume*.

SMYTHE, D. K. and KENOLTY, N. 1975. Tertiary sediments in the Sea of the Hebrides *Journal of the Geological Society of London*, **131**, 227–233.

SRIVASTAVA, S. & TAPSCOTT, C. 1986. Plate kinematics of the North Atlantic. *In*: TUCHOLKE, B. & VOGT, P. R. (eds) *Geology of North America. The Western Atlantic Region, The Decade of North American Geology, Geological Society of America*, Boulder, 379–404.

STECKLER, M. S. 1985. Uplift and extension at the Gulf of Suez: indications of induced mantle convection. *Nature*, **317**, 135–139.

STRIDE, A. H. B., CURRAY, J. R., MOORE, D. G. & BELDERSON, R. H. 1969. Marine geology of the Atlantic continental margin of Europe. *Philosophical Transactions of the Royal Society of London*, **264A**, 31–75.

TUCKER, R. M. & ARTER, G. A. 1987. Tectonic evolution of the North Celtic Sea and Cardigan Bay Basins with special reference to basin inversions. *Tectonophysics*, **137**, 291–308.

WALSH, P. T. 1966. Cretaceous outliers in Southwest Ireland and their implications for Cretaceous palaeogeography *Journal of the Geological Society of London*, **122**, 63–84.

——, BOULTER, M. C., IJTABA, M. & URBANI, D. M. 1972. The preservation of the Neogene Brassington formation of the Southern Pennines. *Journal of the Geological Society of London*, **128**, 518–560.

WILSON, H. E. & ROBBIE, J. A. 1966. Geology of the country around Ballycastle, *Memoir of the Geological Survey of Northern Ireland*. HMSO, London, 370 pp.

WOOD, M. V., HALL, J. & VAN HOORN, B. 1987. Post-Mesozoic differential subsidence in the north east Rockall Trough related to volcanicity and sedimentation. *In*: BROOKS J. & GLENNIE, K. W. (eds) *Petroleum Geology of Northwest Europe*, Graham & Trotman, London, 677–685.

ZIEGLER, P. A. 1982. *Geological Atlas of Western and Central Europe*. Elsevier, Amsterdam, 130 pp.

—— 1983. Inverted Basins in the Alpine Foreland. *In*: BALLY, A. W. (ed.) *Seismic Expression of Structural Styles*, American Association of Petroleum Geologists Studies in Geology Series, **15**, 3.3.3–3.3.12.

—— 1987. Late Cretaceous and Cenozoic intraplate compressional deformations in the Alpine foreland – a geodynamic model, *Tectonophysics*, **137**, 389–420.

D. G. ROBERTS, British Petroleum PLC, Britannic House, Moor Lane, London EC2Y 9BU, UK.

Inversion on the European Continental Shelf

The kinematics of inversion in the Danish Central Graben

J. A. Cartwright

SUMMARY: The Danish Central Graben is a complex mosaic of half-grabens and structural highs that formed during Permo–Triassic and Upper Jurassic rift phases at the confluence of the NW–SE and N–S trending elements of the Central North Sea Rift. The extensional structures forming the margins to the rift were reactivated in a reverse sense in the Late Cretaceous as part of the regional inversion of basins in the southern North Sea. Deposition proceeded throughout the deformation, and thickness variation in the deep marine Chalk succession is directly attributable to compressional deformation of the basin floor. The total shortening across the rift amounts to only a few percent, and the structural expression of the reactivated margins is typically as low amplitude folds with minor associated reverse faulting. Compressional reactivation was restricted to basement faults, and the shallow inclination of the margin faults may have been an important factor in their subsequent reactivation. Zechstein evaporites and overpressured Jurassic shales influenced the development of some of the compressional structures. The orientation of the growth fold axes is consistent with a net convergence in an approximately NE–SW direction. The fold axes are arranged in a segmented pattern which can be related to the underlying basement fabric. This segmentation not only resulted in variations in the amounts and style of shortening along the length of the rift, but seems to have influenced the timing of the deformation. Seismic–stratigraphic analysis of the Upper Cretaceous sequence indicates that the compressional deformation was markedly diachronous. Abrupt changes in the deformational chronology are found to occur across the transverse segment boundaries.

The Danish Central Graben was inverted in Late Cretaceous times against a background of continuous deep marine sedimentation along the axis of a former rift basin. The stratigraphy of the Upper Cretaceous Chalk in this area provides a unique record of the interplay between depositional and deformational processes in force during the inversion. Many of the other inverted rift basins of Northwest Europe were more intensely deformed. Consequently, potential stratigraphic markers of the deformation were either never deposited or were subsequently exhumed and eroded.

The aim of this study was to analyse the style, distribution and timing of deformation through a combination of structural and seismic–stratigraphic interpretation of a regional grid of 5000 km of high resolution reflection seismic data. Particular attention has been paid to the documentation of the structural geometries and evolution of early formed extensional structures. A subsidiary objective was to establish how the compressional deformation influenced Chalk depositional systems. The paper concludes by examining existing dynamic models for the regional inversion of basins in Northwest Europe in the light of observations of the deformation in the Danish Central Graben.

Structural framework

In order to attempt a kinematic analysis of the Late Cretaceous inversion of the Danish Central Graben (DCG), it is important to consider the structural evolution during the preceding extensional phases of basin development. A detailed understanding of the extensional structural geometries allows variations in structural style that relate to regional dynamics to be distinguished from those due to local variations in fault geometry or in the mechanical competence of the graben fill.

The Danish Central Graben is a complex mosaic of half-grabens and structural highs that formed at the confluence of the NW–SE trending UK Central Graben and the N–S trending Dutch Central Graben. The abrupt change in orientation of the Central North Sea Rift has been attributed by Cartwright (1987), to the influence of WNW-trending splays of the Tornquist Zone. The two main Mesozoic depocentres in the Danish Central Graben are

From, Cooper, M. A. & Williams, G. D. (eds), 1989, *Inversion Tectonics*
Geological Society Special Publications No. 44, 153–175.

the Feda Graben and the Tail End Graben (Fig. 1). A third, less important depocentre is the Outer Rough Basin, lying on the NE flank of the Mid North Sea High. The Tail End Graben is a NNW-trending series of half-graben segments which display rotational thickening of Triassic and Jurassic sequences against the Coffee Soil Fault. The Feda Graben is, by contrast, more symmetric and is bounded on both margins by WNW-trending basement normal faults. It is divided into two sub-basins by an axial fault zone, the Gert Ridge (Figs 1 & 2).

Depth conversions of the 7 s two-way travel time seismic data show the major bounding faults of both grabens to be planar at depth and inclined between 30 and 40 degrees to the horizontal. The throws on these faults are large, up to 8000 m in the case of the Coffee Soil Fault. The structural geometry of these large basement structures is simple, and the degree of associated fracturing of the basement and sedimentary fill is minor (Figs 2, 3 & 4). Complications do, however, occur along the Coffee Soil Fault at several positions along its length where the fault plane undergoes abrupt changes of strike (Fig. 1). These changes of strike or offsets of the boundary fault are related to a segmentation of the Tail End Graben by WNW and NE-trending transverse structural zones (Cartwright 1987).

More complex structural geometries have been interpreted from both the Olaf Fault Zone, which forms the NE margin of the Outer Rough Basin, and the Jens–Otto Fault Zone, which forms the western margin of the Tail End Graben (Figs 3 & 4). The complexity of the Olaf Fault Zone arises from the superposition of two *en échelon* normal fault suites of Upper Jurassic and Upper Palaeozoic age which dip in opposite directions. The interference of the two fault trends led to the development of complex crossing conjugate structural geometries (Figs 2 & 3, also see Horsfield 1980).

The Jens–Otto Fault Zone is a fractured monoclinal hinge which developed initially as a series of broadly NNW trending intra-basinal normal faults and subsequently acted as a fulcrum for the eastward rotation of the Tail End Graben. Stretching over the rotational hinge was alleviated by the development of crestal collapse structures. The along-strike arrangement of the conjugate faults that comprise the crestal collapse structures is highly variable. This variation has been shown to be related to the differences in the degree and style of rotational subsidence along the length of the Tail End Graben (Cartwright 1987). Abrupt offsets and changes in structural style or polarity of the dominant 'root' faults are linked to similar sense offsets or changes in the orientation of the Coffee Soil Fault by transverse fault zones of WNW and NE trend (Fig. 1).

Extensional basin development

Rifting in the DCG commenced in the Late Carboniferous (Cartwright 1988). The onset of rift formation was coeval with a major phase of dextral shear across Northwest Europe and a phase of widespread extrusive magmatism (Arthaud & Matté 1977). The depositional limits of the Rotliegend and the Zechstein were determined by the rift-bounding faults that transected the E–W trending Ringkobing–Fyn and Mid North Sea Highs. Although basement faults were active across the crest of this major high trend, the underlying influence of the highs was expressed in the distribution of basinal and marginal Zechstein lithofacies (Gowers & Saeboe 1985). The restriction of the mobile basinal facies Zechstein to the extreme north and south of the DCG had a significant influence on the geometries of the subsequent basin inversion.

Two distinct phases of rifting are recognized in both the Feda and Tail End Grabens. The Permo–Triassic phase was characterized by dispersed and moderate extension on the rift-bounding faults and numerous intra-rift faults, including those along the Jens–Otto Fault Zone (Fig. 1). The floors of the broad rift axis subsided horizontally and slowly and sedimentation rates matched subsidence rates. This pattern of subsidence resulted in the parallel and continuous reflection signatures observed throughout both grabens.

This first phase of rifting ended in the Early Jurassic. However, there was no major uplift of the rift during the tectonic hiatus preceding the second, Middle to Late Jurassic, phase of rifting. The hiatus between the two rift phases is marked by a prominent sequence boundary, known as the Base Middle Jurassic. There was only minor local erosion at this boundary and this occurred mainly over low amplitude salt swells.

The Middle to Late Jurassic phase of rifting was marked by rapid subsidence and pronounced rotation of the graben floors. In the Tail End Graben subsidence was controlled by the Coffee Soil Fault. This resulted in the development of divergent stratal configurations in the Middle and Upper Jurassic sequences. The step-faulted terrace along the Jens–Otto Fault Zone became the fulcrum for the rotation of the

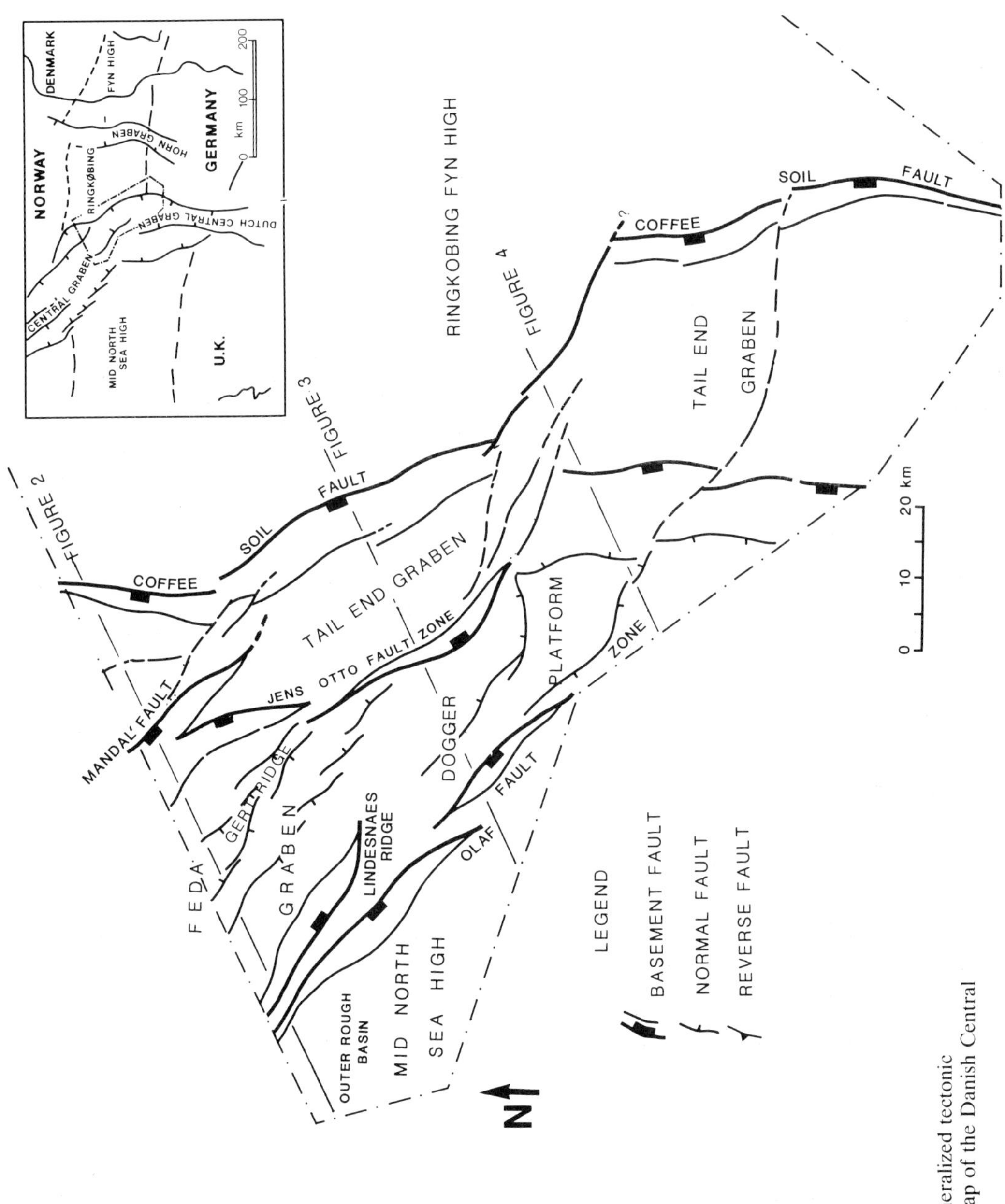

FIG. 1. Generalized tectonic elements map of the Danish Central Graben.

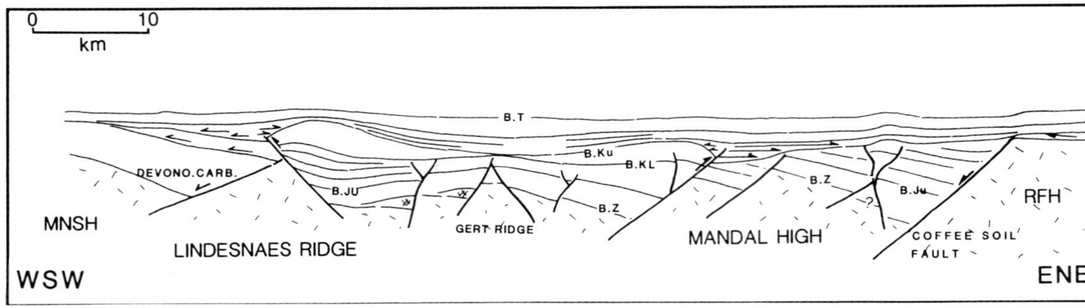

FIG. 2. Schematic cross-section of the Feda Graben. The Upper Cretaceous sequence is amplified to clarify thickness variations across inverted basin margins. Note the relative simplicity of the basement faults. BZ., Base Zechstein; B Ju, Base Upper Jurassic; BV., Base Volgian; BKl., Base Lower Cretaceous; BKu., Base Upper Cretaceous; BT, Base Tertiary. Approximately true scale.

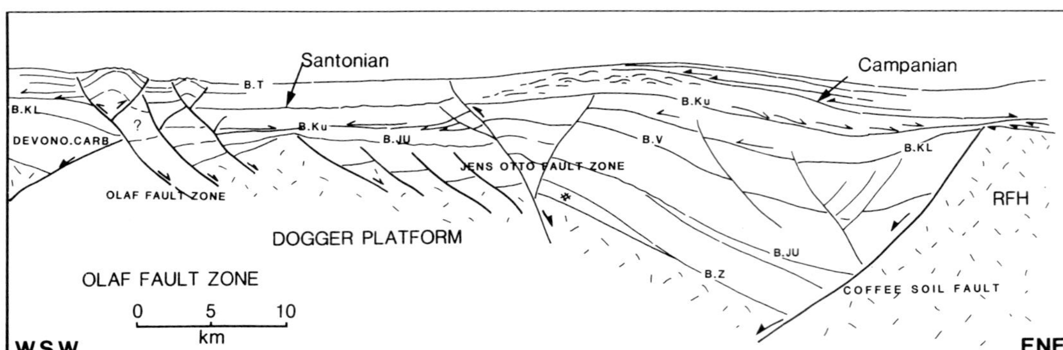

FIG. 3. Schematic cross-section, Outer Rough Basin–Central Tail End Graben. Contrast inversion geometries with Fig. 2. No reactivation of Coffee Soil Fault–Lower Cretaceous and Lower Chalk onlap the fault scarp. Approximately true scale.

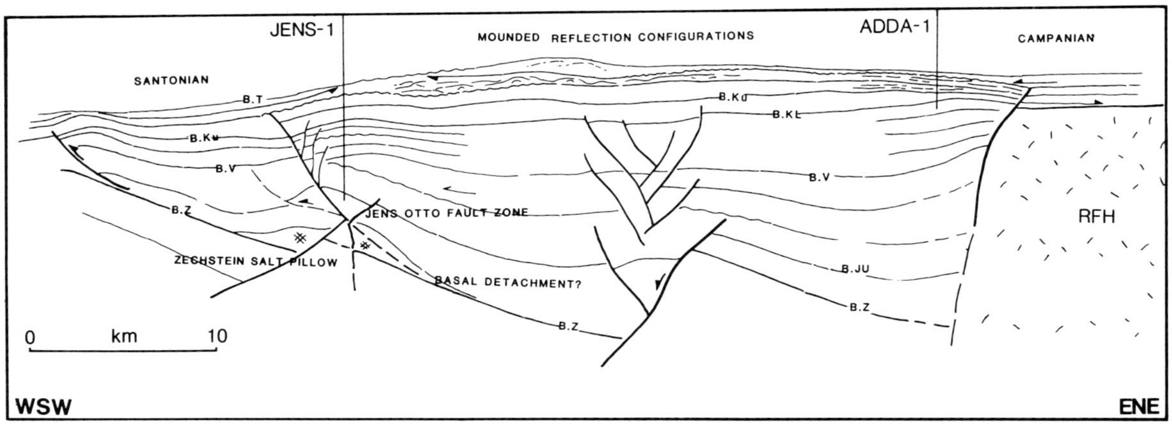

FIG. 4. Schematic cross-section, Central–Southern Tail End Graben. Note possible influence of mobile Zechstein evaporites in reversed polarity of shallow inversion structure on western margin relative to deeper structure. Approximately true scale.

Tail End Graben. It separated the stable Dogger Platform from the rapidly subsiding graben. The Gert Ridge served in a similar capacity to separate the two oppositely rotating segments of the Feda Graben (Fig. 1).

The segmented structure of the Tail End Graben exerted a profound influence on the depositional systems during the Jurassic phase of rifting. Rates of subsidence varied from segment to segment. The combination of WNW and NE-trending faults which formed the transversely oriented segment boundaries functioned to allow the differential rotations at segment boundaries. The torsion expressed at the segment boundaries was greatest at the N and S margins of the central two segments. These two central segments were the locus for extremely rapid subsidence and deposition. Up to 5000 m of Upper Jurassic shales accumulated in approximately twenty million years. One possible consequence of such rapid sedimentation rates of the marine shale sequences may have been the development of abnormal pore fluid pressures. Wells penetrating the Upper Jurassic in the Tail End Graben have encountered only a few permeable units interbedded in the shaly sequences. There would therefore have been only limited avenues for fluid expulsion during compaction and burial. The presence of considerable volumes of undercompacted shale in the graben would have had important consequences for the later compressional phase of basin development in that shortening of the graben fill could have been expressed in the form of tectonic compaction rather than as fault reactivation.

The base of the Lower Cretaceous sequence is a prominent sequence boundary. It displays a strong basal onlap but little or no erosional truncation. This sequence boundary is often referred to as the Late Cimmerian Unconformity in the central North Sea (Rawson & Riley 1982). It has been attributed to the onset of Lower Cretaceous rifting both in this area (Michelsen & Andersen 1983), and over Northwest Europe (Ziegler 1982). In the DCG, however, the widespread onlap of truncated Upper Jurassic fault scarps by Lower Cretaceous marine shales implies that active basement faulting had effectively ceased by the end of the Jurassic. In the DCG at least, it is clear that the Late Cimmerian Unconformity should more properly be regarded as the event marking the termination of rifting. There is no evidence for any regional phase of uplift at the Jurassic/Cretaceous boundary. Minor fault displacements are mapped at Lower Cretaceous structural levels but these are more likely to have resulted from local differential compaction or Zechstein salt withdrawal than from any basement response to the regional rifting. Isopach maps (Cartwright 1987; Frandsen *et al.* 1987) show that Early Cretaceous depocentres are located directly above areas of maximum accumulation of Upper Jurassic shales. This is consistent with a compactional mechanism for the minor faulting. The transition from active rifting in the Jurassic to passive regional subsidence of the rift and its shoulders occurred over a long period. Overstep of the rift margins commenced in the Santonian, some 50 million years after the last definitive basement fault activity.

Late Cretaceous inversion

The Late Cretaceous evolution of the DCG was predominantly influenced by a regional mechanism of subsidence that progressively incorporated greater and greater areas of the rift shoulders. Superimposed on this regional subsidence was a compressional deformation that reactivated elements of the extensional fabric of the rift. This led to the development of reverse faults and broad, low amplitude anticlines along the rift margins.

The scale of the Late Cretaceous inversion of the Danish Central Graben is small when compared with that described from other basins subjected to the same compressional event, such as the Sole Pit or the West Netherlands Trough (Ziegler 1982). The total shortening across the DCG amounted to only a few percent, with the maximum structural relief on inversion related structures only amounting to 500 m. The combination of the moderate degree of compressional deformation and uplift and the prevalent regional basin subsidence centred on the DCG ensured that the stratigraphic record of the Upper Cretaceous Chalk was relatively complete. The stratigraphy therefore holds a comprehensive record of the compressional deformation. The DCG is one of the rare examples of an inverted basin where seismic–stratigraphic techniques can be employed in the structural analysis. Thickness relationships in the Chalk of the DCG (Fig. 6) are the product of three factors: the regional subsidence pattern, the syn-depositional deformation of the sea floor and the relative contributions to total sediment supply of allochthonous and autochthonous chalks.

The basis of the seismic–stratigraphic approach to the structural analysis lies in the extraction of the compressional signature from the convolved response through a filtering of the other parameters. Once the regional subsidence and local depositional patterns are acounted for any features that remain should be primarily related to the deformation. The next section deals with the standard description and analysis of the inversion structures. Thereafter, the emphasis of this chapter shifts to the application of the seismic–stratigraphic interpretation in the refinement of the structural analysis.

Geometries of the inversion structures

The main axes of inversion are simplified from the Upper Cretaceous structural contour and Upper Cretaceous isochore maps (Figs 5 & 6), and illustrated in Fig. 7. Compressional reactivation was localized along the SW and NE margins of the Feda Graben, along the NE margin of the Outer Rough Basin and along the Jens–Otto Fault Zone (Figs 1, 5 & 6). The inversion axes are recognizable as linear arrangements of reverse faults and anticlines expressed at Cretaceous structural levels, which closely follow the orientation and position of the extensional fabric of the rift (c.f. Figs 1 & 7).

Feda Graben

The two basement faults bounding the Feda Graben were reactivated to similar degrees to produce a crude basin-wide symmetry at Cretaceous structural levels (Fig. 2). Growth anticlines developed on both margins. These are low amplitude, asymmetric folds associated with minor thrust faults that propagated from the tips of the extensional faults. The Chalk thins towards the crest of the folds from both limbs. The patterns of thinning and onlap onto the folds indicate that the structures were expressed as topographic features on the sea floor. The structural relief on the folds is greatest at the Base Cretaceous sequence boundary (Figs 8A & B). On both margins, the steeper limbs of the growth folds face away from the basin axis. Reverse throws on the reactivated basement faults are small. Pre-Cretaceous structural markers show a net normal stratigraphic offset from footwall to hanging wall.

The Tail End Graben

Compressional reactivation in the Tail End Graben was also confined to the graben margins. However, in contrast to the symmetric disposition of fold/thrust structures in the Feda Graben, the arrangement of inversion-related structures in the Tail End Graben is highly asymmetric. This is not so surprising when the asymmetry of the extensional structure of the Tail End Graben is taken into account. What is perhaps unusual, is that the most dramatic reactivation took place not along the major basement discontinuity of the Coffee Soil Fault, but along the complexly fractured hinged margin of the Jens–Otto Fault Zone (Figs 3, 4 & 7). The Coffee Soil Fault was only reactivated to a minor extent in the southern part of the Tail End Graben, where low amplitude folds are mapped in association with minor reverse faulting at Cretaceous structural levels (Figs 4, 5 & 6). In the northern part of the Tail End Graben, the structure at the upper tip of the Coffee Soil Fault is a simple compactional monocline (Figs 3 & 5). It is evident that the fault plane remained dormant throughout the regional compressional deformation.

The geometry of the inversion structures along the Jens–Otto Fault Zone is similar to that described for the margins of the Feda Graben (Fig. 9). A representative seismic line showing both the quality of the seismic data base and the typical structural expression of the inversion in the Central Tail End Graben is illustrated in Fig. 10. The shortening across the Tail End Graben was taken up by the development of pop-up growth folds whose spatial arrangement and degree of symmetry were controlled by the segmented fracture patterns of the underlying hinge structure. The segmentation of the hinge was a direct consequence of the segmentation of the graben as whole. The structural relief of the growth folds varies considerably along the strike of the Jens–Otto Fault Zone, reaching a maximum value of some 500 m. The style of the folds also changes along strike, from W-verging asymmetric folds, to symmetric forms and then to E-verging asymmetric forms. Where the pop-up fold has a large amplitude, it is invariably observed that the reverse throw on the reactivated crestal faults of the underlying hinge structure are small. However, where the amplitude of the fold is small, there is a complimentary increase in the amount of reverse throw. In general, the vergence of the inversion structure follows that of the principal 'root' fault of the hinge structure.

An exception to this generalization of the dependent relationship between the extensional and compressional structural geometry can be seen at the southern end of the Jens–Otto

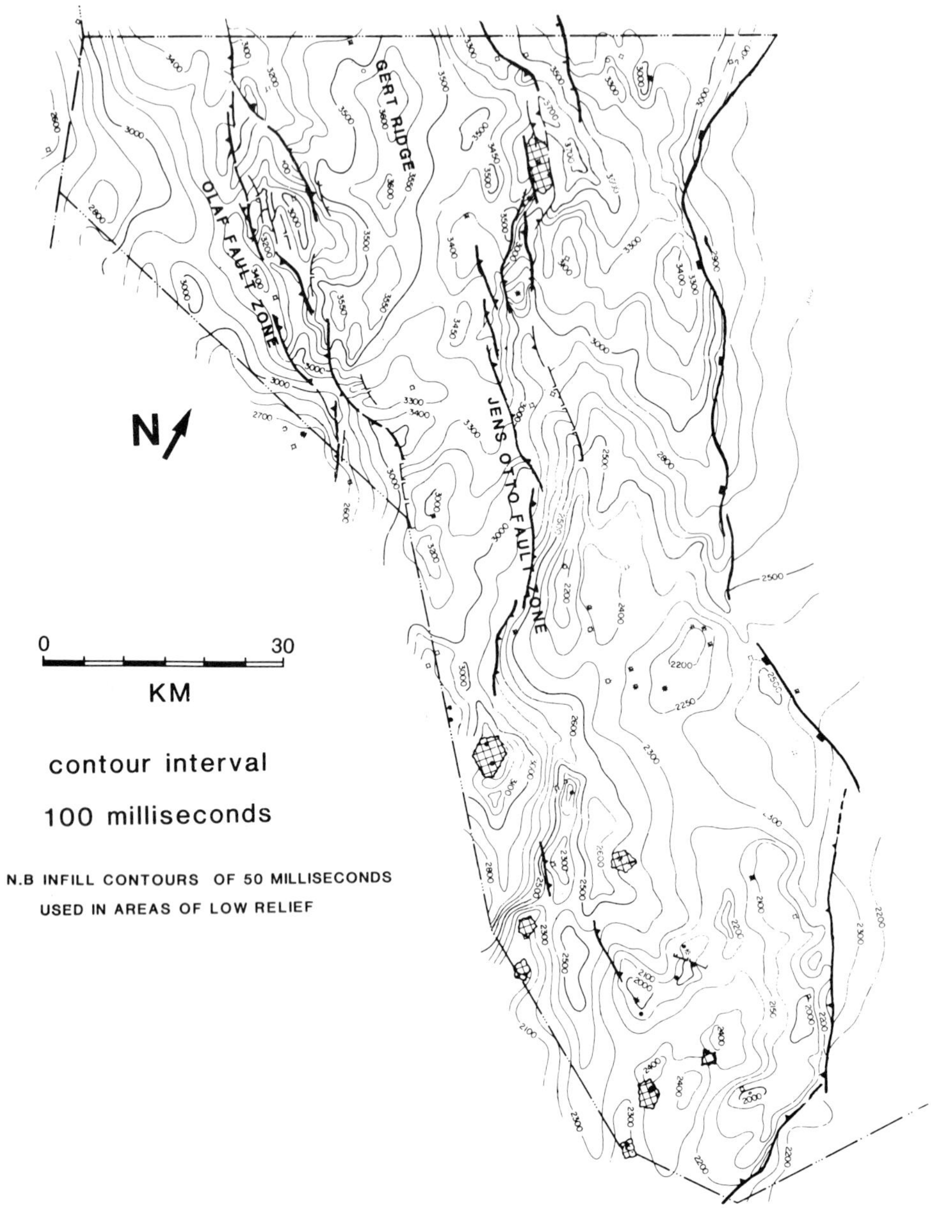

FIG. 5. Base Upper Cretaceous structural contour map. Main inversion axes located along graben margins are characterized by discontinuous arrays of reverse faults and ellipsoidal anticlines.

Fault Zone, where the W-verging asymmetric Late Cretaceous fold represents shortening against the grain of the W-dipping 'root' fault (Figs 4 & 9B). This departure from simple reactivation behaviour is attributed, in this case, to the lubricating effects of a Zechstein salt pillow which straddles the deeper part of the hinge structure. A thrust is interpreted to have cut across the trace of the 'root' fault, and to have soled out into the mobile Zechstein salt layer. The development of a thin-skinned inversion structure in this locality was facilitated by the earlier (Late Jurassic) tilting of the potential detachment horizon into an ideal inclination for

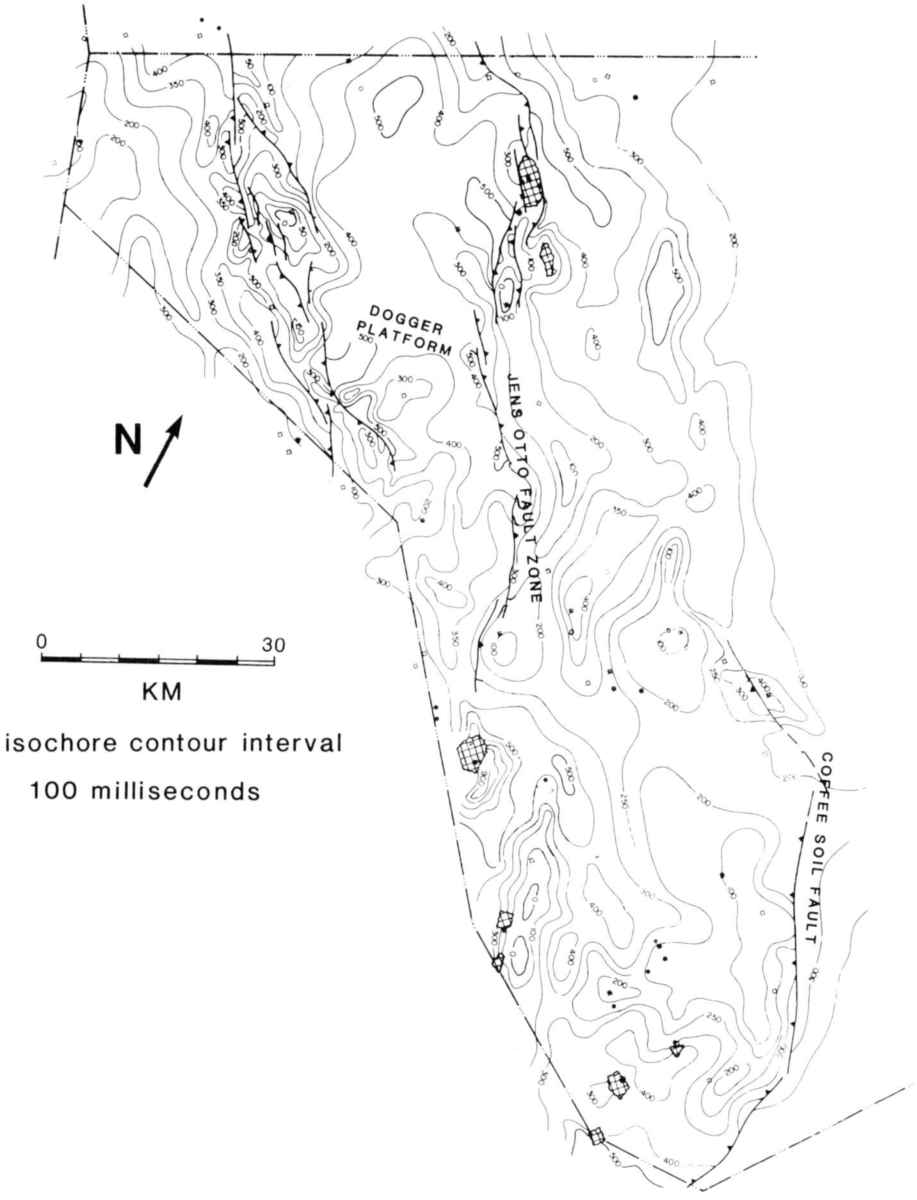

FIG. 6. Upper Cretaceous isochore map (thickness in time measured normal to bedding). Syndepositional fold axes are marked by pronounced contour minima and maxima, often seen to offset abruptly or change strike at zones of transverse contour lineations.

subsequent compressional evisceration of the graben fill.

The position and orientation of fold axes along the Jens–Otto Fault Zone are directly related to the segmented nature of the earlier extensional structure of the graben. The style, polarity and orientation of the hinge structure change abruptly across the transversely intersecting segment boundaries. Abrupt changes in the vergence or amplitude of the pop-up folds along the Jens–Otto Fault Zone are found to occur across precisely the same intersections. The transverse structures were not themselves reactivated at the surface in a compressional or

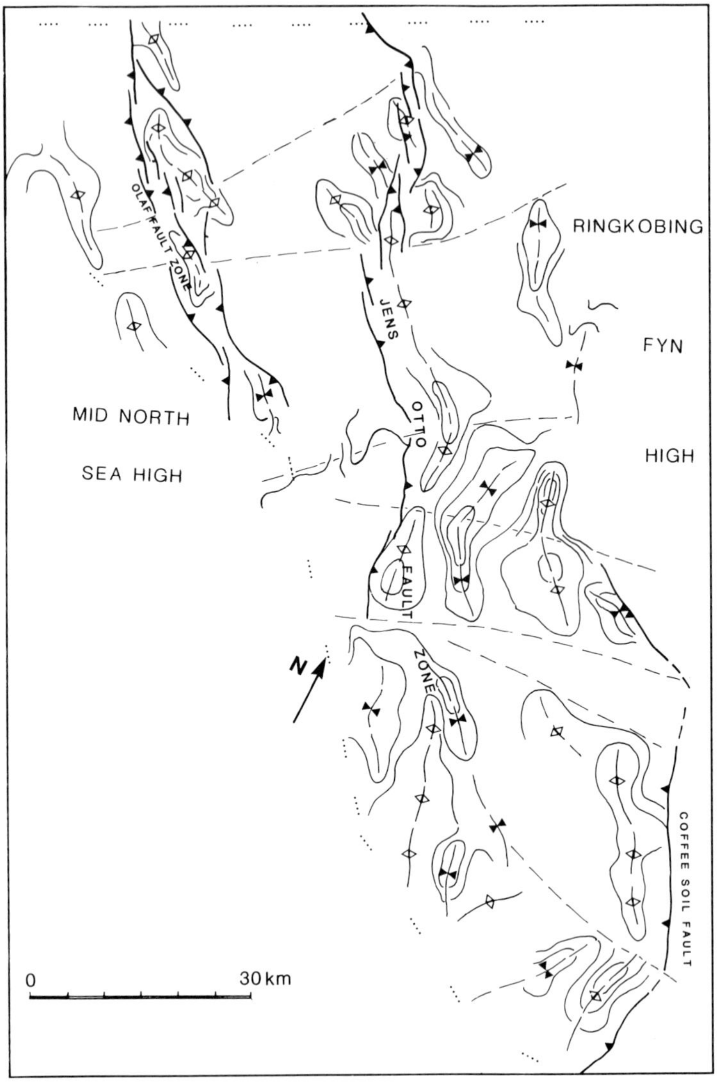

FIG. 7. Transverse lineaments and inversion axes. Upper Cretaceous growth folds form a segmented pattern related to the deeper structural fabric. Within each segment, the axes are aligned uniformly, but change strike or are offset along linear traces recognizable in Fig. 6 and which correspond to transverse fault zones in the basement.

transcurrent sense during the inversion. However, from the distribution of fold axes it is evident that their influence on the Late Cretaceous deformation was nonetheless important (Fig. 7). The segment boundaries acted to partition the compressional strains, although the precise manner by which this partioning was effected is far from clear. The presence of subvertical discontinuities transecting the graben at a high angle to its axis would, nevertheless, represent a severe constraint on the longitudinal distribution of the bulk shortening across the graben. The relationship between the segmentation of the rift structure and the segmentation or partitioning of the compressional reactivation is clarified by a detailed investigation of the timing of the deformation using seismic–stratigraphic interpretation techniques.

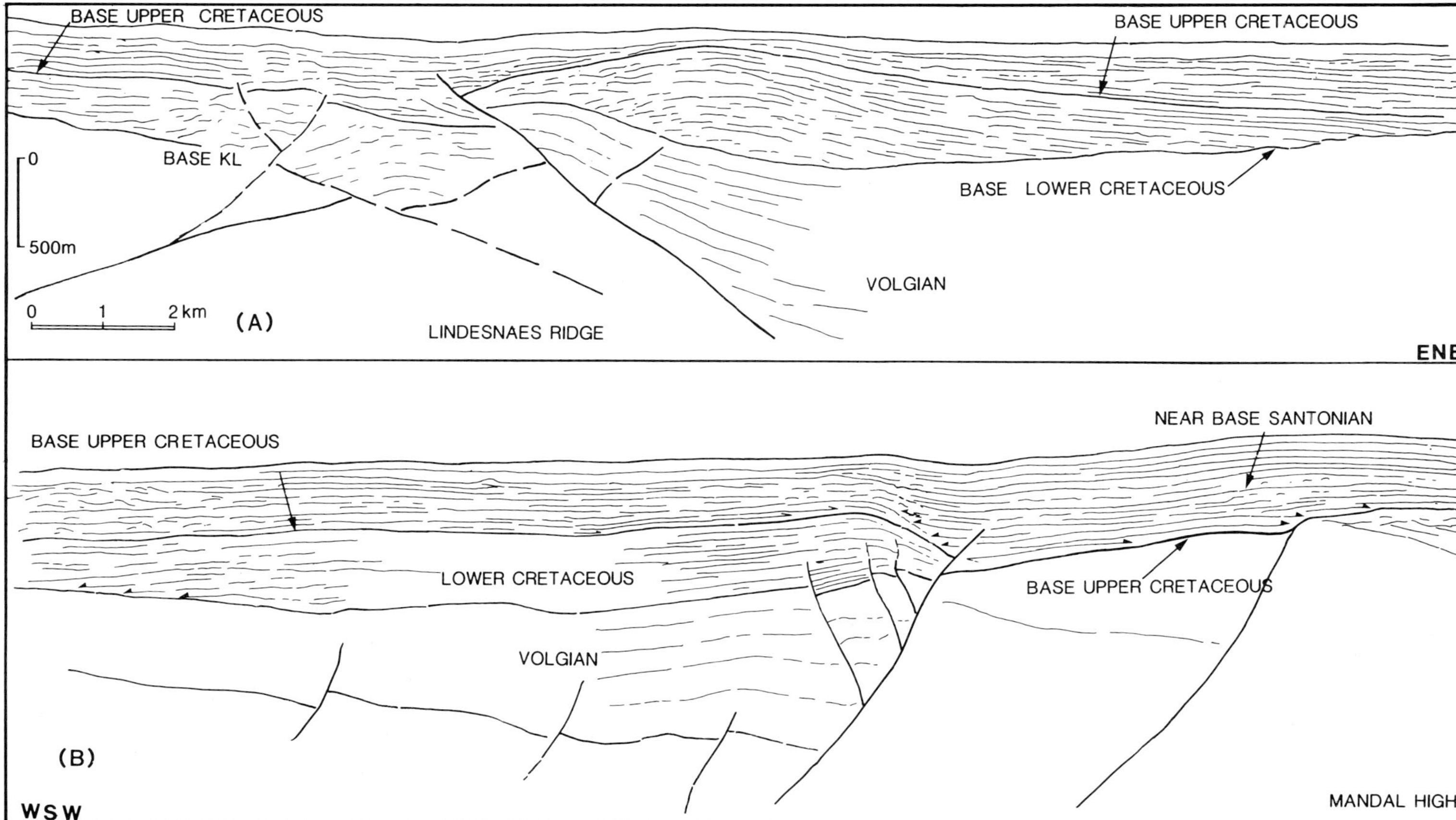

FIG. 8. Line drawings of seismic data across the inverted margins of the Feda Graben. Asymmetric growth folds are developed on both margins as a result of a dominantly dip–slip reversal of motion on the bounding basement normal faults. Note the thinning of the Chalk through a combination of onlap and convergence.

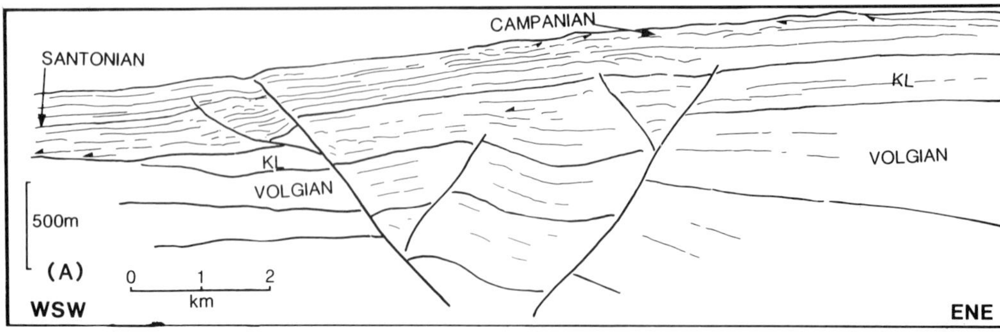

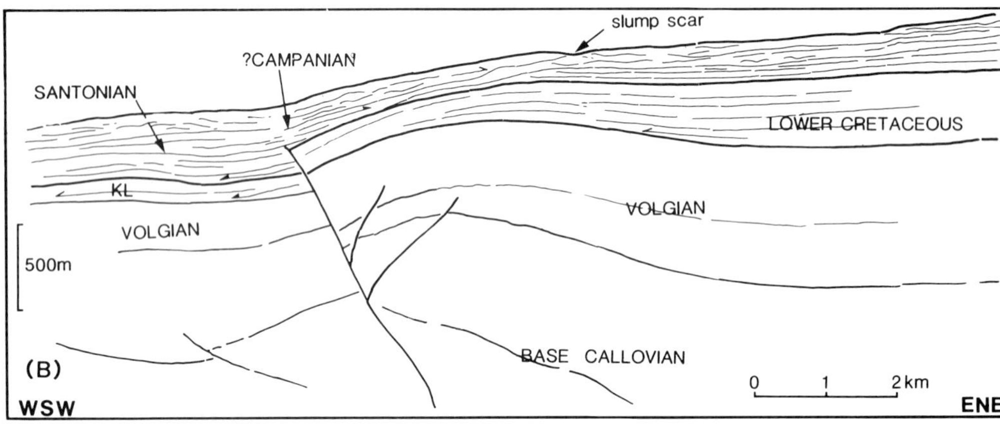

FIG. 9. Line drawings across two neighbouring segments of the Jens–Otto Fault Zone. Note the contrast in both the compressional and earlier extensional structural geometries.

The seismic stratigraphy of the Upper Cretaceous sequence

The aims of the seismic–stratigraphic interpretation were to assess the relative contributions of regional subsidence, tectonic uplift and depositional processes in the overall configuration of the Upper Cretaceous basin fill, to investigate the timing of the deformation on the inverted margins and to consider the interactions between structural development and depositional processes.

In order to satisfy these objectives, the interpretation concentrated on two aspects of the seismic stratigraphy: the accurate correlation and calibration of regional and local unconformities and sequence boundaries within the Upper Cretaceous and the definition and mapping of seismic facies for chronostratigraphic intervals of an appropriate duration within the Upper Cretaceous sequence. There were sufficient released wells to constrain the basin-wide correlation of key markers within the sequence. The widespread occurrence of Chalks in a discontinuous seismic facies proved to be the single most important source of error in the calibration of the seismic data. In all cases, however, local inconformities or sequence boundaries could be correlated to several wells. The reliability of the picking and correlation of the seismic event could thus be checked against the chronostratigraphy of several independent well sections.

By comparing the spread of age determinations for a variety of seismic events in groups of closely spaced wells, it was found that the resolution of the chronostratigraphic calibration of the seismic data was of the order of a stage length. This represents the compounding of errors and the limits of resolution in the seismic correlation, depth conversion, and biostratigraphic determination in the well. The resolution limit of one stage length should be regarded as the absolute upper limit for the method. Even though plus or minus several million years might not seem to be precise enough to draw any useful conclusions about

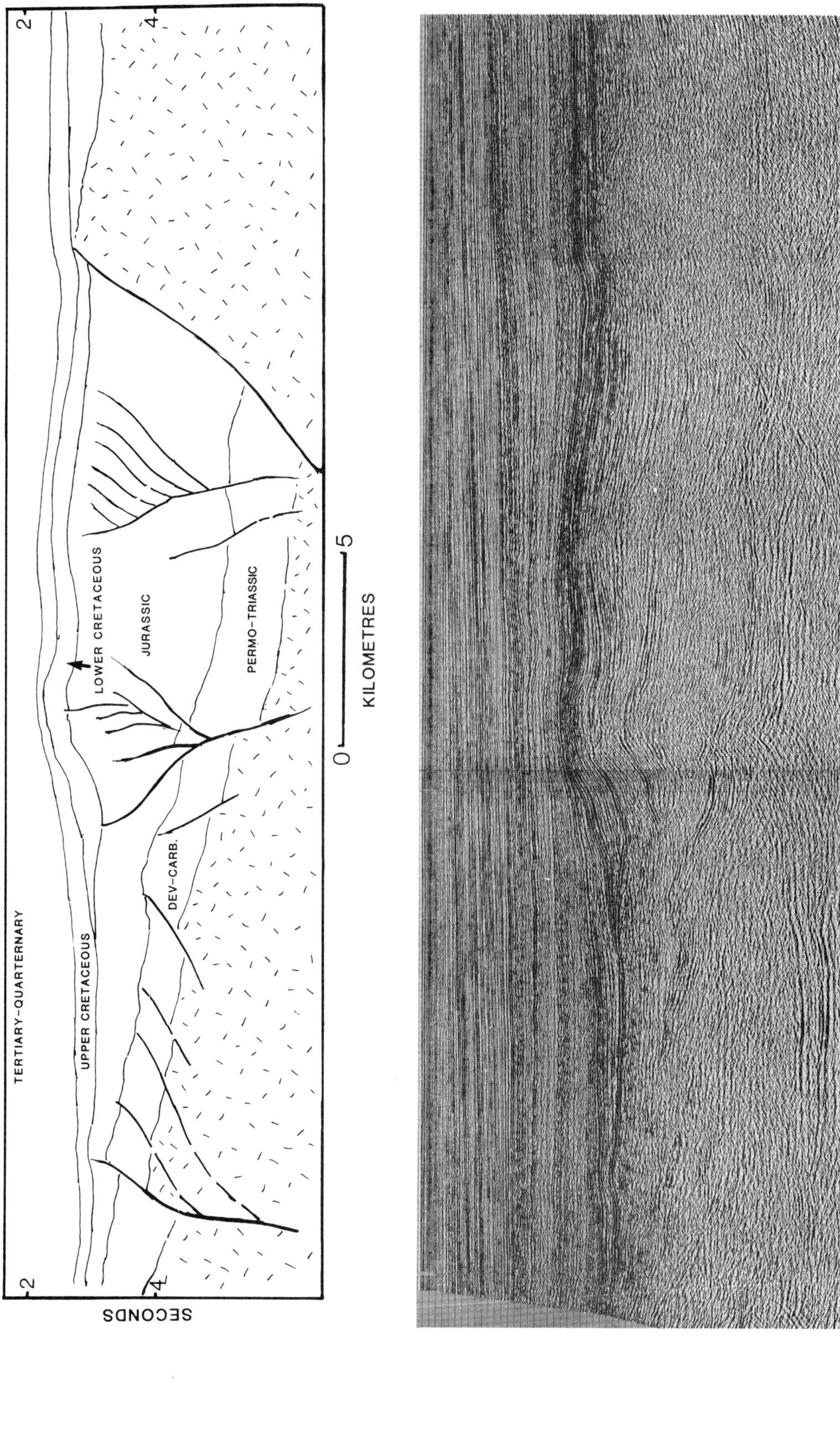

FIG. 10. Representative seismic section and geoseismic interpretation across the Central Tail End Graben.

the temporal distribution of the deformation, it is clearly far more sensitive than the approach of simply analysing bulk isochore or isopach trends. It should be noted that on average, the entire Upper Cretaceous sequence is expressed as 10 to 12 seismic loops. Each reflection event therefore represents several millions of years of deposition.

Timing the deformation

The development of growth folds along the inverted basin margins was noted previously from the isochore data (Fig. 6) and on line drawings of cross-sections (Figs 8 & 9) as a gross thinning of the Upper Cretaceous sequence towards the fold axes. Closer inspection of the reflection geometries shows that this thinning is represented by onlaps and convergence onto the fold limbs. In Figs 8A and B for example, thinning of the Chalk from the axial depocentre of the Feda Graben towards the margins is marked by gentle onlaps and convergence onto the inner facing limbs of the inversion anticlines. The steeper, outward facing limbs display a more pronounced onlap, particularly in the lower part of the sequence. This adds an asymmetry in the stratigraphic relationships associated with the folds to that previously noted for their structure.

The recognition of simultaneous onlap on both limbs of the folds towards their crests is a sure indication of the earliest manifestation of the development of the fold at the depositional surface. The sequence boundary on which the onlapping reflections truncate is thus a convenient chronostratigraphic marker for the onset of the deformation (Fig. 11). However, it should be noted that reversal of motion on faults at some depth in the basin may proceed for a considerable period before there is a response in the near surface.

Since each reflection cycle represents the

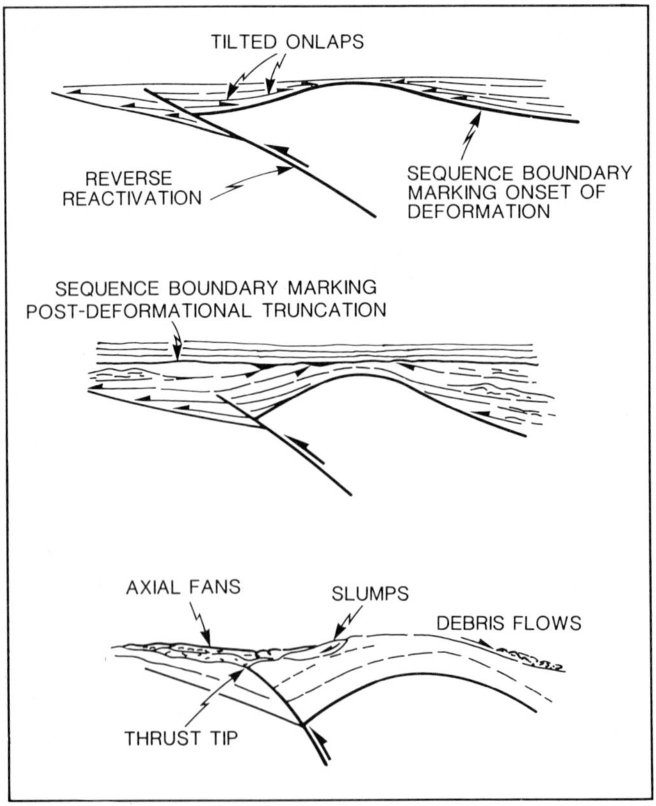

FIG. 11. Schematic representation of some typical reflection configurations across the Late Cretaceous growth folds, showing the form of the stratigraphic markers for the onset and termination of the compressional deformation.

cumulative deposition over a period of several million years, it is not possible to distinguish individual episodes of relief growth beyond this degree of temporal resolution. The growth of relief observed in the progressive rotation of the early onlap units away from the crests of the folds (cf. Figs 8, 9 & 11) offers an additional means for estimating both the growth rate and the magnitude of the early structural relief.

The estimation of the time at which the compressional deformation effectively ceased to influence the topography of the depositional surface is made with reference to truncational reflection configurations observed over the crestal regions of the inversion structures. Truncational sequence boundaries are commonly developed in the crestal areas of the inversion axes. The truncation probably developed by a combination of slope failure on the actively growing structures and erosion by basin floor currents and low-density turbidity currents.

Slump scars are identified along the Jens–Otto Fault Zone (Fig. 9B). Seismic activity was probably an important agent in triggering gravitational collapse features. The truncation surfaces are often observed within the syndeformational sequence. In such cases, the angularity of the truncation surfaces and the degree of removal of material gives a loose indication of the intensity of the deformation. However, when the truncation surface is overstepped or draped by concordant units with parallel and continuous reflection characteristics (Fig. 11), it effectively represents an erosional 'event' which eliminated sea floor relief. The truncation surface can then be taken as a marker for the end of active growth of the folds. The truncational sequence boundaries over the inversion structures along the graben margins correlate laterally into correlative conformities over the more stable adjacent platform areas. Wells located on the Dogger Platform and the Mandal High–Sogne Basin block penetrated stratigraphically complete and thick sections of Chalk. The correlation of the 'deformational markers' into these areas allowed a sharper calibration of the timing of the deformation than would have been possible using wells located over the inversion axes themselves.

Using the method of defining and calibrating deformation onset and termination markers outlined above, it is possible to date the duration of the inversion (as expressed the at surface) on all the principal thrust/fold structures. When this information is overlain on the structural

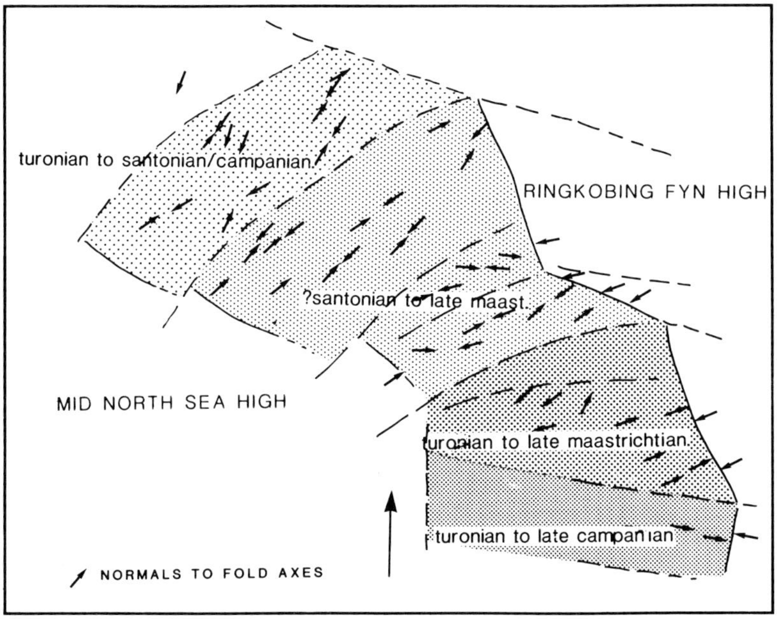

FIG. 12. Kinematic data. Normals to inversion fold axes are oriented between N 20 E and N 80 E suggesting that convergence of the bounding highs was in an approximately NE–SW direction. The compressional deformation was markedly diachronous, and the boundaries between segments with different deformation histories correspond to the transverse lineaments of Fig. 7.

maps, a remarkable pattern becomes apparent (Fig. 12).

The inversion structures can be grouped into several distinct temporal provinces. Within each there is a remarkable synchroneity in the onset and termination of the compressional deformation. However, there is a clear diachroneity in the timing of the deformation from province to province. Along the Jens–Otto inversion axis, for example, the folds in the extreme N and S began to develop synchronously in the Turonian. However, the folds in the central section of the structural trend did not start to develop until the Santonian. The separation of these two onset markers is much greater than the error range in the chronostratigraphic calibration. Therefore, this diachroneity cannot be dismissed as an artefact of the method.

The boundaries between the temporal provinces are found to bear a close relationship with the traces of the transverse lineaments identified from the segmentation of the fold axes (cf. Figs. 7 & 12). The temporal province boundaries are arranged in a fanning pattern with orientations ranging from WNW–ESE to NE–SW. This points to a component of rotation about a vertical axis in the bulk compression of the DCG, almost as if the extensional segmented mosaic was caught in a torsional moment between the closing vices of the rigid, undeformable massifs of the bordering highs. In summary, it is suggested that in addition to partitioning the shortening along the length of the graben, the transverse lineaments forming the segment boundaries were also responsible for confining the deformation to certain particular segments for certain specific periods.

Seismic facies analysis

Two seismic facies were distinguished:

Facies A: parallel, sub-parallel, or wavy reflection configuration with good reflection continuity.

Facies B: hummocky, lenticular, mounded or occasionally progradational reflection configuration, with frequent downlap event terminations and poor reflection continuity.

The reflection character of facies A indicates that the lithofacies were uniform over considerable areas and depositional systems were low energy. Facies A is observed in tectonically stable areas on the flanking Mid North Sea and Ringkobing–Fyn Highs and the Dogger Platform. The seismic facies is interpreted to be composed mainly of pelagic (autochthonous) chalks with admixtures of distal turbiditic chalks along the axis of the stable, subsiding graben floor (Fig. 13).

In contrast, the reflection configurations of facies B are indicative of higher energy depositional processes. This facies is equated with the gravity flow deposits (allochthonous chalks) as described by Watts *et al.* (1980), Nygaard *et al.* (1983), and Hatton (1986) from nearby areas in the Central North Sea. Facies B is closely associated with tectonically active areas adjacent to the inversion axes (Fig. 13). In the Norwegian sector allochthonous chalks are associated with zones of tectonic instability either along inverted fault margins or along the flanks of salt structures. Reworking of pelagic chalks took place in an analogous manner in the Danish sector (Nygaard *et al.* 1983). This relationship between sourcing of allochthonous chalk deposits and inversion axes is supported by the common observation of slump scars and erosion surfaces on the crests and limbs of many of the growth folds (e.g. Figs 8 & 9).

Progradational configurations are observed along the axis of the Tail End Graben (Fig. 13). These are interpreted as prograding fan lobes. Many of the lobate or mounded forms have no recognizable progradational geometry and are often chaotic in internal configuration. These units are mapped on the flanks of the growth folds immediately adjacent to the break of slope on the fold margins. They are interpreted as admixtures of slump units and debris or mud flows.

In order to build up a representative palaeogeography at different stages of the basin inversion, these two seismic facies were mapped for two stratigraphic intervals, the Turonian–Coniacian and the Maastrichtian. The maps presented in Fig. 13 therefore represent generalized patterns of deposition for early and late stages in the basin inversion. They emphasize the effects of the diachronous 'switching' of compression between segments on the spatial arrangement of the depositional systems (cf. Figs 12 & 13).

Deposition of the Lower Chalk (Fig. 13A) was largely confined to the axis of the former rift. Both margins of the rift were onlapped at topographic features corresponding to the underlying rift-bounding faults (e.g. Fig. 10). The Lower Chalk is thin or absent along the Olaf Fault Zone, the northern and southern extremities of the Jens–Otto Fault Zone and the inverted margin of the southern segment of the Coffee Soil Fault (Fig. 6). Facies A predominates on the Dogger Platform, the Feda Graben, and the northern and southern seg-

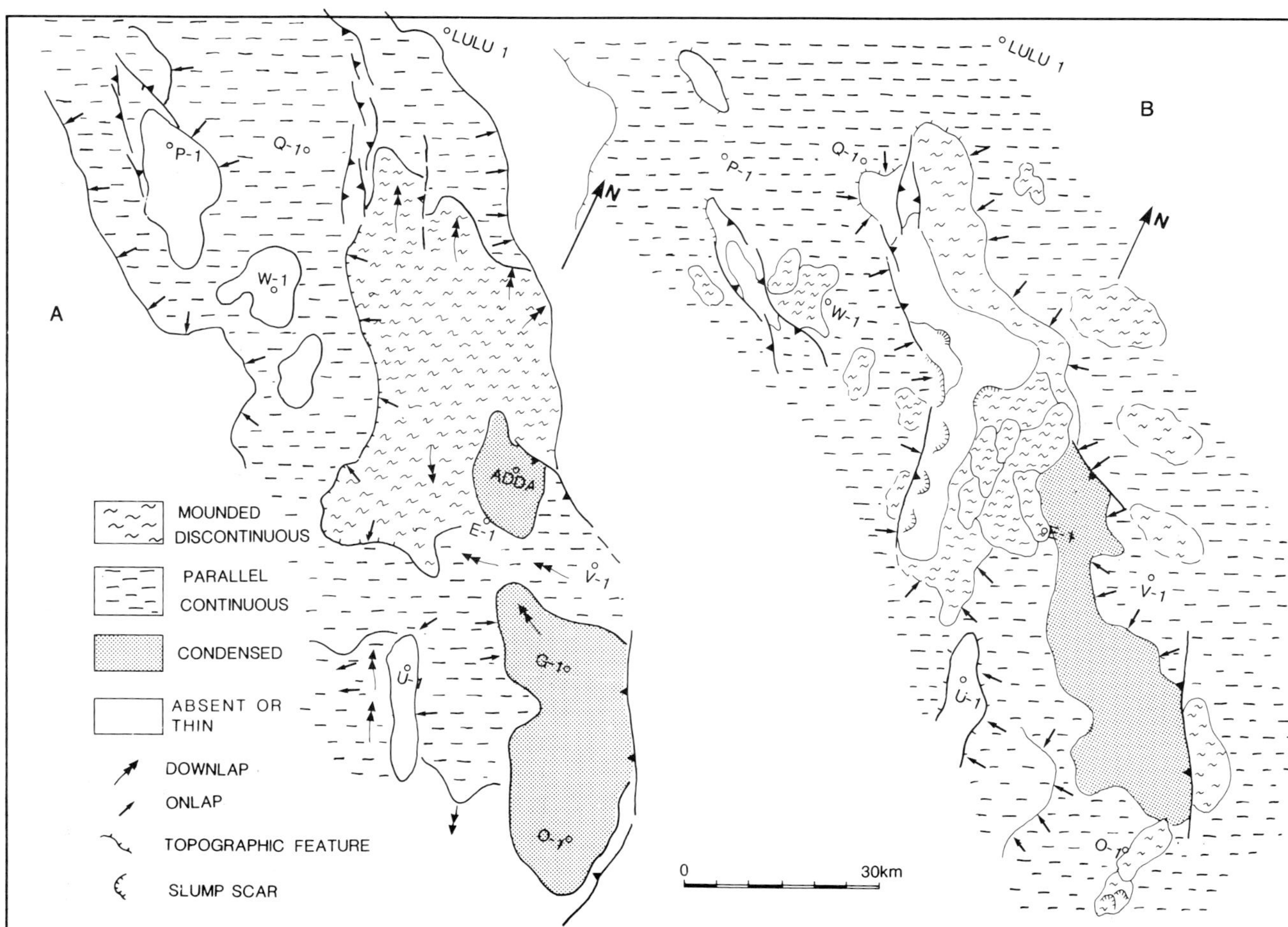

FIG. 13. Seismic facies maps of representative sequences of the Lower and Upper Chalk. A, Turonian–Coniacian, and B, Maastrichtian.

ments of the Tail End Graben. Facies B is confined to the central segment of the Tail End Graben, which was an area of pronounced and continuous subsidence until the Santonian. This area was in effect a sub-basin, delimited by topographic features on the sea bed that were controlled by the dormant marginal faults and by the transverse lineaments. The influence of deeper structure on the morphology of the basin floor is clearly seen in the onlap relationships illustrated in Fig. 13A.

The sourcing and flow paths of the allochthonous chalks were markedly influenced by the actively emerging topography along the inversion axes. Downlaps observed in Lower Chalk units indicate that the pattern of sediment dispersal was influenced by depressions in the sea floor between the inverting margin structures. In the vicinity of the V–1 well, for example, progradational reflection configurations are observed from E to W into the graben axis from the shelf area on the adjacent Ringkobing–Fyn High. The progradational units are confined to a valley-type feature occupying a saddle between two offset margin inversion structures (Figs 7 & 13A). Gravity flow chalks triggered by slope instabilities associated with the reactivation of the Coffee Soil Fault in the vicinity of the shelf edge may have been focused into the topographic low of the Central Tail End Graben through this gap in the inverting margin. It is likely that downlapping and progradational units observed parallel and adjacent to other inversion axes are also indicative of the deflective or confining role of the structurally controlled relief on the sediment dispersal patterns.

By the late Santonian the flanking highs were incorporated in a regional phase of subsidence. As a result of the broader base of subsidence and the rise in absolute sea level, the shelf margins retreated away from their earlier fault-controlled positions on the rift margins. The most easily mobilized source of allochthonous chalk material was thus removed far from the potential repository on the basin plain. Nevertheless, facies B is widely recognized in the Upper Chalk and is closely associated with inversion axes (Fig. 13B). By Campanian and Maastrichtian times, the main locus of compressional reactivation had shifted towards the central area, along parts of the Olaf and Jens–Otto Fault Zones. Sea floor relief amounted to over 100 m along these growth folds. The reworking from the crestal regions to immediately adjacent areas is a testament to the strong morphological control of the basin floor by the reactivated structures.

The distribution of the two seismic facies thus shifted progressively in accordance with changing patterns of regional subsidence and inversion. The dominantly pelagic chalks of facies A accumulated mainly on the tectonically stable areas, whilst the allochthonous chalks display patterns of sourcing, dispersal and accumulation that were strongly influenced by topographic features corresponding to the axes of inversion. A picture thus emerges of the control exerted on the depositional systems of the Upper Cretaceous by both the dormant and the reactivated elements of the structural framework.

Discussion

The most important observations presented in the preceding sections of this paper are summarized below:

1 Late Cretaceous basin inversion in the Danish Central Graben is manifested in the compressional reactivation of extensional faults controlling the margins of the earlier rift.

2 The compressional structures are simple. Symmetric and asymmetric growth folds developed at the tips of minor thrust faults that propagated towards the depositional surface from a point of nucleation on the pre-existing fault planes.

3 The orientation and position of the inversion axes mimics directly the extensional fabric of the rift. Offsets and changes in the structural geometry along the strike of the reactivated rift margins are related to the influence of a suite of transverse lineaments that served to segment the rift.

4 The transverse lineaments were responsible not only for a partitioning of the compressional strain into segments of different degrees and styles of shortening, but were also instrumental in the diachronous fashion in which individual segments were reactivated.

5 The deformation proceeded amidst a background of regional basin subsidence and deep marine sedimentation. The development of growth folds had profound effects on the sourcing, transport and accumulation of the Upper Cretaceous Chalk, principally through the interaction between structurally generated basin floor topography and gravity flow processes.

Previous analyses of the inversion of the Danish Central Graben consider that the deformation was essentially transpressional (Michelsen & Andersen 1983; Vejbaek & Andersen 1987). This view follows the doctrine of transpressional or strike–slip deformation of the Alpine Foreland basins resulting from Alpine collisional orogenesis propounded by Ziegler (1982). Ziegler likened the behaviour of the previously thinned Mesozoic basins of Northwest Europe to shear pins collapsing under tangential stresses generated at the collisional front. This dynamic model appears to have been derived in part from Ziegler's (1982) observations that inversion movements were more or less synchronous in widely separated basins, and that 'sub-Hercynian and Laramide compressional and wrench deformations of the Alpine foreland coincide with major orogenic events in the Alpine domain.' Although some authors have argued strongly that the European platform was a medium for long-range and relatively uniformly oriented compressional stress transmission related to Alpine collisional orogenesis (Bergerat 1987; Letouzey 1987), the dynamics of the inter-relationship between continental collision and foreland stress transmission are far from being completely understood. Indeed, recent studies of the relative plate motion vectors of Africa and Europe and kinematic data from the circum-Mediterranean orogenic belts show that there is no straightforward correlation between the bounding plate motions and the detailed kinematic response of the collisional belt (Dewey *et al.*, *in press*). Given the problems in establishing a satisfactory kinematic framework for the SE boundary of the European plate, it is perhaps presumptuous both to relate ambient intraplate stress fields to Alpine orogenesis and to isolate the dynamics of that particular boundary in a causative sense with regard to the intraplate compression.

Ziegler (1982) suggested that compressional stresses in the foreland in the Late Cretaceous 'had an essentially N–S orientation'. In the Danish sector, Michelsen & Andersen (1983) concluded from the apparent *en échelon* arrangement of fold axes, that inversion was 'caused by a compressional or strike–slip stress regime'. However, they gave no indication of the orientation of this causative stress field beyond relating it to a regional context with the statement that 'the inversions are believed to be caused by sub-Hercynian or possible Laramide orogeny.' Vejbaek & Andersen (1987) are more specific in that they refer to a likely stress system for the Late Cretaceous as 'governed by N–S to NNW–SSE directed right lateral strike–slip motions' but their dynamic interpretation of fault kinematics shows a NE–SW orientation for the principal compressive stress (their Figure 17).

Interpretations of stress field orientations in areas such as the DCG are hampered by the strong influence that pre-existing lines of mechanical weakness exert on the distribution of deformation. Indeed, so little is known of the refractive effects of crustal anisotropy that any discussion of 'causative stress fields' is only of limited worth. However, the ubiquitous development of approximately NW–SE oriented fold axes is difficult to interpret in any other way than as the result of a net NE–SW convergence. From the several percent shortening measured across the graben this amounts to a relatively trivial horizontal motion of the order of one kilometre. In a strictly kinematic sense, the inversion of the Danish Central Graben was the result of a minor NE–SW convergence of the rift shoulders (the rigid flanking massifs). This tells us nothing about any causative stress field and nothing about the mechanics of the deformation. There is certainly no justification for deducing any principal stress configurations from the orientation of inversion axes, particularly in view of the fact that the orientation of the inversion axes was conditioned by the orientations of the earlier extensional structures.

The basis for Vejbaek & Andersen's (1987) and Michelsen & Andersen's (1983) contention of a strike–slip component in the deformation is founded on the apparently *en échelon* arrangement of the fold axes and the flower-type structural geometries developed on the reactivated graben margins. It has been demonstrated here, however, that the arrangement of the fold axes was conditioned exclusively by the extensional basement fabric. In every case, the fold axes are parallel to the strike of the root fault underlying the structure. If the term *en échelon* is applied at all, and that is highly debatable, then it relates not to the compressional reactivation structures but to their extensional antecedents.

Flower structures are commonly assumed to be the result of strike–slip deformation. As noted originally by Harding (1985), similar structures can also be produced under a range of predominantly dip–slip conditions. Unfortunately, the flower structure is now taken as *prima facie* evidence for strike–slip motion in sub-surface studies almost with the same degree of certainty as striae on slickensides are taken as kinematic indicators in outcrop-based studies. Certainly, the inverted Olaf and Jens–Otto fault Zones could be described as flower structures, in that they are characterized by pop-up structures bounded by downward converging fault strands which merge at depth into steep basement faults. In addition, there are rapid variations in stratigraphic and structural expression along the fault zones, a facet which is often taken as diagnostic of strike–slip deformation. However, the complex geometries exhibited by these two inverted basin margins are equally explicable as the product of the superposition of several phases of dip–slip fault movements, in both extensional and compressional senses. In the case of the Jens–Otto Fault Zone, for example, the flower-type geometry was the product of a phase of step faulting in the Triassic, followed by hinge development and crestal collapse faulting during the rotational subsidence of the Tail End Graben in the Mid-Late Jurassic, then by salt flowage and minor diapiric intrusion in the Early Cretaceous (with its corollaries for longitudinal thickness variations). This culminated in the buckling and reversal of motion in the Late Cretaceous.

It is suggested, therefore, that the complexity and lateral variability in the geometry of the inversion structures can be adequately explained as a simple dip–slip reactivation along an approximately NE–SW convergence direction by reference to the lateral variation in the extensional geometries and by taking account of additional factors such as halokinesis and the mechanical properties of the graben fill. Due caution should be exercised, therefore, in attaching a particular kinematic significance to a term that is, after all, merely a geometric description.

In view of the arguments presented above, it may be more appropriate to consider the compressional deformation not so much in terms of regional horizontal stress fields, but rather in terms of relative displacement of crustal blocks on a regional scale in Northwest Europe. The Mid North Sea and Ringkobing–Fyn Highs were evidently displaced relative to one another in a convergent sense during the course of the inversion. It now remains to be considered whether this convergence resulted from passive reactivation of crustal discontinuities under compressive stress field or whether some alternative mechanism is more applicable.

The European Plate is a mosaic of Precambrian, Caledonian and Variscan massifs, which are delimited by ancient and often reactivated lithospheric fault lineaments (Fig. 14). Most of the Mesozoic basins of the European Platform lie at intersections or terminations of these major continental lineaments. Some of these lineaments connect to the plate boundaries. It follows, therefore, that the network of continental lineaments offers a plausible route for the translation of plate boundary motions (as opposed to forces) deep into the interior of the plate. Since the intra-plate rift basins link into this network, it also follows that motions translated along the lineaments could be expressed in a reactivation of the fault structures bounding the rift basins.

How then might such differential plate boundary motions occur? The two most obvious sources of differential motion imparted to the plate boundary are those arising from a collisional impact along only a portion of a convergent boundary and those arising from a jump in the location of ocean floor spreading across a transform fault. Both these conditions appear to have applied in the case of the rhombohedral European Plate during the Late Cretaceous and Tertiary. This can be seen from schematic reconstructions of the plate boundary processes (Fig. 15). This diagram illustrates some possible connections of incipient North Atlantic transforms with 'Tornquistian' (NW–SE) lineaments in the interior of the plate, via offsets or dog-legs along the network of Mesozoic rift basins. One of the most striking alignments is that between the Tornquist Zone and the Faeroe–Iceland Ridge (Pegrum 1984). Adjustments of local plate kinematics as the zones of active spreading propagated into the Norwegian–Greenland Sea might therefore have been communicated into the North Sea region by the oceanic transform–continental lineament link. On the opposite side of the plate, the major lineament bordering the SW margin of the Bohemian Massif projects into the Alpine belt at a point of marked change in orientation, coincident with the boundary between the Western and Eastern Alps. It is plausible that this major NW trending lineament accommodated the differential and diachronous convergence in the Western and Eastern Alps and translated the 'buffer' motions into the Polish–Danish Trough and the Southern North Sea (see the Late Cretaceous reconstruction in Fig. 15).

The inversion of basins in the NE half of the plate (bounded to the SW by a NW–SE trending set of fractures linking from the SW border of the Bohemian Massif, through the Sole Pit Basin, and offset via the Pennine Faults and Cheshire Basin to the kinked margin of the Western Hebridean Platform) pre-dates that of the group of basins in the Western Approaches, the Irish Sea and the Channel. The NE sector was inverted mainly in the Late Cretaceous and Early Tertiary, whereas the more SW portion of the plate was inverted mainly in the

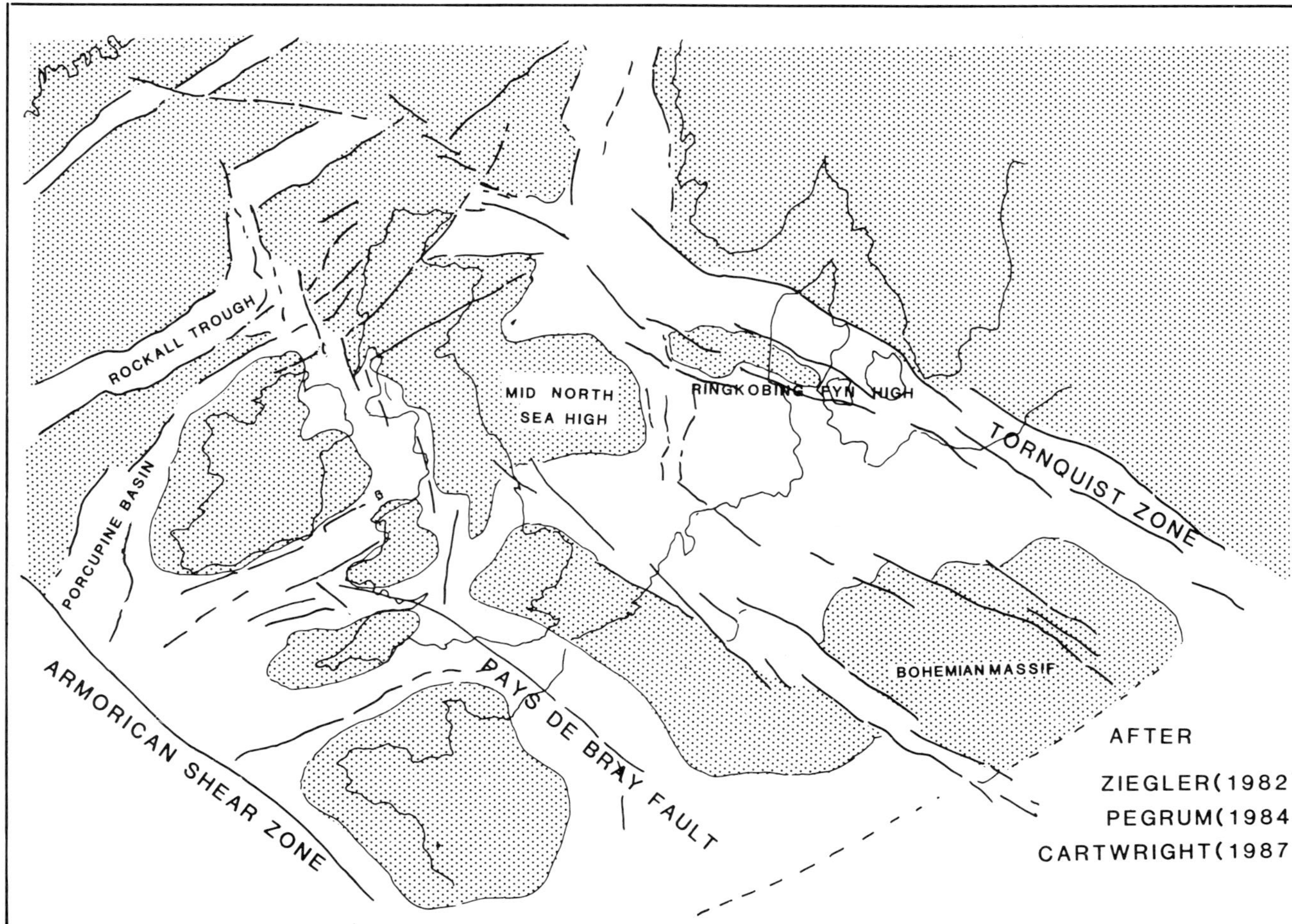

FIG. 14. Fundamental fault lineaments, basins and massifs of Western and Central Europe.

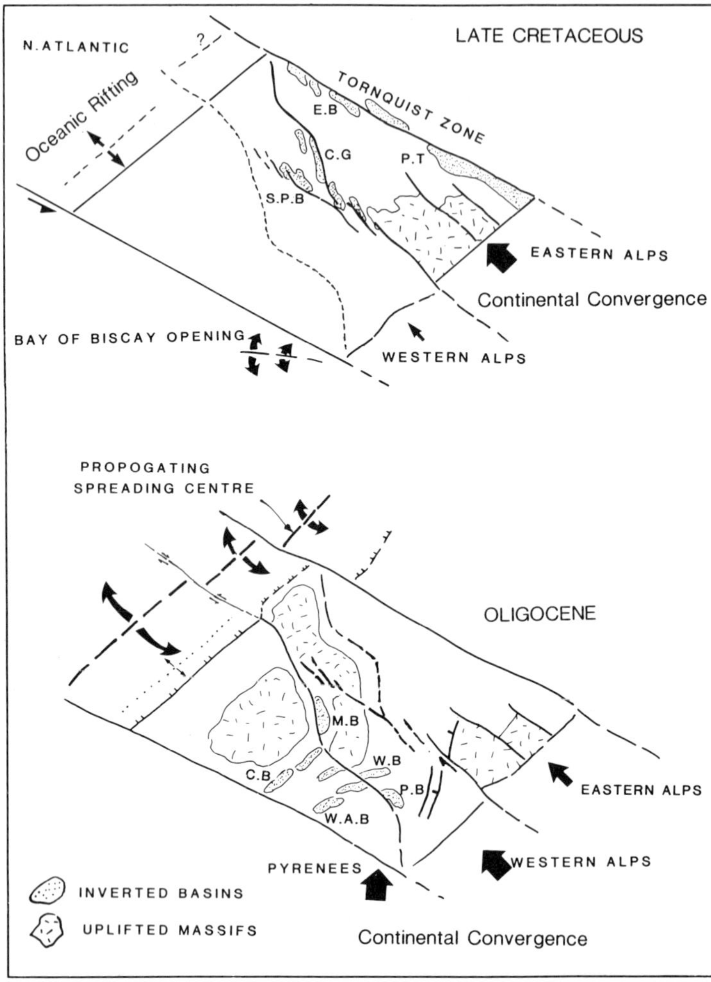

FIG. 15. Schematic reconstruction of the plate boundary configurations of the European Plate during the Late Cretaceous and the Mid-Tertiary. The rhomb-shaped plate was subjected to a diachronously imparted vice-like compression arising from the NE propagation of the North Atlantic Ocean on the NW boundary, and continental collision on the S and SE boundary.

Oligocene–Miocene. It may well be that this gross scale diachroneity across the breadth of the European plate is the response to a diachronous opening of the North Atlantic and a diachronous convergence along the SE margin which is younging from the Eastern Alps to the Pyrenees. The larger scale partitioning of the net convergent motions between the double couple of the divergent and convergent margins may thus have been achieved through the dislocative function of the major NW–SE trending lineaments.

The massifs of the Central North Sea are connected to the plate boundaries via splay lineaments of the Tornquist Zone (Cartwright 1987). If a mechanism as outlined above was in force during the Late Cretaceous, it would seem entirely reasonable that the fan-like arrangement of segment boundaries could have relayed the minor horizontal translations into the interior of the Danish Central Graben. The modest amount of shortening exhibited across the area would only have necessitated horizontal displacements of the order of several kilometres along the Tornquistian splays. This is easily conceived as a minor jostling of continental massifs throughout the plate mosaic.

Diachroneity on the scale of a single basin, as described here, or from basin to basin can thus be understood in terms of the vagaries of the connections of the lineament network. Ziegler's shear pin analogy is particularly apt, but rather than appealing to 'tangential stresses', here the emphasis is on communication of *all* the surrounding plate boundaries with the interior basins through the network of lithospheric lineaments. An analagous process of pull-apart through differential shear on the same network of lineaments applied for the formation of the intra-plate rift basins (Arthaud & Matté 1977). The intra-plate compression can be visualized simply as a reversal of this dynamic process. In conclusion, therefore, the inversion of rift basins on the European Platform is attributed to the interactions between the convergent SE and divergent NW plate boundaries with the network of fundamental lithospheric discontinuities transecting the plate. The lack of rigidity exhibited by continental plates under bulk compression or extension may be mainly due to the mechanical influences of these ancient and durable fundamental continental lineaments.

ACKNOWLEDGMENTS: Thanks are due to Petrofina (UK) for their financial assistance, to Merlin Geophysical, for the loan of their excellent seismic data, and to Drs H. G. Reading and M. Helman for their help in improving the manuscript.

References and further reading

ARTHAUD, F. & MATTÉ, P. 1977. Late Paleozoic strike–slip faulting in southern Europe and northern Africa: result of a right-lateral shear zone between the Appalachians and the Urals. *Bulletin of the Geological Society of America*, **88**, 1305–1320.

BERGERAT, F. 1987. Stress fields in the European Platform at the time of the Africa–Eurasia Collision. *Tectonics*, **6**, 99–132.

CARTWRIGHT, J. A. 1987. Transverse structural zones in continental rifts — an example from the Danish Sector of the North Sea. *In*: BROOKS, J. & GLENNIE, K. W. (eds) *Proceedings of the Third Conference on the Petroleum Geology of NW Europe*, Graham & Trotman, London, 441–452.

—— 1988. *A seismic interpretation of the Danish North Sea*. D.Phil. Thesis. University of Oxford, UK.

DEWEY, J. F., HELMAN, M. L., TURCO, E., HUTTON, D. H. W. & KNOTT, S. D. Kinematics of the Western Mediterranean. *In*: COWARD, M. P., DIETRICH, D. & PARK, R. G. (eds) *Alpine Tectonics. Special Publication of the Geological Society of London*. (in press).

FRANDSEN, N., VEJBAEK, O. V., MOLLER, J. J. & MICHELSEN, O. 1987. A dynamic geological model of the Danish Central Trough during the Jurassic–Early Cretaceous. *In*: BROOKS, J. & GLENNIE, K. W. *Proceedings of the Third Conference on the Petroleum Geology of the North Sea*, Graham & Trotman, London, 453–468.

GOWERS, M. B. & SAEBØE, A. 1985. On the structural Evolution of the Central Trough in the Norwegian and Danish sectors of the North Sea. *Marine and Petroleum Geology*, **2**, 298–318.

HANCOCK, J. M. 1983. The setting of the Chalk and its initial accumulation. *In*: *Chalk in the North Sea*. JAPEC Course Notes, No. 20, Geological Society of London. A1–31.

HARDING, T. P. 1985. Seismic characteristics and identification of negative flower structures, positive flower structures, and positive structural inversion. *Bulletin of the American Association of Petroleum Geologists*, **69(4)**, 582–600.

HATTON, I. R. 1986. Geometry of allochthonous Chalk Group members, Central Trough, North Sea. *Marine & Petroleum Geology*, **3**, 79–98.

HORSFIELD, W. T. 1980. Contemporaneous movement along crossing conjugate normal faults. *Journal of Structural Geology*, **2**, 305–310.

LETOUZEY, J. 1987. Cenozoic paleo-stress pattern in the Alpine Foreland and structural interpretation in a platform basin. *Tectonophysics*, **132**, 215–231.

MICHELSEN, O. & ANDERSEN, C. 1983. Mesozoic structural and sedimentary development of the Danish Central Graben. *Geologie en Mijnbouw*, **62**, 93–102.

NYGAARD, E., LIEBERKIND, K. & FRYKMAN, P. P. 1983. Sedimentology and reservoir parameters of the Chalk Group in the Danish Central Graben. *Geologie en Mijnbouw*, **62**, 177–190.

PEGRUM, R. M. 1984. The extension of the Tornquist Zone in the Norwgian North Sea. *Norsk Geologisk Tidsskrift*, **64**, 39–68.

RAWSON, P. F., & RILEY, L. S. 1982. Latest Jurassic–Early Cretaceous events and the 'Late Cimmerian Unconformity' in the North Sea area. *Bulletin of the American Association of Petroleum Geologists*, **66**, 2628–2648.

VEJBAEK, O. V. & ANDERSEN, C. 1987. Cretaceous–Early Tertiary Inversion tectonism in the Danish Central Trough. *Tectonophysics*, **137**, 221–238.

WATTS, N. L., LAPRE, J. F., VAN SCHIJNDEL-GOESTER, F. S. & FORD, A. 1980. Upper Cretaceous and Lower Tertiary Chalks of the Albuskjell area, North Sea: deposition in a slope and base of slope environment. *Geology*, **8**, 217–221.

ZIEGLER, P. A. 1982. *Geological Atlas of Western and Central Europe*. Elsevier, Amsterdam. 130 pp.

J. A. CARTWRIGHT, Department of Geology, Royal School of Mines, Imperial College, Prince Consort Road, London SW7 2BP, UK.

The Permian to Cretaceous structural evolution of the Western Approaches Basin (Melville sub-basin), UK

T. J. Chapman

SUMMARY: The tectonic–sedimentary evolution of the western part of the Western Approaches Basin (Melville sub-basin) was initiated during early Permian times when synrift clastic sedimentation and volcanism commenced within an ENE trending half-graben. This sequence thickened to the south and onlapped a high to the north. During the Permian an early NNW–SSE basin extension was accommodated on an array of ENE–WSW trending normal faults linked by NNW–SSE orientated transfer faults. This was followed by a period of ENE–WSW extension along the basin axis when the normal and transfer faults of the previous faulting episode swapped roles. This was possibly related to early rifting in the Bay of Biscay. Both sets of faults appear to be controlled to some extent by Variscan thrusts and strike–slip faults within the basement. During the Upper Triassic the ridge in the north subsided allowing a thick sequence of salt to be deposited where there had previously been a topographic high. During Jurassic times sedimentation extended across the Permian–Triassic half-graben into the French sector. This was followed by an episode of rifting and wrench deformation (and inversion?) during which the Permian to Jurassic basin was domed and uplifted. The Cretaceous to Tertiary sedimentation which followed probably resulted from postrift thermal subsidence related to cooling of the continental platform adjacent to the opening Atlantic Ocean. Steep normal and/or strike–slip faults date from the earliest part of this postrifting stage.

Introduction

The Western Approaches, or South Western Approaches Basin is a large ENE trending Permian/Cenozoic sedimentary basin lying between the western end of the English Channel and the edge of the Atlantic continental margin (Fig. 1). It is separated from the South Celtic Sea Basin to the north by the Scilly–Cornubian high (BGS 1985). The basin contains up to 7 km of Permian to Cenozioc sediments (Ziegler 1982, 1987b) and is in part divided by a high, into two basins. The W basin is termed the Melville Basin (BGS 1985). Because, historically the whole area has been referred to as the Western Approaches Basin and the two parts are to some extent part of a single ENE trending basin, the western area will be described hereafter as the Melville sub-basin. Earlier accounts of the Western Approaches Basin and adjacent areas include those of Smith & Curry (1975), Avedik (1978), Hamilton (1979) and Montadert *et al.* (1979). Ziegler (1987a; 1987b) has discussed the evolution of the basin, with emphasis on Cretaceous wrenching and Tertiary inversion episodes. This chapter documents the earlier Permian to Cretaceous history of the basin, with emphasis on the Melville sub-basin in the W part of the Western Approaches.

Although a large portion of the Western Approaches Basin lies within UK waters, an area of much thicker Jurassic strata is preserved within the French sector to the south (Ziegler 1982; 1987b).

The information on the structure of this basin comes from a speculative seismic survey shot by Geco UK Ltd in the Western Approaches in 1978 and 1979. The line layout is shown in Fig. 2. Fig. 1 illustrates the overall structure of the pre-Cretaceous (Permian to Jurassic) basins in the area. Cretaceous and Tertiary strata overstep and cover these basins and the intervening highs.

Stratigraphy

The stratigraphy of the Melville sub-basin and the eastern part of the Western Approaches Basin is illustrated by the stratigraphic correlation of a number of released wells (Fig. 3). A detailed stratigraphy of the Melville sub-basin is outlined by Bennet *et al.* (1985) in their description of well 72/10–1A. Evans *et al.* (1981) described IGS stratigraphic test wells mainly in the eastern part of the Western Approaches Basin. In the Melville sub-basin Permian–Triassic sediments rest unconformably on a basement of probable Devonian and Carboniferous aged rocks. These are proved on the Cornubian High to the north to include slates (well 83/24–1, Evans *et al.* 1981). In BP well 87/12–2, Middle Devonian Limestones are

From COOPER, M. A. & WILLIAMS, G. D. (eds), 1989, *Inversion Tectonics*
Geological Society Special Publications No. 44, 177–200.

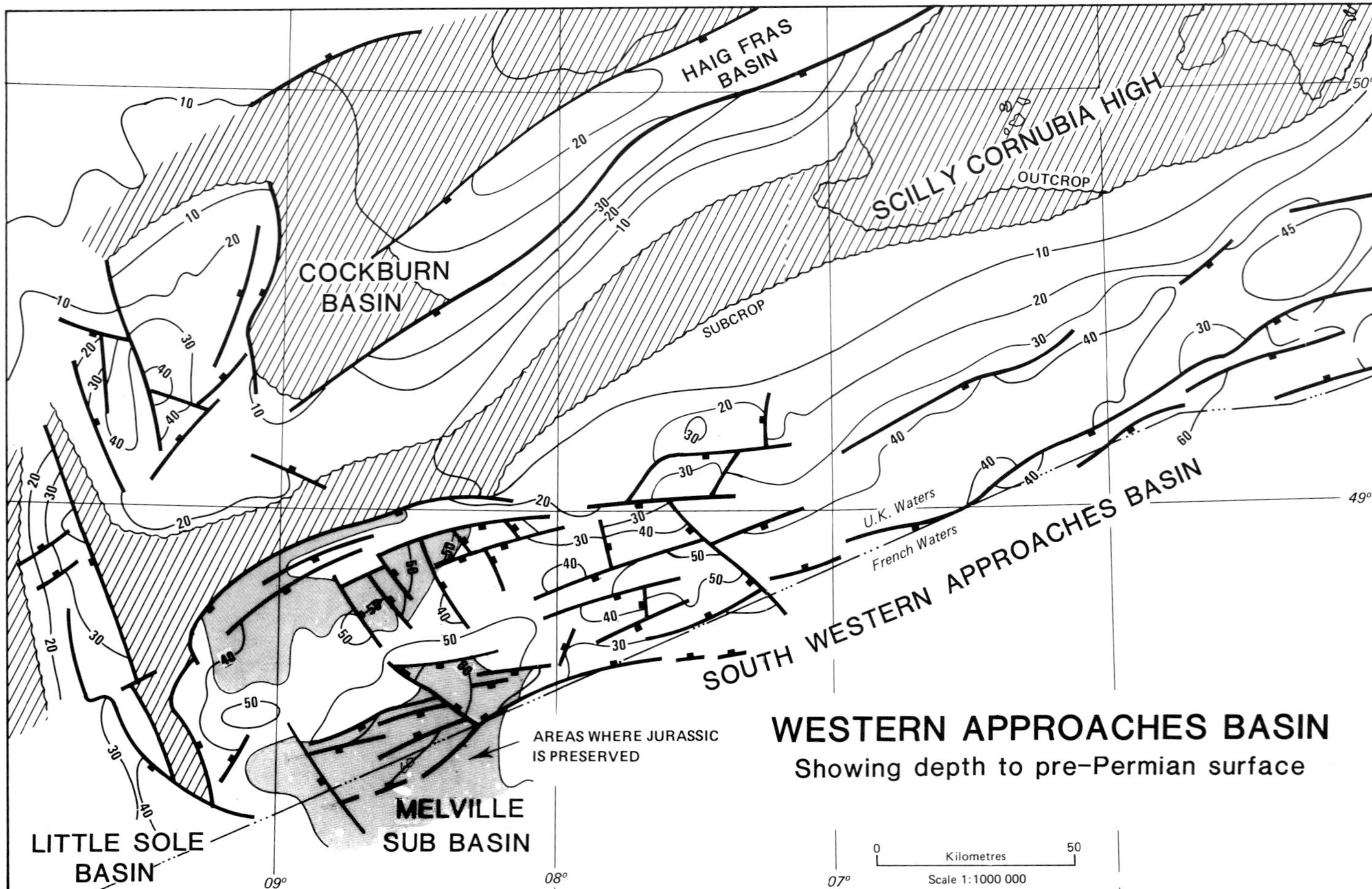

FIG. 1. Simplified geological structual map of the Western Approaches Basin (UK part only). Diagonal shading shows pre-Permian basement highs subcropping beneath the Cretaceous. Dark shading shows extent of Jurassic within the Melville sub-basin. Contours, in hundreds of metres, to top basement modified after the BGS 1:10 000 000 pre-Permian basement contour map (1985).

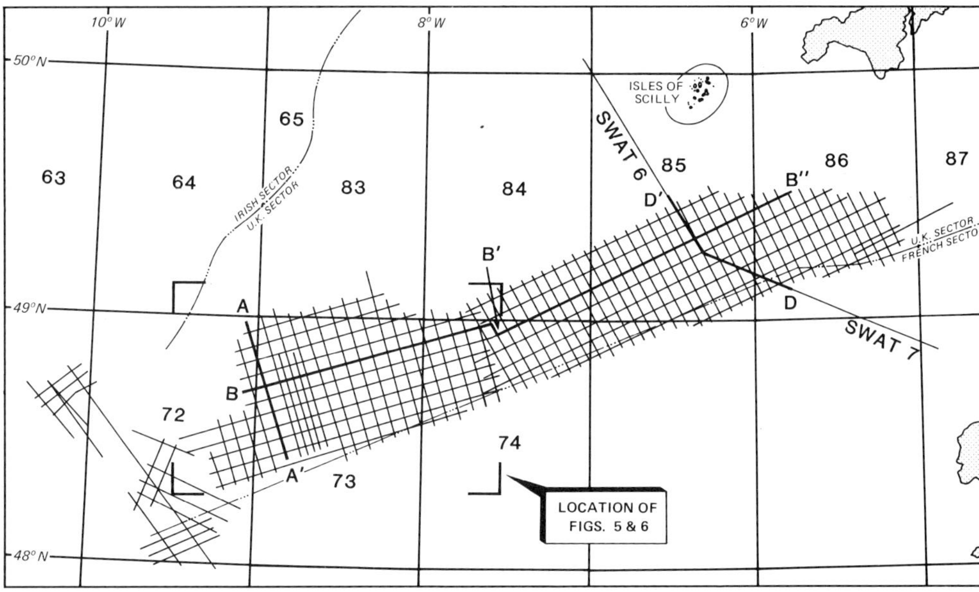

FIG. 2. Plan of the GECO UK Ltd Western Approaches Seismic survey showing the position of SWAT 6 and 7. Sections A–A′ are shown on Figs 5 & 8; B–B′ on Figs 9 & 10; B–B′–B″ on Fig. 13.

found. In some areas possible post-Variscan late Carboniferous or early Permian sediments are preserved (Britoil well 86/18–1). The seismic evidence indicates that such deposits are now restricted to the eastern part of the Western Approaches Basin, below the basal Permo–Triassic unconformity. Cambrian ultramafic rocks were recorded in Monsanto/Amerada Hess well 73/2–1. Considering the prominent zones of southward dipping seismic reflectors below the basin, it seems likely that the basement is composed of a complex of thrust sheets of Variscan and pre-Variscan rocks such as those exposed in the Amorican Massif of Northwest France.

Permo–Triassic deposits form the bulk of the sedimentary fill within the Western Approaches Basin below the regional Upper Cretaceous and Tertiary cover and are generally much thicker than in the Celtic Sea basins further north. Biostratigraphical dates are only available for the Upper Triassic with the exception of one determination in the Lower Permian (Britoil well 86/18–1). By comparison with the Permo–Triassic sequences exposed in South Devon it is possible to suggest a division into Triassic and Permian. However, it should be appreciated that the boundary in both areas is a purely lithostratigraphical one (see Warrington *et al.* 1980).

The base of a sandstone facies (Sherwood Sandstone Group) below a proven Triassic mudstone facies (Mercia Mudstone Group) is taken as the base of the Triassic (Fig. 3). Below this a further fine grained sequence coarsens downwards through mudstones to siltstones. In some places there is a high proportion of sandstones. The sequence then passes down into a basal breccia-conglomerate unit comparable with the Permian deposits in South Devon.

Within the Melville sub-basin the lowest unit comprises a volcanic sequence proved in wells, and mappable as a series of irregular, high amplitude reflectors (See BGS map of Little Sole Bank). These Permian volcanics were no doubt extruded during the earliest phase of rifting in the Western Approaches Basin.

The Upper Triassic typically consists of grey to red claystones with thin siltstones and anhydrite. Within the northern Melville subbasin massive halite of Norian–Carnian age reaches 1700 m thick in well 72/10–1A (see Fisher & Jeans 1982; Bennet *et al.* 1985). It thins away in all directions to form a discrete basins (Fig. 4). The salt thickens across both NW and ENE trending fault sets. Although this is partly due to salt flowage, it is probably also the result of syntectonic deposition. Salt swells are developed NW of the Melville Basin. Within the northern part of the Melville subbasin a

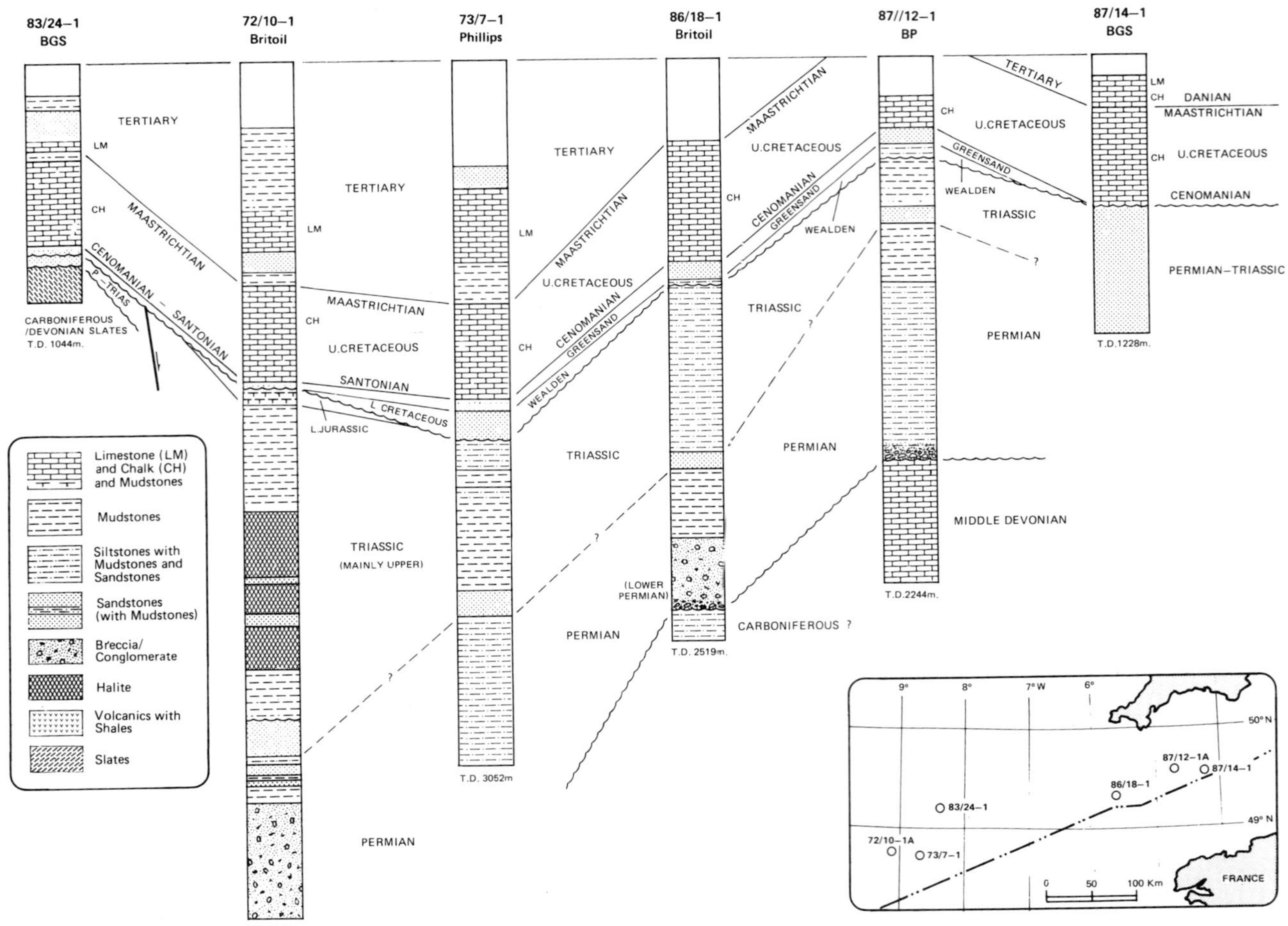

FIG. 3. Stratigraphic correlation of released commerical and BGS wells in the Western Approaches. The position of the Permian–Triassic boundary is based on a tentative comparison with the lithostratigraphy of South Devon.

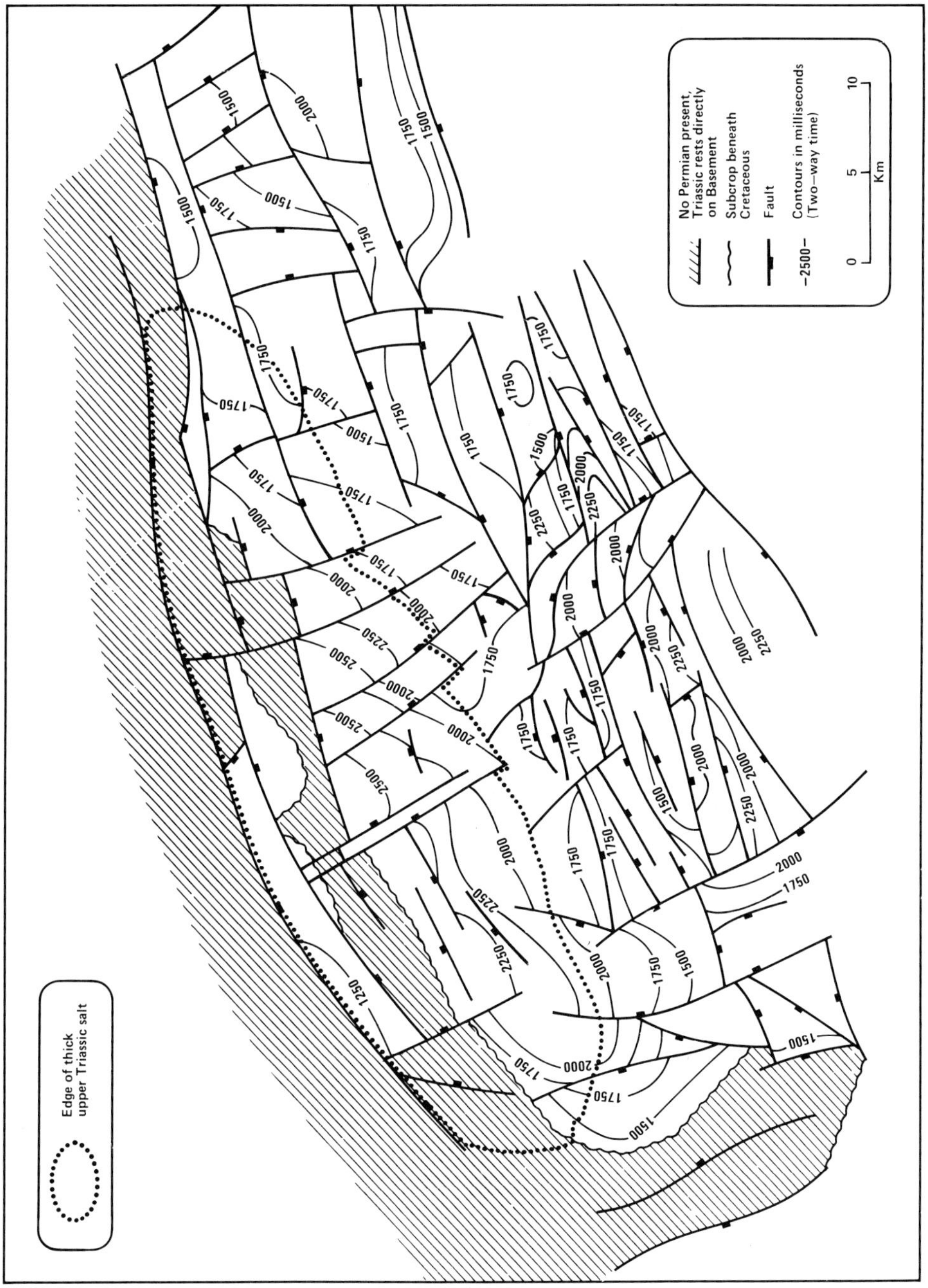

FIG. 4. Structural map of the Melville sub-basin with contours to a near base Triassic reflector in ms (two-way time).

local, but important, unconformity is developed some way below the base of the salt. In the area of well 72/10–1A, this unconformity places Upper Triassic Carnian and younger rocks directly on the Bunter/Sherwood Sandstone Group at the base of the Lower Triassic or on older Permian sandstones (Fig. 5 and see Fig. 9). Bennet *et al.* (1985) infer an unconformity near this position (their Figure 4).

Above the salt further red-grey claystones and siltstones with anhydrite pass up into Rhaetic dark grey claystones with thin limestones. Up to about 1000 m of Lower Jurassic sediments are preserved in the Melville sub-basin. Well 72/10–1A penetrated only the lowest part of sequence (22.5 m), which consisted of dark grey claystones and limestones (Bennet *et al.* 1985).

These Lower Jurassic and Triassic strata are truncated by a major unconformity below a cover of Aptian–Albian rocks. In some areas there is also a thin Wealden sequence. In the Melville sub-basin there is about 50 m of probable Aptian aged calcareous claystones and altered volcanics (Bennet *et al.* 1985). Wealden sandstones and claystones are locally preserved in the south (Fig. 3).

The Upper Cretaceous is between 400 and 500 m thick. It consists of chalk, the lowermost part of which (Cenomanian to Campanian) is absent in the Melville sub-basin (Bennet *et al.* 1985) but is present further east (Fig. 3). The Tertiary comprises Eocene limestones overlain unconformably by Oligocene and Miocene claystones with thin limestones and sandstones (Bennet *et al.* 1985).

Structure and tectonic evolution of the Melville sub-basin

Figs 6 & 4 show time structure contour maps of base Permo–Triassic/Top basement and a near base Triassic reflector (near base Sherwood Sandstone Group) respectively. These maps illustrate the very complex fault pattern in the basin. The pattern is dominated by ENE–WSW and NNW–SSE trends. Many of these faults had a complex history and acted as both normal dip–slip and transfer faults at different times. Fig. 4 also shows a Permian ENE-trending ridge, defined by the onlap limit of the near base Triassic reflector, and lying directly beneath the axis of the younger Upper Triassic salt basin. This ridge and the deepest part of the salt basin both lie S of the present Cornubian High.

The main feature of the UK part of the basin is the pronounced NNW tilt of the pre-Cretaceous sequences which are truncated and overstepped by the Cretaceous strata (Figs 7 & 5). A prominent steep normal fault forms the northern boundary to the pre-Cretaceous part of the Melville sub-basin. The Upper Cretaceous chalk (including a thin Greensand equivalent) oversteps onto Devonian or Carboniferous basement to the north. In the French sector there is a general southwards tilt giving the overall geometry of an inverted Permian–Jurassic basin with the inversion axis lying within the UK sector (see Fig. 12). Jurassic strata are now preserved in two distinct areas of the Melville sub-basin (Figs 1 & 7). The largest area is in the south within an EW-trending graben-like basin. Fig. 5 shows part of an interpreted seismic section of a line trending NNW–SSE across the W part of the Melville sub-basin (line A–A′ on Figs 2 & 6). This section highlights the pronounced truncation and overstep of the pre-Cretaceous beds which are tilted towards the NNW. Also evident from this section is the marked onlap of the Permian and Lower Triassic sequence northwards onto the prominent ENE trending high in the N.

From the section it is evident that faults were active during the Permian and the early Triassic, because those at the northern end of the section pre-date and, in general, neither cut the Upper Triassic salt/claystones sequence nor detach upwards into it. The Upper Triassic sequence thickens northwards as it cuts down below the top of the Lower Triassic sandstone unit. The boundary is an erosional unconformity. This demonstrates that the topographic high of Permian and early Triassic times had become an area of localized subsidence in the Upper Triassic. The Permian and Lower Triassic sequences accummulated within a major half-graben. The bounding fault lies close to the UK/French boundary (see Figs 1, 4, 6 & 12).

Fig. 8 shows a depth converted interpretation of a seismic line. Fig. 5 is a part of this section. This balanced section has been sequentially restored (Dahlstrom 1969; Gibbs 1983), although no allowance for decompaction has been made. The restored sections demonstrate the following sequence of events:

1 An early Permian rifting phase with synrift clastic sedimentation and associated volcanism. NNW–SSE extension probably occurred within a major half-graben. The southern bounding fault (north dipping) lay a short distance south of the section close to UK/French boundary (see also Fig. 12).

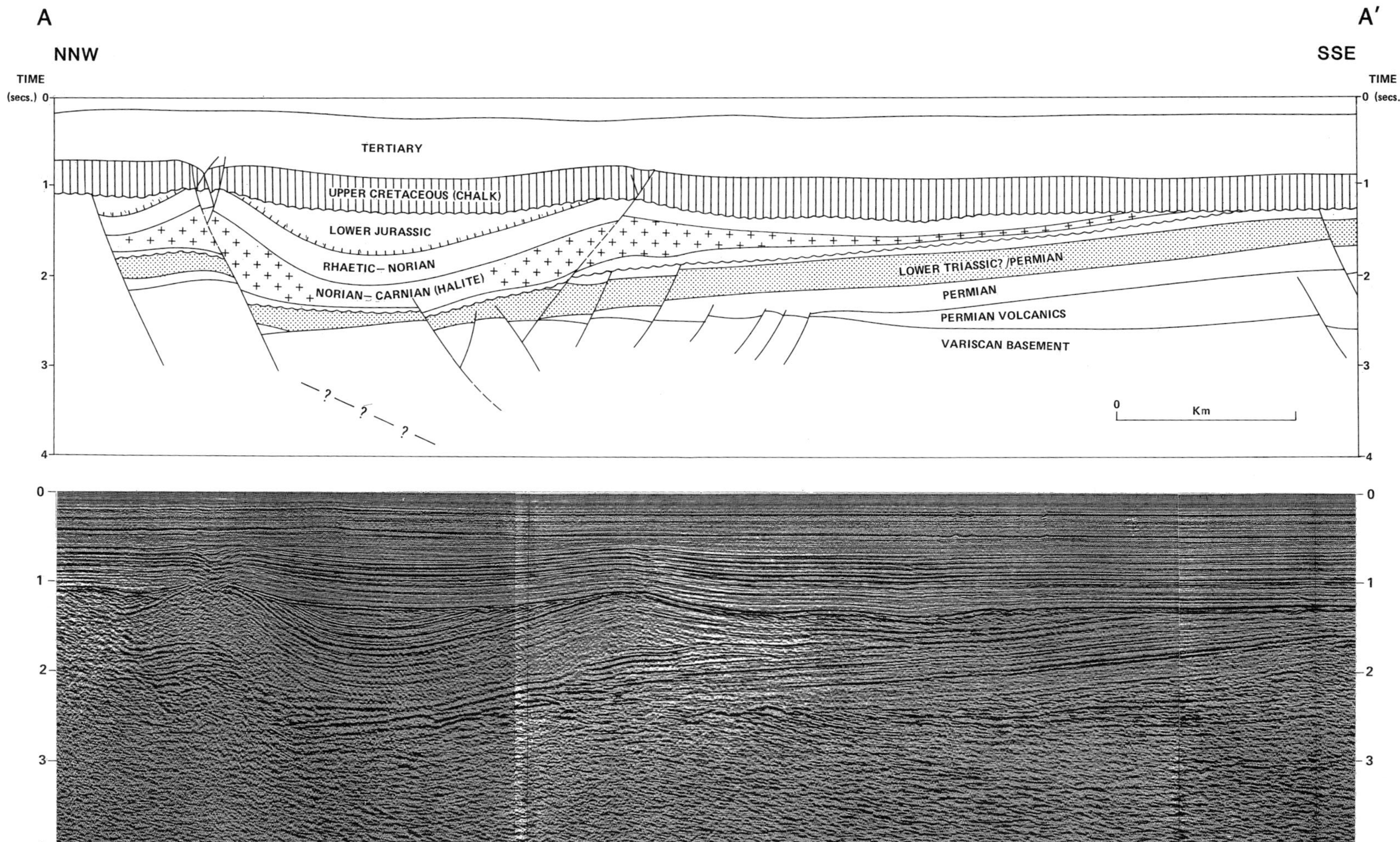

FIG. 5. Part of an interpreted seismic profile along line A–A′ (Fig. 6). Whole of section A–A′ shown in Fig. 8.

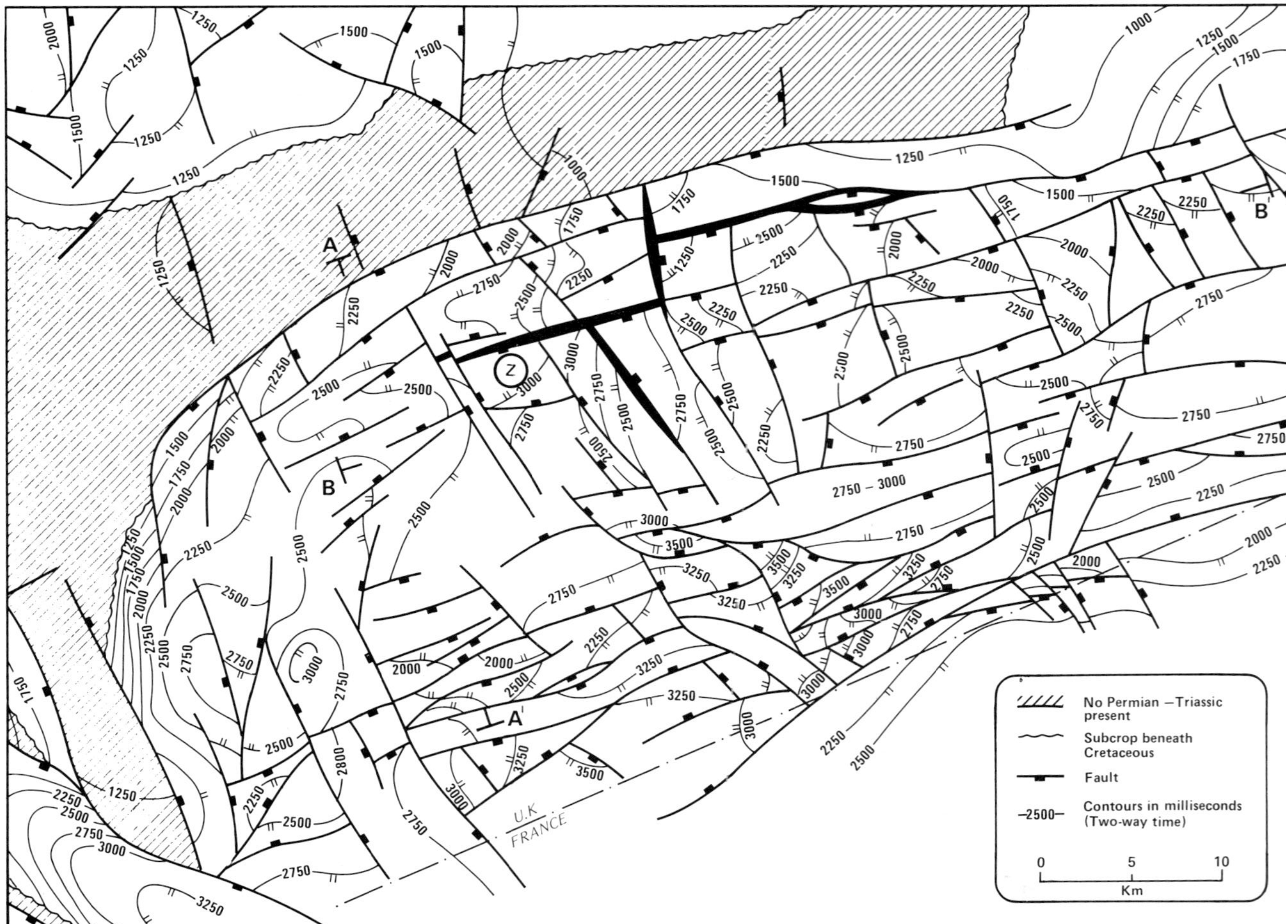

FIG. 6. Structural map of the Melville sub-basin with contours to base Permo–Triassic in ms (two-way-time). Z, typical quarter-graben controlled by a reactivated Variscan thrust and transfer fault. A–A′ and B–B′ are the cross-sections shown in Figs 8 & 10, respectively.

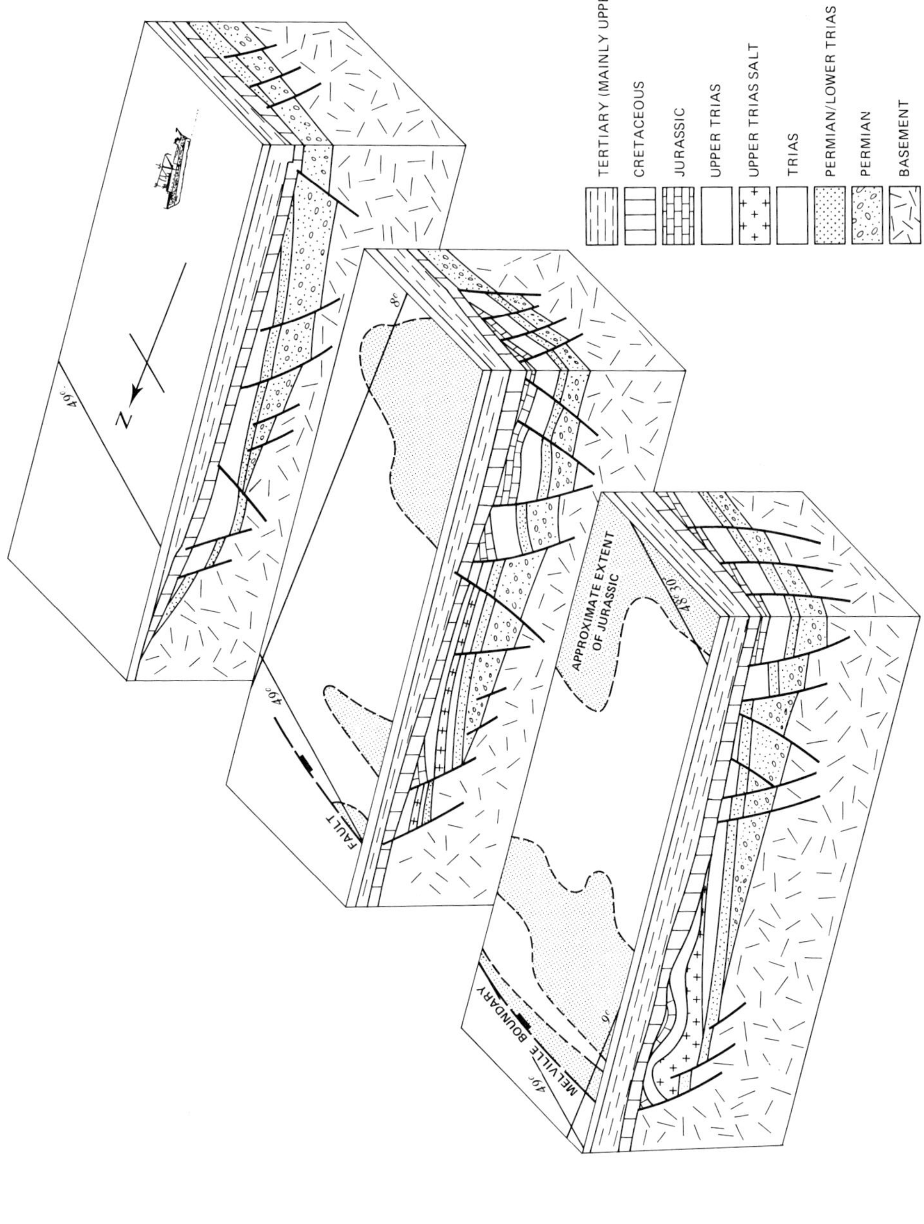

FIG. 7. Block diagram to illustrate general structural and stratigraphical features of the Melville sub-basin. Vertical scale approximately ×3 horizontal scale.

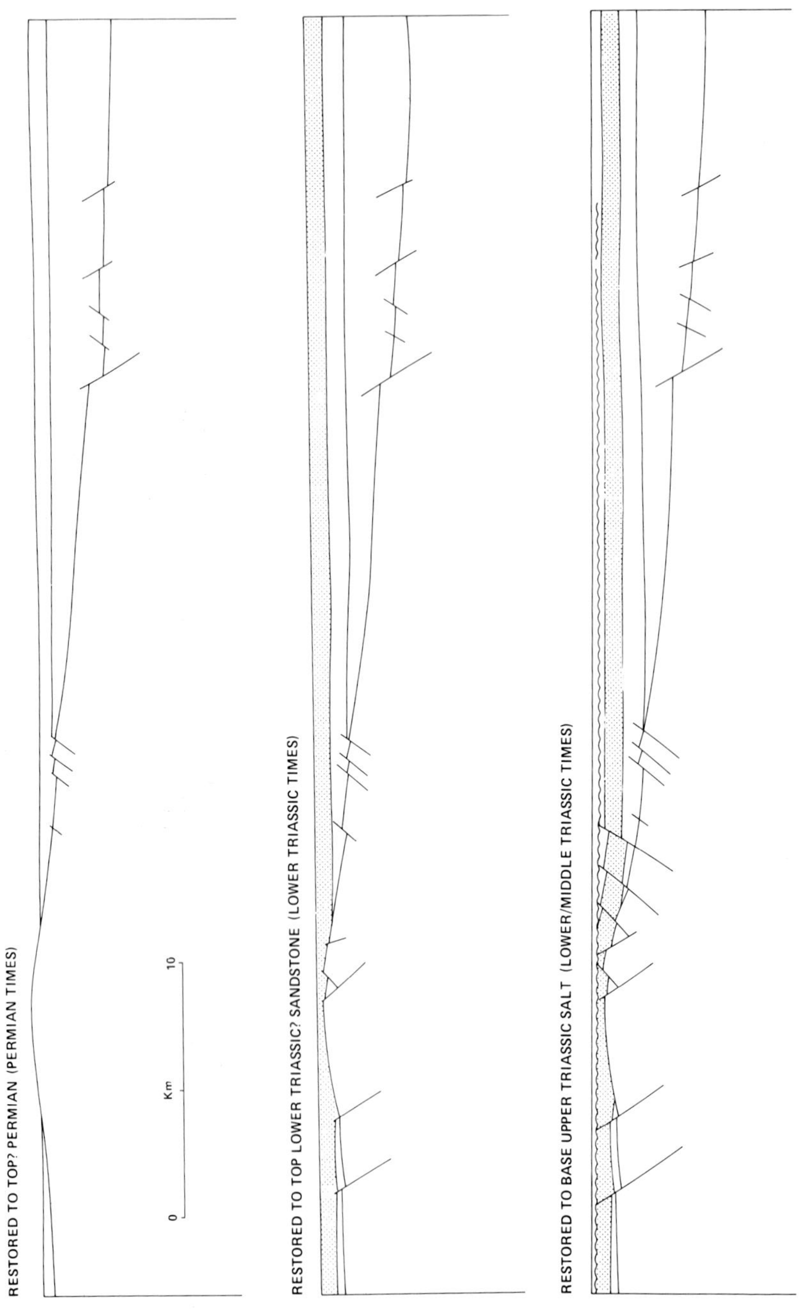
RESTORED TO TOP? PERMIAN (PERMIAN TIMES)
0
Km
10
RESTORED TO TOP LOWER TRIASSIC? SANDSTONE (LOWER TRIASSIC TIMES)
RESTORED TO BASE UPPER TRIASSIC SALT (LOWER/MIDDLE TRIASSIC TIMES)

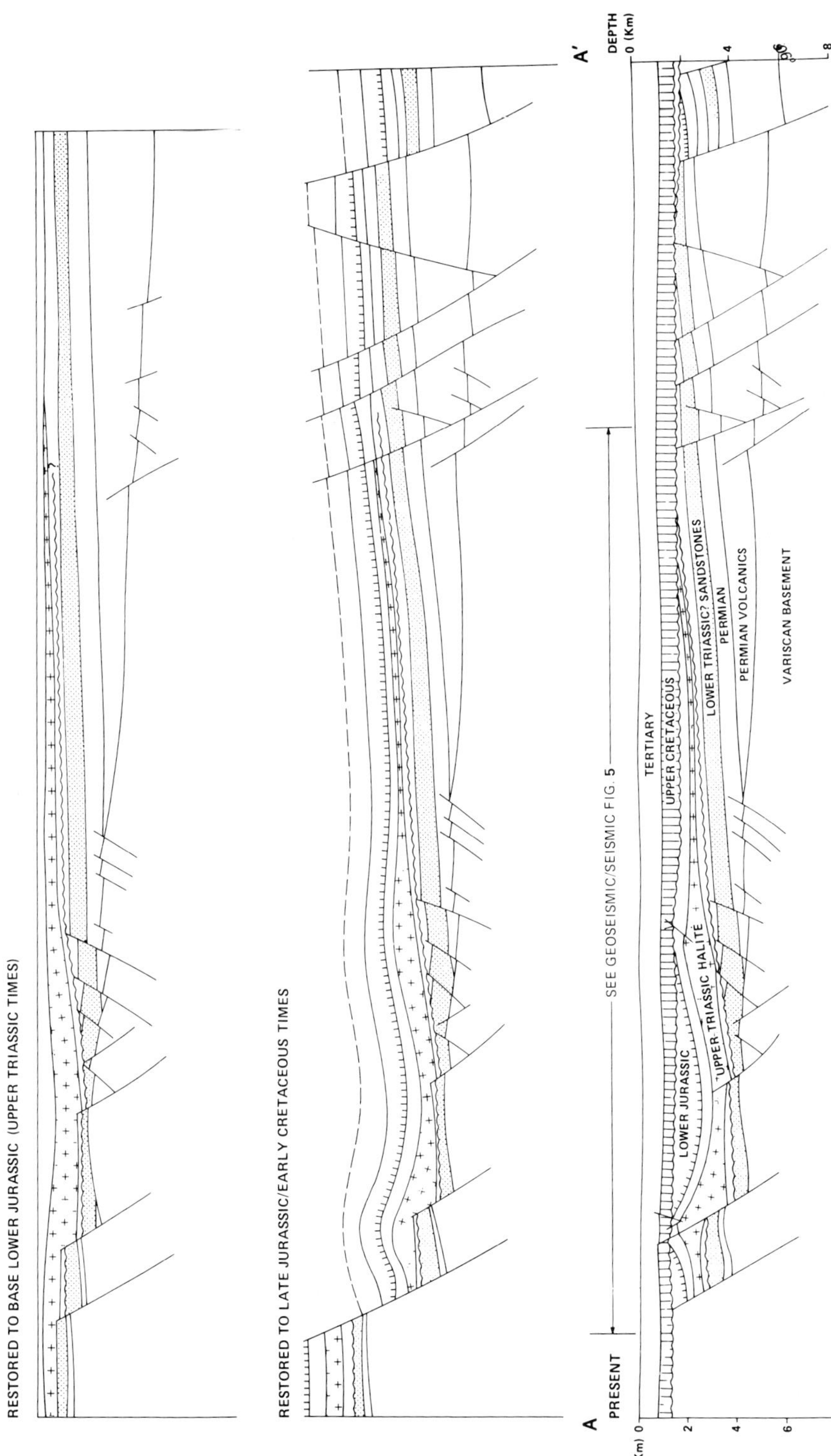

FIG. 8. Sequential structural restorations of line A–A′, based on the depth conversion of a seismic profile. See text for details.

2 Progressive northwards onlap took place onto the high to the north. This high was actually a ENE trending ridge with a further (minor?) area of subsidence to the N. The subsidence was S of the main Cornubian High which may not have come into existence until later.
3 In the early Triassic local uplift and doming occurred on an axis located on the high in the north. This doming was accommodated in part by the development of inward dipping faults. This gives the appearance of an uplifted rift axis, although the faulting was only a relatively local development. There was little or no extension across the basin at this time.
4 Accompanying, or immediately following the up-doming event was a major period of erosion which bevelled off the rift axis and cut down to the Lower Triassic sandstone sequence.
5 During the Upper Triassic the ENE-trending high subsided, allowing the deposition of a thick salt, claystone and anhydrite sequence as well as thinning and onlapping away from the centre of this local basin (see Fig. 4).
6 Sedimentation continued unbroken(?) into the Jurassic. Later a major faulting occurred. However, severe erosion has removed evidence of the exact nature and timing of this tectonic episode. Elsewhere the evidence of base Callovian and regional intra-Oxfordian unconformities preserved in the Western Channel Basin and Brittany Trough reflect tectonic instability at this time (Ziegler 1987b).

It appears that at some time in the late Jurassic an episode of extensional rifting ($\beta=1.05$ as estimated from Fig. 8) combined with wrench movements and a major episode of basin uplift and doming affected the area. The apparent 'inversion' axis lay just north of the French/UK international boundary (See Figs 11D & 12). Footwall uplift and erosion was at its maximum immediately north of main rift axis.

Fig. 9 shows part of seismic section with an interpretation of line B–B′ running ENE, approximately perpendicular to the NNW–SSE-trending faults (see Figs 2 & 6). It is clear that the majority of these faults pre-date the Upper Triassic sequence which includes salt. There is a pronounced erosional unconformity cutting down to the Lower Triassic Sandstone unit, which is most severe on the upthrown sides of the faults. Seismic data reveals that at least one of these faults had a pronounced fault scarp (towards the western end of Fig. 10). Sequential structural restoration of the whole section (Fig. 10) indicates an erosional bevelling off after faulting and prior to the deposition of the Upper Triassic sequence.

These pre-Upper Triassic faults are clearly listric. They sole out towards the east. Reflections are produced by some of the fault planes (Fig. 9).

Fig. 10 reveals the following sequence of events:

1 In the early Permian synrift clastic sedimentation was related to extension by rifting approximately out of the plane of the section. Despite the development of a slight high at the eastern end of the section, most of the NNW–SSE trending faults visible on Figs 9 & 10 were acting as transfer faults at this time. The motion was dominantly strike–slip.
2 The next major episode (Lower or Middle? Triassic) involved ENE–WSW extension along the section (along the basin) by rifting on the NNW–SSE-orientated faults with an extension factor of 1.05 (estimated from Fig. 10). These faults, which had previously acted as transfer faults, now became normal faults. Uplift and footwall erosion affected many of them.
4 In the Upper Triassic deposition of claystones, evaporites and a thick salt unit commenced. These beds thicken on the hanging wall side of the faults. In at least one case this is due to the infilling of a topographic depression below a fault scarp. On other faults, however, thickening may have been due to syntectonic sedimentation on active growth faults.

Fig. 11 is a series of fault activity maps which illustrate the development of the Melville sub-basin area and show which faults were active at a particular time. Of interest is the switch from NNW extension during the Permian to extension along the basin (ENE) at the end of Lower/Middle Triassic times. This could be related to the formation of the Little Sole Permian–Triassic half-graben further west. This half-graben trends NW–SE and runs parallel to the present continental slope, which is now separated from the Melville sub-basin by a prominent high (Fig. 1). There was certainly some relationship between this ENE extension and the similarly orientated extension across the NNW trending Permian–Triassic Bay of Biscay rift (Ziegler 1982).

The evolution of the Western Approaches Basin should be considered with reference to both the UK and French parts of the basin. The schematic sequence of events shown in Fig. 12

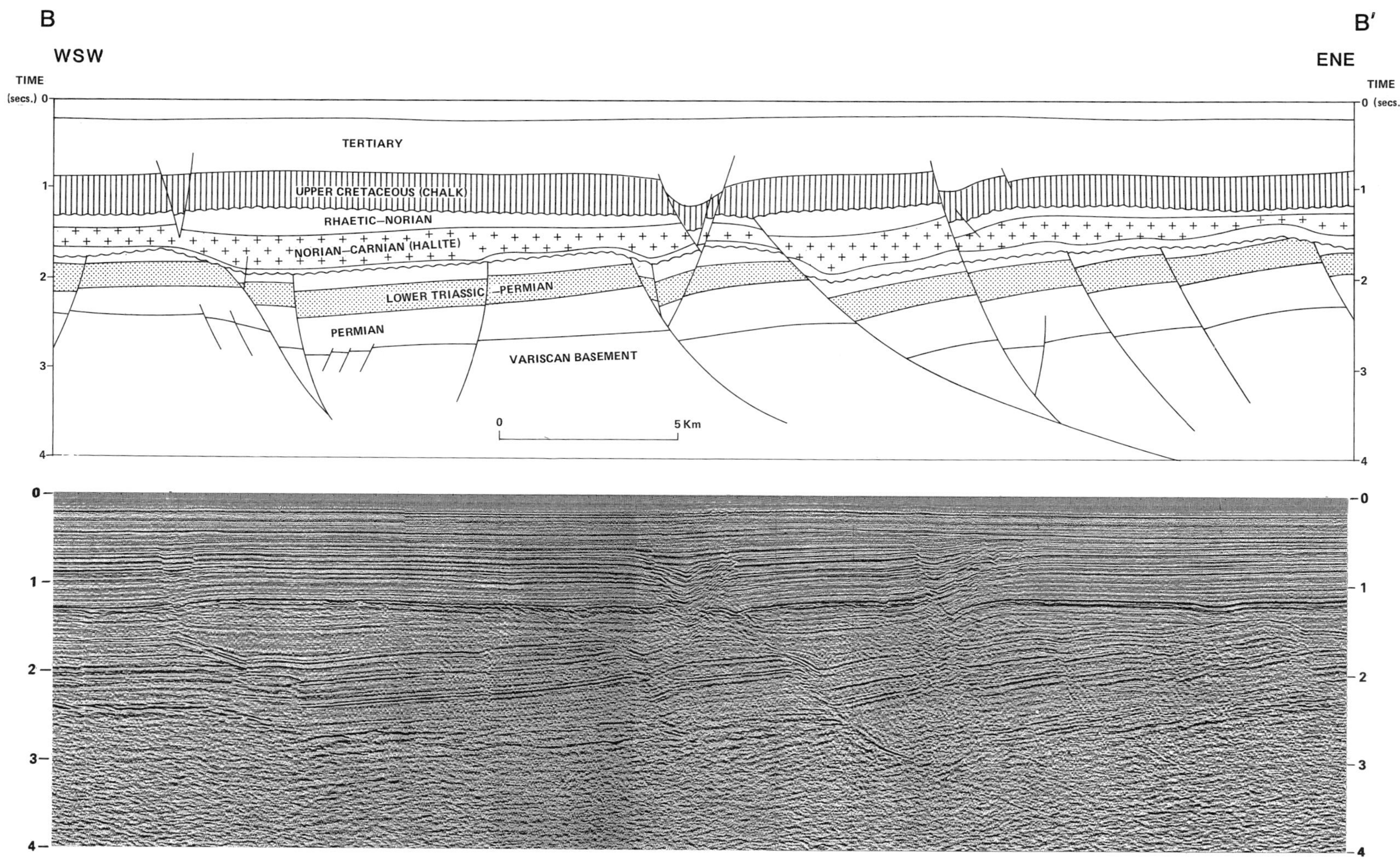

FIG. 9. Part of an interpreted seismic profile along line B–B′ (Fig. 6). Whole section B–B′ shown in Fig. 10.

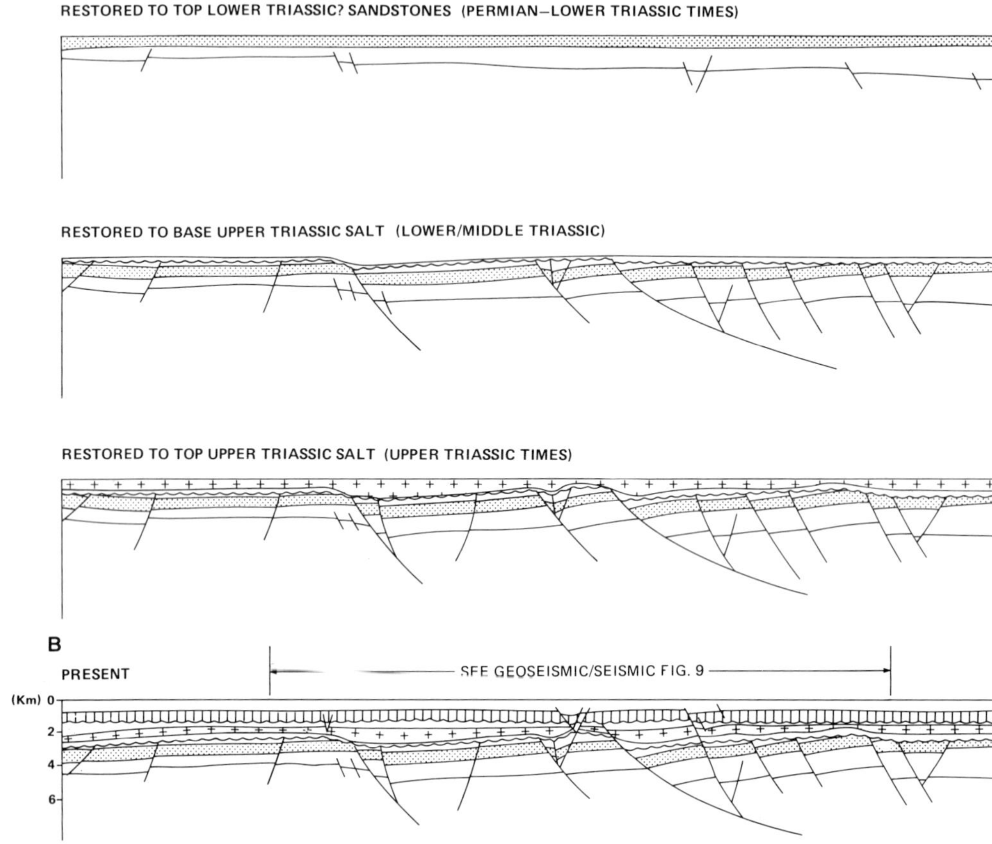

FIG. 10. Sequential structural restorations of line B–B′ based on the depth conversion of a seismic profile. See text for details. Ornamentation as in Fig. 9.

illustrates the Permian to Cretaceous history of the area. These reconstructions are based on a generalized section across the Melville sub-basin into the French sector, which is drawn approximately to scale. The areas of the stratigraphic units are only approximately maintained. It should be noted that some strike–slip movements have occurred. Therefore, the section may not balance precisely (Gibbs 1983). It is evident (Fig. 12) that more Jurassic section is preserved in the French sector but the Permian–Triassic sequence is much thinner (Ziegler 1982, 1987b). It appears that the major part of Permian–Triassic basin lay within the UK sector. The later Jurassic (including Upper Jurassic?) basin was much wider and extended across the whole area. It probably became deeper within the French sector.

During the early Permian a major half-graben developed within the UK sector. The southern bounding fault to this half-graben was a northward dipping normal fault with a downthrow to the N. This contrasts with the eastern part of the Western Approaches Basin where the faults on the S side of the basin all dip S (See Ziegler 1982 & 1987b). Volcanics and clastics were followed by the deposition of mudstones and sandstones. These sediments onlapped a local high to the north. The overall geometry of this high resembles a dome-like axis with a further (minor?) half-graben depositional area to the north (Fig. 12). This axis is characterized by inward dipping faults. In places some of these south dipping faults sole onto pre-existing Variscan basement thrusts (Fig. 12). During the Lower Triassic an up-doming of this axis occurred, at least its W end (Fig. 8). This was followed by erosion. A phase of subsidence in the Upper Triassic produced a local basin in which a salt–claystone sequence was deposited.

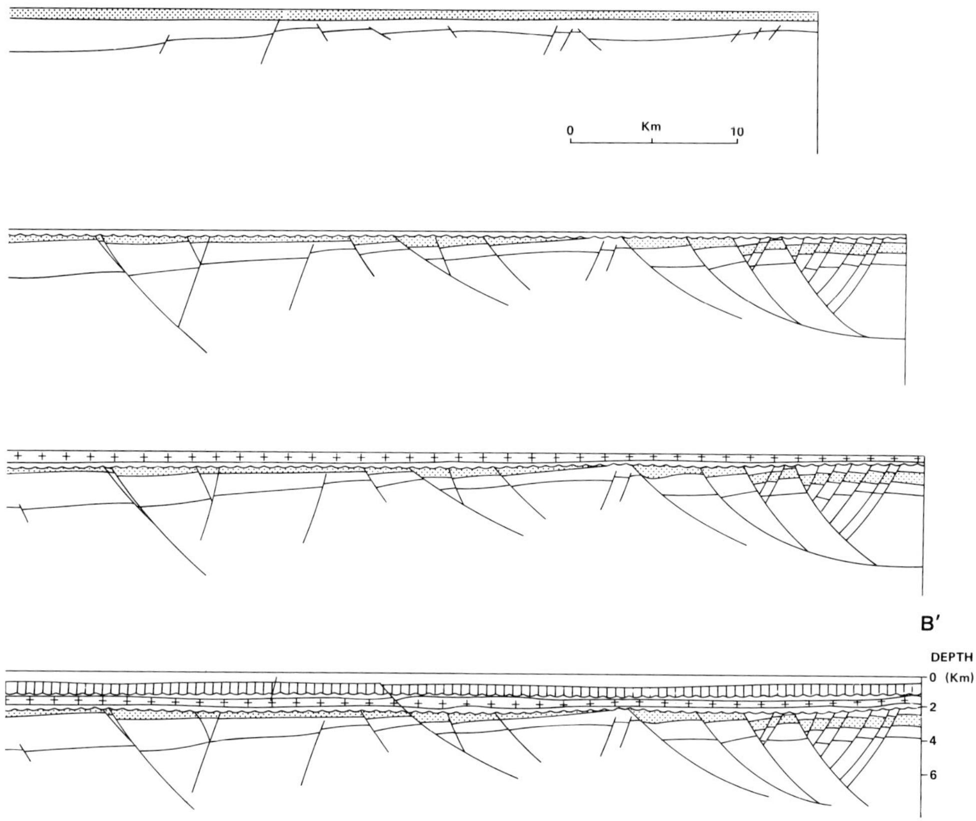

Jurassic deposition covered a much greater area than that occupied by the present Western Approaches Basin. The present N boundary fault to the Melville sub-basin is late Jurassic or early Cretaceous in age. It is difficult to know if the Cornubian High was in existence prior to this. Up to 2000 m of Jurassic sediments were deposited. Less than 1000 m of Lias are generally preserved in the UK sector.

In late Jurassic–early Cretaceous times a major doming, rifting and regional uplift event occurred which inverted the basin. In places all of the Jurassic and some of the underlying Triassic strata were removed. The maximum uplift was on the northern shoulder of the rift in the centre of the dome (Figs 5, 11D & 12). A complicated series of ENE-trending dip–slip extensional faults are linked by NNW trending transfer faults. It is possible that the latest Jurassic sequences were deposited syntectonically with this faulting, doming episode.

This doming and regional uplift of the W part of the Western Approaches Basin resulted in the overall appearance of an inverted basin. However, the lack of evidence of compressional stresses at this time implies that inversion is perhaps not an appropriate term. The late Jurassic doming and rifting reflects crustal extension which probably reflects mantle processes. This, combined with a regional intra-Berriasian low stand in sea level (Ziegler 1987b), resulted in a major truncation of the Jurassic strata and bevelling off of the dome.

It is also likely that there were strike–slip movements at this time. Compressional folds within the Jurassic can be mapped (Fig. 11D). This local compression (or transpression) appears to have resulted from an episode of left-lateral strike–slip movement.

These movements and the uplift and erosion

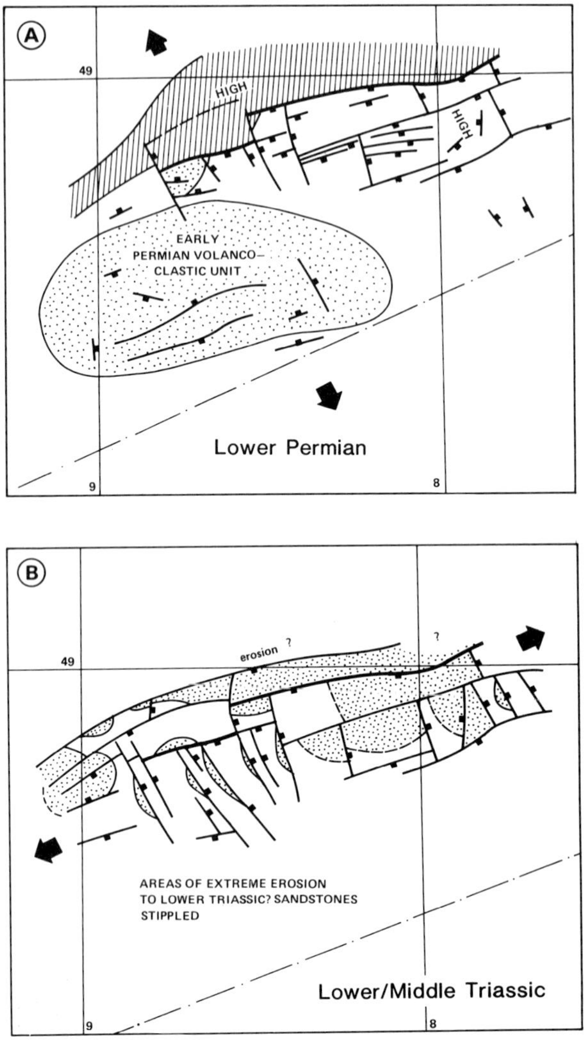

FIG. 11. Fault activity maps for the Melville sub-basin. Arrows show direction of extension or sense of wrenching. Fold axes are also shown in D.

occurred prior to Wealden times. This is seen in the south of the Melville sub-basin where a thin Wealden sequence overlies truncated faulted and folded Jurassic or older strata. This unconformity may mark the end of the rifting phase in the basin. The left-lateral strike–slip movements contrast with the younger (Lower Cretaceous) right-lateral movements described by van Hoorn (1987) in the South Celtic Sea.

Discussion

The most important points relating to the Permian–Cretaceous evolution of the Melville sub-basin are the development of an ENE-trending Permian–early Triassic graben or half-graben; a major ENE extensional episode along the basin during the Lower–Middle Triassic and a major rifting and doming event in the later Jurassic, combined with a regional uplift, which effectively inverted the Permian–Jurassic basin (Fig. 12).

The first two of these events clearly relate to the Permian–Triassic rifting of the Western Europe/North American part of Pangea which in the Bay of Biscay, Western Approaches and Celtic Sea areas was underlain by Variscan/Hercynian orogenic basement (Debyser *et al.* 1971; Ziegler 1982; Gardiner & Sheriden 1983).

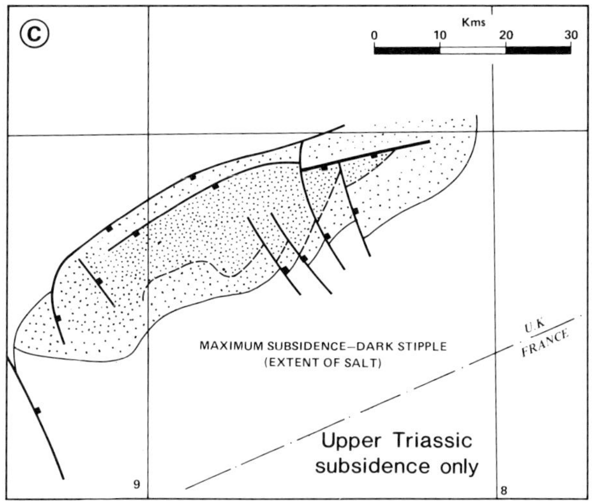

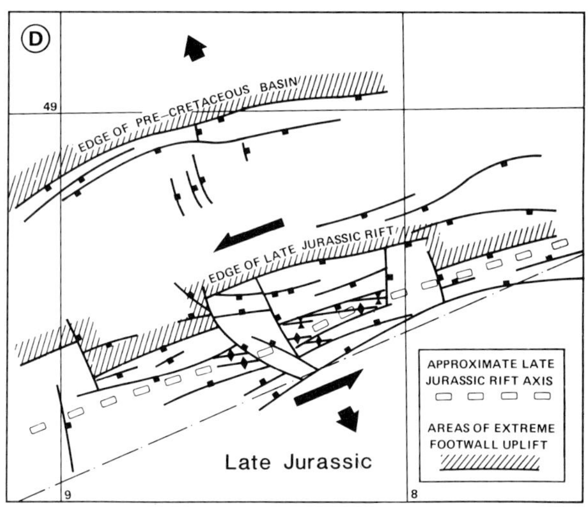

The major rifting, doming and uplift event that effectively inverted the Western Approaches Basin is seen to be post Lower Jurassic and pre-Berriasian in age. Other major Cimmerian unconformities are present in the North Atlantic region. A late Jurassic–early Cretaceous unconformity is present in the Bay of Biscay (Roberts *et al.* 1980), the Galica Bank area (Sibuet *et al.* 1980), and on the western Scotian shelf on the western side of the present Atlantic Ocean (Sibuet *et al.* 1980). In the Bay of Biscay this event is earliest Lower Cretaceous in age (intra-Berriasian; Ziegler 1982).

Further NW in the South Celtic Sea and Bristol Channel Basins, the most prominent unconformity is post-Wealden–pre-Greensand, although there is an unconformity at the base of the Wealden. A thin cover of Aptian clastics overlying both basement and folded Triassic to Jurassic gives evidence of this deformation/inversion event which is late Berriasian–pre-Aptian in age (Kamerling 1979; Naylor & Shannon 1982; van Hoorn 1987).

In contrast, in the Western Approaches Basin the major unconformity is pre-Wealden in age (pre-Berriasian). The late Jurassic or early Cretaceous rifting and deformation event is usually related to the beginning of sea-floor spreading in the Atlantic and Bay of Biscay (Roberts *et al.* 1980; Ziegler 1982; Kamerling

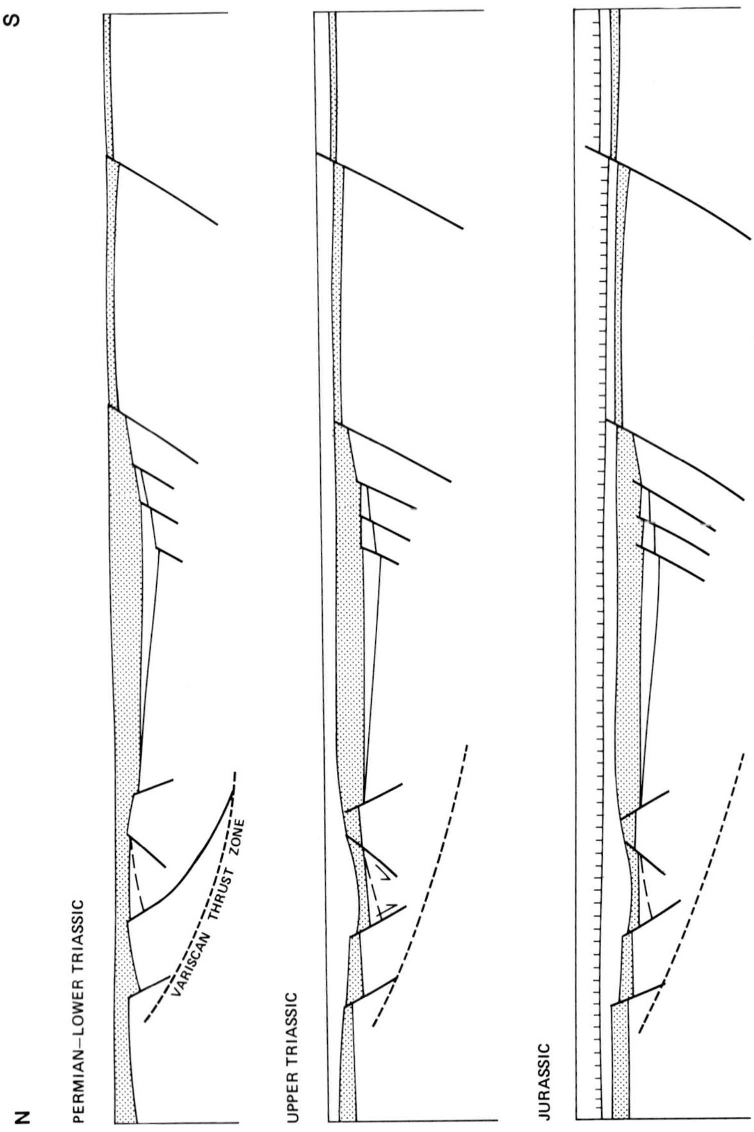
S
N
PERMIAN–LOWER TRIASSIC
VARISCAN THRUST ZONE
UPPER TRIASSIC
JURASSIC

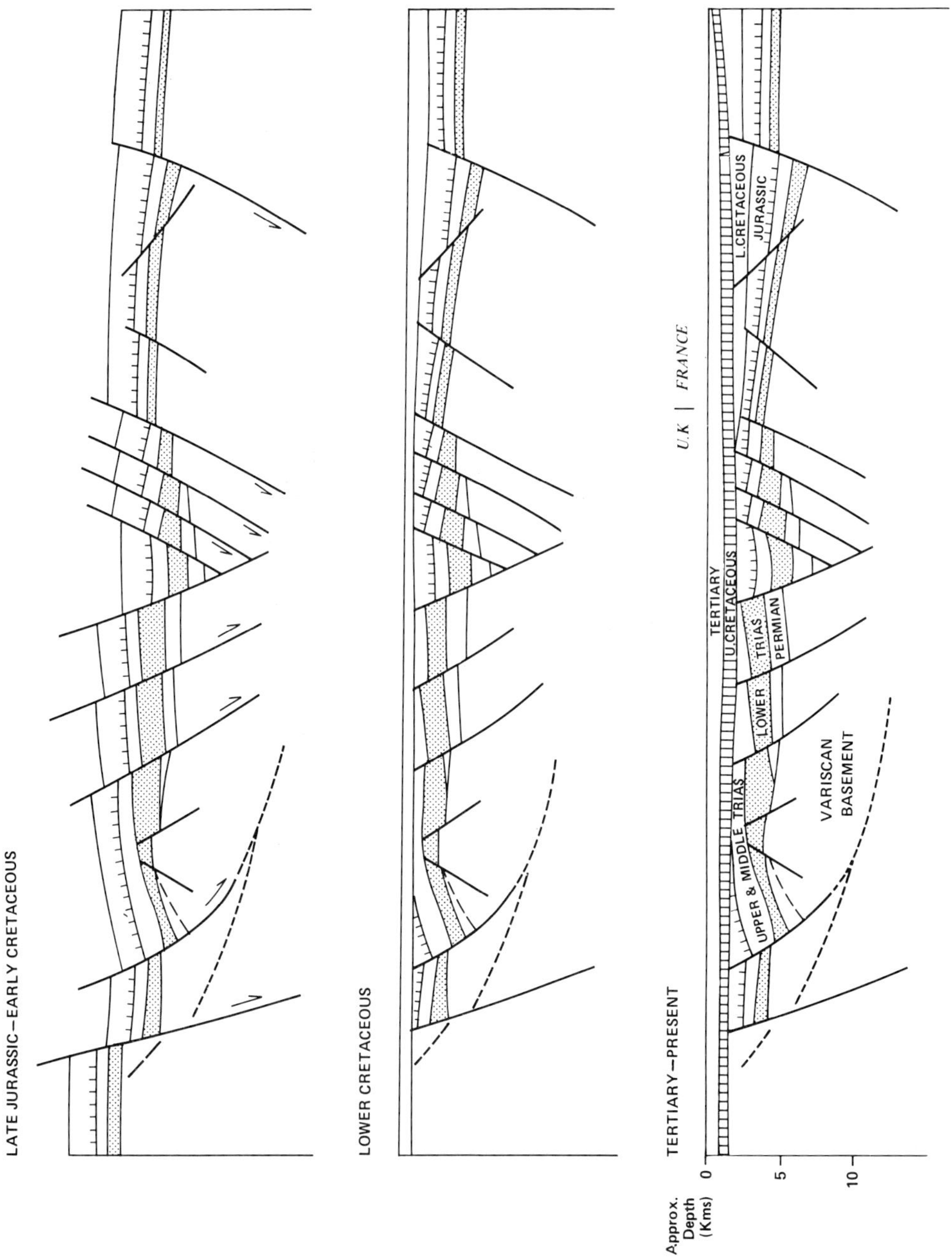

Fig. 12. Generalized evolutionary sequence across the Melville sub-basin into the French sector based on seismic profiles. Approximately true scale. The thickness of the Permian and Triassic in the French sector is speculative but is probably thin as shown (see Ziegler 1982; 1987b).

1979). This may have migrated northwards with time, hence the contrast in age between the major unconformities in the Western Approaches and South Celtic Sea Basins.

The considerable uplift on the northern shoulder of this rift in the Melville sub-basin removed up to 2000 m of Jurassic strata (Fig. 11D). Such a large uplift could have been caused by a variety of mechanisms including a thinning of the lithospheric part of the mantle. This could be due to a greater amount of stretching compared with the crust or a genuine asthenosphere upwelling. The overall doming of the rift (Fig. 12) could indicate that the latter possibility is more likely than an inhomogenous stretching (Royden & Keen 1980; Coward 1986; Badley *et al.* 1988).

With the commencement of spreading, any uplifted areas along the margins of the pre-Atlantic–Bay of Biscay rift would have begun to subside when the adjacent continental areas begin to cool as the accreting ocean ridge moved further away westwards (Roberts *et al.* 1980; Sibuet *et al.* 1981). Possibly the Lower Cretaceous was sourced from this uplifted area to the west (Allen 1981) prior to any overall thermal subsidence. Roberts *et al.* (1980) point out that the cooling phase in the northern Bay of Biscay, SW of the Western Approaches Basin, started at about 120 Ma. In contrast, the South Rockall Plateau further N did not subside until 52 Ma. Thus the northward migrating inception of Atlantic spreading may have influenced rifting and deformation of the adjacent continental margin basins, the timing of which also became later northwards. This is dated as pre-Wealden (early Berriasian?) in the Western Approaches Basin and pre-Greensand (post-Berriasian to pre-Aptian) in the South Celtic Sea Basin.

The postrift thermal subsidence phase (McKenzie 1978; Dewey 1982) probably resulted from cooling following lithospheric stretching of the Western Approaches Basin, the Bay of Biscay and the North Atlantic rift to the W. There is no simple relationship between extension, as estimated from seismic sections, for the rifting phase and the subsequent thermal subsidence phase across the Western Approaches and Celtic Sea basins because of their proximity to the early North Atlantic Ocean. For this reason there has been no attempt to relate stretching during the tectonic extension phase to footwall uplift (Barr 1987) or thermal subsidence during the flexural phase (McKenzie 1978; Dewey 1982).

Deep structure and the basin origin

Although the commerical GECO seismic survey is processed only down to 5 s two way time, it does reveal a series of dipping events within the basement similar to those seen on deeper, non-commerical lines (BIRPS & ECORS 1986). These reflectors have been mapped out by Hillis & Day (1987) in the Western Approaches. They described a series of planar events dipping at 20–35° (after depth conversion) to the SE which may represent thrusts with a similar trend to the Variscan structural grain of Southwest Cornwall (see also Day & Edwards 1983). Hillis & Day point out that some of these Variscan thrusts locally exert a control on the Permian–Mesozoic basis. Others clearly do not.

In the Melville sub-basin the two prominent sets of faults, ENE and NNW (Figs 6 & 4) may be controlled by structures within the Variscan basement. In view of the fact that a well developed set of NNW trending strike–slip faults are developed with a fairly constant orientation across the Variscan orogenic belts in SW

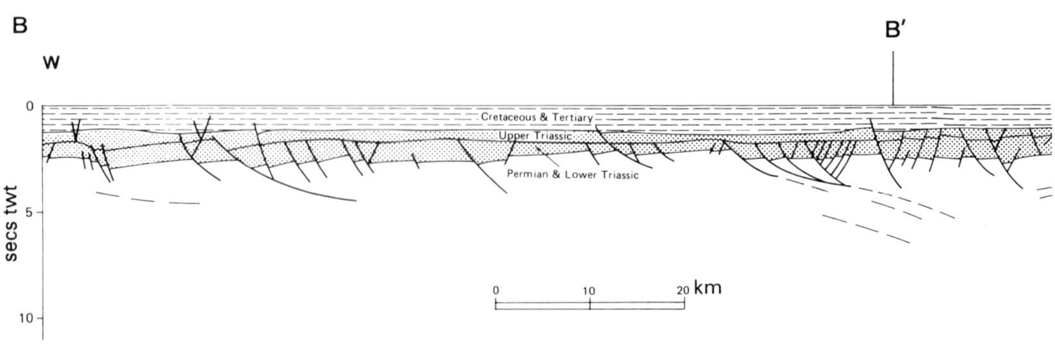

FIG. 13. ENE seismic profile along the axis of the Western Approaches Basin (see Fig. 2) showing events within the basement. Position of reflection Moho is from SWAT 6.

England, South Wales and Southern Ireland (e.g. see Dunning 1985; Cooper *et al.* 1986), it seems likely that the NNW fault trend in the Melville sub-basin is controlled by Variscan basement strike–slip faults. Brooks *et al.* (1988) suggest a similar origin for the highly compartmentalized Triassic–Jurassic Bristol Channel Basin. In the Melville sub-basin the orientation of the ENE fault set could have been controlled to some extent by reactivated Variscan thrusts as indicated by the basement reflectors (Day & Edwards 1983; Hillis & Day 1987). An extension approximately perpendicular to the Variscan trend could have reactivated some Variscan faults. Seismic evidence suggests reactivation of Variscan thrusts, as indicated by the small half-graben at the N end of Fig. 12.

The Lower or Middle Triassic extension along the basin axis (Fig. 11B) was accommodated on listric faults which dip mainly towards the ENE. Some of these faults appear to flatten downwards towards shallow reflectors in the basement as shown at the western end of Fig. 13 which shows a section along the centre of the Western Approaches Basin (Section B–B′B″ on Fig. 2). At the eastern end of this section planar extensional faults also appear to be controlled by a basement reflective zone. These basement events actually dip SSE–ESE obliquely out of the plane of Fig. 13 and clearly do influence the Mesozoic normal faults.

These same basement reflectors are well displayed on SWAT 6 (see Fig. 2), a deep seismic line shot by BIRPS (BIRPS & ECORS 1986). Part of SWAT 6 (unmigrated) is shown in Fig. 14. The line is interpreted in part using a coincident migrated GECO line, which has much better resolution. The deep reflectors (may be both real events and diffractions produced by, for example, anastomosing shear zones (Reston 1987). In particular, it is possible to demonstrate the position of a steeply dipping fault in the centre of the profile which displaces both the Permian–Mesozoic basin and a package of deeper reflectors in the basement and also produces diffractions. On the unmigrated line SWAT 6 the reflectors just above 5 s to the right of this fault (Fig. 14) are difficult to separate from the diffractions. However, they are clearly visible on the GECO line.

The purpose of drawing attention to this steep fault is that two recent interpretations of the SWAT 6 line show major faults dipping along the southward dipping reflective zones. These faults control the Permo–Triassic basin geometry (Cheadle *et al.* 1987; Beach 1987), if real. It is possible that early basin development was controlled by reactivated Variscan thrust zones but later steep normal and/or strike faults such as that in Fig. 14 cut through these zones. Beach *et al.* (1987) have proposed a similar two-phase evolutionary model for the Viking Graben of the North Sea. The suggestion of Badley *et al.* (1988) that early rifting extension may be accommodated on shallow listric normal faults but later thermal subsidence could involve some steeper planar faults is also relevant. The steep fault shown in Fig. 14, is laterally continuous, as are others in the eastern part of the Western Approaches (see Fig. 1).

These may be strike–slip faults, or oblique–slip with both normal dip–slip and strike–slip displacements. According to Ziegler (1987b) strike–slip faults were active during a phase of wrench deformation in the earliest Aptian which coincided with crustal separation in the Bay of Biscay. This appears to be coincident with the earliest part of the postrift subsidence phase in the Western Approaches and is later than the phase of left-lateral strike–slip deformation, which occurred in the late Jurassic towards the end of the rifting phase. It is therefore conceiv-

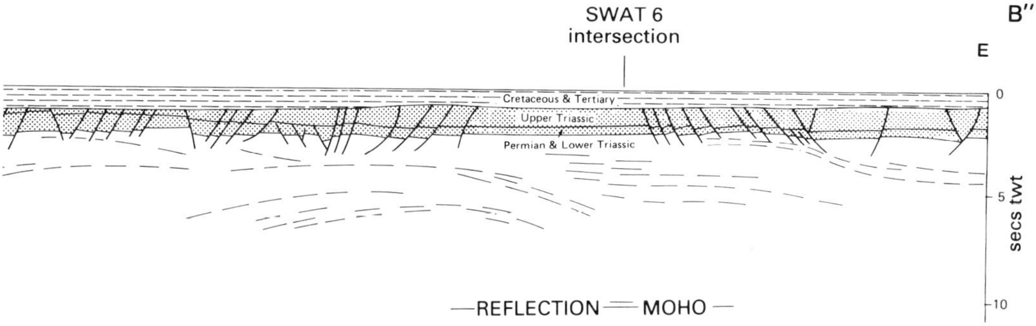

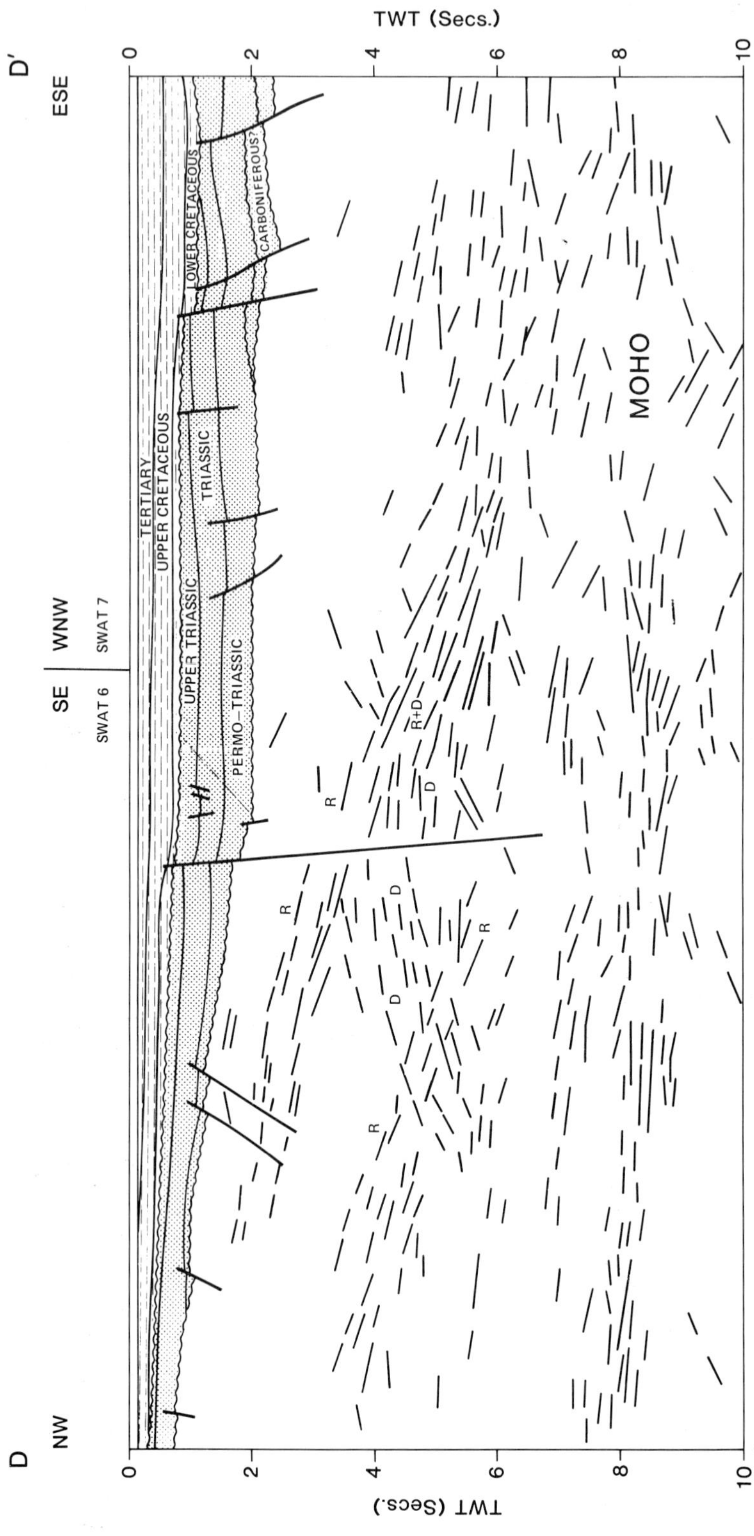

FIG. 14. Part of SWAT 6 interpreted using a coincident migrated GECO line for the upper 5 s. R, reflection; D, diffraction.

able that early rifting extension was accommodated on normal faults, some of which were listric and detached onto reactivated Variscan thrusts. This may have been followed at a later stage by the activation of steeper normal and/or strike–slip faults related to, or coincident with, the thermal subsidence phase. However, it should be noted that other apparently older (Permo–Triassic) faults also cut through basement reflection packages in Fig. 14. This may indicate little or no influence by reactivated Variscan thrust zones on the development of the Permian–Mesozoic basin.

This paper does not provide a preferred model for the origin and deep structure of the Western Approaches Basin. In agreement with Cheadle *et al.* (1987) it is felt that the elongate shape of the Western Approches, Haig Fras, South and North Celtic Sea Basins is consistent with a mainly crustal extension origin. Wrench-induced subsidence is of a minor nature only. However, the subsidence and inversion history of the Western Approaches Basin was complex. Shallow detaching planar and listric faults and steep planar faults developed at various times. Although basement structure had some control on the basin geometry, the full extent of its influence is difficult to ascertain at present.

ACKNOWLEDGMENTS: The management of Merlin Geophysical Ltd and GECO Geophysical Company Ltd. are thanked for permission to publish confidential seismic sections. Thanks are also due to the drafting department of GECO Exploration Services for producing figures. The author is grateful to Murphy Petroleum Ltd. and Amerada Hess Ltd. for access to confidential information.

References

ALLEN, P. 1981. Pursuit of Wealden models. *Journal of the Geological Society of London*, **138**, 375–405.

AVEDIK, F. 1975. The seismic structure of the Western Approaches and the South American continental shelf and its geological interpretation. *In*: WOODLAND A. W. (ed.) *Petroleum and the Continental Shelf of North-West Europe, Vol. 1: Geology*. Applied Science Publications, London, 29–44.

BADLEY, M. E., PRICE, J. D. RAMBECH DAHL, C. & AGDESTEIN, T. 1988. The structural evolution of the northern Viking Graben and its bearing upon extensional modes of basin formation. *Journal of the Geological Society of London*, **145**, 455–472.

BARR, D. 1987. Lithospheric stretching, detached normal faulting and footwall uplift. *In*: COWARD, M. P., DEWEY, J. F. & HANCOCK, P. L. (eds) *Continental Extensional Tectonics. Special Publication of the Geological Society of London*, **28**, 75–94.

BEACH, A. 1987. A regional model for linked tectonics in North-West Europe. *In*: BROOKS, J & GLENNI, K. W. (eds) *Petroleum Geology of North-West Europe*. Graham & Trotman, London, 43–48.

BEACH, A., BIRD, T. & GIBBS, A. 1987. Extensional tectonics and crustal structure; deep seismic reflection data from the northern North Sea Viking Graben. *In*: COWARDS, M. P., DEWEY, F. J. & HANCOCK, P. L. (eds) *Continental Extensional Tectonics. Special Publication of the Geological Society of London*, **28**, 467–76.

BENNET, S., COPESTAKE, P. & HOOKER, N. P. 1985. Stratigraphy of the Britoil 72/10–1A Well, Western Approaches. *Proceedings of the Geologists Association, London*, **96**, 225–261.

BRITISH GEOLOGICAL SURVEY (BGS) 1985. Pre-Permian 1 : 1 000 000 contour map (compiled by N. S. P. SMITH).

BIRPS & ECORS, 1986. Deep seismic reflection profiling between England, France and Ireland, *Journal of the Geological Society of London*, **143**, 45–52.

BROOKS, M., TRAYNER, P. M. & TRIMBLE, T. J. 1988. Mesozoic reactivation of Variscan thrusting in the Bristol Channel area, U.K. *Journal of the Geological Society of London*, **145**, 439–444.

CHEADLE, M. J. MCGEARY, S., WARNER, M. R. & MATTHEW, D. H. 1987. Extensional structures on the Western UK continental shelf: A review of evidence from deep seismic profiling. *In*: COWARD, M. P., DEWEY, J. F. & HANCOCK, P. L. (eds) *Continental Extensional Tectonics. Geological Society Special Publication*, **28**. 445–465.

COOPER, M. A., COLLINS, D. A., FORD, M. A., MURPHY, F. X., TRAYNER, P. M. & O'SULLIVAN, M. 1986. Structural evolution of the Irish Variscides. *Journal of the Geological Society of London*, **143**, 53–61.

COWARD, M. P. 1986. Heterogenious stretching, simple shear and basin development. *Earth and Planetary Science Letters*, **80**, 325–336.

DAHLSTROM, C. D. A. 1969. Balanced cross-sections. *Canadian Journal of Earth Science*, **6**, 743–757.

DAY, G. A. & EDWARDS, J. W. F. 1983. Variscan thrusting in the basement of the English Channel and SW Approaches. *Proceedings of the Ussher Society*, **5**, 432–436.

DEBYSER, J., LE PICHON, X. & MONTADERT, L. 1971. *Histoire Structurale du Golfe de Gascogne, Vols I & II*. Technip, Paris.

DEWEY, J. F. 1982. Plate tectonics and the evolution of the British Isles. *Journal of the Geological*

Society of London, **139**, 317–412.

DUNNING F. W. (ed.) 1985. Geological structure of Great Britain, Ireland and surrounding seas. *Geological Society of London* Map Chart.

EVANS, C. D. R., LOTT, G. K. & WARRINGTON, G. (Compilers) 1981. The Zephyr (1977) wells, South Western Approaches and Western English Channel. *Report of the Institute of Geological Science*, No. 81/8.

FISHER, M. J. & JEANS, C. V. 1982. Clay mineral stratigraphy in the Permo–Triassic Red Bed Sequences of BNOC 72/10–1A. Western Approaches and the South Devon Coast. *Clay Minerals*, **17**, 79–89.

GARDINER, P. R. R. & SHERIDAN, D. J. R. 1981. Tectonic framework of the Celtic Sea and adjacent areas with special reference to the location of the Variscan Front. *Journal of Structural Geology*, **3**, 317–331.

GIBBS, A. D. 1983. Balanced cross-section construction from seismic sections in areas of extensional tectonics. *Journal of Structural Geology*, **5**, 471–482.

HAMILTON, D. 1979. The geology of the English Channel, South Celtic Sea and continental margin, South Western Approaches *In*: BANNER, F. T., COLLINS, M. B. & MASSIE, K. S. (eds). *The North-West European Shelf Seas: 1 Geology and Sedimentology*, Elsevier, Amsterdam, 61–87.

HILLIS, R. R. & DAY, G. A. 1987. Deep events in U.K. South Western Approaches. *Geophysical Journal of the Royal Astronomical Society*, **89**, 243–250.

VAN HOORN, B. 1987. The South Celtic Sea/Bristol Channel Basin; origin, deformation and inversion history. *Tectonophysics*, **137**, 309–334.

KAMERLING, P. 1979. The Geology and hydrocarbon habitat of the Bristol Channel Basin. *Journal of Petroleum Geology*, **2**, 75–93.

MCKENZIE, D. 1978. Some remarks on the development of sedimentary basins. *Earth and Planetary Science Letters*, **40**, 25–32.

MONTADERT, L. 1977. Rifting and subsidence on passive continental margins in the North East Atlantic. *Nature*, **268**, 305–309.

MONTADERT, L., ROBERTS, D. G., DE CHARPAL, O. & GUENNOC, P. 1979. Rifting and subsidence of the northern continental margin of the Bay of Biscay. *In*: *Initial Reports of the Deep Sea Drilling Project*, Vol. 48, US Government Printing Office, 1025–1060.

NAYLOR, D. & SHANNON, P. 1982. *Geology of Offshore Ireland and West Britain*. Graham & Trotman, London, 161 pp.

RESTON, R. J. 1987. Spatial interference reflection character and the lower crust under extension. Results from 2-D seismic modelling. *Annales Geophysical*, **5B(4)**, 339–348.

ROBERTS, D. G., MONTADERT, L., THOMPSON, R. W., AUFFRET, G. A., LUMSDEN, D. N., KAGAMI, H., TIMOFEEV, P. P., CARLA MÜLLER, BOCK, W. D., DUPEUBLE, P. A. SCHNITKER, D., HAILWOOD, E. A., HARRISON, W. and THOMPSON, T. L. 1980. Geological setting and principal results on the margins of the Bay of Biscay and Rockall Plateau during Leg 48. *Philisophical Transactions of the Royal Society of London*, **A294**, 65–75.

ROYDEN, L. & KEEN, C. E. 1980. Rifting processes and thermal evolution of the continental margin of eastern Canada determined from subsidence curves. *Earth and Planetary Science Letters*, **51**, 343–361.

SIBUET, J. C., RYAN, W. B. F., ARTHUR, M., BARNES, R., BLECHSMIDT, G., DE CHARPAL, O., DE GRACIANSKY, P. C., HABIB, D., IACCARINO, S., JOHNSON, D., LOPATIN, B. G., MALDONADO, A., MONTADERT, L., MOORE, D. G., MORGAN, G. E., MOUNTAIN, G., REHAULT, J. P., SIGAL, J. and WILLIAMS, C. A. 1980. Deep drilling results of Leg 47b. (Galica Bank area) in the framework of the early evolution of the North Atlantic Ocean. *Philosophical Transactions of the Royal Society of London*, **A294**, 51–61.

SMITH, A. J. & CURRY, D. 1975. The structure and geological evolution of the English Channel. *Philosophical Transactions of the Royal Society of London*, **A279**, 3–20.

WARRINGTON, G. M., AUDLEY-CHARLES M. G., ELLIOT, R. E., EVANS, W. B., IVIMEY-COOK, H. C., KENT, P. E., ROBINSON, P. L., SHOTTON, F. W. & TAYLOR, F. M. 1980. A Correlation of Triassic rocks in the British Isles. *Geological Society of London, Special Report*, **13**, 78 pp.

WERNICKE, B. 1985. Uniform-sense normal simple shear of the continental lithosphere. *Canadian Journal of Earth Sciences*, **22**, 108–125.

ZIEGLER, P. A. 1982. *Geological Atlas of Western and Central Europe*. Elsevier, Amsterdam, 130 pp. & enclosures.

—— 1987a. Celtic Sea–Western Approaches area; an overview. *Tectonophysics*, **137**, 285–289.

—— 1987b. Evolution of the Western Approaches Trough. *Tectonophysics*, **137**, 341–346.

T. J. CHAPMAN, GECO Exploration Services, Merlin House, Boundary Road, Woking, Surrey, GU21 5BX, UK.

Inversion, reactivated faults and related structures: seismic examples from the southern North Sea

M. E. Badley, J. D. Price & L. C. Backshall

SUMMARY: This study examines the geometry and timing of inversion features in offshore Quadrants 53/54 of the UKCS. The area studied lies south of the Zechstein salt basin and is dominated by two major and temporally-persistent structural elements, the Wales–Brabant Massif (in the SW of the area) and the South Hewett Fault (in the centre of the area). The Wales–Brabant Massif is a stable basement block, flanked to the north by an inverted Triassic basin. This Triassic basin in turn forms the footwall block to the South Hewett Fault. NE, and locally SW, of the South Hewett Fault lies a series of Jurassic tilted fault blocks symmetrically disposed about a central 'keystone'. These central blocks contain no synextension sediment fill and are interpreted as outer-arc extensional structures parasitic on a large wavelength uplift produced by an underlying thermal anomaly. All major structural features lying to the NE of the Wales–Brabant Massif were inverted during the late Cretaceous. This inversion produced forced folds in the cover rocks above deeper basement faults, and harpoon-shaped structures at the sites of earlier rotational half-grabens. This account considers the effects of both this late Cretaceous inversion and a succeeding period of inversion during the Miocene on earlier extensional structures.

Introduction

Examples of fault reactivation, inversion structures and related features from UKCS Quadrant 53, are illustrated using seismic data (Fig. 1). This area is a part of the southern North Sea which lies south of the Upper Permian Zechstein salt basin. It is thus free from the effects of halokinesis which frequently mask basement–cover relations elsewhere in the southern North Sea and make analysis of fault style and timing difficult (cf. Glennie & Boegner 1981 and Walker & Cooper 1987, who discuss the Sole Pit Inversion). Here it is possible to study the inter-relationships between subsidence, stratigraphy, inversion, regional uplift, and associated unconformities, from the Carboniferous to the Recent.

A brief summary of the geological history and structural evolution of the area is given. A detailed description of geology of the area and interpretation of the tectonic processes which have influenced and controlled the area's structural evolution will appear elsewhere.

Well-imaged seismic examples of inversion features, reactivated faults and related structures are accompanied by line drawing interpretations. The line drawings are offered as a guide, mainly to stratigraphy, and are not intended as detailed interpretations of the seismic data, which would require illustration at a much larger scale. The examples are taken from a non-exclusive seismic survey in the southern North Sea, acquired and processed by Merlin Geophysical in 1983–1984. The seismic data, shot in a 4 × 4 km grid, are of extremely good quality and resolution. All of the examples have been migrated. Details of the acquisition and processing of the survey are given in Table 1.

A N–S geoseismic cross-section is shown in Fig. 2. The cross-section is slightly oblique to the main structural trends (dominantly NW–SE to NNW–ESE) but the relatively steep dips (generally in the range of 50° to 70°) of the main, higher-level, penetrative structures (faults and fold axes), determined from detailed interpretation of seismic data, indicate that the structural style and dips of the N–S sections are representative. The cross-section shows the main structural elements (see Fig. 1 also) and the locations of the seismic examples off-line. Names for the principal structural elements follow van Hoorn (1987).

Stratigraphic nomenclature in this paper (Fig. 3) is deliberately informal, and is based

From COOPER, M. A. & WILLIAMS, G. D. (eds), 1989, *Inversion Tectonics*
Geological Society Special Publications No. 44, pp. 201–219.

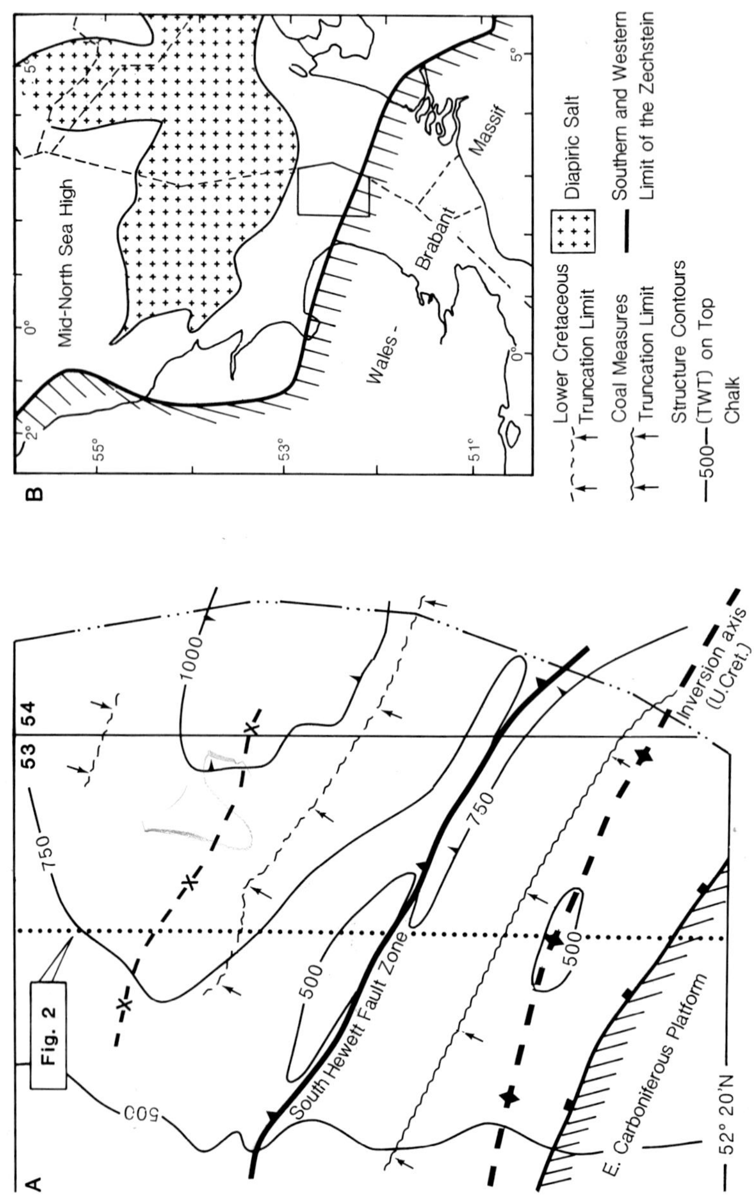

FIG. 1. A, Location map for UKCS Quadrant 53 study area showing principal structural elements. (B, After Glennie 1984.)

TABLE 1. Acquisition, processing and display parameters for the data (survey prefix SP53N) used in this paper

ACQUISITION:		
SHOT BY:	SEISMIC PROFILERS A/S (LIV PROFILER, NOV DEC 1983)	
ENERGY SOURCE:		
type	wide airgun array	
shot point interval	25m	
source depth	5m	
source energy	3640 cu.in.	
RECEIVING ARRANGEMENT:		
fold of recording	48	
no. of groups	96	interval 25m
cable length	2375m	depth 11m
near trace	1	offset 0m
INSTRUMENTATION:		
recording system	DFS V	
gain type	I.F.P.	
filters: low cut	5.3 Hz.	slope 18 dB/octave
high cut	180Hz.	slope 72 dB/octave
record format	segB. 1600 bpi. 96 channels	
record length	4s	
sample rate	2ms	
POSITIONING SYSTEM:		
primary: hifix–6	secondary: satnav/sonar	

PROCESSING:		
PROCESSED BY:	MERLIN GEOPHYSICAL CO. LTD.,	
1. DEMULTIPLEX		
2. SOURCE SIGNATURE COMPENSATION:		
input	calculated far–field signature	
desired output	fitted minimum phase Gaussian filter	
3. RESAMPLE:		
subsampling	2ms to 4ms	
anti–aliasing filter	90(48) Hz (dB/oct) lowpass	
4. STATIC CORRECTIONS:		
source and receiver depth	+11ms	
5. FK MULTIPLE ATTENUATION:		
pre–attenuation inverse scaling	+16dB/s exp. to 2.5s	
6. DECONVOLUTION BEFORE STACK:		
type	minimum phase least squares inverse	
autocorrelation window length	1700ms	
max. prediction lag	120ms	
min. prediction lag	4ms	
post–deconvolution inverse scaling	−16dB/s exp. to 2.5s	
7. NMO CORRECTION:		
velocity derivation	contoured semblance spectra	
offset dependent mute		
8. STACK:		
type	standard mean amplitude CDP	
coverage	4800%	
space variant geometrical divergence compensation		
9. FK COHERENCY FILTER:		
attenuation fan	±8 to ±40 ms/trace	
10. DECONVOLUTION AFTER STACK:		
type	minimum phase least squares inverse	
	zone I	zone II
autocorrelation window length	1200ms	1800ms
max. prediction lag	120ms	120ms
min. prediction lag	16ms	16ms
11. WAVE EQUATION MIGRATION:		
finite difference solution	48ms depth step	
pre–migration filter	6(24)–90(36) Hz (dB/oct)	
migration velocity derivation	interval velocity model	

12. SPACE TIME VARIANT FILTER:			
at sp 101	time (ms)	low cut Hz (dB/oct)	high cut
	50	10(24)	90(36)
filters linearly interpolated	500	8(24)	90(36)
in space and time.	1700	8(24)	90(36)
cuts and slopes specified	1800	7(24)	45(36)
at −3dB point.	4000	7(24)	30(36)

13. TWO DIMENSIONAL FILTER:	
number of adjacent traces	7
passband (4–550ms)	+1.5 to −1.5 ms dip per trace
passband (550–4000ms)	+5 to −5 ms dip per trace
percentage input feedback	50
14. AMPLITUDE BALANCE:	
a) general amplitude trend analysis and compensation	
b) trace equalisation type	Robust AGC
DISPLAY:	
system	SCIITEX laser plotter
horizontal scale	1:25,000 (20 traces/cm)
vertical scale	10 cm/sec
gain	2 dB
bias	5 %
polarity	compression : positive : trough
datum plane	sea level
shotpoint location	source position

on data from released wells and inferences from the seismic data. Thicknesses, either taken from well data or derived from seismic data are present-day, and have not been corrected for either compaction or erosion.

Geological history and structural evolution

A number of recent publications (Badley 1984, papers in Glennie & Boegner 1981; Crittenden 1982; Glennie 1984; Brooks *et al*. 1986; Milsom 1986; van Hoorn 1987; van Wijhe 1987 and Walker & Cooper 1987) describe aspects of the geology of the area, although most of these references concentrate on the gas fields which lie to the N and E. The following account describes only briefly the main features of the area's geology and geological evolution and concentrates on illustrating distinctive features attributable to structural inversion. The description uses the cross-section in Fig. 2 to illustrate stratigraphic and structural relationships that can be determined from seismic and well data. Four major truncational unconformities can be recognized in the seismic data (Fig. 2). These are referred to in the figures as: BP (Base Permian), BC (Base Cretaceous) and IUC (Intra-Upper Cretaceous) and MU (Intra-Miocene). There is also a major intra-Westphalian unconformity, and minor unconformities are developed locally at the Base Albian (Fig. 4), Base Chalk (Fig. 6), in

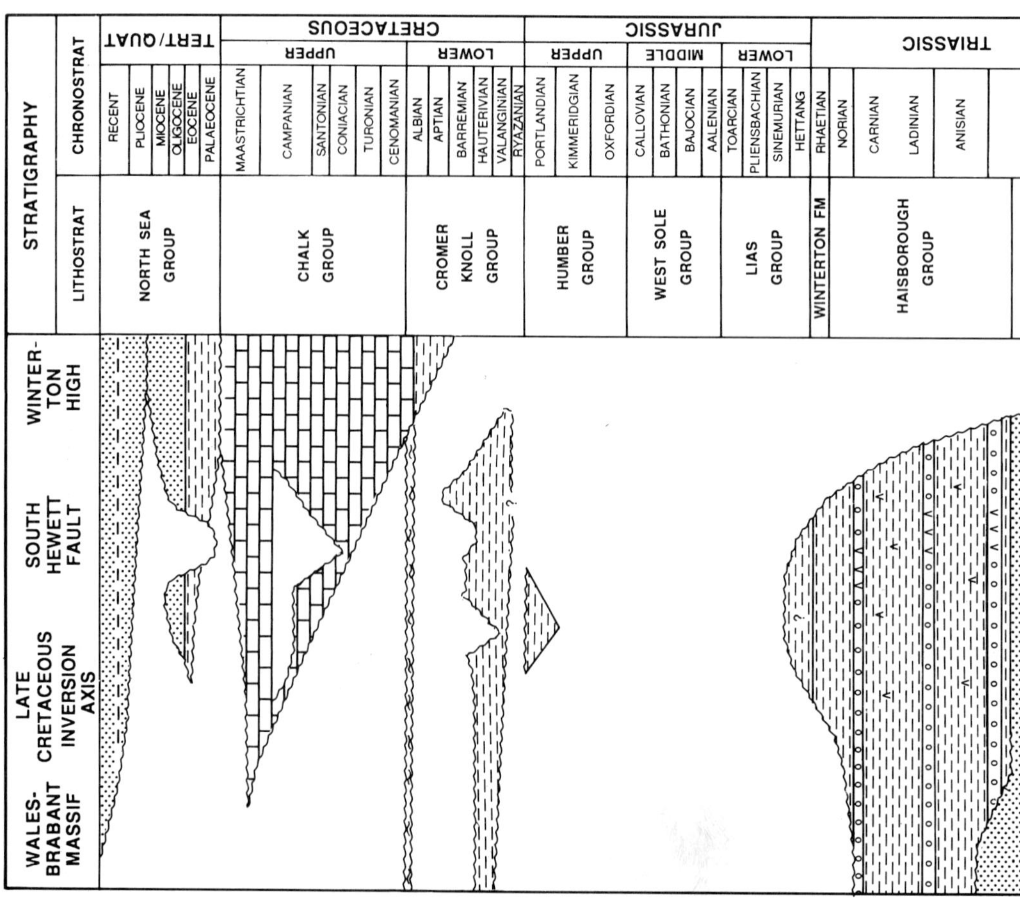
WALES-BRABANT MASSIF
LATE CRETACEOUS INVERSION AXIS
SOUTH HEWETT FAULT
WINTERTON HIGH
STRATIGRAPHY
LITHOSTRAT
CHRONOSTRAT
NORTH SEA GROUP
CHALK GROUP
CROMER KNOLL GROUP
HUMBER GROUP
WEST SOLE GROUP
LIAS GROUP
WINTERTON FM
HAISBOROUGH GROUP
TERT/QUAT
CRETACEOUS
UPPER
LOWER
JURASSIC
UPPER
MIDDLE
LOWER
TRIASSIC
RECENT
PLIOCENE
MIOCENE
OLIGOCENE
EOCENE
PALAEOCENE
MAASTRICHTIAN
CAMPANIAN
SANTONIAN
CONIACIAN
TURONIAN
CENOMANIAN
ALBIAN
APTIAN
BARREMIAN
HAUTERIVIAN
VALANGINIAN
RYAZANIAN
PORTLANDIAN
KIMMERIDGIAN
OXFORDIAN
CALLOVIAN
BATHONIAN
BAJOCIAN
AALENIAN
TOARCIAN
PLIENSBACHIAN
SINEMURIAN
HETTANG
RHAETIAN
NORIAN
CARNIAN
LADINIAN
ANISIAN

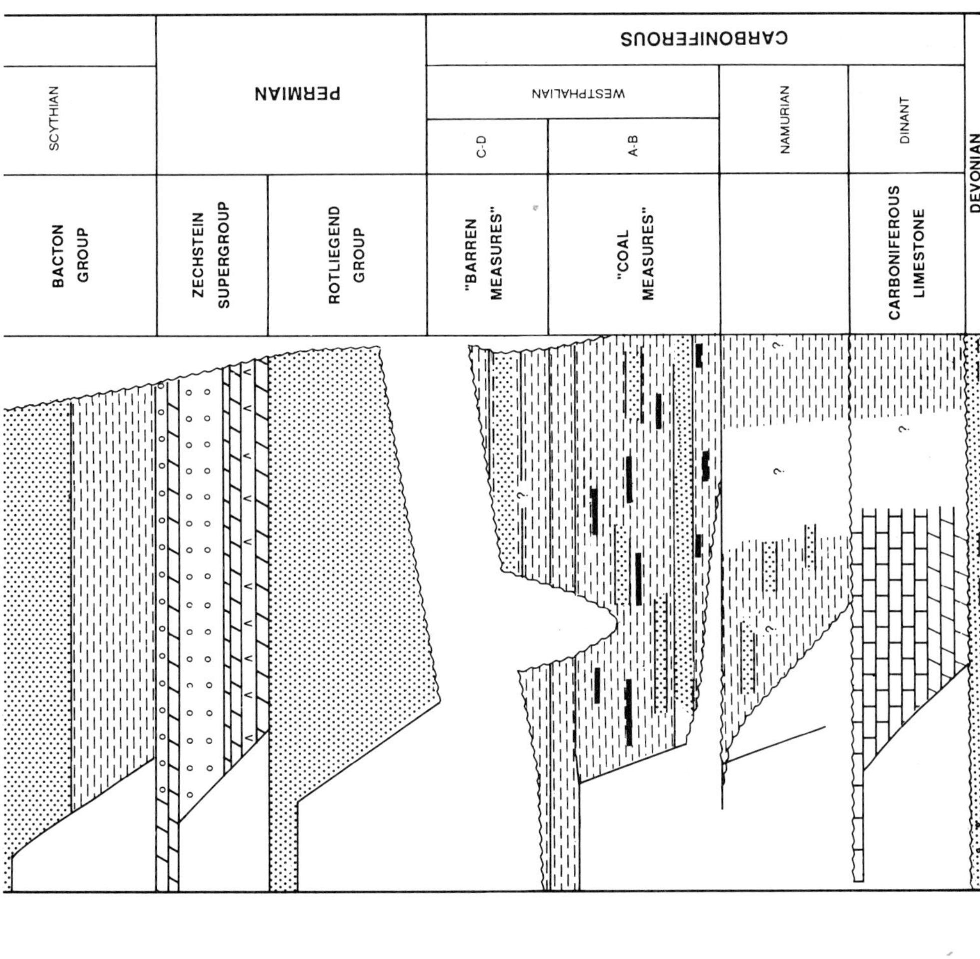

FIG. 3. Generalized stratigraphic column for UKCS Quadrant 53 (after Walker & Cooper 1987).

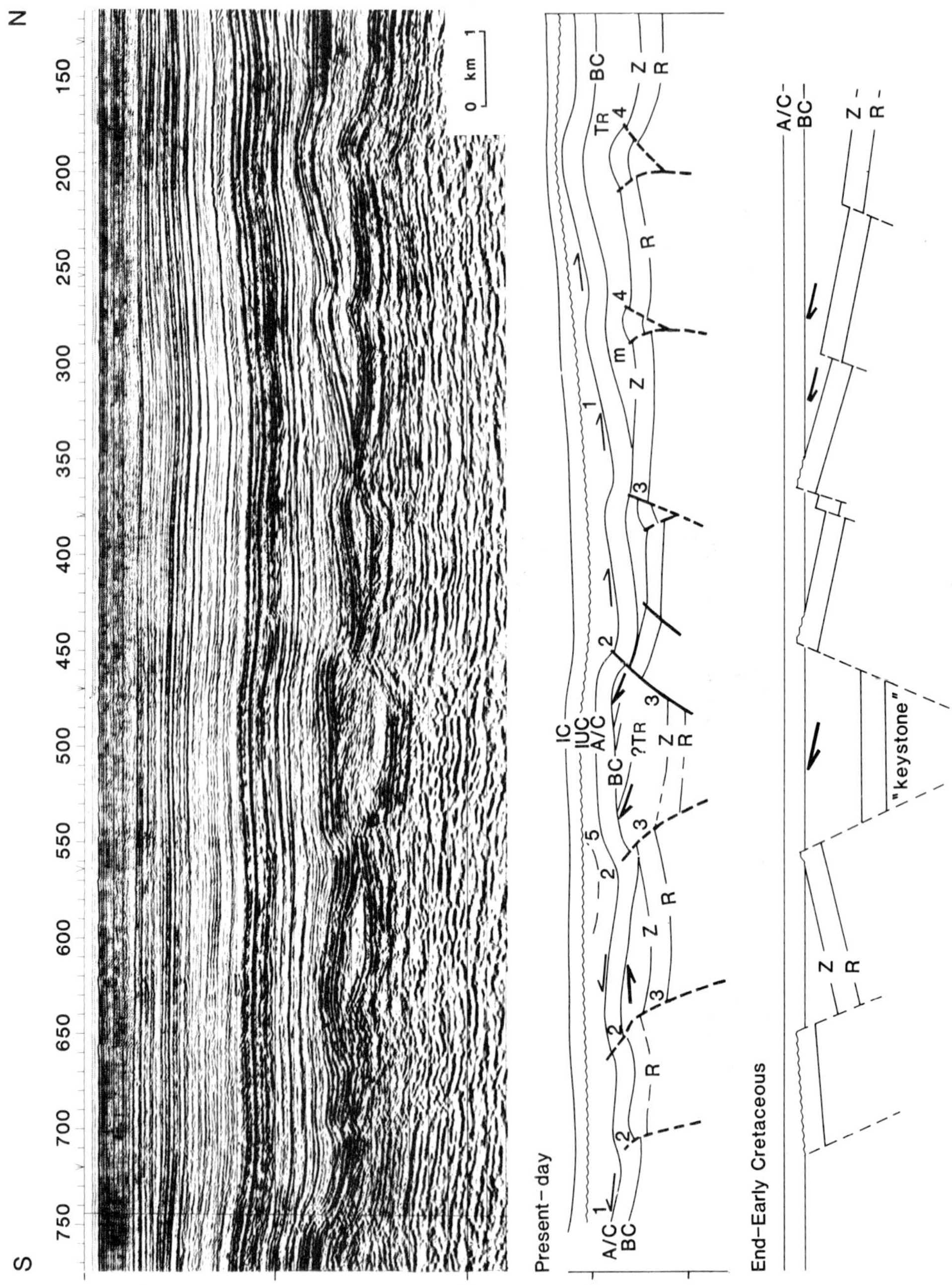
N
S
0 km 1
TWT(s)
0.0
1.0
2.0
1.5
Present-day
End-Early Cretaceous
"keystone"
IC
IUC
A/C
BC
TR
?TR
Z
R
m

FIG. 4. Seismic section across the axis of the now partially subsided Late Jurassic uplift NE of the South Hewett Fault (see Fig. 2 for location and key). Line drawings show an interpretation of end early Cretaceous and present day configurations. End early Cretaceous reconstruction: the most prominent features are Jurassic extensional faults and fault blocks. These contain no synextension sediments. Reflections showing apparent baselap are in multiples. The base Cretaceous unconformity everywhere erodes into the preextension sequence. The fault blocks are parasitic on a more regional uplift centred on the 'keystone' block. The elevation of fault block crests increases towards the 'keystone' from both NE and SW as a result of this uplift. The faults are interpreted as outer-arc extensional structures around a large wavelength anticline. The anticline is interpreted as the product of a possible underlying Jurassic thermal anomaly. The Albian sequence shows relatively uniform thickness across the periclinal axis, indicating a period of uniform subsidence across the area. Present day configuration: the striking present day structure is interpreted to be primarily the result of a relatively short-lived ?post-Cenomanian to pre-Campanian compressional event. Before the compressional event occurred, however, differential subsidence had begun to affect the area following deposition of the 'Albian' sequence. Subsidence was greatest above the 'keystone' marking the previous periclinal axis. Onlap to the SW and NE, away from the periclinal axis, onto the top 'Albian' is clearly evident (1). The subsidence was interrupted by the compressional event. The compression caused reactivation of, and reverse movement on, most pre-existing normal faults. The overlying cover rocks, including the 'Albian' sequence and overlying Chalk, were thrown into forced asymmetric folds above the reversed faults. In some cases the faults propagated up into the folded cover (2). Reverse movement on many of the faults was insufficient to cancel the original normal throw (3). In other cases it produced net reverse movement and 'flower' structures that are particularly evident at top Zechstein level (4). The burial depth of the pre-existing normal faults immediately prior to the compressional event is unlikely to have been more than several hundred metres. The relatively low values of deviatoric stress (probably in the range of 10–30 mPa) required for fault reactivation at such shallow burial depths is in accord with the observation that a majority of pre-existing faults were reactivated. The compressional event was short-lived and no compressional features affect the intra-upper Cretaceous (IUC) and younger reflections. Terminations of reflections beneath this reflection at (5) date the cessation of uplift and the return to subsidence. The early return of subsidence to this part of the area is shown by the conformable nature of reflection IC. Further to the S reflection IC is a major truncational unconformity (Figs 6 & 8).

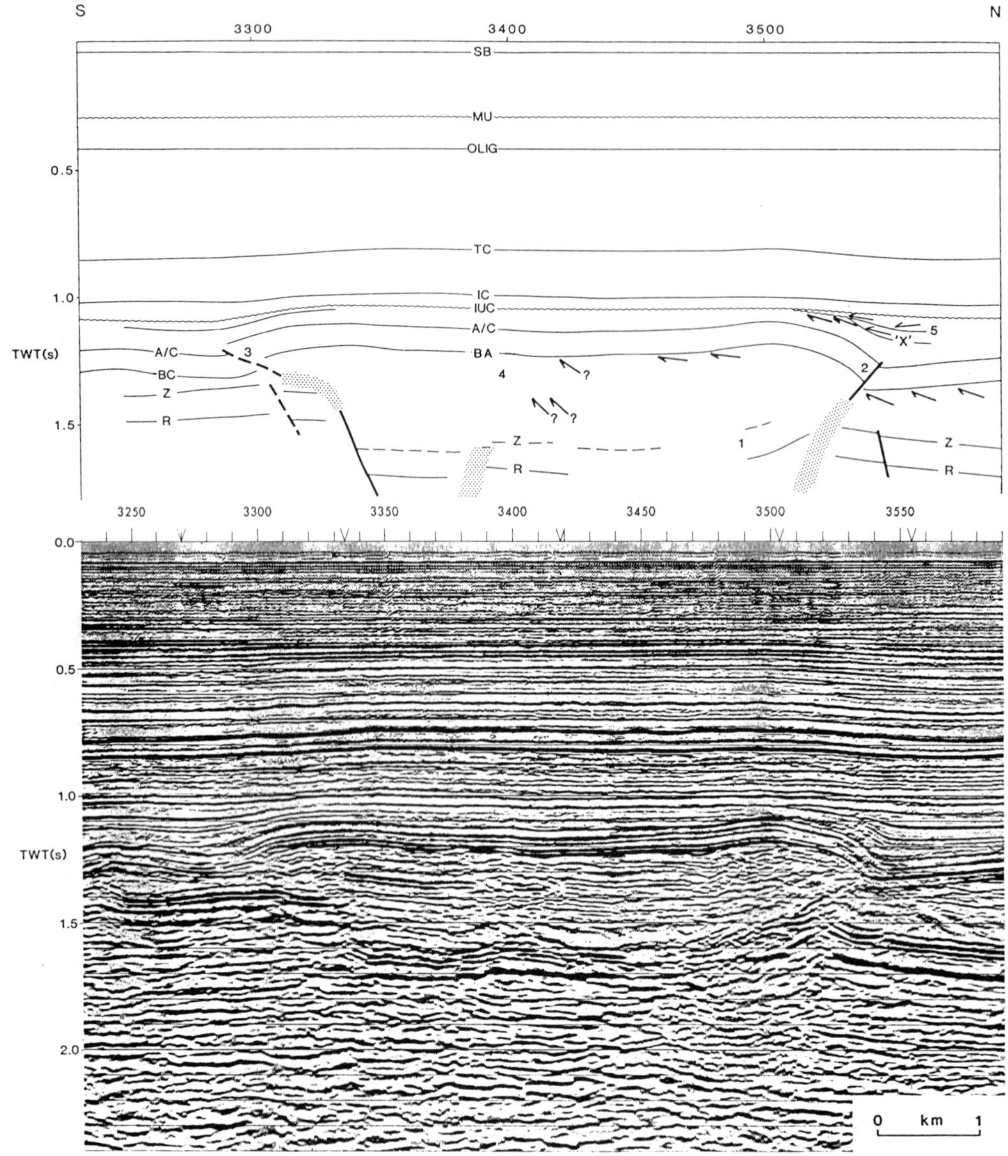

FIG. 5. Seismic section and line drawing interpretation showing the detailed structure of the 'keystone' block at the graben axis (see Fig. 2 for location and key). This illustration can be compared closely with the 'graben axis' illustrated in Fig. 4, which is located approximately 5 km to the NW. The general description given in Fig. 4 also applies to this figure. There are several specific points of interest. For example, poorly-imaged, tectonically-disturbed Zechstein reflections (1). In the fold affecting the Albian sequence, the underlying normal fault has cut-through (2). The Albian itself is offset by a low-angle reverse fault (3). Extension of Zechstein reflections beneath the fold affecting the Albian indicates a lower dip on the upper portion of the reverse fault. Koopman *et al.* (1987) describe sand-box experiments which show similar flattening of reactivated normal faults as they propagate into previously unfaulted cover. Triassic reflectors are truncated beneath reflection BA (4). Reflection BC cannot be picked within the axial graben. Onlap onto the folded reflection X indicates the onset of compression and initiation of folding. The pattern of onlap, flexuring and truncation in the vicinity of (5) documents the growth of the fold during the compressional event. Available stratigraphic control does not permit a detailed evaluation of the duration or rate of fold growth. Reflection IUC is the unconformity marking the cessation of folding and uplift and the resumption of sedimentation. Reflection IC, which, further SW, is a major unconformity developed as a consequence of the compressional event, is here conformable with underlying reflections.

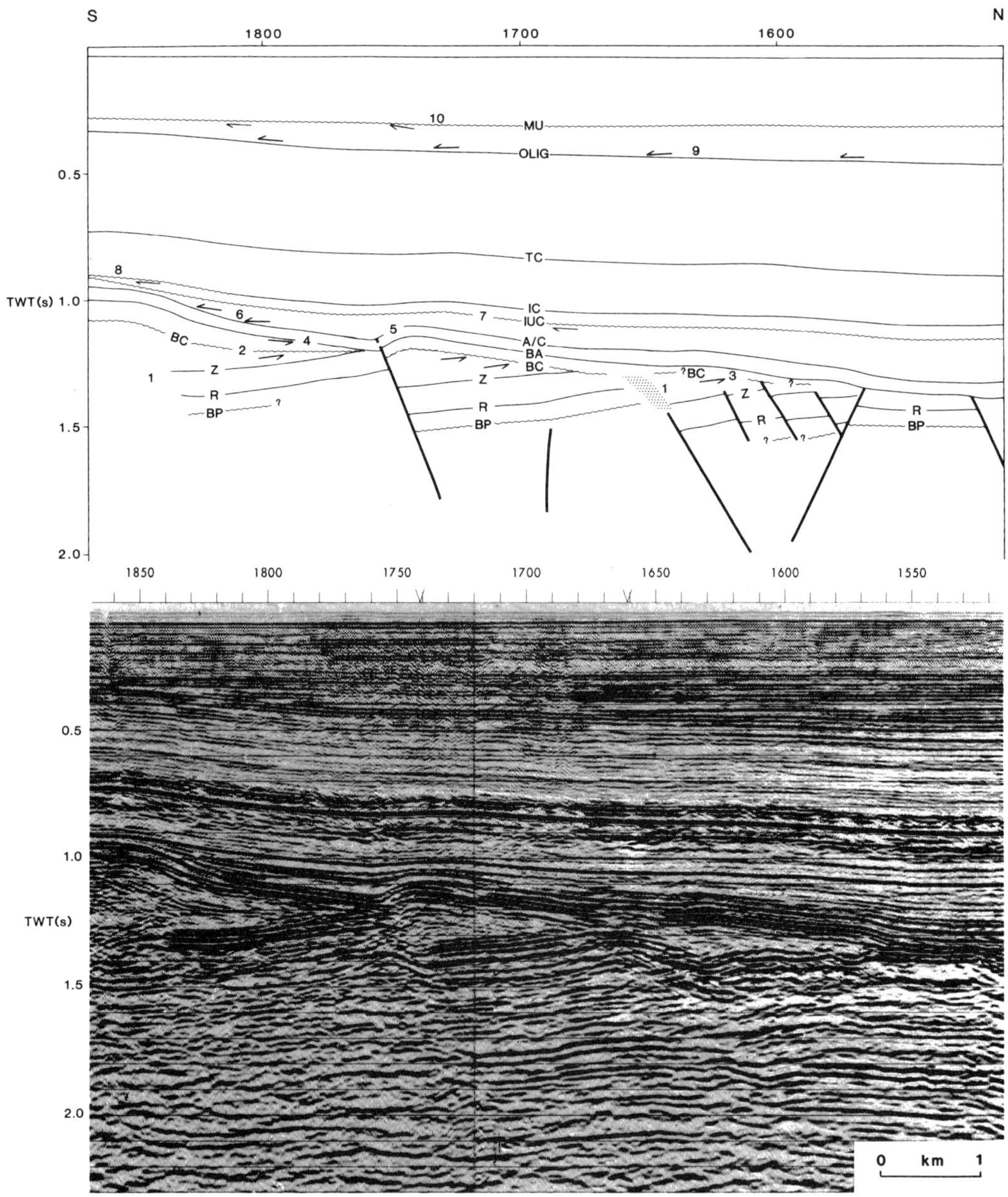

FIG. 6. Seismic section and line drawing interpretations showing detail of the SW crest of the now subsided Late Jurassic 'periclinal high' adjacent to the graben axis (see Fig. 2 for location and key). Features of interest in the seismic section include a poorly-imaged fault zone. This is possibly a consequence of the oblique trend of the fault relative to the seismic section (1). A truncational unconformity (BC) lies beneath Early Cretaceous sequences (2), which are themselves overstepped (3) or ?truncated (4) by the Albian sequence BA–A/C. Normal faults, reactivated during the Late Cretaceous show reverse movements and there is a large fold at (5). Intra-Chalk reflections onlap complexly onto the NE-dipping top Albian sequence, and are truncated beneath the unconformity IUC (7). This truncation surface marks the close of a Late Cretaceous compressional event, which resulted in uplift and folding. To the SW the reflection IUC merges with the surface IC which, in this direction, becomes a major truncational unconformity (8). An Oligocene basin with low-angle onlap to the SW developed above the graben axial area (9). The intra-Miocene truncational unconformity (10) developed in response to localized uplift and the growth of a major fold.

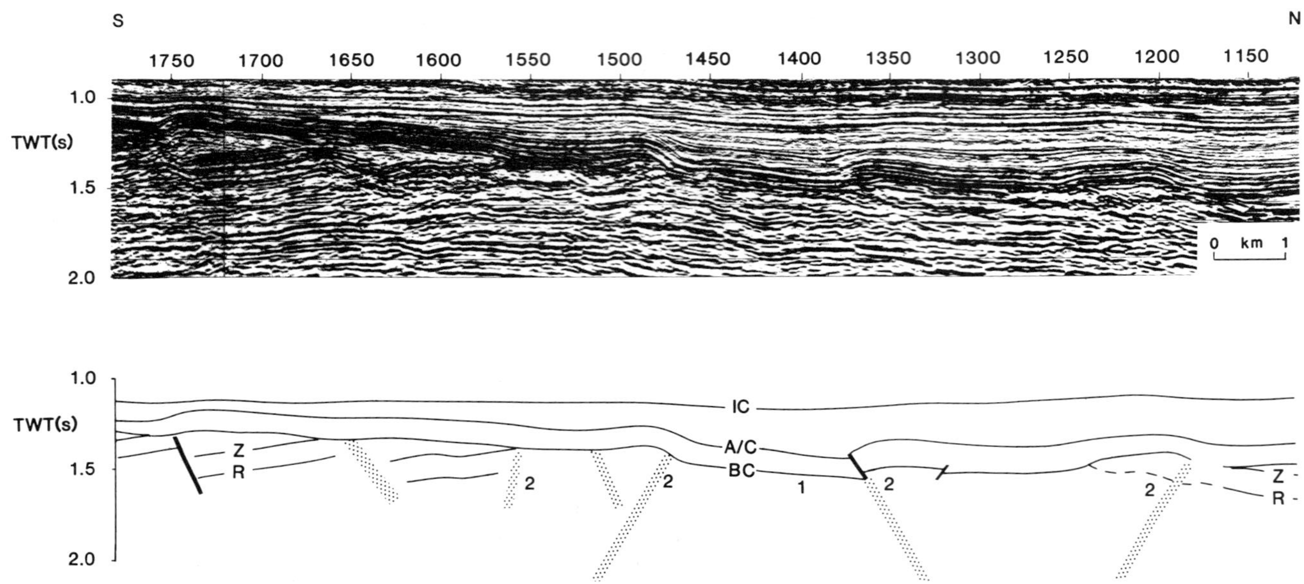

FIG. 7. A N–S seismic section and line drawing interpretations showing detail of the 'graben axis' area approximately 5 km SE of the seismic section shown in Fig. 4 (see Fig. 2 for location and key). Erosion associated with the late Jurassic periclinal uplift increases inwards towards the graben axis and also to the SE along the axis of the dome. The seismic section shows Albian resting on Carboniferous in the centre of the seismic section (1). The seismic grid at this locality is too coarse to allow detailed mapping of the pre-Cretaceous. The characteristic reflections of the Zechstein sequence can be seen clearly, however, to the immediate S and N where it exhibits opposing dips. The asymmetry of folds affecting the Albian sequence can be used to indicate the hade of reactivated faults in the pre-Cretaceous (2).

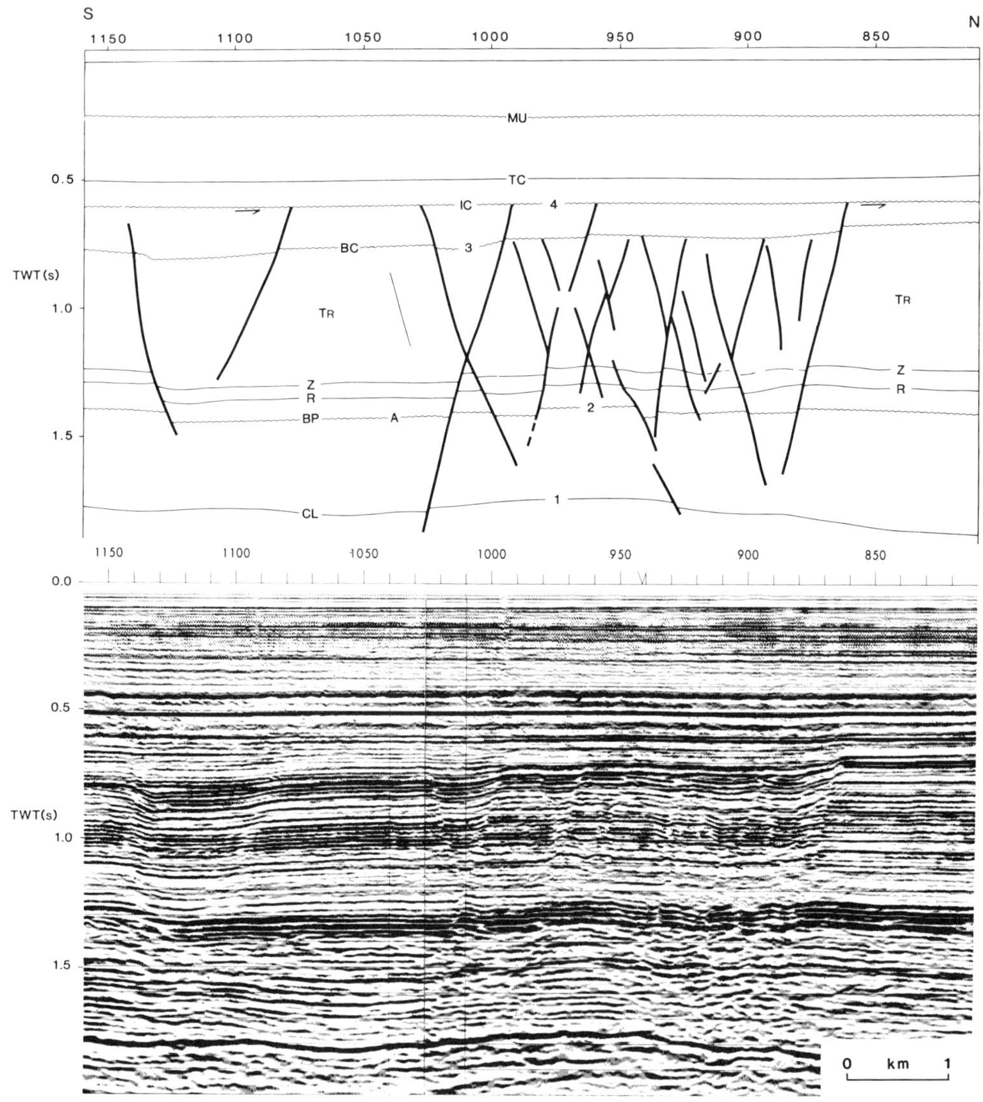

FIG. 8. A N–S seismic line across the inverted footwall to the South Hewett fault (see Fig. 2 for location and key). The striking feature is the set of conjugate faults, which are interpreted to have formed during the Late Cretaceous compressional event which resulted in periclinal uplift. The faults are interpreted to have formed contemporaneously and show geometries similar to those described by Horsfield (1977). Conjugate faulting is a relatively common style seen in seismic data both here and elsewhere. The faults are characterized by a graben in the upper half, a cross-over zone, outside of which, at the same depth as the cross-over zone, reflections appear to show no displacement, and a horst beneath the cross-over. Other features of interest in the seismic section are: The top lower Carboniferous limestone forms a prominent reflection, characterized by a positive reflection coefficient, within the footwall of the South Hewett Fault (1). The Basal Permian unconformity is approximately parallel with underlying Carboniferous reflections and no obvious angular relationship is evident here (2). Similarly the Base Cretaceous unconformity here is not marked by significant angularity (3). Slightly more angularity can be observed beneath the intra-Campanian unconformity (IC), beneath which the larger conjugate faults also terminate (4).

the early Late Cretaceous (Fig. 5) and Intra-Campanian (Fig. 8).

Pre-Permian

The oldest sequences imaged clearly by seismic data are the Carboniferous and Devonian over the Wales–Brabant Massif (WBM), and the Carboniferous only over the remainder of the area. Detailed Permian (and Cretaceous) subcrop maps showing the regional distribution of Carboniferous sequences, have been published by van Hoorn (1987). On the Wales–Brabant Massif a Carboniferous Limestone (Dinantian) sequence, approximately 1 m thick (see well 53/16–1), overlies Devonian red beds. At the northern margin of the Wales–Brabant Massif the Carboniferous Limestone exceeds 400 m in thickness (well 53/12-2) and thickens considerably further N. However, in this area it has not been penetrated by any of the released wells. There is evidence seen in the seismic data for thickening of the Lower Carboniferous Limestone into a fault zone marking the northern margin of the Wales–Brabant Massif, although the top Lower Carboniferous Limestone reflection is generally unfaulted in its passage from the basinal area to the N onto the Wales–Brabant Massif. The high-amplitude and continuous top Lower Carboniferous reflection cannot be traced N of the South Hewett Fault. A possible explanation of this would be provided if the South Hewett Fault had been active in the Lower Carboniferous, separating a carbonate platform to the S from a more basinal area to the N. The younger Carboniferous can be divided into three sequences: a presumed, late Early Carboniferous (?Namurian) basinal sequence (not penetrated by any released wells) which onlaps the Lower Carboniferous Limestone, and which is overstepped; a Coal Measures (Westphalian A–B) sequence, characterized by high-amplitude but poorly continuous reflections, which thickens to the NE and, a red-bed sequence (Westphalian C–D) free of high amplitude reflections, which onlaps into the Coal Measures and equates with the Barren Measures onshore.

Permo–Triassic

Uplift, faulting and folding, on a trend parallel with the South Hewett Fault, occurred after deposition of the Barren Measures. Widespread erosion, which followed uplift in the Late Carboniferous/Early Permian, led to the development of a regional unconformity which bevelled the underlying Carboniferous sequences and structures (unconformity BP; Figs 2 & 8). Sedimentation resumed in the Permian with the deposition of relatively tabular units of Rotliegend clastics and Zechstein carbonates, which thicken gradually from 200 m to 500 m over the 70 km S–N transect (Fig. 2). In the south of the area onlap of Zechstein carbonates onto the depositional topography of a Rotliegend aeolian dune belt sequence, and large scale internal structures within the Rotliegend, can be observed on seismic sections. Triassic sediments concordantly overlie the Permian.

Jurassic

No Jurassic sediments are recognized with certainty in the study area, although they are present immediately to the N and E (Glennie & Boegner 1981, their Figure 3b). The main activity in the area during the Jurassic appears to have been a regional uplift, centred about an axis 20 km NE of the South Hewett Fault (Fig. 1). This uplift was accompanied by extensional faulting (Fig. 4). This is an apparent paradox, which requires explanation.

The Jurassic extensional fault blocks are most clearly seen NE of the South Hewett fault (Fig. 4), where they form the northern part of the Winterton High. They are also locally present to the SW. Thus the south western Hewett Fault was not the master fault to the Jurassic extension. On first inspection these fault blocks appear to be classic, domino-style, tilted blocks, but in detail they differ from typical tilted blocks formed during basin formation. Firstly, there is no 'synrift' sediment-fill within the hanging walls of any of the faults. The post-extension (Base Cretaceous) unconformity everywhere truncates tilted prerift sediments (Figs 4, 6 & 7). Secondly, the tilted blocks are symmetrically disposed about a central 'keystone' fault block (Fig. 4). Thirdly, although the fault blocks are all of approximately the same size and spacing, the post-extension unconformity erodes progressively deeper into fault block crests as the 'keystone' block is approached from both the NE and SE. In places Lower Cretaceous sits on Westphalian adjacent to the 'keystone'. The absence of sediment-fill within the hanging walls indicates that no net subsidence accompanied extension.

The regional uplift centred on the 'keystone' block is a large structure with an observable half-wavelength of approximately 50 km. The blocks filled synchronously with this uplift are parasitic on this larger structure but because of their abundance appear to be the dominant

Jurassic structures in the area. It is possible that this regional uplift could be the result of long-wavelength, low-amplitude late Jurassic compression. However, at this time the remainder of the North Sea was in extension. We therefore favour the argument that the uplift may have been thermally, rather than mechanically, induced. In that case the tilted blocks would be outer-arc extensional structures to the uplift.

It is the erosion products derived from this probably thermal uplift which were redeposited elsewhere in the southern North Sea as the Volgian–Ryazanian Spilsby Sandstone and its correlatives. This erosive event produced a near-planar, truncational unconformity across the area (event BC, Figs 2 & 6).

Locally, erosion has completely removed the ?Jurassic, Triassic, and Permian (Figs 4 & 7). The regional pattern of erosion associated with the uplift along the extension axis defines an elongate NW–SE-trending periclinal structure. The amount of erosion increases along the uplift axis, to the SE, so that ultimately Albian sediments overlie Carboniferous Coal Measures (Fig. 7). The approximate total extent of the uplift coincides with, and was the precursor to, the Winterton High (van Wijhe 1987). To the south of the South Hewett Fault, erosion beneath the unconformity reaches a minimum along the inversion axis running between the WBM and the fault. Along this trend, in the NW of the studied area, the thickness of Triassic sediment preserved beneath the unconformity exceeds 1 km.

Early Cretaceous

Sedimentation resumed in the Early Cretaceous and the truncational unconformity associated with the Jurassic uplift was onlapped. This indicates that continued net uplift of the periclinal axis continued into the Early Cretaceous such that its crest probably did not become the site of sedimentation until the Albian. Movement on the majority of the smaller faults had ceased by the Early Cretaceous and they terminate at the truncational unconformity. Elsewhere, the Early Cretaceous sediments show a more concordant relationship to the unconformity, with very low-angle onlap or parallelism of reflections.

Late Cretaceous–Oligocene

Cessation of differential uplift by the Albian–Cenomanian is indicated by the preservation north of the South Hewett Fault of a unit of approximately constant thickness (300 m). A low-angle, truncational relationship between this unit and the underlying Lower Cretaceous sediments on the flanks of the periclinal area (Fig. 6, location 4) may mark the final decay of uplift. Immediately following deposition of this unit differential subsidence commenced, centred above the area of deepest erosion and the extension axis. Intra-Chalk reflections onlap onto the edges of the subsiding area to the S and NE of the subsidence axis.

The pattern of uplift followed by differential subsidence, centred on the area of earlier periclinal uplift, is interpreted to be the result of a thermal event accompanied by limited crustal extension. Uplift of the WBM may have occurred as an isostatic response to lateral heat flow into areas unaffected by crustal extension. The widespread occurrence of late Jurassic–early Cretaceous volcanically-derived clays across the southern North Sea, and onshore in the UK and north Germany (Jeans *et al.* 1982) and the presence of the late Jurassic Zuidwal volcanic vent in the Vlieland Basin (Perrot & van der Poel 1987) and other volcanic and intrusive activity in the Broad Fourteens Basin (van Wijhe 1987) during the latest Jurassic–Early Cretaceous, provide evidence of a regional thermal event.

A short-lived, compressional event, of probable Turonian–Coniacian age, resulted in reverse movements on the South Hewett Fault and on many faults both in its hanging wall and to the NE of it. Previous movement on these faults had been of normal sense (Fig. 2). These reverse movements generated forced folds in the Cretaceous cover rocks, producing the characteristic harpoon-shaped structures of inverted half-grabens, and the symmetric inversion above the former extension axis (Figs 4–7; cf. Koopman *et al.* 1987, their figures 3 & 4). The fact that many of the minor, as well as the major, faults were reactivated, may be a result of their shallow burial depth which probably ranges from only a few tens to hundreds of metres. Consequently, only a small deviatoric stress (probably in the range of 10–30 mPa) would be required for fault reactivation (Ragan 1985, p. 141).

Compressional effects were not confined to the hanging wall of the South Hewett Fault because the area between the South Hewett Fault and the WBM experienced its first inversion at this time (Fig. 2). A second inversion along the same axis occurred in the Miocene. The development of conjugate fault sets accompanied the Late Cretaceous inversion

(Fig. 8). The conjugate fault sets are developed throughout the inversion feature but are mostly concentrated in its axial region. Line-length estimates indicate an average 3% shortening across the area. The Late Cretaceous inversion event has been interpreted to be the product of transpressional deformation (Ziegler 1981; van Hoorn 1987 and van Wijhe 1987). The pattern of folding of the Albian sediments and its direct relationship to the underlying previously normal faults indicates, however, that in the block 53/4 area movement at shallow structural levels was dominantly dip–slip.

An intra-Chalk unconformity, developed during the Campanian, bevelled both the uplifted area between the South Hewett Fault and the WBM (Fig. 8) and the inversion-related fold associated with reactivation of the South Hewett Fault (Fig. 9). Previous inversion structures lying further to the NE, in an area then subsiding at a greater rate which had returned to a regime of net deposition, escaped the erosional effects producing the unconformity. Sedimentation had resumed across the entire area by the Late Campanian–Maastrichtian. A chalk unit, thickest above the site of the periclinal uplift (Fig. 5), was deposited. This indicates a continued greater subsidence rate. This pattern of subsidence continued into the Oligocene.

Post-Oligocene

A further compressional event affected the area in the Miocene. The South Hewett Fault was reactivated as a reverse structure for a second time, and a large fold developed in the hanging wall block (Fig. 9). The area to the SW was also inverted for a second time. To the NE of the South Hewett Fault, however, none of the structures inverted during the earlier Late Cretaceous event were reactivated. At an average burial depth of 1–2 km the deviatoric stress required for fault reactivation would have been considerably greater than in the Late Cretaceous, and was probably not achieved. The crest of the fold above South Hewett Fault and the inversion structure to the SW, were eroded following the uplift and folding. The overlying Pliocene sedimentation pattern bears no relationship to the amount of erosion affecting the subcropping sequences, a depositional characteristic of unconformities associated with crustal thickening.

Conclusions

Seismic examples of inversion structures

The series of seismic sections and line drawings presented in the Figures, show examples of the structures and stratigraphic relationships produced during the sequence of geological events described above. The seismic data are presented at a large scale, in order to enable the reader to form an independent opinion of the geology. The accompanying line drawings are provided only as a guide to the stratigraphy and obvious aspects of the geology. No attempt has been made to project or predict fault trajectories, where they are not clear in the seismic data. When interpreting the seismic data the points listed in the Appendix were considered. They are essential to ensuring correct interpretation of the data.

The principal structural elements of Quadrant 53 are, from SW to NE:

1 The Wales–Brabant Massif, a stable Palaeozoic platform with a thin veneer of onlapping Mesozoic and Tertiary sediments.
2 A basin with a thick Triassic–?Jurassic sequence forming the footwall to the South Hewett Fault. This basin was inverted during the late Cretaceous to form an eroded pericline, and inverted again during the Miocene.
3 The South Hewett Fault, a major ?pre-Mesozoic structure NE of which no pre-Permian reflectors are imaged. After an early extensional history the fault was reactivated in a reverse sense during the Late Cretaceous (Turonian–Coniacian) and Miocene to form a subsequently eroded, monoclinal fold.
4 A zone of SW-dipping tilted fault blocks, formed by localized Late Jurassic extension, possibly on the outer arc of a thermally-induced periclinal structure. The faults were reactivated in a reverse sense during the Late Cretaceous when forced-folds and dramatic harpoon structures formed in the cover sequence.
5 The Late Jurassic extension axis over which the tilted fault blocks change dip direction to NE inclined. Locally, over the crests of the most axial tilted fault blocks, Lower Cretaceous post-extension sediments directly overlie the Westphalian at a marked erosional unconformity.
6 The pre-Mesozoic history of Quadrant 53 can partially be deduced from the seismic data discussed in this paper. Devonian red bed sediments are known from the Wales–

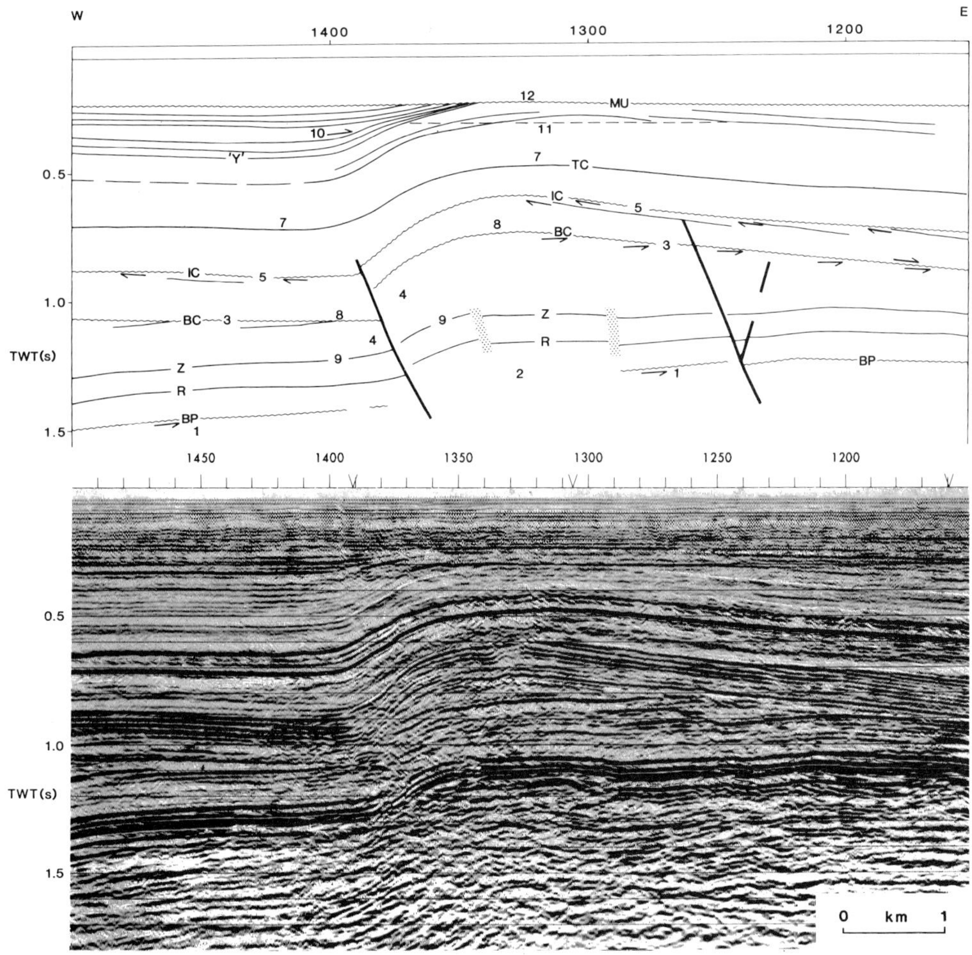

FIG. 9. Seismic section and line drawing interpretation of an E–W seismic line across the South Hewett Fault and associated hanging wall fold. The line is orientated at approximately 50° to the regional dip direction. The Basal Permian unconformity in this area is not characterized by a strong or consistent reflection coefficient. The unconformity can only be located in seismic sections where angularity between Carboniferous and truncating Rotliegend reflections can be demonstrated. Terminations of Carboniferous reflections mark the unconformity. Angular discordance between Carboniferous reflections and the Rotliegend can be seen at (1). Poor data quality in other areas precludes any interpretation (2). The prominent unconformity at the base of the Cretaceous sequences, here truncating SW-tilted Triassic beds, is readily located by both reflection terminations and a reflection at the unconformity surface, characterized here by a positive reflection coefficient (3). The original (pre-Cretaceous) NW downthrow on the South Hewett Fault can be recognized by the greater thickness of Triassic beds in the hanging wall than in the footwall (4). The truncational nature of the intra-Campanian unconformity is very obvious (5). Lower Cretaceous sequences have NE dips imparted by the tilting associated with differential subsidence towards the 'graben axial' area and later folding related to the Upper Cretaceous compressional event. The unconformity in the area imaged by the seismic section juxtaposes Campanian chalk and Lower Cretaceous sands and shales. The resulting negative reflection coefficient produces a clear trough/peak (white above black) response in the seismic section. (Note how the truncated reflections terminate beneath the peak (black of the unconformity reflection, at least 20 ms beneath the actual subsurface truncation position. The South Hewett Fault was reactivated during the Miocene compressional event, resulting in a second phase of reverse movement. The amplitude of the fold affecting the top Chalk provides a measure of the magnitude of the Miocene reverse movement at this location (7). The offset of the Base Cretaceous unconformity (8) and top Zechstein reflection represents the minimum aggregate value of the Upper Cretaceous and Tertiary reversals. The smaller offset of the top Zechstein reflection (9) gives a measure of the pre-Cretaceous downthrow to the NE. The growth history of the fold developed above the South Hewett fault is recorded in the complex sequence beneath the Miocene unconformity (10). Reflections beneath Y are subparallel. Those above indicate sequences thinning towards the fold. Reflection Y within the late Eocene–Oligocene marks the onset of folding. Reverberations (11) within the first 350 m of seismic section make interpretation of the shallow sequences difficult. The regional truncational intra-Miocene unconformity (12) can, however, be followed throughout the area.

Brabant Massif, where they are overlain by a thin veneer of Carboniferous limestone.

7 The Carboniferous limestone thickens considerably to the NE, although it cannot be traced beyond the South Hewett Fault. It is onlapped by a Namurian basinal sequence. The Westphalian, developed across the whole area, forms two sequences, an underlying coal measure sequence and an onlapping red bed sequence, which became truncated beneath the Permian. It is present only in the SW.

The Rotliegend extends across the whole area, and is onlapped from the NE by Zechstein carbonates. A thick Triassic basin formed over the central part of the quadrant, although elsewhere substantial parts of the Triassic and any trace of presumed Jurassic sediments, were eroded during the periclinal uplift accompanying Late Jurassic extension. Subsidence during the Cretaceous and Tertiary, attributable to thermal relaxation, was punctuated by two short compressive events, which are responsible for giving the area its dramatic structural style. During the earlier ?Turonian–Coniacian event normal displacement on the South Hewett Fault was reversed and a large monoclinal fold formed in the overlying cover rocks. To the SW the whole footwall was periclinally uplifted and a set of conjugate faults developed, largely in pre-Cretaceous sequences. To the NE of the South Hewett Fault all of the Late Jurassic extensional faults were reversed. This formed forced folds and harpoon structures locally in the overlying cover sequence.

During Miocene inversion only the South Hewett Fault was reactivated, and a second monoclinal fold developed above it. Its footwall again experienced periclinal uplift.

ACKNOWLEDGMENTS: This paper has benefited greatly from discussion with Professor N. Kusznir and Alan Roberts, to whom we are grateful. Thanks are due also to the staff at Badley, Ashton & Associates who were responsible for the drafting and typing.

References and further reading

BADLEY, M. E. 1984. *Petroleum Potential of UK Quadrant 53 Area*. Unpublished Report, Merlin Exploration Services.

——, 1985. *Practical Seismic Interpretation*. IHRDC, Boston.

BROOKS, J. & GLENNIE, K. W. (eds) 1987. *Petroleum Geology of Northwest Europe*. Graham & Trotman, London.

——, GOFF, J. C. & VAN HOORN, B. (eds) 1986. *Habitat of Palaeozoic Gas in NW Europe. Special Publication of the Geological Society of London*. **23**.

CRITTENDEN, S. C. 1982. Lower Cretaceous lithostratigraphy NE of the Sole Pit area in the UK Southern North Sea. *Journal of Petroleum Geology*, **5**, 191–202.

GLENNIE, K. W. (ed.) 1984. *Introduction to the Petroleum Geology of the North Sea*. Blackwell Scientific Publications, Oxford.

—— & BOEGNER, P. L. E. 1981. Sole Pit inversion tectonics. *In*: ILLING, L. V. & HOBSON, G. D. (eds) *Petroleum Geology of the Continental Shelf of North-west Europe*. Heyden & Son, London, 110–120.

HORSFIELD, W. T. 1977. An experimental approach to basement controlled faulting. *Geologie en Mijnbouw*, **56**, 363–370.

JEANS, C. V., MERRIMAN, R. J., MITCHELL, J. G. & BLAND, D. J. 1982. Volcanic clays in the Cretaceous of southern England and Northern Ireland. *Clay Minerals*, **17**, 105–156.

KOOPMAN, A., SPEKSNIJDER, A. & HORSFIELD, W. T. 1987. Sandbox model studies of inversion tectonics. *Tectonophysics*, **137**, 379–388.

MILSOM, J. 1986. Small inversion structures on seismic sections from the Southern North Sea. *Proceedings of the Geologists Association*, **97**, 379–382.

PERROT, J. & VAN DER POEL, A. B. 1987. Zuidwal — a Neocomian gas field. *In*: BROOKS, J. & GLENNIE, K. W. (eds) *Petroleum Geology of Northwest Europe*. Graham & Trotman, London, 325–335.

RAGAN, D. M. 1985. *Structural Geology. An Introduction to geometrical techniques*. John Wiley & Sons, Chichester.

SHERRIF, R. E. 1985. Aspects of seismic resolution. *In*: BERG, O. R. & WOOLVERTON, D. G. (eds) *Seismic Stratigraphy II. American Association of Petroleum Geologists Memoir*, **39**, 1–10.

TUBB, S. R., SOULSBY, A. & LAWRENCE, S. R. 1986. Palaeozoic prospects on the northern Flanks of the Wales–Brabant Massif. *In*: BROOKS, J., GOFF, J. C. & VAN HOORN, B. (eds), *Habitat of Palaeozoic Gas in NW Europe. Special Publication of the Geological Society of London* **23**, 55–72.

WALKER, I. M. & COOPER, W. G. 1987. The structural and stratigraphic evolution of the north-east margin of the Sole Pit Basin. *In*: BROOKS, J. & GLENNIE, K. W. (eds) *Petroleum Geology of Northwest Europe*. Graham & Trotman, London 263–275.

VAN HOORN, B. 1987. Structural evolution, timing and

tectonic style of the Sole Pit inversion. *Tectonophysics*, **137**, 239–284.

VAN WIJHE, D. H. 1987. Structural evolution of inverted basins in the Dutch offshore. *Tectonophysics*, **137**, 137–219.

ZIEGLER, P. A. 1981. Evolution of sedimentary basins in north-west Europe. *In*: ILLING, L. V. & HOBSON, G. D. (eds), *Petroleum Geology of the Continental Shelf of North-west Europe*. Heyden & Son, London. 3–39.

M. E. BADLEY, Badley, Ashton & Associates Ltd, Winceby House, Winceby, Horncastle, Lincolnshire, LN9 6PB, UK.

J. D. PRICE, Badley, Ashton & Associates Ltd, Winceby House, Winceby, Horncastle, Lincolnshire, LN9 6PB, UK.

L. C. BACKSHALL, Merlin Exploration Services, Merlin House, Boundary Road, Woking, Surrey, GU21 5BX, UK.

Appendix

Interpretation of the seismic data

The following points are essential to interpretation of the data discussed here.

Seismic display

The seismic data have been processed to produce a minimum phase wavelet response to an acoustic impedance contrast and are displayed with reverse polarity following the Society of Exploration Geophysicists (SEG) convention. This implies that an isolated (interference-free) reflector with a positive reflection coefficient (an increase in acoustic impedance) produces a response of a peak followed by a trough (Fig. 10), e.g. the top Zechstein or top Lower Carboniferous limestone. Reflectors with negative coefficients (a boundary across which acoustic impedance decreases) produce a trough followed by a peak. The majority of the reflections on the seismic sections, however, are interference composites resulting from reflectors being spaced more closely than half the wavelength of the seismic wavelet at a given depth.

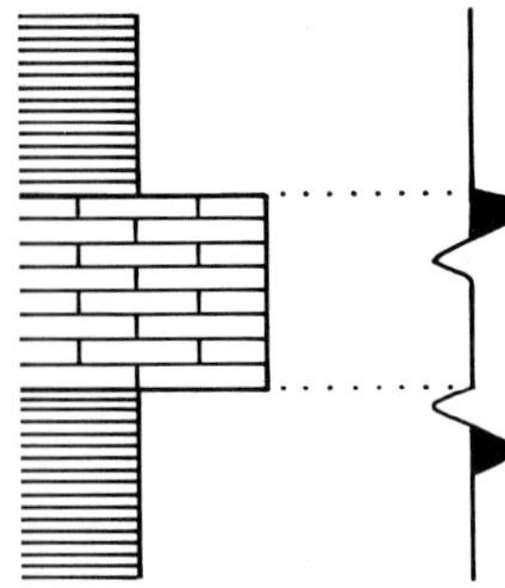

FIG. 10. Idealized seismic response from interference-free reflectors (RC = reflection coefficient).

Scale

The seismic sections are displayed at a horizontal scale of 1 : 50 000. The vertical scale is displayed conventionally in two-way-time (TWT). The relation between TWT and depth varies with velocity. The variation in velocity from SW to NE for the main stratigraphic units is shown in Table 2 together with its interval velocity expressed as corresponding metres per millisecond. As a guide to vertical exaggeration for seismic displayed at a horizontal scale of 40 shot points per km, an interval velocity of 5000 ms^{-1} produces a true scale time section and an interval velocity of 2500 m s^{-1} produces a vertical exaggeration of 2 : 1, etc.

Vertical resolution

Vertical resolution is related to the spacing of reflectors relative to the wavelength of the seismic wavelet at a particular depth. Units thicker than half-wavelength are interference-free. Intervals around a quarter wavelength thick exhibit either tuning (for reflection coefficients of opposite sign at the interval's top and base) or dimming (for similar-sign reflection coefficients at an interval's top or base). Intervals thinner than an eighth of a wavelength cannot be distinguished from a single interface (Badley 1985, Sherrif 1986). The variation in the approximate wavelength (or dominant

TABLE 2. *Tabulation of characteristic interval velocity variation for the main lithostratigraphic intervals*

Interval	Interval velocity m/s SW		NE	Thickness per millisecond m/ms SW		NE
Tertiary	1750		1900	0.875		0.95
Chalk	2700		3700	1.35		1.85
Lower Cretaceous	2200		3000	1.1		1.5
Triassic		3000			1.5	
Zechstein	3500		5000	1.75		2.5
Rotliegend		3500			1.75	
Coal and Barren Measures	3300		4000	1.65		2.0

frequency) of the seismic pulse can be estimated from the time duration of interference-free reflections from reflectors over a range of travel times.

The approximate dominant frequencies of the Merlin seismic data are:

0–700 ms	70 Hz
700–1200 ms	60 Hz
1200–2000 ms	40 Hz
2000–3000 ms	30 Hz

Wavelength is related to dominant frequency:

$$\text{Wavelength} = \frac{\text{Interval velocity (m s}^{-1}\text{)}}{\text{Dominant frequency (Hz)}}$$

Interpreting unconformities and sequence boundaries

The seismic response of discordant boundaries and their interpretation depends on the acoustic impedance contrasts across the boundary and the nature of the seismic wavelet. In a seismic section processed to produce a minimum phase wavelet, onlapping reflections terminate at the unconformity. Reflections from truncated reflectors terminate beneath the second half-cycle of the unconformity reflection, i.e. the terminations in the seismic for a truncational unconformity are actually some depth/time below the true unconformity.

Inversion in Other Geological Environments

The Lewis Thrust/Rocky Mountain trench fault system in Northwest Montana, USA: an example of negative inversion tectonics?

C. M. Powell & G. D. Williams

SUMMARY: In the Rocky Mountains of Northwest Montana, USA, Oligocene basins occur within the Laramide thrust system. During emplacement of the Laramide Lewis thrust sheet in the Glacier National Park area, a penetrative cleavage was developed with an asymmetry suggestive of NE directed movement. This cleavage can be used as a pre-extensional marker to reinterpret the geometry of the Rocky Mountain Trench extensional fault system. The cleavage is confined to a zone of constant thickness above the thrust. Thus its upper limit can be considered a mappable surface horizon from which the depth to the Lewis Thrust can be interpolated. Sequentially restored sections from the Rocky Mountain Trench to the leading edge of the Lewis Thrust illustrate the interpreted geometries and the interaction of the extensional and contractional fault systems.

Regional setting

Glacier National Park in Northwest Montana, USA forms part of the Rocky Mountain Front Ranges between longitudes 115°30′ W and 113°30′ W and latitudes 48°15′ N and 49°00′ N (Fig. 1).

Geological reconnaisance of the area dates back to Willis (1902) who described the major Lewis Thrust to the east of the Park and recognized the stratigraphy in its hanging wall as Precambrian. Willis also documented an extensive lineament running NW–SE along the Flathead River. He named it the North Fork Fault and speculated upon its having an extensional origin. Childers (1963) was the first worker to consider the control upon this extensional fault by pre-existing structures but concluded that the Lewis Thrust had had only a minor influence upon the geometry of the extensional system. In the past two decades there has been a concentration of effort north of the Canadian border, inspired by the classic works of Bally, Gordy & Stewart (1966) and Dahlstrom (1970) with structural models considering the contractional and extensional fault systems as having a common decollement at shallow depth. Constenius (1982) applied this model of reactivation to the extensional fault system in Montana. He concluded that in Glacier National Park the extensional faults were directly controlled by the position of pre-existing thrusts.

In this paper, a model is presented which considers this interaction of the Laramide contractional and the Rocky Mountain Trench extensional fault systems in Northwest Montana in terms of negative inversion tectonics. Emphasis is placed on the use of small scale structures to model hanging wall geometries and related fault profiles.

The major lineaments of the Glacier National Park area are the Lewis Thrust of the late Cretaceous Laramide Orogeny and the North Fork Fault of Willis (1902), now renamed the Flathead Fault in Canada and the Blacktail Fault in Montana (Fig. 1). In Canada, all the extensional displacement is taken up on the Flathead Fault and only one tilted fault block exists. In contrast, in Northwest Montana, multiple extensional faults and associated tilt blocks occur. The most westerly of these extensional faults is the Misson Fault which runs along the eastern shore of Flathead Lake and controls the Rocky Mountain Trench (Fig. 1). This fault system is a reflection of considerable supracrustal extension and demonstrates the effects of Basin and Range style deformation as far north as Canada (Constenius 1982).

The stratigraphy within Glacier National Park is dominated by the Precambrian 'Belt' Series in the hanging wall of the Lewis Thrust. The entire sequence is comprised of maroon and green siltstones with subordinate felspathic quartzite (the Appekunny, Grinnel, Snowslip, Mount Shields and McNamara Formations) or massive dolomitic grey limestones (the Altyn,

From Cooper, M. A. & Williams, G. D. (eds), 1989, *Inversion Tectonics*
Geological Society Special Publications No. 44, 223–234.

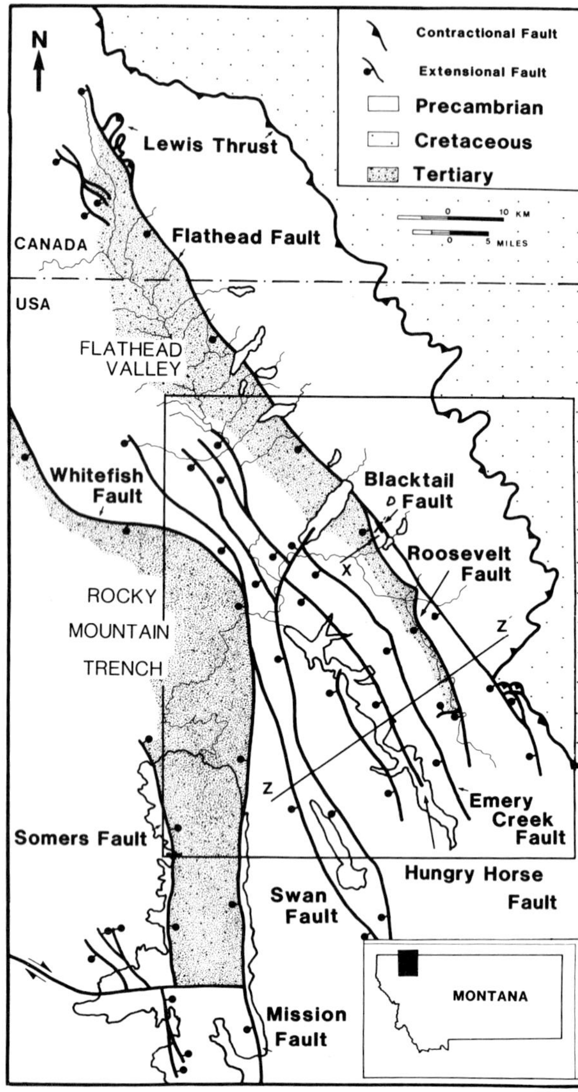

FIG. 1. Schematic geological map of the Glacier National Park area showing the positions of major contractional and extensional lineaments. (Inset box shows location of Figs 2, 3, 10A & 10B.)

Helena and Shepard Formations) (Figs 2 & 3). Both the Empire Formation, a coarse dolomitic sandstone, and the Purcell Lava, a pale blue/grey basalt within the Snowslip Formation, form distinctive marker horizons throughout Glacier National Park and adjoining Whitefish Range (Whipple *et al.* 1984). From stratigraphic logged sections taken throughout the Glacier National Park area (Fig. 2), it is clear that the lowest four formations in the Lewis Thrust sheet have a remarkably consistent thickness across the entire area. This facilitates the restoration of cross-sections with a broadly layer-cake stratigraphic template as a realistic assumption.

Lewis Thrust contractional system

The most obvious feature of the Laramide contractional event is the Lewis Thrust, which extends laterally for 250 km from Alberta to

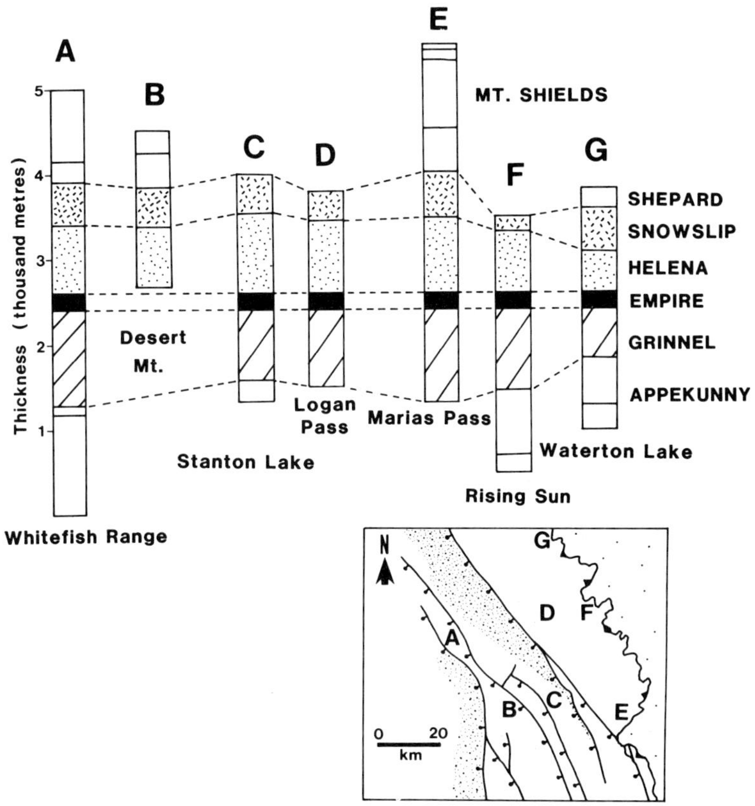

FIG. 2. Fence diagram demonstrating the consistency of thickness of Precambrian stratigraphic units throughout Glacier national Park. (Ornament of inset box as for Fig. 1)

Central Montana (Fig. 3). Laramide deformation within the Glacier National Park area has been bracketed as between 72 and 56 Ma (Hoffman *et al.* 1976). Where the Lewis Thrust is exposed in the cliffs on the eastern side of Glacier National Park, it trends NNW–SSE and appears to be a planar structure, subparallel to bedding and dipping SW at 15°. The Precambrian strata of the Lewis Thrust sheet are essentially undeformed on a large scale. However, on a smaller scale deformation features are evident. The Appekunny Formation directly above the Lewis Thrust has a strong mylonitic foliation and the Altyn limestones are considerably deformed by duplex development.

Mudge (1982) estimated movement on the Lewis Thrust in Glacier Park as 65 km with a stratigraphic throw of 6500 m. Slip vectors indicate a movement direction to 050° (Fig. 4B).

Small scale contractional structures include asymmetric folds with NE vergence and minor forethrusts and backthrusts. Thrust related tip folds exhibit fanning cleavages and the existence of finite neutral points which are usually related to buckle folding (Ramsay & Huber 1983) (Fig. 4A). Non-orthogonal penetrative cleavage is developed in incompetent beds, with an orthogonal spaced cleavage in competant units. The asymmetric penetrative cleavage consistently dips to the SW and its angular relationship with bedding may be used as an indicator of the bed parallel shear direction, coincident with the Lewis Thrust sheet emplacement direction. The cleavage becomes less intense up stratigraphy within the thrust sheet, and dies out completely 100 m below the boundary of the Helena and Snowslip Formations. This horizon is a mappable entity throughout the Glacier National Park area, and can be used to predict the geometry of the Lewis Thrust at depth.

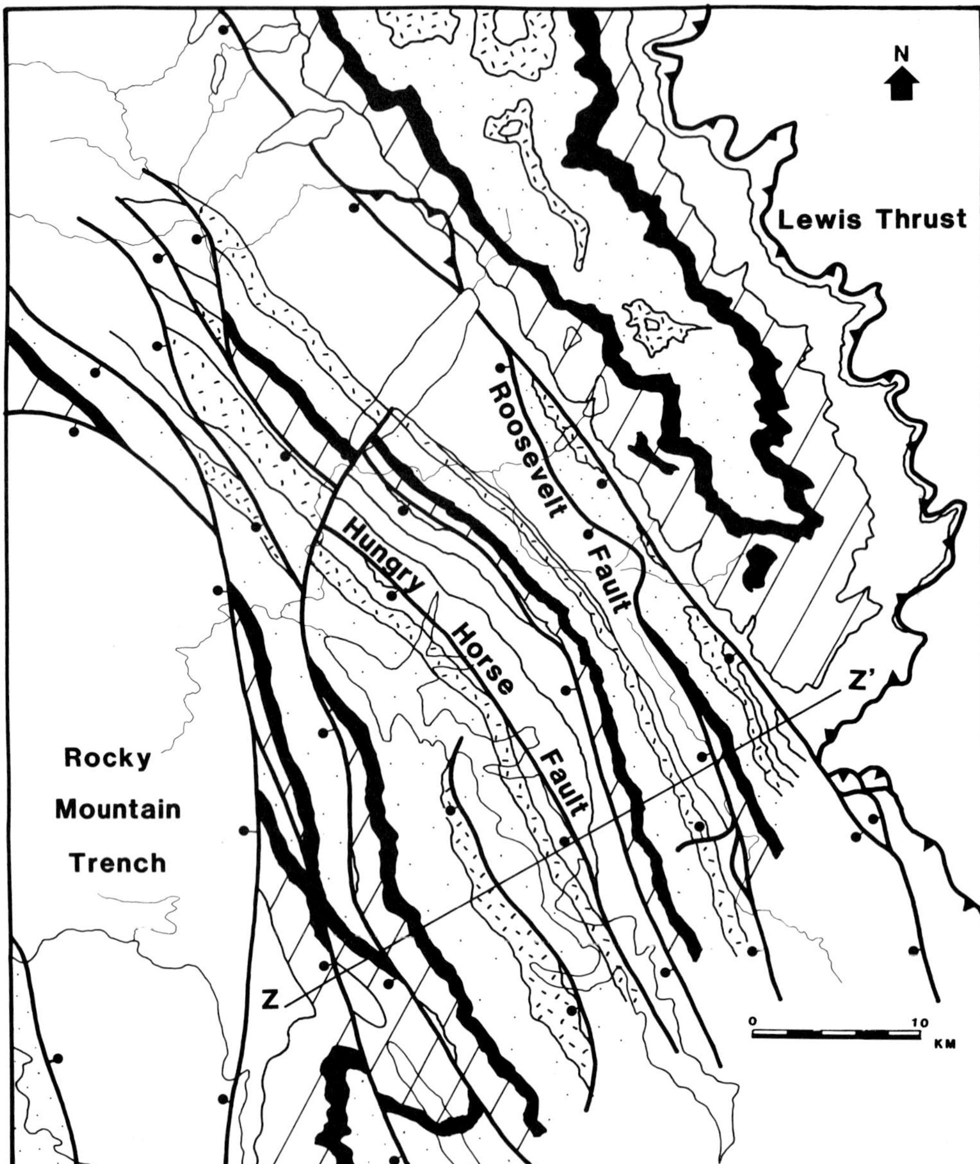

FIG. 3. Generalized geology of the Lewis Thrust sheet. (Ornament as for stratigraphic columns of Fig. 2).

The Rocky Mountain Trench extensional system

Superimposed upon the contractional Lewis Thrust system is a series of extensional fault blocks comprising the Rocky Mountain Trench system (Fig. 3). Tertiary synrift sediments unconformably overlying the Precambrian strata are preserved in the hanging walls of some of the fault blocks. This date the extension at 33 Ma (Middle Oligocene) (Constenius & Dyni 1983). Between the leading edge of the Lewis Thrust in Montana and the Rocky Mountain Trench there are seven major extensional faults (Fig. 3). All the extensional faults strike sub-parallel to the trace of the Lewis Thrust and all downthrow to the southwest. The stratigraphic separation on the faults is extremely large; 6100 m and 2100 m on the Blacktail and Roosevelt faults respectively.

Associated small scale extensional structures include faults with graphic hanging wall geometries of high angle antithetics and low angle synthetics cutting bedding at 80° and 15°, respectively (Fig. 5A). Sets of conjugate shear fractures exposed in the footwall of the Blacktail Fault, together with extension fractures, indicate an extension direction of 052°–242° (Fig. 5B). The Laramide contractional and the Tertiary extensional systems have coincident movement directions orthogonal to the strike of both major fault systems. Thus any reactivation of Laramide thrusts by the later system can be considered as dip–slip with no transpressional or transtensional component.

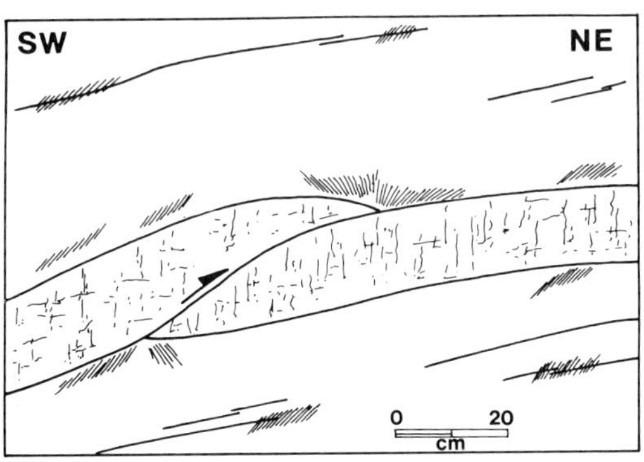

A

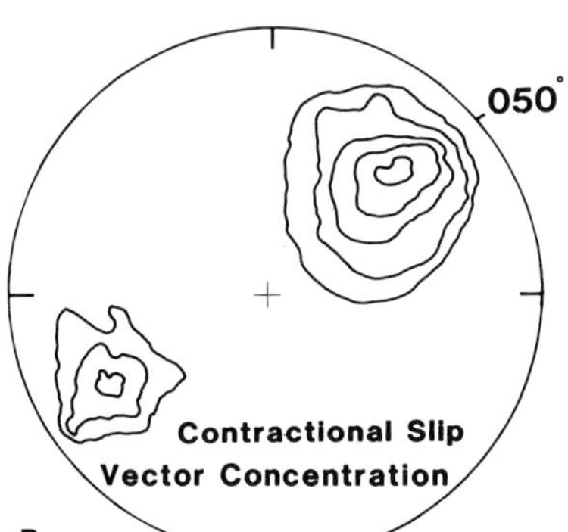

B

FIG. 4. A, Minor Laramide upward facing thrust with cleavage fanning about a finite neutral point. B, Equal area lower hemisphere projection of slickenside data on minor thrusts. Contour intervals of 1, 3, 5, 10 and 15%.

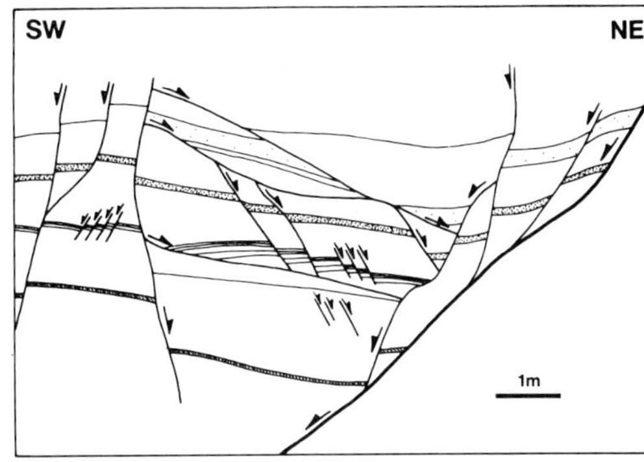

A

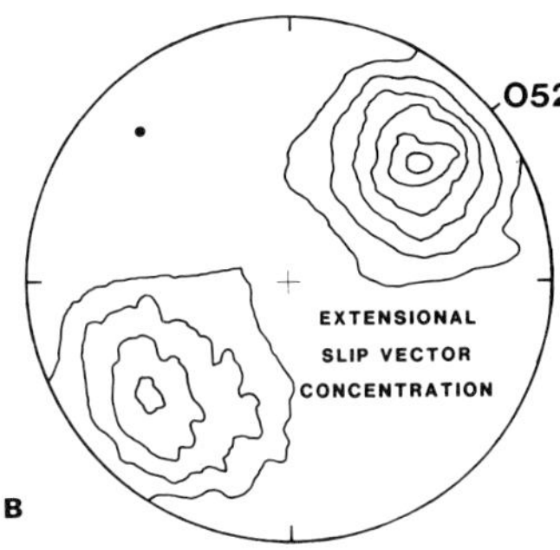

B

FIG. 5. A, Minor extensional fault system on Going-to-the-Sun road. B, Equal area lower hemisphere projection of slickenside data on minor extensional faults in Glacier National Park. Contour intervals of 1, 3, 5, 10 and 15%. Solid dot indicates intersection lineation of conjugate shear fractures.

Interaction of contractional and extensional systems

Previous workers (Bally *et al.* 1966, Constenius 1982) have believed the extensional faults of the Rocky Mountain Trench system to have listric geometries (Fig. 6). Each fault was considered to sole into and reactivate the Lewis Thrust. The Lewis Thrust then acts both as a sole thrust and as a decollement for the extensional system. A large ramp was envisaged in the Lewis Thrust at the position of the Blacktail Fault near the eastern margin of Glacier Park (Constenius 1982) and at the position of the Flathead Fault in Canada (Bally *et al.* 1966) (Fig. 6).

An alternative model is presented here. Although extensional faults of the Rocky Mountain Trench system are controlled to some degree by the pre-existing contractional structures, the reactivation of these contractional

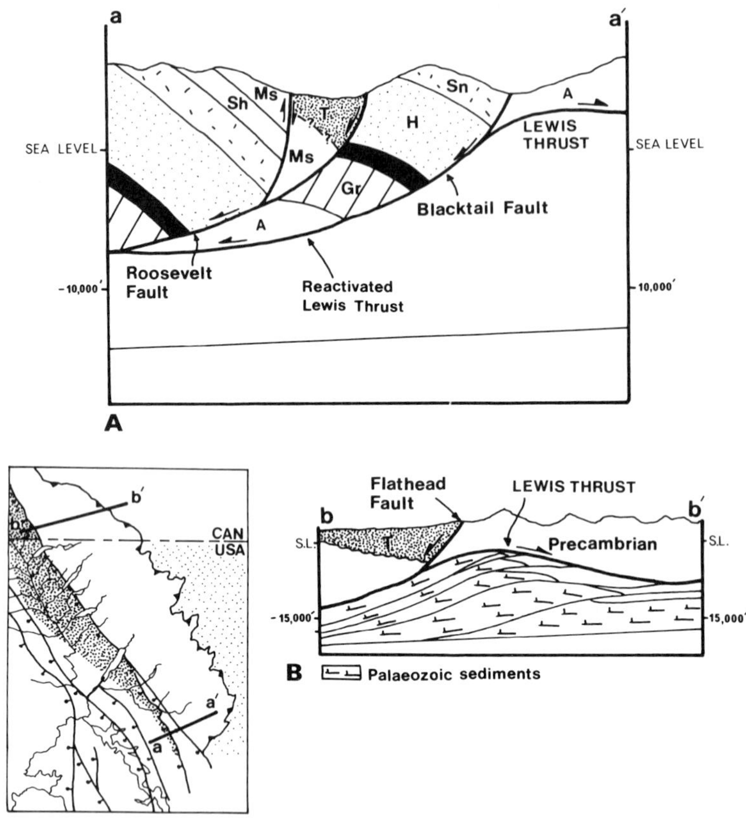

FIG. 6. Previous models for the structural evolution of the area. A, Constenius 1982; B, Bally *et al.* 1966.

fault surfaces is localized, with only part of the later system being controlled by earlier structures.

If existing models are applicable to the Glacier National Park area, the Lewis Thrust would have a relatively constant depth beneath the Park at the present day with a large ramp in the thrust plate marking the position of the most easterly fault of the Rocky Mountain Trench system. As the shear zone containing the intense penetrative cleavage has a limited thickness above the Lewis Thrust the cleavage would not be seen throughout the Park (Fig. 7B).

However, the horizon at which the cleavage dies out is a mappable entity throughout the area (Fig. 8A). The cleavage is present within every fault block, specifically in the footwall directly to the NE of each major fault. This surface distribution of the cleavage is indicative of the Lewis Thrust attaining near surface levels at the leading edge of each extensional fault block (Fig. 8A). Structural contour maps for the Lewis Thrust plane constructed from exposed stratigraphy and the elevation of the cleavage horizon illustrate the shallow depths of the thrust within the footwall of each fault block (Fig. 8B).

A model is thus required that accounts for the changing depth of the Lewis Thrust across the fault blocks. The Lewis Thrust must be displaced by the later extensional faults and remain subparallel to bedding within each fault block. The thrust would then dip northeast within each fault block and the depth of the thrust from the surface would increase to the NE within the individual blocks. The extensional fault profiles would have a planar geometry rather than being listric in form (Fig. 7D).

In order to verify the geometry of any individual fault in the extensional system, the deviation of the penetrative cleavage from its pre-extensional sheet dip can be used. On first examination, the hanging wall of the Roosevelt

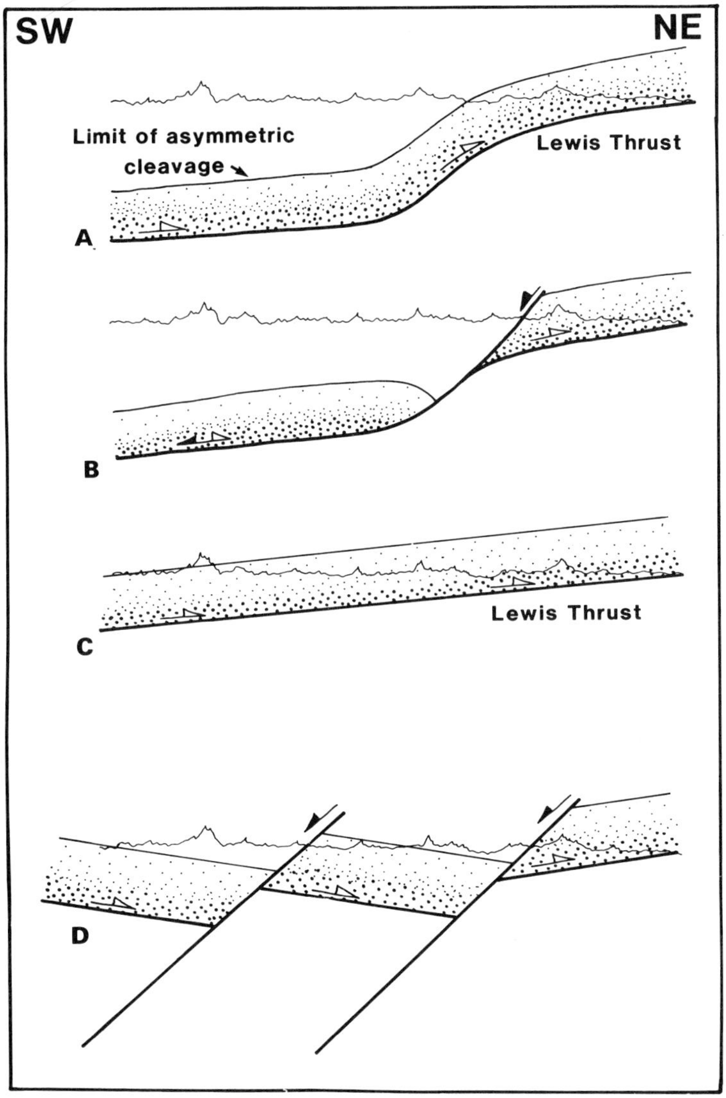

FIG. 7. Alternative models for the relationship of contractional and extensional fault systems. A, B, Pre-extensional fold in the Lewis Thrust sheet with reactivation of the steep portion of the Lewis Thrust by listric, shallowly detaching extensional faults. C, D, Deep throughgoing extensional faults with no reactivation of the Lewis Thrust.

Fault appears to be deformed into an extensional 'rollover' geometry (Hamblin 1965). The beds dip steeply to the NE at a distance of 2 km from the fault trace. However, although the bedding dip is increasing towards the fault, the cleavage maintains a constant dip angle (Fig. 9). Small folds contain the upward facing cleavage in their axial planes. The cleavage is axial planar to a medium scale asymmetric Laramide fold; the bedding is deviating from its regional dip due to a contractional structure rather than indicating an extensional rollover.

Approximately 200 m to the NE, the bedding flattens but the cleavage remains upward facing and axial planar to the Laramide fold. One kilometre nearer to the Roosevelt fault trace, both the bedding and the cleavage are rotated with the cleavage maintaining a constant angle to bedding. The cleavage is rotated through the horizontal to become downward

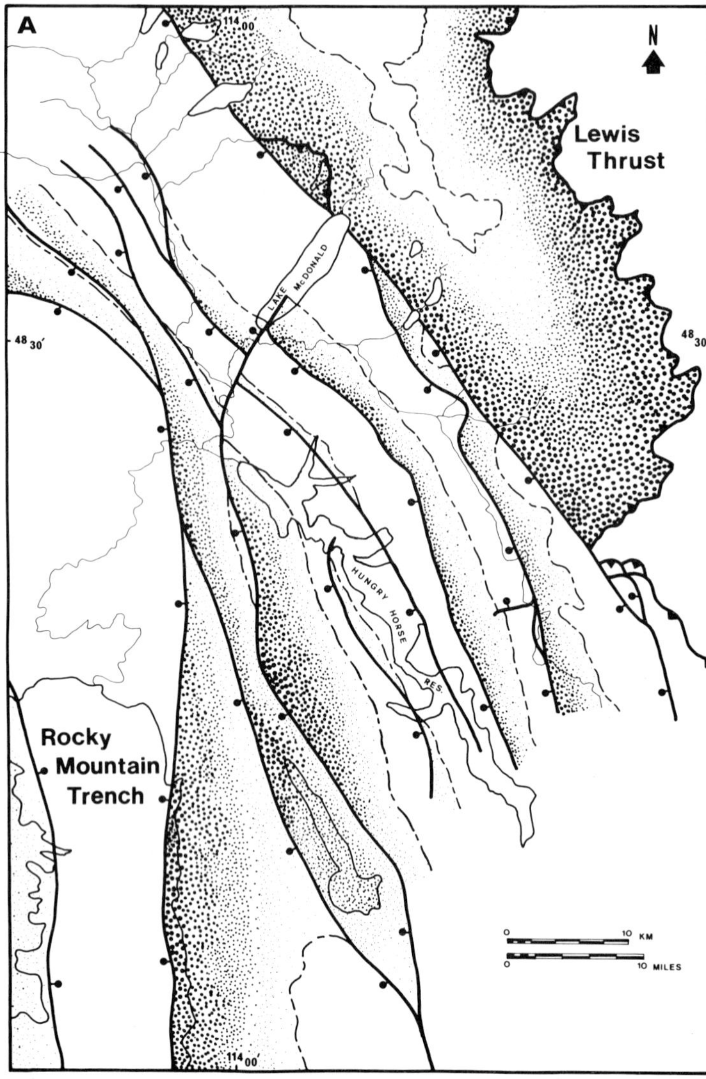

FIG. 8. A, Distribution of asymmetric cleavage within the fault blocks of the Rocky Mountain Trench extensional system. (Intensity of stipple equates to asymmetric cleavage intensity.) B, Structural contour map for the Lewis Thrust.

facing (Fig. 9). This is the true extensional hanging wall geometry, defined not only by the bedding but by the departure of the cleavage from its regional dip.

Upon application to every extensional fault block in the Rocky Mountain Trench system, pre-existing small scale contractional features define two types of hanging wall geometries with two corresponding types of fault profiles constructed using the 'slip line' Chevron construction of Williams & Vann (1988). The two types of fault profiles include listric fault profiles with rollover geometries in the hanging wall and planar faults with uniform hanging wall dips.

Only two major faults with a listric geometry are recognized; the Roosevelt Fault and the Hungry Horse Fault, which sole into and reactivate the Lewis Thrust at shallow depths. The majority of the fault profiles are planar in geometry. They cut through and displace the Lewis Thrust. The Chevron construction predicts these

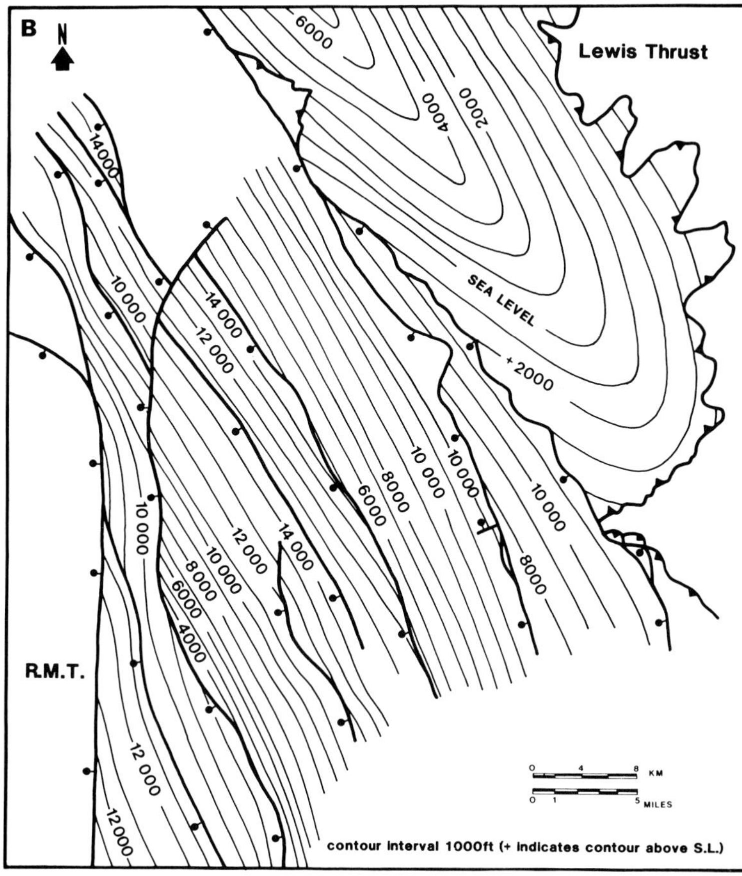

faults to be throughgoing to at least 15 km depth. They are not expected to form part of a shallow linked system of extensional faults (Gibbs 1983). The zone of ductile deformation (Fig. 10A), below which the faults cannot be accurately constrained, is equated to the 'mid-crustal detachment' at the depth of the brittle–ductile transition in continental crust. The thrust has no apparent control upon the spatial distribution of these planar faults.

A section from the Rocky Mountain Trench to the leading edge of the Lewis Thrust along the southern margin of Glacier National Park illustrates the extensional fault geometries, the distribution of penetrative cleavage and the partial reactivation of the Lewis Thrust (Fig. 10A).

The position of the listric Roosevelt Fault is controlled by a fold in the Lewis Thrust located approximately 14 km W of the ramp postulated by earlier workers. The fold in the thrust sheet is caused by a deep level backthrust, manifested at the surface as a zone of intense northeasterly dipping cleavage and numerous minor backthrusts. Partial restoration of the section shows a probable sequential development of the extensional faults with the early development of listric faults controlled by pre-existing structures (Fig. 10B). The planar faults appear to form at a later stage as they are documented cutting through the reactivated thrust plane and displacing it. Thus, the deep throughgoing planar faults, such as the Blacktail Fault develop with no influence from the contractional system.

Discussion

There appears to be a direct relationship between listric fault geometries and the presence of Oligocene basin fill sediments. The Roosevelt

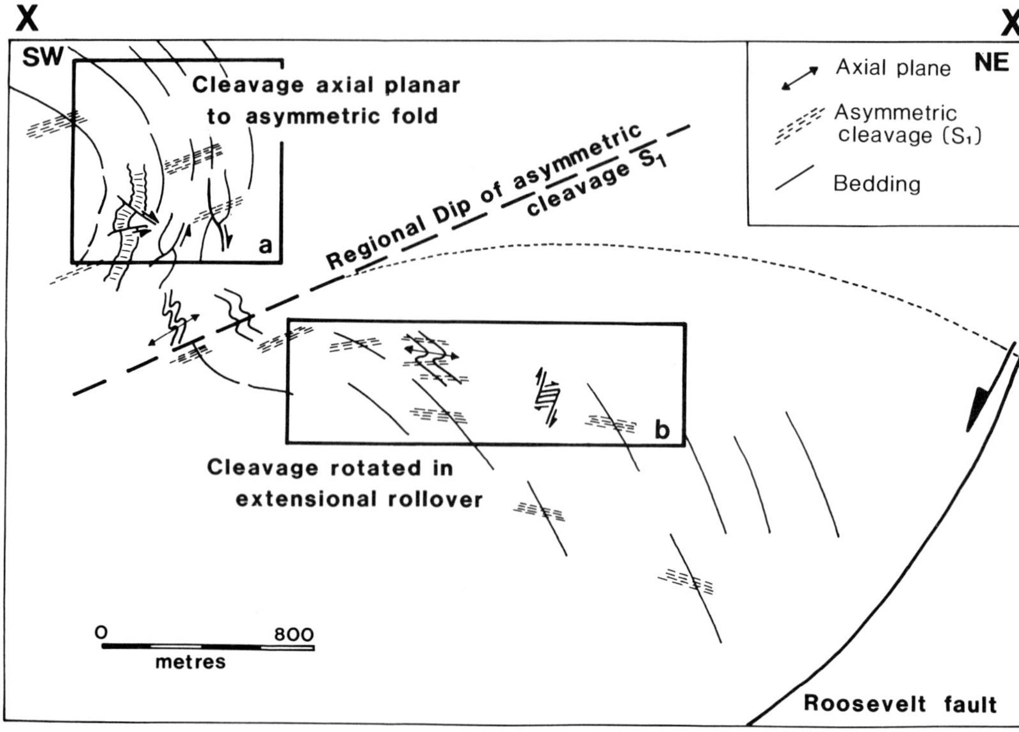

FIG. 9. Sketch section through the hanging wall of the Roosevelt Fault demonstrating the use of Laramide cleavage rotation to determine large scale extensional geometry.

Fault contains the Kishenehn Basin in its hanging wall and the Hungry Horse Fault appears to control a very narrow Tertiary basin in the valley of the Hungry Horse reservoir (Fig. 1). It may be postulated that as the Roosevelt Fault and the Blacktail Fault merge within Glacier National Park, the Flathead Fault in Canada has a listric geometry controlling the northern part of the Kishenehn Basin (McMechan & Price 1980). The Misson Fault is somewhat enigmatic because it is modelled as having a planar geometry and yet contains Tertiary sediments in its hanging wall. Further detailed work is needed to determine the precise geometry of the Rocky Mountain Trench and its controlling fault.

In conclusion, it can be stated that in the Glacier National Park area of Northwest Montana the presence of an existing fault system has had a localized and specific influence on those structures produced during a later deformation episode. The extent of this influence appears to relate to the pre-extensional dip of the thrust. The Lewis Thrust has a regional sheet dip of only 15° and is not reactivated by any of the extensional planar faults with a surface dip of 50–65°. Where it is seen to influence the geometry of the Roosevelt Fault, backthrust folding has rotated the thrust plane into a more preferential angle for reactivation in an extensional system. This can be termed negative inversion tectonics in as much as the Oligocene basinal areas and their master faults, for example the Roosevelt Fault, are controlled by the geometry of the pre-existing contractional system. However, the majority of the faults are planar and cut through the Lewis Thrust with no reactivation. Hence, the negative inversion that occurs has to be considered as localized. It relates directly to the geometry of the pre-existing contractional system.

ACKNOWLEDGMENTS: The authors would like to thank colleagues at University College Cardiff and Dr M Cooper for much useful discussion during the preparation of this manuscript. C.M.P. gratefully acknowledges receipt of a Shell International Petroleum Development PhD Scholarship, and extends thanks to Pete Fermor of Shell Canada and Ken Glennie of Shell UK for advice throughout the duration of this work.

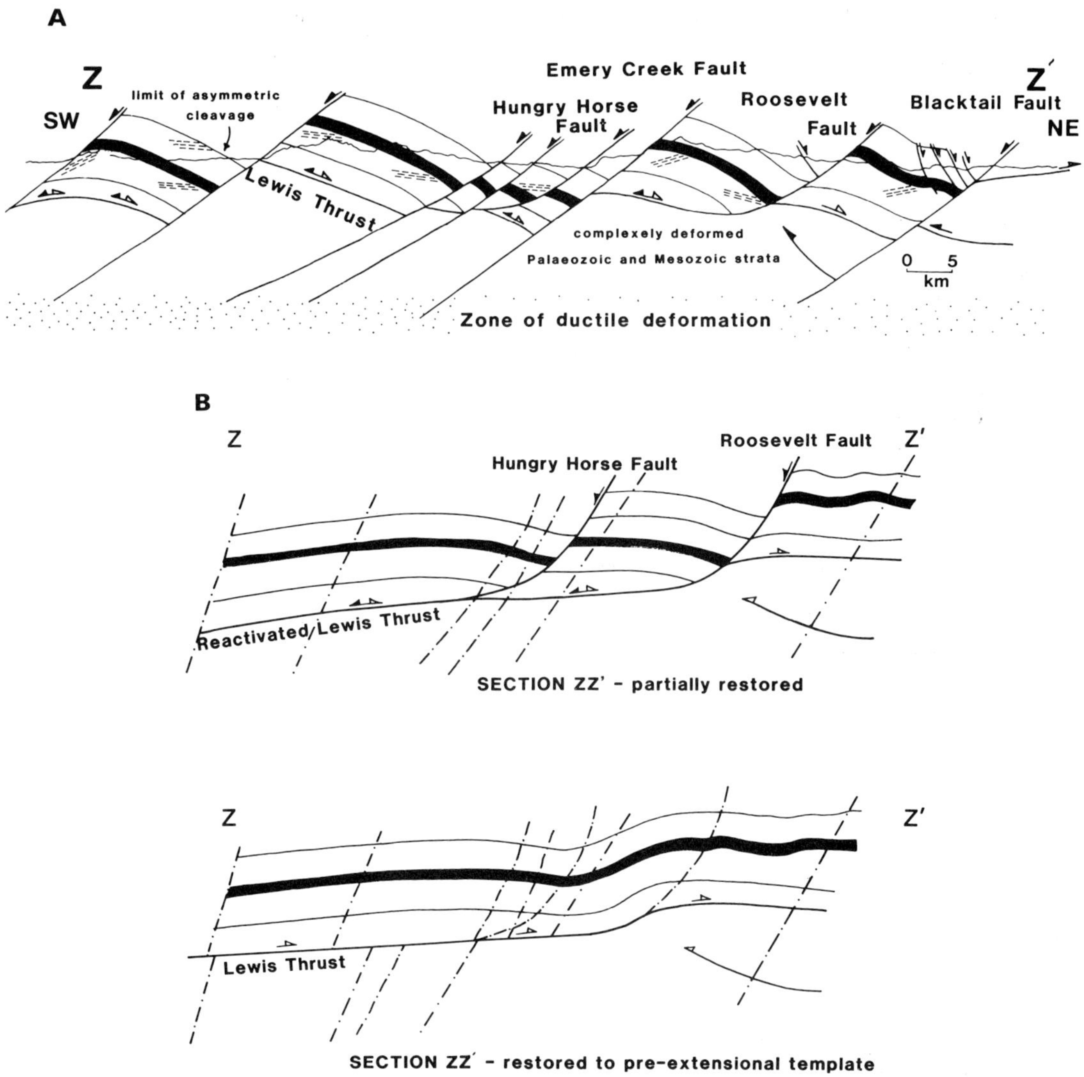

FIG. 10. A, Structural cross section Z–Z′ showing the interaction of contractional and extensional systems. B, Pre-extensional restoration of cross section Z–Z′.

References and further reading

BALLY, A. W., GORDY, P. L. & STEWART, G. A. 1966. Structure seismic data and orogenic evolution of Southern Canadian Rocky Mountains. *Bulletin of Canadian Petroleum Geologists*, **14**, 337–381.

CHILDERS, M. D. 1963. Structure and stratigraphy of the southwest Marias Pass area, Flathead County, Montana, *Bulletin of the Geological Socciiety oof America*, **74**, 141–164.

CONSTENIUS, K. N. 1982. Relationship between the Kishenehn Basin, and the Flathead listric normal fault system and Lewis Thrust salient. *In*: POWERS, R. B. (ed.) *Geologic Studies of the Cordilleran Thrust Belt*: Rocky Mountain Association of Geologists, 817–830.

—— & DYNI, J. R. 1983. Lacustrine oil shales and stratigraphy of part of the Kishenehn Basin, N. W Montana. *Mineral and Energy Resources*, **26**, 1–16.

DAHLSTROM, C. D. A. 1970. Structural geology of the eastern margin of the Canadian Rocky Mountains. *Bulletin of Canadian Petroleum Geologists*, **18**, 332–406.

GIBBS, A. D. 1983. Balanced cross-section construction from seismic sections in areas of extensional tectonics. *Journal of Structural Geology*, **5**, 153–160.

HAMBLIN, W. K. 1965. Origin of 'reverse drag' on the downthrown side of normal faults. *Bulletin of the Geological Society of America*, **76**, 1145–1164.

HOFFMANN, J. & HOWER, J. & ARONSON, J. L. 1976. Radiometric dating of time of thrusting in the disturbed belt of Montana, *Geology*, **4**, 16–20.

MCMECHAN, R. D. & PRICE, R. A. 1980. Reappraisal of a reported unconformity in the Palaeogene (Oligocene) Kishenehn Formation; implications for Cenozoic tectonics in the Flathead Valley graben, Southeastern British Columbia, *Bulletin of Canadian Petroleum Geologists*, **28**, 37–45.

MUDGE, M. R. 1982. A resume of the structural geology of the Northern Disturbed Belt, Northwestern Montana. *In*: Powers, R. B. (ed.) *Geologic Studies of the Cordilleran Thrust Belt*. Rocky Mountain Association of Geologists, 91–121.

RAMSEY, J. G. & HUBER, M. I. 1983. *The Techniques of Modern Structural Geology*, Academic Press, London.

WHIPPLE, J. W., CONNOR, J. J., RAUP, O. B. & MCGIMSEY, R. G. 1984. *Preliminary Report on the Stratigraphy of the Belt Supergroup, Glacier National Park and Adjacent Whitefish Range, Montana. 1984 Field Conference to NW Montana*. Montana Geological Society 59–52.

WILLIAMS, G. D. & VANN, I. D. The geometry of listric normal faults and deformation in their hanging walls. *Journal of Structural Geology*, (in press).

WILLIS, B. 1902. Stratigraphy and structure, Lewis and Livingstone Ranges, Montana, *Bulletin of the Geological Society of America Bulletin*, **13**, 305–352.

C. M. Powell, Department of Geology, University of Keele, Keele, Staffordshire ST5 5BG, UK.

G. D. Williams, Department of Geology, University of Keele, Keele, Staffordshire ST5 5BG, UK.

Inversion of the Kechika Trough, Northeastern British Columbia, Canada

K. R. McClay, M. W. Insley & R. Anderton

SUMMARY: The Kechika Trough, in the western Rocky Mountains of northern Canada, is a NW-trending fold and thrust belt of Hadrynian through Mississippian strata within the telescoped passive margin of ancestral North America. The Kechika Trough underwent major crustal extension in the Mid–Late Devonian with the deposition of coarse to fine 'Antler' type black clastics in an asymmetrical basin consisting of a terraced half-graben system to the west and a roll-over structure to the east. Inversion and contraction of the Kechika Trough occurred during the Jurassic–Cretaceous Columbian–Laramide orogeny. The geometries of the resultant fold and thrust systems were strongly controlled by the half-graben extensional architectures and by the geometries of the synrift sediment packages. Inversion structures include contractionally reactivated extensional faults, rotated and re-utilized extensional faults, and thrust faults pinned by extensional faults as a result of buttressing against rigid sediment wedges. Inversion of the synrift sediments is associated with strongly developed chevron folding, pervasive cleavage development and imbrication by out-of-half-graben thrust faults. These thrust faults may locally climb down stratigraphic section in the direction of transport as a result of cutting through strata tilted by Mid–Late Devonian extensional faults. Major stratigraphic contrasts between adjacent thrust packages have been resolved in terms of lateral facies differences produced by extension, rather than by over-complex thrust geometries. Although syndepositional extensional faults are not preserved in the thrust belt, their position and geometry may be inferred from an analysis of structural, stratigraphic and sedimentological relationships.

Introduction

The Kechika Trough, in the western Rocky Mountains of northeastern British Columbia (Fig. 1), is an inverted and contracted extensional basin system that now forms a NW-trending, complex fold and thrust belt of Palaeozoic miogeoclinal strata within the telescoped continental margin of ancestral North America. This research on the extension and inversion history of the Kechika Trough forms part of a multidisciplinary study of the tectonics, stratigraphy, sedimentation and mineralization of this region and is based upon detailed structural and stratigraphic mapping (at a scales of 1:20 000 to 1:10 000), of approximately 3200 km^2 in the Gataga area that encompasses the central part of the Kechika Trough (Fig. 1). Here the fold and thrust belt is at least 90 km long but only 15–18 km wide. Detailed section measuring, palaeontological, structural and sedimentological studies were used to determine the tectonic history, depositional environments, tectonic controls on sedimentation and to constrain section construction. Strata range in age from Late Precambrian (Hadrynian) through to Mississippian. They outcrop in a 180 km long fold and thrust belt that contains widespread stratiform Ba–Fe–Zn–Pb mineralization in Late Devonian synrift 'black clastic' sediments.

Previous work in the Gataga area has chiefly been reconnaisance style 1:250 000 scale mapping (Gabrielse 1962; Taylor & Stott 1973; MacIntyre 1981; 1983) and detailed exploration for mineralization (Carne & Cathro 1982). This research programme began in 1985 with detailed mapping of the Driftpile Creek area (McClay & Insley 1986) and continued in 1986 with a regional mapping programme (McClay *et al.* 1987; 1988).

The Kechika Trough is the southern extension of the mid–late Palaeozoic Selwyn Basin in the Yukon (Fig. 1). Through mid-Proterozoic to Jurassic times the western edge of cratonic North America was a passive margin accumulating shallow-marine clastic and carbonate assemblages with a dominant source from the cratonic area to the east. The stratigraphic record contains evidence of episodic extension throughout this period. Thompson *et al.* (1987) have identified four distinct passive margin sequences each associated with extensional (rift) episodes and subsequent subsidence. The fourth

From Cooper, M. A. & Williams, G. D. (eds), 1989, *Inversion Tectonics*
Geological Society Special Publications No. 44, 235–257.

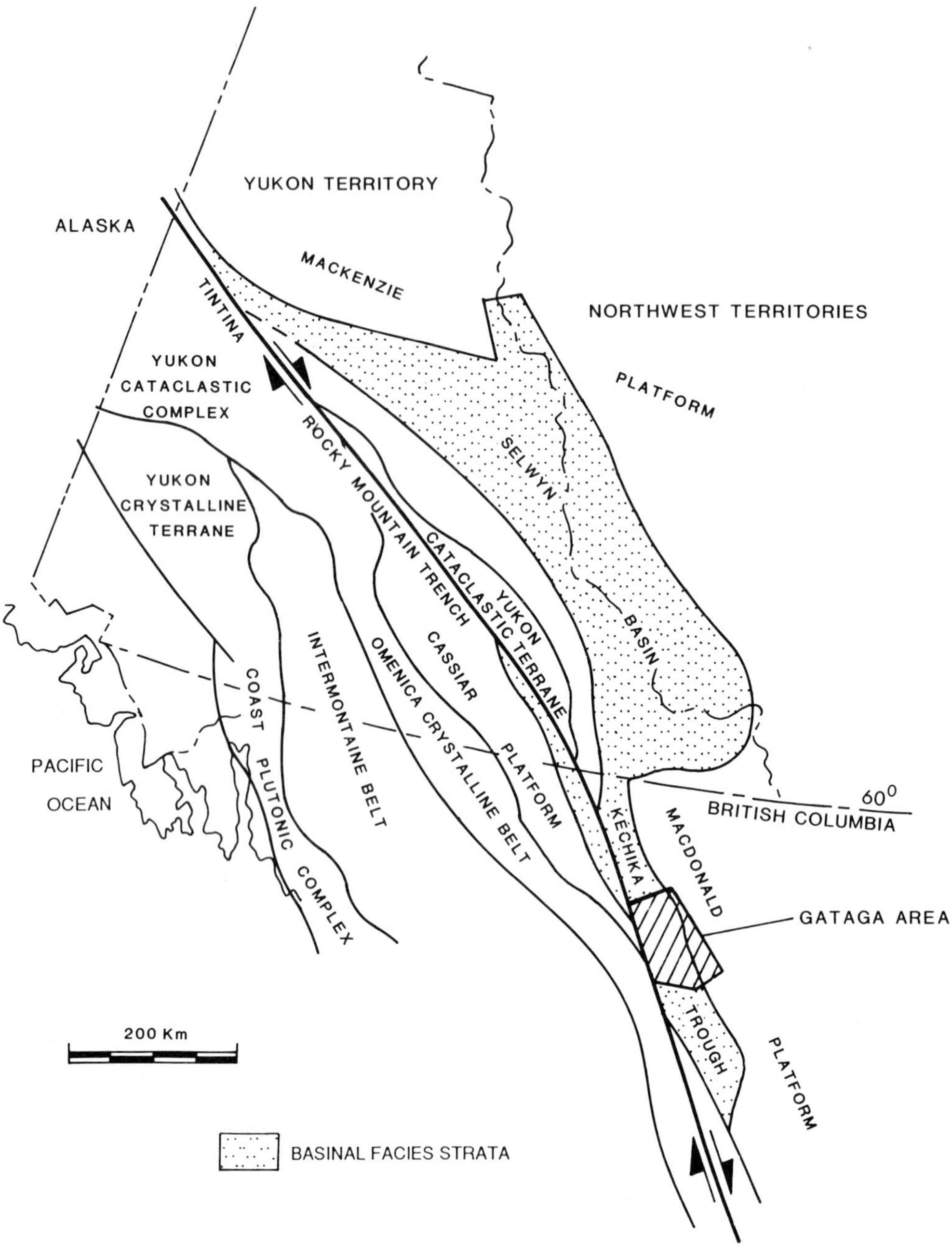

FIG. 1. Location of the Gataga map area within the Kechika Trough, NE British Columbia, Canada.

passive margin sequence 370 Ma to 180 Ma (upper Palaeozoic and lower Mesozoic sequence) appears to have been initiated in Mid–Late Devonian when the environment changed dramatically with the influx of large volumes of coarse to fine-grained 'Antler black clastics' which flooded across the continental margin largely from the NW (Gordey *et al.* 1987). These black clastics make up the Lower Earn Group (Gordey *et al.* 1982) and are found throughout the Selwyn Basin and Kechika Trough. They appear to have been deposited within extensional graben and half-graben systems that developed within the passive margin (Gordy *et al.* 1982; McClay 1984; McClay & Insley 1986; McClay *et al.*, 1988).

This paper describes the extensional evolution of the Kechika Trough half-graben system and its subsequent contraction and inversion within the Columbian–Laramide fold and thrust belt. We demonstrate that the geometries of syndepositional Late Devonian extensional faults and the architecture of the sediment bodies and graben systems strongly influenced and locally controlled the contraction and inversion. In particular there is a strong interplay between the synrift basin architecture and the later thrust fault geometries and fault sequences.

Tectono-stratigraphic setting of the Gataga area

The Gataga area is bounded to the west by the Rocky Mountain Trench dextral strike–slip fault system, with at least 750 km cumulative displacement (Gabrielse 1985), and to the E by folded and thrust, Hadrynian age siliciclastics Taylor & Stott 1973; McClay *et al.* 1988, (Figs 2 & 3).

The Gataga area of the Kechika Trough forms part of the Late Jurassic–Cretaceous Columbian–Laramide NE verging fold and thrust belt of the western Rocky Mountains. The study area comprises five distinct, narrow (3–5 km wide) but elongate (50–70 km long), tectono-stratigraphic assemblages (I–V in Fig. 2) which are bounded by steep SW dipping, NE verging thrust faults. Assemblage I contains only highly folded and thrust Kechika Group strata and forms the western margin of the Kechika Trough (Figs 2 & 3). Assemblage II consists of Kechika through Upper Earn Group strata (Fig. 4) that are strongly folded and imbricated on thrust faults (Figs 2 & 3). Assemblage III contains a distinct series of imbricates in Road River and Lower Earn Group rocks. The Driftpile sub-basin (Figs 2 & 3) constitutes tectono-stratigraphic unit IV which is dominated by a thick sequence of Lower Earn Group synrift, fine-grained 'black clastic' sediments. The more broadly folded Hadrynian through Cambrian strata on the eastern margin of the Kechika Trough form assemblage V (Figs 2 & 3). Within each assemblage the strata generally young westwards.

The Mid–Late Devonian 'black clastics' of the Kechika Trough formed in sediment starved extensional basins and host widespread stratiform barite and barite-sulphide deposits (cf. Figs 4 & 5) (Carne & Cathro 1982; MacIntyre 1983; McClay & Insley 1986). Deposits of this type in the Selwyn Basin–Kechika Trough are considered to have formed from metalliferous brines which discharged along active growth faults and ponded in local extensional basins during crustal extension in the Middle–Late Devonian (Gordey *et al.* 1982; MacIntyre 1983; McClay 1984; McClay & Insley 1986).

General descriptions of the stratigraphy and the stratigraphic nomenclature used in this paper are given in Gabrielse (1962), Taylor & Stott (1973), Fritz (1980), MacIntyre (1983) and McClay *et al.* (1987; 1988). Stratigraphic columns incorporating measured sections from both the eastern and western parts of the map area are shown in Fig. 4. Locations of measured sections MS1–MS4 are shown on Fig. 2. Accurate thickness determinations are, in places, hampered by the intense deformation and cleavage development.

The pre Mid–Late Devonian rift strata (Fig. 4) consist of a minimum 1.2–1.5 km cumulative thickness of Hadrynian phyllites and grits; 1.1 kms of Cambrian shallow water platform clastics and carbonates; a westwards thickening wedge (150–800 m) of Cambrian–Ordovician calcareous phyllites and thin limestones of the Kechika Group; and Ordovician–Lower Devonian black argillites and cherts, orange weathering siltstones (Silurian) and silver grey weathering cherts and argillites of the Road River Group with a total thickness of approximately 200–350 m (Fig. 4). The Lower Devonian cherts and argillites are distinctive lithologies that mark the top of the Road River Group and can be found in most thrust panels (Fig. 4). They are the uppermost units in the pre-Mid–Late Devonian rift sequence and are unconformably overlain by the synrift Lower Earn Group strata (Fig. 4).

The stratigraphy and sedimentology of the Lower Earn Group 'black clastics' of Mid–Late Devonian age are described in detail because these synrift sediments permit determination of the extensional basin architecture.

Stratigraphy of the Earn Group

Lower Earn Group (Mid–Late Devonian)

The Mid–Late Devonian Lower Earn Group of the Selwyn Basin–Kechika Trough is characterized by a flood of W and NW derived coarse

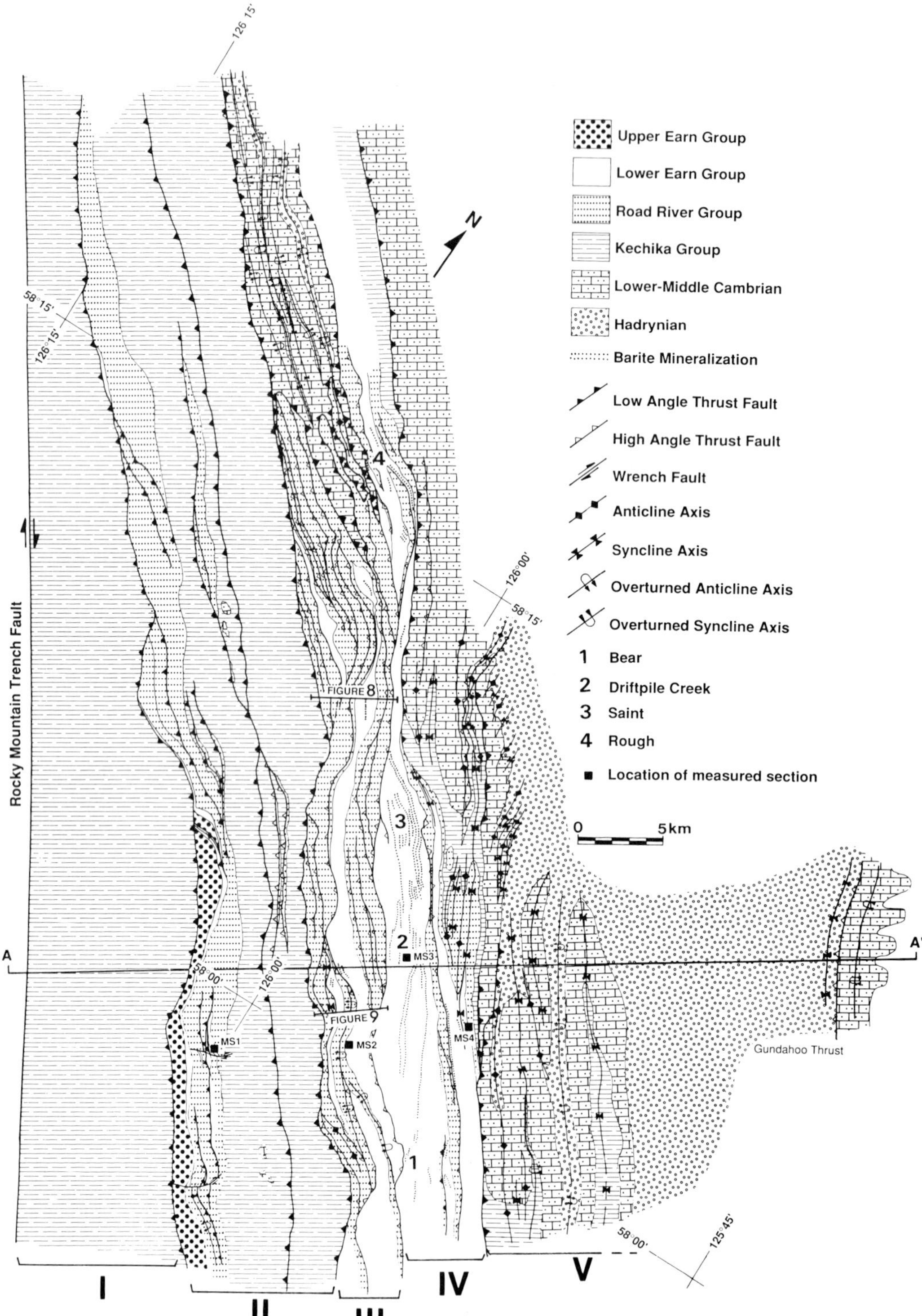

FIG. 2. Regional geology of the Gataga area compiled from detailed 1 : 10 000 and 1 : 20 000 mapping. I to V are major thrust bounded tectono-stratigraphic assemblages discussed in the text. MS1 to MS4 are the locations of measured sections shown in Fig. 4. Localities 1 to 4 are occurences of significant stratiform Ba-Zn-Fe-Pb mineralization.

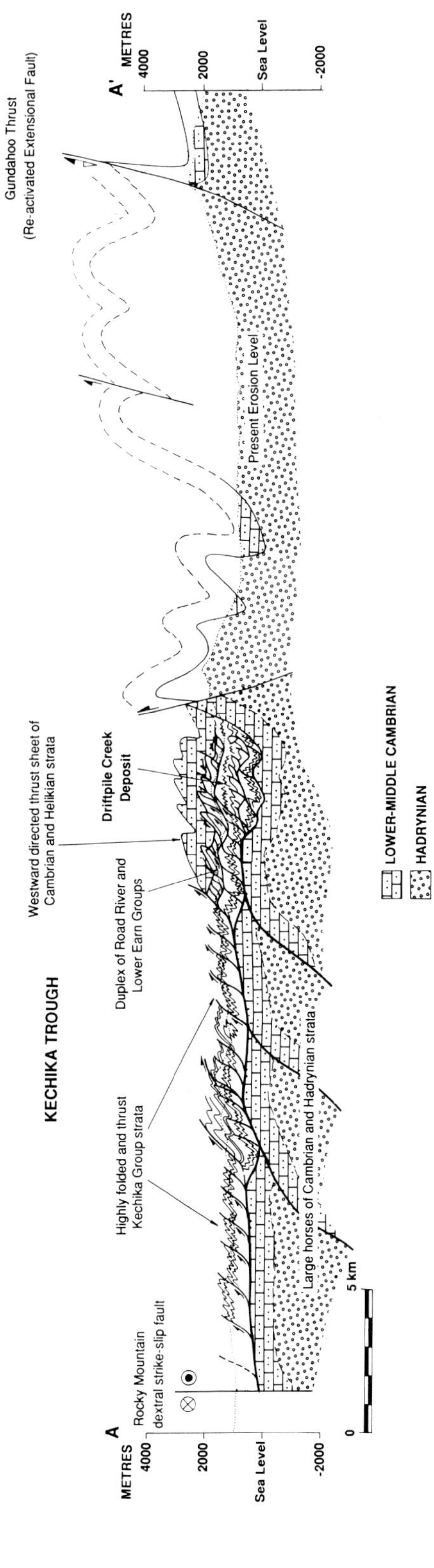

FIG. 3. Regional cross-section of the Gataga area along line A–A′ in Fig. 2.

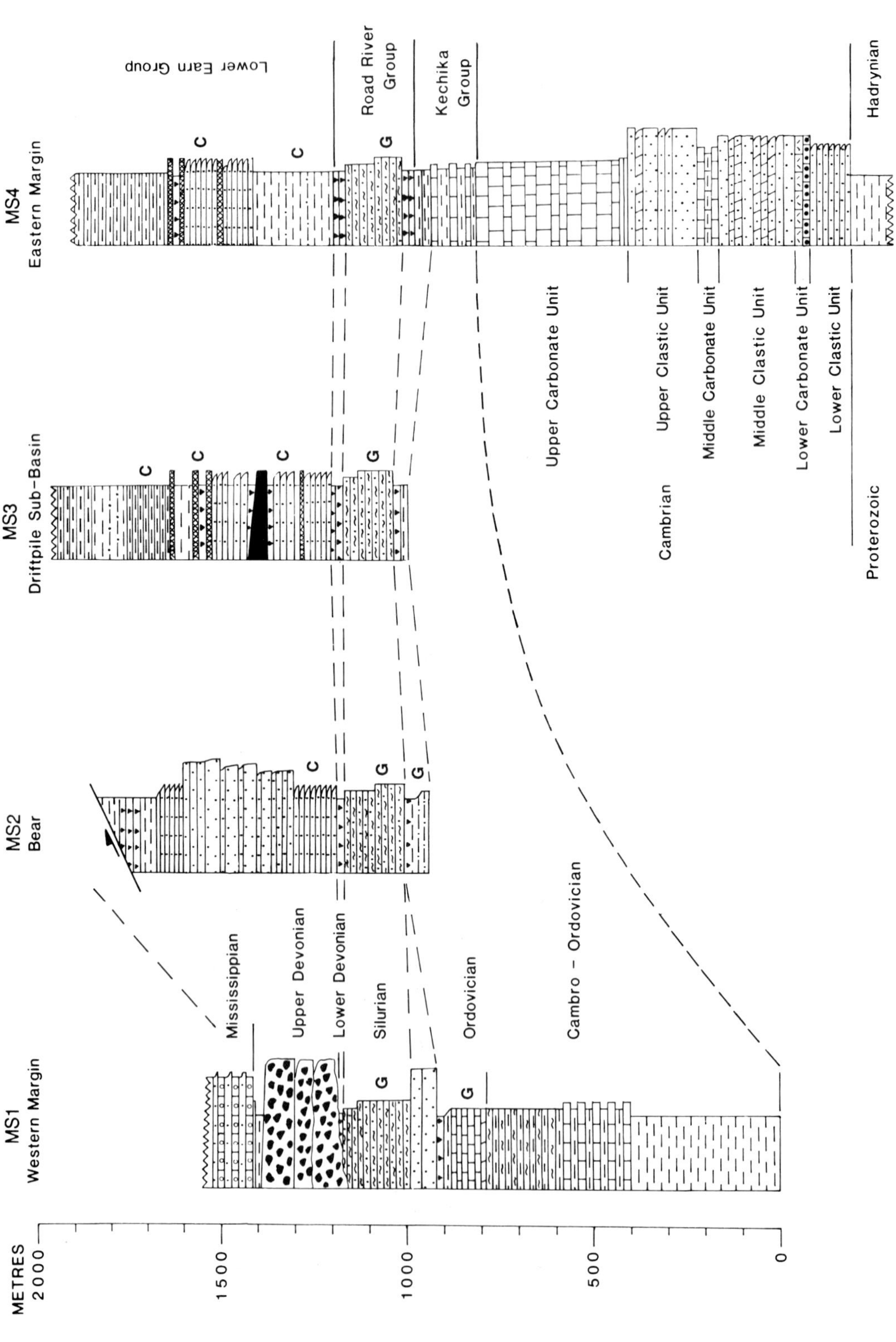
MS1
Western Margin
MS2
Bear
MS3
Driftpile Sub-Basin
MS4
Eastern Margin
METRES
2000
1500
1000
500
0
Mississippian
Upper Devonian
Lower Devonian
Silurian
Ordovician
Cambro - Ordovician
Lower Earn Group
Road River Group
Kechika Group
Upper Carbonate Unit
Upper Clastic Unit
Middle Carbonate Unit
Middle Clastic Unit
Lower Carbonate Unit
Lower Clastic Unit
Cambrian
Proterozoic
Hadrynian
G
C

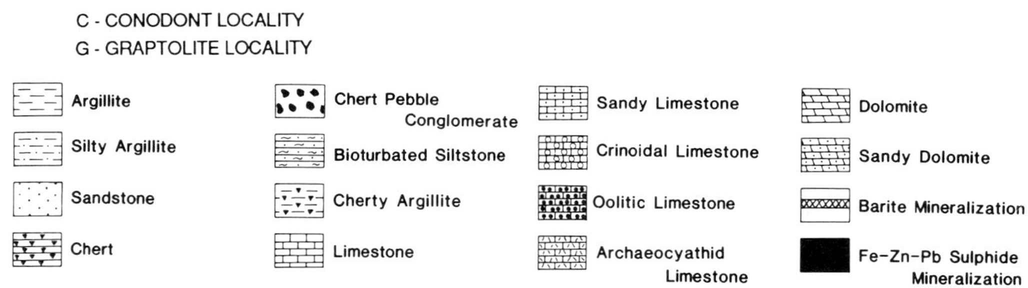

FIG. 4. Regional stratigraphic sections from west to east across the Kechika Trough showing thickness variations in the Mid–Late Devonian synrift Lower Earn Group. Location of measured sections MS1 to MS4 are shown on Fig. 2. Key stratigraphic boundaries are linked by tie lines. The datum for the pre-Mid–Late Devonian rift sequence is taken as the distinctive Lower Devonian cherty argillite sequence at the top of the Road River Group.

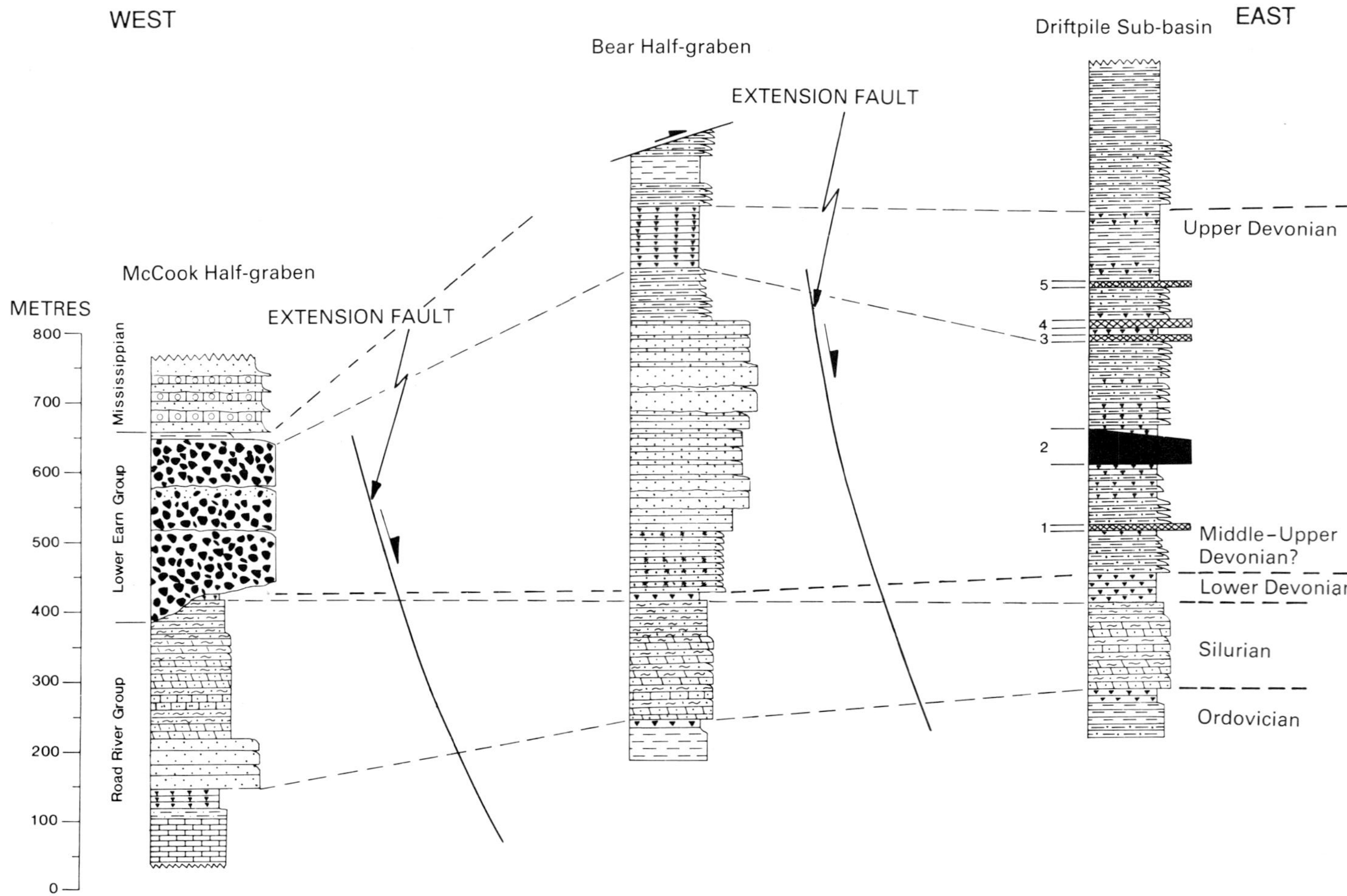

FIG. 5. Detailed measured sections through the McCook half-graben (MS1, Fig. 2), the Bear half-graben (MS2, Fig. 2), and the Driftpile sub-basin (MS3, Fig. 2). These illustrate the changes in Mid–Upper Devonian Lower Earn Group sediment character and thickness due to deposition in extensional fault bounded basins. Key as in Fig. 4.

grained chert pebble conglomerates, chert grits, sandstones and siltstones with rapid thickness changes, proximal submarine fans and fault controlled deposition (Gordey *et al.* 1982; 1987; McClay *et al.* 1987; 1988). These Lower Earn Group 'black clastics' have been interpreted as a synrift sequence (Gordey *et al.* 1987; McClay *et al.* 1988) and represent a dramatic change in depositional environment from the relatively quiet sedimentation conditions of the underlying Road River Group.

Within the Gataga map area detailed mapping, section measurement and sedimentological analysis has permitted the identification of three distinct, narrow (probably only 8 to 12 km wide) but elongate (at least 50 to 70 km long), extensional fault bounded basins. These include McCook half-graben in the west (MS1 in Fig. 2), the Bear half-graben (MS2 in Fig. 2) and the Driftpile sub-basin (MSE and locality 2 in Fig. 2). Detailed measured sections through these basins are shown in Fig. 5.

In the western part of the Gataga area (Figs 2 & 3), the base of the Lower Earn Group in the McCook half-graben is characterized by sheets and lenses of resistant, thick bedded chert pebble conglomerates and chert grits which thin and interdigitate eastwards with thin bedded, laminated siltstones and silt banded argillites (Fig. 5, MS1 in Fig. 4). The thick bedded to massive chert pebble conglomerates (Fig. 5) overlie the Road River strata with a strongly erosive and downcutting base. These coarse conglomeratic westernmost exposures of the Lower Earn Group are overlain by only a few tens of metres of fine-grained black argillites which are in turn overlain by Mississippian age medium to thick bedded crinoidal grainstones and sandstones of the Upper Earn Group (Fig. 5).

The Bear half-graben (MS2 in Figs 2 & 4) contains Lower Earn Group strata that are stratigraphically and sedimentologically distinct from the McCook half-graben. The base of this sequence is characterized by medium to thin bedded chert pebble grits, sandstones and local chert pebble conglomerates (Fig. 5). The conglomerates and grits are overlain by several hundred metres of recessive, unlaminated to thinly laminated silver-grey weathering black argillites, cherty argillites and cherts which range in age from Frasnian to Fammenian (M. Orchard, personal communication, 1985). Palaeocurrent evidence indicates that these sediments prograded towards the N–NE.

The Driftpile sub-basin in the central–eastern part of the map area (MS3 and locality 2 in Fig. 2) is larger and more extensive than the two eastern half-graben systems. It contains a minimum 1 km thick sequence of fine-grained carbonaceous black argillites and cherty argillites and thin distal turbiditic siltstones (Fig. 5). Palaeocurrent measurements indicate sediment transport towards the NW and NNW parallel to the basin axis. This sequence, dated as Fammenian using conodonts (M. Orchard, personal communication, 1986), contains at least three horizons of stratiform barite–lead–zinc mineralization and probably a further two horizons of stratiform barite mineralization (Fig. 5). The Lower Earn Group strata within the Driftpile sub-basin thin eastwards and are sedimentologically distinct from those in the Bear half-graben system to the SW.

The distinctive coarse grained, eastward thinning sediment wedges, combined with sedimentological analysis and mapped structural/stratigraphic relationships permit the inference that each of the three Lower Earn Group basins within this part of the Kechika Trough was generated by active Mid–Late Devonian extensional faulting with hanging wall movement down to the E (Fig. 5). This produced distinctive NW-trending synrift black clastic basins each with a characteristic stratigraphy. Postrift deposition (Fig. 5) is characterized by laminated black argillites and siltstones and probably passes upwards into the Mississippian Upper Earn Group.

Upper Earn Group (Mississippian)

This consists of a minimum of 70 m of grey-black, medium-thick bedded crinoidal grainstones, sandstones and siltstones with abundant shell debris which yield a Mississippian fauna (H. Gabrielse, personal communication, 1986). This unit is only found in the footwall of the westernmost major thrust fault in the map area (Fig. 2). Its exact tectono-stratigraphic position is unclear (McClay *et al.* 1988).

Tectonics of the Gataga Area

The Gataga area comprises five major tectono-stratigraphic packages (I–V, Fig. 2) which exhibit complex polyphase folding and thrust faulting (Figs 2 & 3). Detailed structural analysis of the Gataga area has indicated three deformation phases, which are all probably Columbian–Laramide in age, and which are common to all tectonic units (McClay & Insley 1986). These are:

1 An early, pre-main cleavage phase of folding

on NE-trending fold axes. A local early cleavage is found around phase one fold hinges.

2 Dominant NW-striking, NE verging folding and thrusting with the accompanying development of a penetrative cleavage. The folds are tight to chevron style and generally plunge gently either to the NW or to the SE. All thrust faults are now generally steeply SW dipping and have been rotated into this attitude by movement on underlying thrusts.

3 Late NE–SW to E–W striking dextral reverse kink folding has produced minor folds, dilatant vein systems and minor reorientation of earlier structures.

In the W part of the map area the folds and thrusts show a NE transport direction (i.e. vergence NE) whereas at the E boundary both NE and SW verging folds and thrust faults are found (Figs 2 & 3). In the northern part of the area large thrust sheet of Cambrian strata has been thrust southwestwards over Driftpile subbasin. Subsequent to emplacement, the thrust sheet was folded by movement on underlying thrusts thus giving rise to the complex map patterns in the northern part of the Gataga area (Fig. 2). Total tectonic shortening is difficult to measure because individual thrust sheets acted independently (see cross-section in Fig. 3) and because of the intense cleavage development. A minimum shortening of 100% has been estimated from some sections in the thrust belt.

Anomalous thrust relationships are found within panels II, III and IV (Fig. 2) where duplexes and imbricate fans incorporate progressively older strata towards the foreland and in the direction of tectonic transport. This implies that either these thrusts develop progressively into the hinterland or that some of the thrusts climb down section in the direction of transport and therefore do not follow the simple rule that thrusts climb up stratigraphic section in the direction of tectonic transport. In subsequent sections of this paper we develop models to explain these anomalies.

Extension in the Kechika Trough

In common with other parts of the Canadian Cordilleran margin (e.g. Struik 1987 and Thompson *et al.* 1987) the Gataga area contains a shallow water siliciclastic and carbonate passive margin sequence that extends at least from the Upper Proterozoic to the Lower Mesozoic. Such an extended sequence cannot be simply explained by a single continental rifting and separation event which would be expected to reach thermal equilibrium and zero subsidence potential in 200 Ma or less (McKenzie 1978; Thompson *et al.* 1987). The detailed structural, stratigraphic, palaeontological and sedimentological analysis carried out during this research programme has indicated that the Kechika Trough has undergone episodic extension from at least the Eocambrian through to the Late Devonian. In particular, the identification of distinct half-grabens and sub-basins combined with definition of the geometries of the sediment packages has permitted an interpretation of the positions and geometries of extensional faults within the now complex fold and thrust belt.

Eocambrian–Middle Cambrian extension

In the Gataga area, the latest Hadrynian sediments (Fig. 4) are interpreted to have been deposited on a storm-dominated shelf and are overlain by the Cambrian shelf clastics and carbonates (Fig. 4). The former consist of tidal-shelf sandstones which, on a regional scale, probably interdigitate westwards with a carbonate succession that dominated the outer part of the Cambrian passive continental margin. The Gataga area lies in an intermediate position between the inner detrital and middle carbonate belts and probably occupied a slightly protected shelf basin between source lands to the E on the craton and a complex of shelf-edge carbonate ridges to the W. This basin may have been bounded by syndepositional faults throughout the Cambrian, for example, the Gundahoo Fault (Taylor & Stott, 1973). However; subsidence never significantly outpaced sedimentation and shallow to marginal marine sediments accumulated until late in the Cambrian (Figs 2 & 6A). Within the western part of the Gataga area, the spectacular development of Mid-Cambrian algal bioherms also indicates tilted fault blocks at this time. Just east of the Gataga area in the MacDonald Platform (Fig. 1) Taylor & Stott (1973) have described Lower Cambrian syndepositional extensional faults with as much as 1500 m of displacement and overlain by unaffected Ordovician strata. Taylor & Stott (1973) also document and angular unconformity of 40° between Lower Cambrian and Helikian strata. These factors indicate a period of Early Cambrian extension with the development of rifted and rotated fault blocks. This agrees with the prediction by Bond & Kominz (1984) of a 550 Ma rifting and separation event based upon tectonic subsidence curves in the southern Canadian Cordillera and with the interpretations of Struik (1987) for the E-central

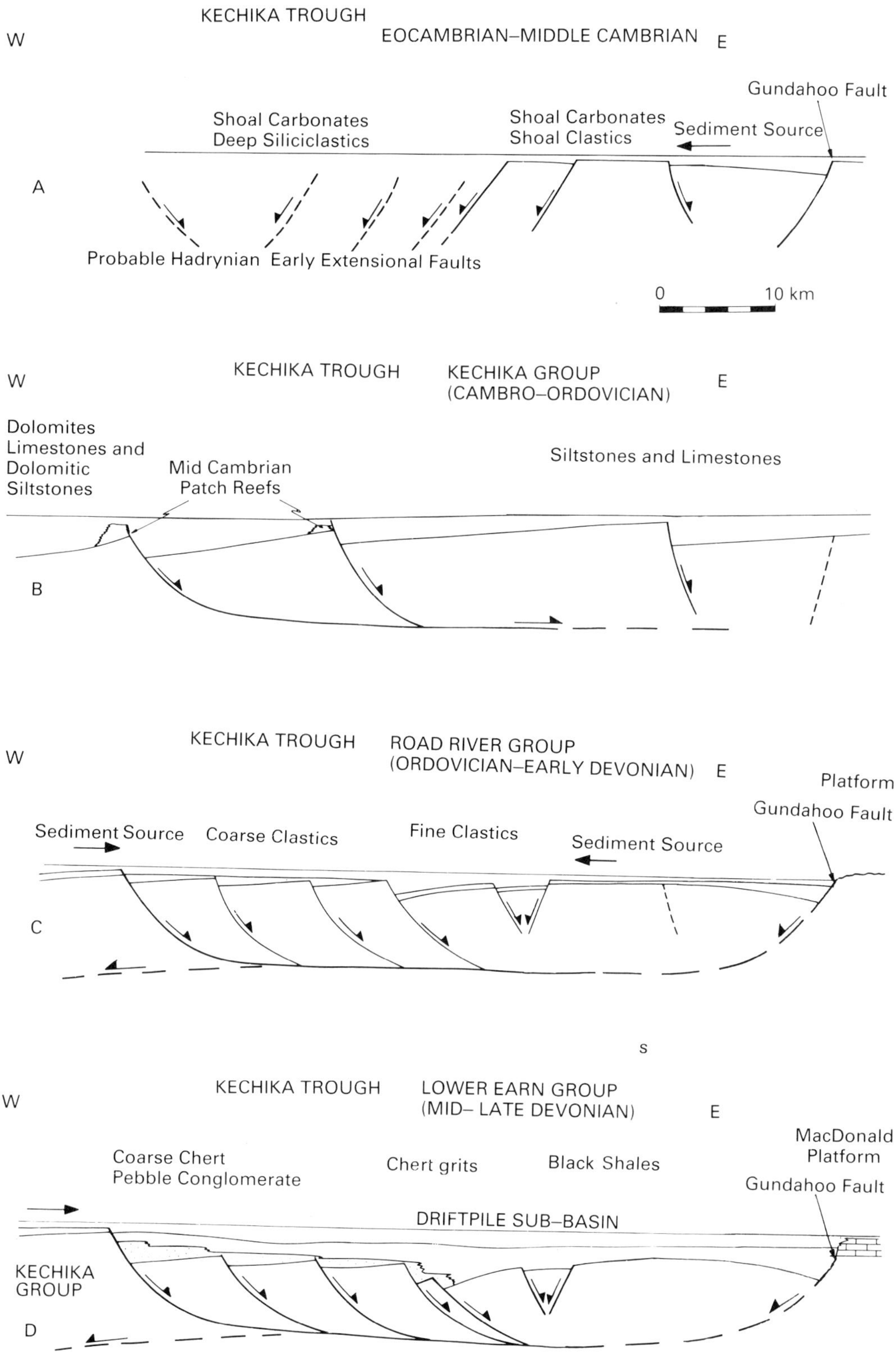

FIG. 6. West to east schematic sections illustrating the pre-contractional tectono-stratigraphic evolution of the Kechika Trough from the Eo–Middle Cambrian through to the Mid–Late Devonian. See text for details.

Canadian Cordillera and those of Thompson *et al.* (1987) for the northern Canadian Cordillera.

Cambro–Ordovician Kechika Group deposition

In late Cambrian–Ordovician times the Gataga area underwent what may be interpreted to be postrift, passive margin thermal subsidence with the development of a thick sequence of fine grained siltstones, shales and thin limestones of the Kechika Group (Fig. 6B). The Kechika Group thickens westwards (Figs. 4 & 6B) but detailed stratigraphic relationships are obscure by the intense Columbian–Laramide penetrative deformation.

Ordovician–Early Devonian Road River Group deposition

During Road River Group times probable thermal subsidence continued with slow deposition of anoxic, carbonaceous rich muds (Figs 4 & 5). In late Ordovician, turbidites containing pelagic carbonate and chert detritus derived from a sediment starved offshore ridge to the west were deposited in the W part of the Gataga area. This suggests that a deep water trough with a fault-bounded western margin, had come into existence by then (Fig. 6C). A similar tectonic framework influenced the Silurian sedimentation of the Road River Group (Figs 4 & 6C), although during this period fault margin uplift produced an influx of more sand and the end of anoxic sea bed conditions. Small submarine fans built into the deep western side of the trough, with shallow water sands accumulating on the more gently tilted eastern side (Fig. 6C). Lower Devonian (Upper Road River Group) sediment starvation throughout the area is indicated by the deposition of thinly bedded pelagic cherts (Figs 4 & 6C). Road River Group sedimentation patterns may be interpreted to indicate the initiation of a fault bounded Kechika Trough with extensional faults ping eastwards but possibly linked at depth to a detachment extending to the continental margin (Fig. 6C).

Mid–Late Devonian extension

In Mid–Late Devonian the Kechika Trough underwent marked extension and rapid subsidence forming a system of deep half-grabens and sub-basins (Fig. 6D) produced by activation of a linked fault system which had possibly been in existence since the Ordovician (Figs 6B & C). The Kechika Trough is asymmetric with the western margin probably bounded by a series of east dipping faults and maximum sediment accumulation in the Driftpile sub-basin in the east (Fig. 6D). At least three basin systems have been recognized in the Gataga area: the McCook half-graben, the Bear half-graben and the Driftpile sub-basin (Figs 5 & 6D). They were probably 8 to 12 km across and at least 50–70 km long but precise estimates of their dimensions are difficult due to the intense cleavage and thrust development in the area.

Coarse conglomerates, derived by subaerial erosion of uplifted fault blocks to the west (Gordey *et al.* 1987), and dominated by Lower Palaeozoic chert and quartz-arenitic sandstone clasts, poured eastwards into the Kechika Trough and were deposited in submarine fans in the western half-grabens of the Gataga area (cf. the McCook half-graben, Figs 5 & 6D). The lack of mud matrix and coarse-clast size of the conglomerates (probably fed directly from alluvial fan channels) indicates that they were rapidly deposited in narrow fans dominated by unstable channels. In the eastern part of the Kechika Trough, Mid–Late Devonian clastic deposition is characterized by a thick sequence of thin bedded distal turbidites (cf. the Driftpile sub-basin, Fig. 5). The widespread occurrence of stratiform Ba and Ba–Zn–Fe–Pb mineralization also supports the fault bounded basin interpretation.

Inversion

The Mid–Late Devonian basin architecture of the Kechika Trough has exerted a profound influence on the Columbian–Laramide contractional tectonics in the region. In the Gataga region the NW strike of the thrust faults (Fig. 2) most likely reflects the trend of the margins of the extensional basins (cf. NW palaeocurrents parallel to the basin axes). This feature has been noted elsewhere in the Cordillera (e.g. Taylor & Stott 1973; Thompson *et al.* 1987; and others). In detail, the geometries of the Mid–Late Devonian extensional systems control the detachment levels, fault trajectories and apparent fault sequences in this part of the Kechika Trough. The following examples, taken from the three different basins identified in this study, demonstrate the extent to which the early basin architecture and corresponding synrift sediment characteristics shaped this section of the Cordilleran Fold and Thrust Belt.

Inversion of the McCook half-graben

One of the simplest inversion structures in the Kechika Trough is exhibited by the McCook half-graben where Road River and Lower Earn Group clastics are preserved in the footwall of a major northeast verging thrust fault (Panel II in Fig. 2 & Fig. 3). Similar geometries, where Lower Earn Group clastics are only preserved in the footwalls to major thrusts, are commonly found elsewhere in the Kechika Trough (MacIntyre 1983). In the case of the McCook half-graben, stratigraphic and sedimentological evidence indicates that the thrust fault now occupies the position of a Mid–Late Devonian extensional fault. Chevron folded and imbricated Road River and Lower Earn Group strata are preserved in the footwall of this major thrust sheet which consists of strongly deformed Kechika Group rocks (Fig. 3).

A model that illustrates the sequential contraction and inversion of the McCook half-graben system is shown in Fig. 7. Initial contraction and inversion involved pinning of a NE propagating thrust at the extensional fault and against the massive chert pebble conglomerates of the Lower Earn Group (Fig. 7B). The Kechika Group underwent tight folding and the half-graben began to contract with the development of an 'out-of-the-half-graben thrust'. Continued contraction tightened the folding and rotated the Mid–Late Devonian extensional fault (Fig. 7C) so that it became re-utilized as a thrust. With increased deformation, the out of the half-graben fault developed into an imbricate fan/duplex system as the original thrust overcame the buttressing effect of the original half-graben (Fig. 7D). This style of deformation sequence produced an imbricate fan in which each successive foreland propagating fault slice contains progressively older strata (Fig. 7D). Although individual imbricate thrusts cut up stratigraphic section, the floor thrust to the imbricate fan must gradually cut down stratigraphic section in the direction of transport i.e. toward the NE. This relationship may be explained by superimposing a relatively flat lying thrust trajectory upon strata previously tilted by extensional faults. In this example, westerly dipping Road River and Lower Earn Group strata are related to a rollover structure developed in the hanging wall above an easterly dipping extensional fault (Fig. 7).

Chevron folding and imbrication of the half-graben synrift sediments is attributed to the space problems generated by shortening of the half-graben during inversion. Similar geometries are found in experimental models are described by McClay (this volume). In addition, high fault cut-off angles suggest that much of the imbrication occurred after folding.

An example of a duplex structure developed in a half-graben system from the northern part of the map area (see Fig. 2) is shown in Fig 8. Here, progressively older strata are found in the thrust horses towards the foreland of the thrust system. This geometry, in which the floor thrust cuts down stratigraphic section in the direction of tectonic transport, is commonly found in the footwall package beneath major thrust faults in the Gataga area. We interpret this geometry to indicate that the thrusts are cutting wedges of strata that were tilted towards the SW on Mid–Late Devonian extension faults.

Inversion of the Bear half-graben

Detailed mapping on the Bear mineral claims in the southern part of the Gataga area (locality 1, Fig. 2) has permitted unravelling of the complex geology of this area. Stratiform sulphides and barite occur adjacent to an inferred syndepositional fault. Fig. 9 shows a model for the sequential deformation and inversion of the Bear half-graben. The present day structure comprises a W panel of tightly folded Kechika Group strata thrust NE against a series of SW directed thrust faults that form the W margin of a small pop-up structure of Silurian siltstones (Fig. 9D). These thrusts have been rotated and steepened by further shortening after emplacement. The pop-up structure was itself thrust NE over a coarsening-upward submarine fan sequence of medium to thick bedded sandstones and grits that from a NE verging anticline (Fig. 9D). Although the eastern limb of this fold is partially thrust out, an entire Lower Earn Group sedimentary section is exposed below the thrust. This thrust bounded panel contains bedded sandstones and grits that thin northeastwards and interdigitate with thin bedded siltstones and shales that are deformed into a series of upright to overturned folds and thrusts.

In its pre-contractional configuration, the half-graben consisted of a steep NE dipping extensional fault with approximately 200 m downthrow to the NE and a corresponding increase in thickness of sandstones and grits in the hanging wall (Fig. 9A). In the early stages of contraction a NE directed thrust became pinned against the extensional fault and an adjacent wedge of competent, thick bedded sandstones and grits (Fig. 9B). Continued deformation in the hanging wall to this thrust eventually led to the formation of the pop-up structure and

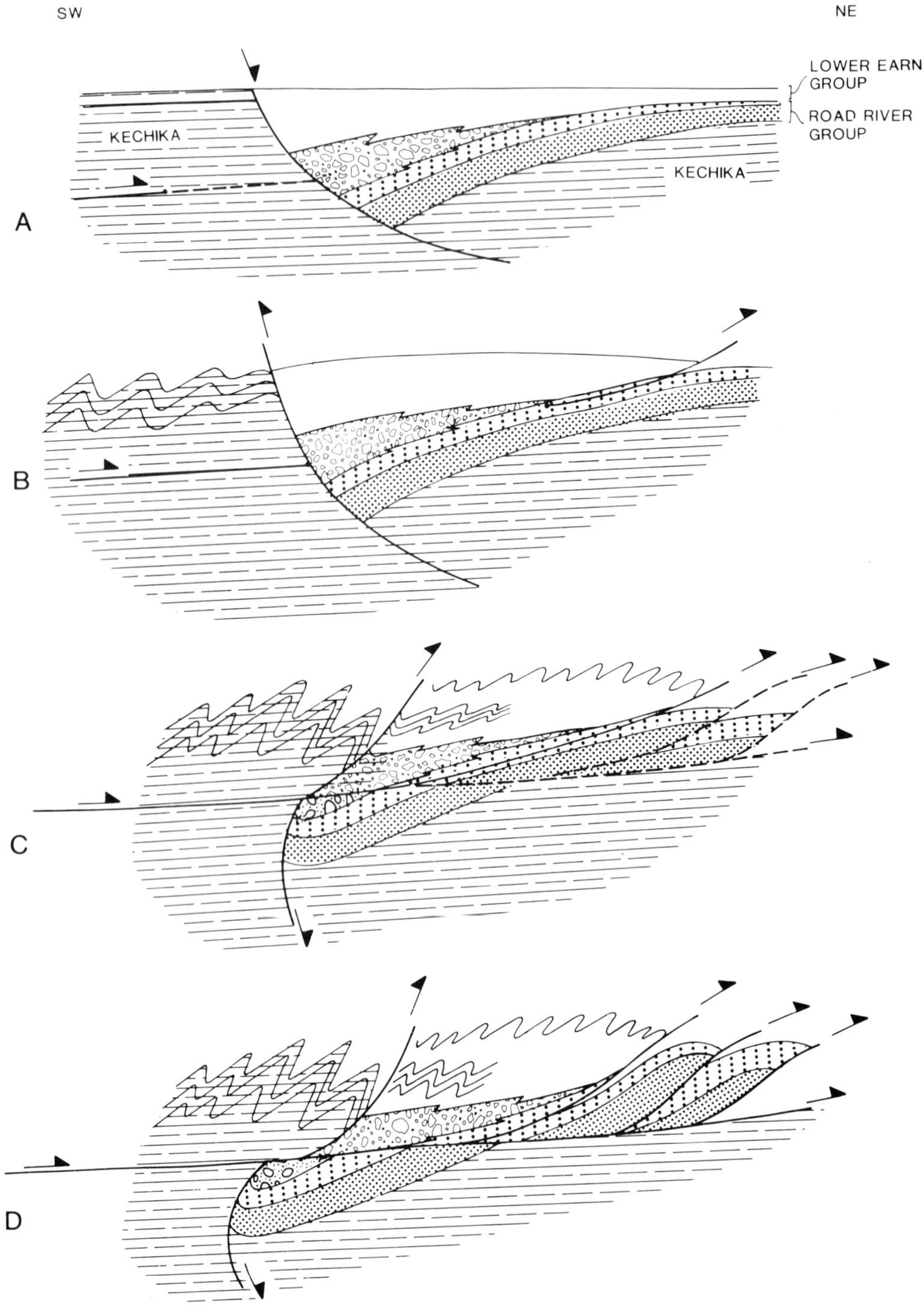

FIG. 7. Schematic model (not to scale) for the inversion of the McCook half-graben (MS1 in Fig. 2). A, Pre-inversion geometry of a down to the NE half graben with a wedge of synrift Lower Earn Group sediments. B, Initial inversion showing reactivation of the half-graben bounding fault and an 'out of the half-graben thrust'. A northeastwards propagating thrust in the Kechika Group in the footwall to the half-graben became pinned against the extensional fault by the wedge of competent chert pebble conglomerates in the hanging wall of the extensional fault. C, Continued deformation with folding in both the footwall and hanging wall of the extensional fault; break-through of the thrust fault; rotation of the extensional fault so that it becomes re-utilized as a thrust and incipient imbrication of the synrift sediments by out of the half-graben faults. D, Translation and imbrication the half-graben system on through going thrust faults. This is the present day configuration of the McCook half-graben as determined from detailed mapping together with structural and sedimentological analysis.

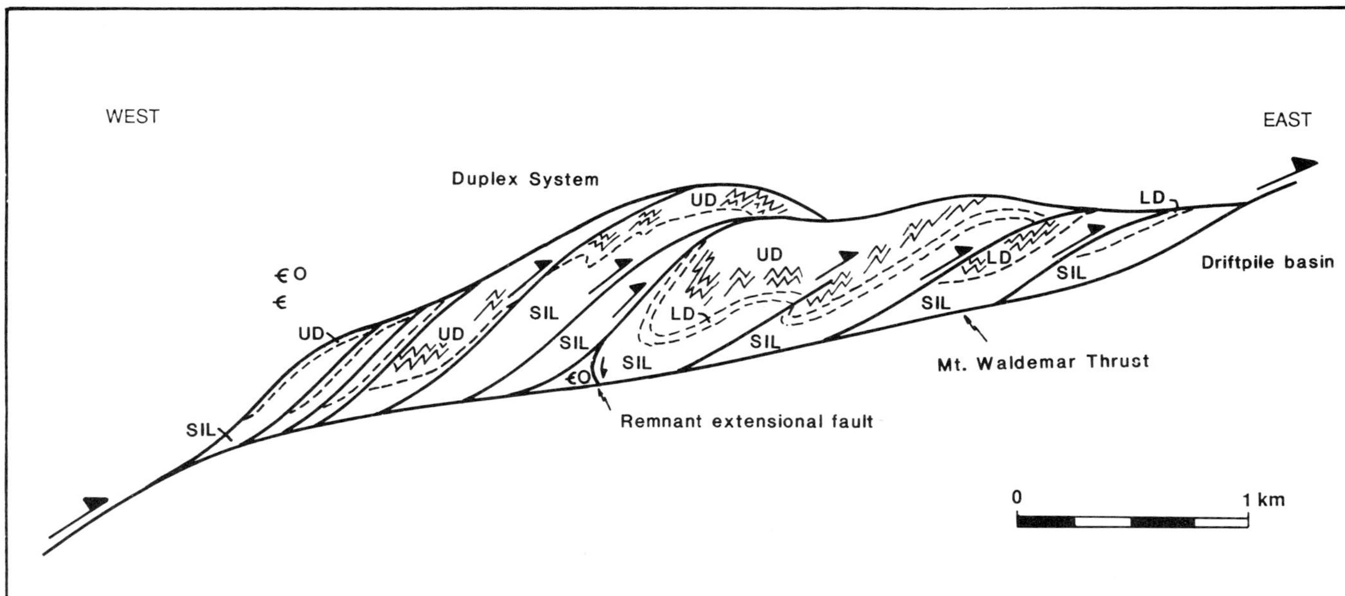

FIG. 8. Detailed cross-section through the Mount Waldemar duplex (location of section on Fig. 2). The floor and roof thrusts of this duplex cut down section and the duplex horses contain variable stratigraphy as a result of both the floor thrust and the roof thrust cutting through a wedge of Mid–Upper Devonian sediments formed in an extensional half graben system.

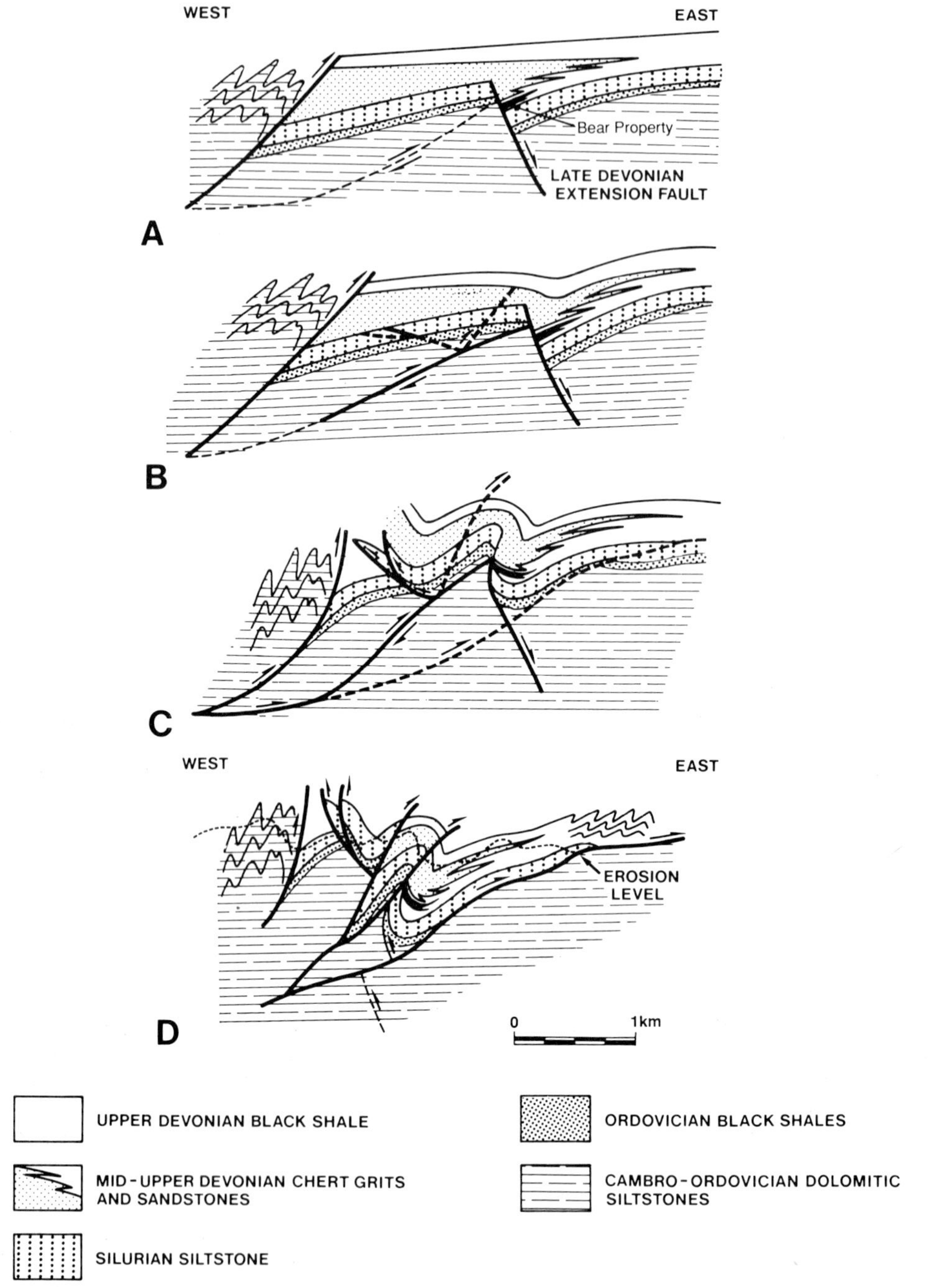

FIG. 9. Development of the Bear structure (location of section is shown on Fig. 2). A, Initial extensional geometry of the Bear half-graben showing the location of the Bear Property mineralization in the hanging wall of the extensional fault. The western most thrust, depicted at the start of inversion as a thrust, is probably a reactivated extensional fault. B, An early stage in the inversion showing the thrust fault in the footwall Cambro–Ordovician shales of the Kechika Group pinned by the Bear extensional fault and by the wedge of Mid–Upper Devonian grits and sandstones in the half-graben. The trajectories of incipient back thrusts are shown. C, Back thrusting with the development of a pop-up structure and a syncline–anticline–syncline fold system. D, Rotation of the Late Devonian extensional fault so that it was re-utilized as a thrust fault. Folding and thrusting of the synrift sequence and forward propagation of a through-going thrust system that carried the half-graben system to the NE. This is the present day configuration of the Bear structure determined from 1 : 10 000 scale mapping, drill hole data and detailed structural analysis.

rotation and folding of the synrift sediments into an anticline–syncline pair (Fig. 9C & D). Amplification of this fold pair caused the extensional fault to become rotated until its orientation became favourable for re-utilization as a thrust fault which then cut through the eastern limb of the anticline (Fig. 9D). At high strains the structure locked-up and further shortening was accommodated by chevron folding, cleavage development, and imbrication of the more easterly parts of the half-graben infill.

Inversion of the Driftpile sub-basin

The Driftpile sub-basin (locality 3 Fig. 2) contains a thick sequence of tightly folded and strongly thrust faulted Lower Earn Group black argillites and cherty argillites (Fig. 5). It is interpreted to occupy a position near to the crestal collapse of a rollover structure (Fig. 6D). The sub-basin is bounded to the W by a NE verging duplex of predominantly Road River strata whereas the Lower Earn Group black clastics thin onto the eastern margin of the basin (Fig. 6). The broad folds at this E margin of the sub-basin (Fig. 2) are controlled by large scale folds and thrust horses in the thick underlying succession of Cambrian and Hadrynian strata (Fig. 3).

A schematic model for the Driftpile sub-basin is shown in Fig. 10. Initial inversion involved reactivation of the half-graben and crestal collapse extensional faults (Fig. 10B). With increased contraction the crestal collapse graben inverts and becomes a pop-up structure and the major half-graben fault rotates and is re-utilized as the footwall duplex of Road River strata are thrust over the Driftpile sub-basin (Fig. 10C). The Lower Earn Group synrift black clastics are strongly chevron folded and intensely cleaved. At higher strains, shortening within the synrift sediments was accomplished through imbricate out-of-the-half-graben faults.

Discussion

The Kechika Trough is a long, narrow extensional basin that developed within the passive margin of ancestral western North America. In the Mid–Late Devonian the Kechika Trough experienced a major extensional deformation that resulted in fault controlled sedimentation with a massive influx of the easterly prograding, coarse to fine-grained 'black clastics' of the Lower Earn Group (Gordey *et al.* 1982; Gordey *et al.* 1987). In the Gataga area of the Kechika Trough these 'black clastics' were deposited in long, narrow NW–SE trending extensional half-graben systems. During the Columbian–Laramide Orogeny, the half-graben systems were inverted and incorporated into the thrust sheets of the northwestern/Rocky Mountain fold and thrust belt. The inversion and thrust geometries were controlled by the architecture of the extensional systems and by the shapes of the sediment bodies in the basins. This analysis of the structure of the Gataga area in terms of inversion and thrusting allows the major stratigraphic contrasts seen between adjacent thrust packages to be resolved in terms of lateral facies differences caused by Mid–Late Devonian extension rather than by overcomplex thrust geometries.

Mid–Late Devonian extension

Detailed structural, sedimentological and stratigraphic analysis of the Gataga area, coupled with an interpretation in terms of inversion and telescoping into the Rocky Mountain fold and thrust belt, has permitted a reconstruction of the Mid–Late Devonian architecture of this part of the Kechika Trough as a series of linked extensional half-grabens (Fig. 11A).

In this model, the Kechika Trough is interpreted to be an asymmetrical half-graben system some 30–40 km wide and at least 90 km long (probably 150–180 km long). It developed within the passive margin and consisting of a terraced W edge with a probable rollover anticline linked with the Gundahoo fault to the E (Fig. 11A). During the Mid–Late Devonian, deposition of the Lower Earn Group black clastics, sourced from the west, was controlled by extensional faults. The large volume of synrift sediments suggests that there existed a substantial area to the W, representing the outer portion of the passive margin. It must have been uplifted and undergoing active erosion at this time. This continental sliver is interpreted to have been the Cassiar Platform (Figs 1 & 11) which has since been offset by dextral strike–slip along the Rocky Mountain Trench–Tintina Fault system (Fig. 1). This sliver of outer platform would have been bounded by a westerly facing Atlantic-type continental margin (Tempelman-Kluit 1979; Evenchick *et al.* 1984; Devlin *et al.* 1985) and would perhaps have been a crustal mega-boudin or thinned continental ribbon of the type proposed by Lister *et al.* (1986).

The geometries and linkages of the Mid–Late Devonian extensional fault systems on crustal and lithospheric scales must be speculative at this stage because of the intense contractional

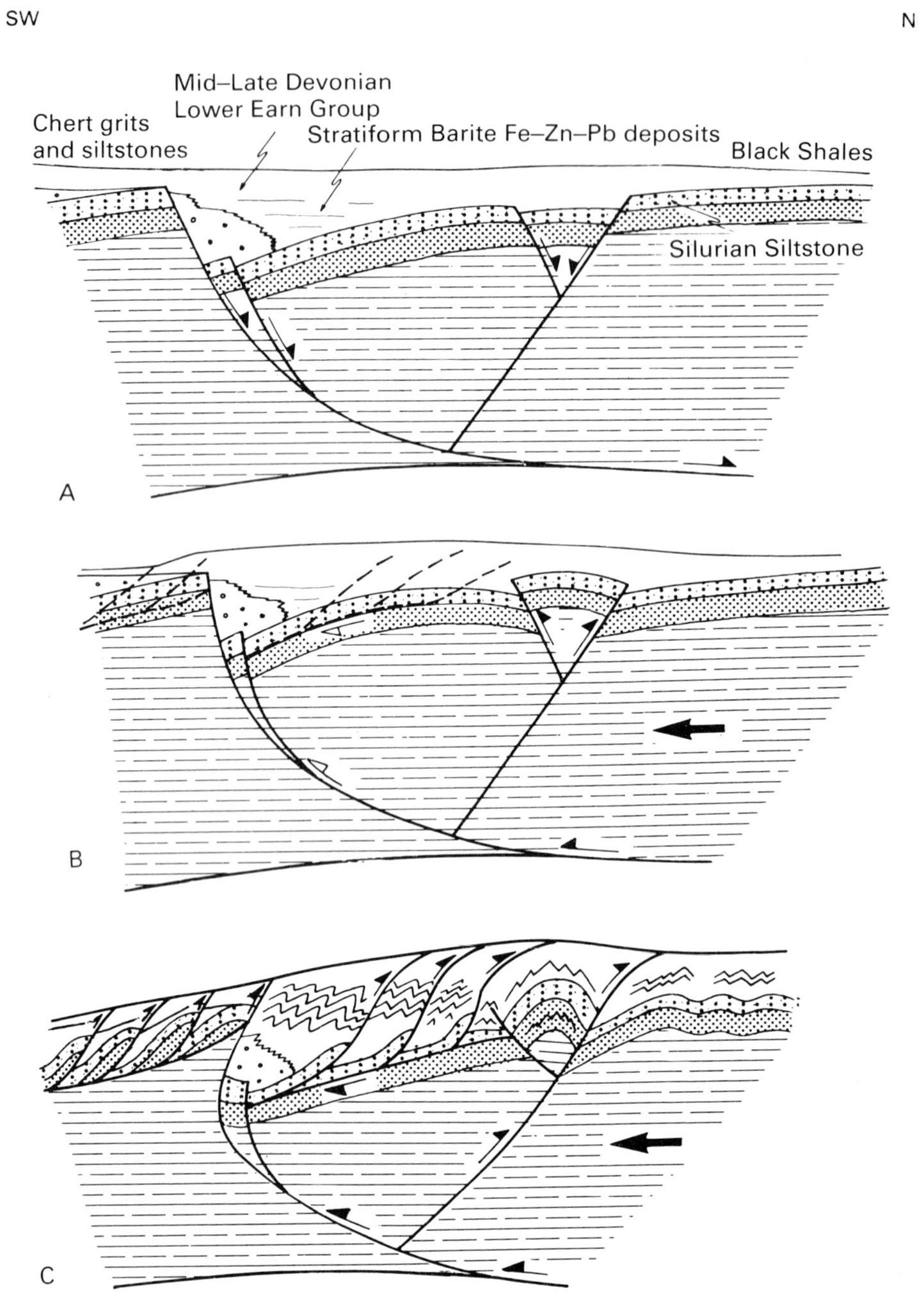

FIG. 10. Inversion of the Driftpile sub-basin (location 3 in Fig. 2). A, Pre-inversion geometry of an extensional half-graben system with a roll-over and crestal collapse graben system. Drill hole and stratigraphic data indicates that the synrift Lower Earn Group sediments thin to the NE. B, Initial inversion showing reactivation of the extensional faults, formation of a 'pop-up' from the crestal collapse structure and bulk shortening of the synrift sequence. The trajectories of incipient thrust faults are shown. C, Inversion and thrusting of the Driftpile sub-basin showing duplexing in the footwall of the original extensional fault, rotation of this fault so that it is re-utilized as a thrust fault and imbrication of the synrift Lower Earn Group by 'out-of-the-half-graben' thrust faults. This is the present day configuration of the Driftpile sub-basin determined from detailed 1 : 5000 and 1 : 10 000 scale mapping and drill hole data combined with structural and sedimentological analysis.

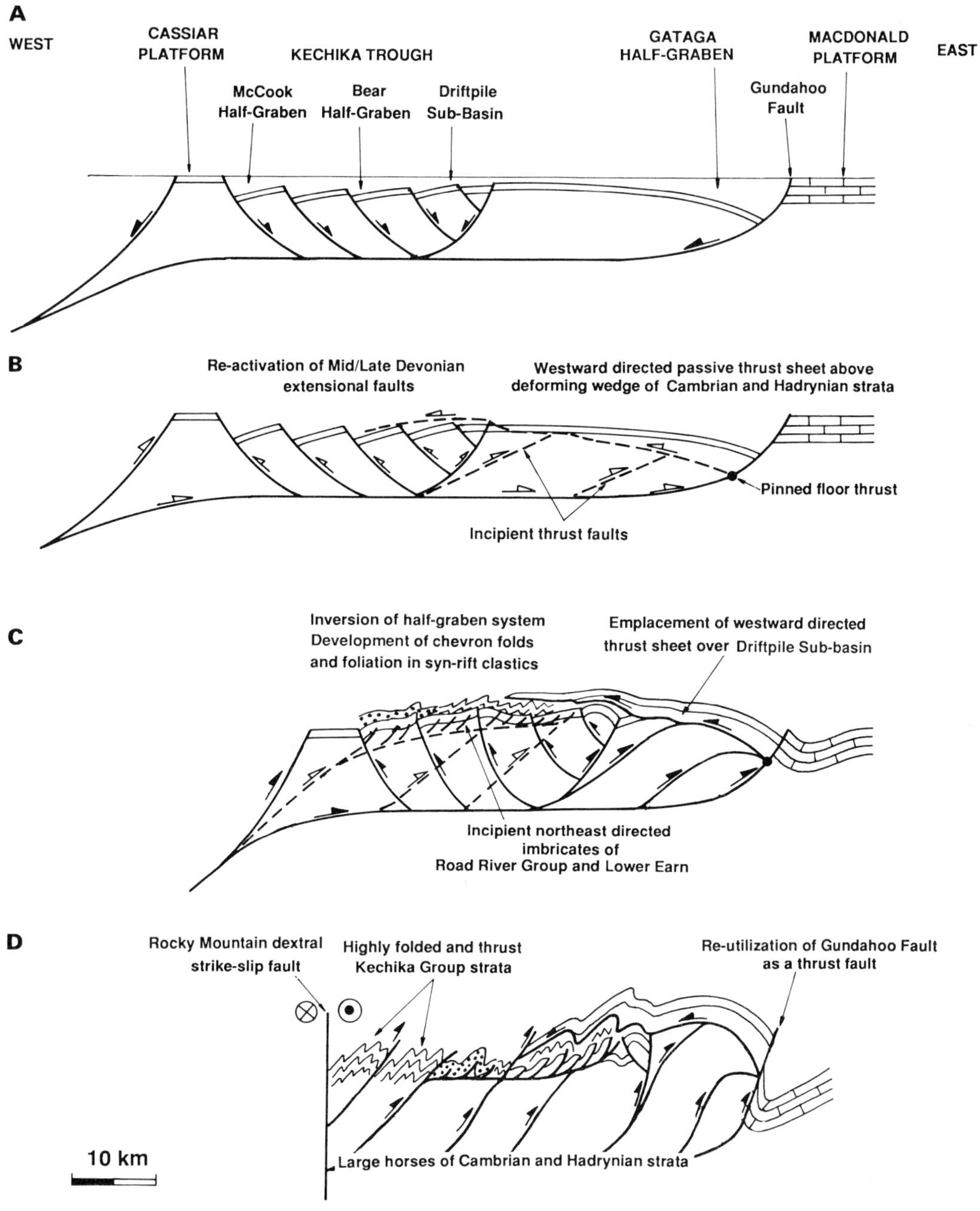

FIG. 11. Regional tectonic model for the inversion and contraction of the Kechika Trough. A, Mid–Late Devonian structure showing the linked system of extensional faults responsible for the evolution of the Kechika Trough and the locations of basins systems discussed in this paper. B, Initial inversion, probably during the Jurassic, with contractional reactivation of extensional faults. Pinning of the floor thrust and incipient wedging. The trajectories of incipient thrust faults are shown. C, Inversion of the half-graben systems during the Cretaceous and westwards emplacement of a thrust sheet over the Driftpile sub-basin. Folding and cleavage developed in the synrift clastics. D, Late Cretaceous. The basic architecture of the fold and thrust belt was established. The Kechika, Road River and Lower Earn Groups had been incorporated into highly deformed distinct tectonostratigraphic assemblages (Fig. 2). The W margin of the Kechika Trough was translated northwards by the dextral strike–slip displacement on the Rocky Mountain Trench Fault system.

deformation, resultant uplift and erosion which has obscured and removed much of the stratigraphic record. It is likely that the extensional faults linked in a ramp–flat system, similar to the models proposed by Gibbs (1984); Wernicke (1985) and Lister *et al.* (1986), and in which essentially crustal-scale down to the west simple shear or combined simple and heterogeneous shear (Coward 1986) generated the western continental margin of ancestral North America. A similar extensional model has been proposed by Struik (1987) for the evolution of the passive margin phase of the E-central Canadian Cordillera S of the Gataga area.

Inversion and contraction of the Kechika Trough

Inversion and contraction of the Kechika Trough in the Columbian–Laramide Orogeny initiated in the Jurassic with the reactivation of the extensional fault system in contraction (Fig. 11B). The westward emplacement of a thrust sheet of Cambrian and Hadrynian strata over the Kechika Trough (Figs 3 & 11B) is interpreted to have resulted from the pinning of the floor thrust and the development of a passive back thrust by wedging and delamination (Fig. 11B & C) similar to structures found in the southern Cordillera (Price 1986). The emplacement of this thrust sheet may be responsible for localized phase 1 fold development within the synrift Lower Earn Group sediments.

During the Cretaceous, inversion and uplift of the Kechika Trough was associated with major NE directed thrusting and folding that generated phase 2 folds and cleavage (Fig. 11C). The half-graben systems were inverted and the extensional faults rotated and re-utilized (Fig. 11C). The synrift clastics were deformed by intense chevron folding and the development of a pervasive cleavage. The formation of out-of-the-half-graben thrusts, duplexing, rotation and northeastwards transport on underlying 'blind' thrusts (Thompson 1981) resulted in the five distinct tectonic packages within the Gataga area (Fig. 2). The exact detachment levels of these underlying thrusts are difficult to determine because they do not outcrop in the Gataga area but they probably lie in the more phyllitic units of the Proterozoic strata as found further to the east by Taylor & Stott (1973).

By the Late Cretaceous the essential architecture of the thrust belt had been established. At this time, a segment of the western margin of the Kechika Trough and the source area for the Mid–Upper Devonian Lower Earn Group were translated in excess of 750 km northwards by dextral strike slip (Gabrielse 1985) along the Rocky Mountain Trench Fault system (Fig. 11D). The phase 3 deformation structures in the Gataga area have been interpreted in terms of a dextral shear couple associated with this major dextral strike–slip faulting event.

Inversion geometries

Inversion of the Kechika Trough is characterized by thrust fault bound packages each with a distinctive stratigraphy and sedimentology that can be related to individual extensional half-grabens. In particular, within each package the thrust faults cut down stratigraphic section towards the NE in the direction of transport as a result of the wedge shaped stratigraphy and tilting on Mid–Late Devonian extensional faults. The detail of inversion structures within the half grabens in the western part of the Kechika Trough are distinctive and include:

1. Contractionally reactivated extensional half graben bounding faults (Figs 7B & 10B);
2. Rotation and reutilization of previous extensional faults by thrust faults (Figs 7C, 9C & 10C);
3. Pop-up structures and 'triangle' zones related to the pinning of thrust faults by the buttressing effect of extensional faults and adjacent competent half graben sediment infill (Figs 9C & 10B);
4. Out-of-half-graben thrust faults, imbrication and intense folding of synrift sediments as a result of space problems in the half-graben as it is shortened (Figs 7D & 10C);
5. Foreland propagating thrust faults that locally cut down stratigraphic section in the direction of propagation, e.g. when a relatively flat trajectory thrust cuts through strata previously tilted on Mid–Late Devonian extensional faults (Figs 7D & 8).

By comparison, eastern margin of the Kechika Trough is characterized by an easterly tapering wedge of Cambro–Ordovician to Upper Devonian strata overlying relatively uniform stratigraphy of the Lower–Middle Cambrian and Proterozoic. The pre-contractional geometry (Fig. 11A) of a rollover anticline results in a less complicated inversion and thrust pattern. Deformation is dominated by large thrust sheets of Cambrian and Hadrynian strata with a simple internal structure and large wavelength folds. Inversion structures (Figs 8 & 9) in the synrift sequence are similar to the first four geometries listed above. Type 5 geometries have not been identified.

Syndepositional Mid–Late Devonian extensional faults have not been found preserved in this part of the Kechika Trough but this may be expected because of the intense contractional deformation and because of the rotation and re-utilization of extensional faults by later thrusting. The geometry and sedimentology of the synrift Lower Earn Group clastic sediments does, however, permit the location and asymmetry of the extensional faults to be determined.

Many of the structures suggested for the inversion of the Kechika Trough have been found in the experimental models of inversion described by McClay, this volume. In particular, rotation of extensional faults, pinning by extensional faults, buttressing and out-of-the-half-graben thrusts have been observed.

Conclusions

The Kechika Trough is an inverted and contracted, narrow elongate extensional basin that now forms part of the highly deformed and telescoped passive margin of ancestral western North America. It forms part of a long lived passive margin that experienced episodic extension from the Proterozoic through to the Early Devonian. The Kechika Trough underwent a major phase of crustal extension and subsidence in the Mid–Late Devonian and subsequent contraction and inversion in the Jurassic–Cretaceous Columbian–Laramide orogeny.

Detailed sedimentological and structural analysis has allowed the Mid–Late Devonian architecture of the Kechika Trough to be defined as an narrow, elongate asymmetrical basin comprising a terraced half-graben system to the west and roll-over structure to the E (Fig. 11). The distinctive synrift 'black clastics' of the Lower Earn Group were probably derived from the erosion of a continental sliver to the W of the Kechika Trough. The patterns of deposition were strongly controlled by the half-graben systems.

Within the Kechika Trough, three phases of deformation have been identified. These can be related to the inversion and contractional history of the area. Inversion was initiated in the Jurassic with the emplacement of westerly directed thrust sheets and continued through the Cretaceous when the basin was incorporated into the fold and thrust belt of the northern Rocky Mountains.

Major stratigraphic contrasts between adjacent thrust packages have been resolved in terms of lateral facies differences produced by extension, rather than over-complex thrust geometries. The position of syndepositional faults has also been established. The extensional fault systems together with the geometries of the basins and bodies of sediment exerted a distinctive influence on the later deformation patterns of the fold and thrust belt.

Detailed analysis of the Gataga area has enabled a progressive series of inversion structures to be proposed. At low bulk contractional strains, extensional faults were reactivated as contractional faults and buckling of the synrift sediments began. Syndepositional faults and the adjacent sediment wedges also behaved as semi-rigid buttresses causing the pinning of thrust faults and lead to the formation of pop-up structures and 'triangle zones'. At intermediate strains, the half-graben margin extensional faults became rotated until they become favourably orientated to be re-utilized by thrust faults. Chevron folding of the basin fill intensified and out-of-graben thrusts developed and imbricated the synrift sediments. At still higher strains the half-graben infill became penetratively deformed by chevron folding and cleavage development. Anomalous thrust relationships occur where thrusts climb down stratigraphic section when they cut through strata previously tilted by extensional faults.

The Kechika Trough is an extensional basin that has undergone extreme inversion and contraction. Nevertheless the control of thrust geometries by the extensional fault systems and by the basin architecture can be clearly recognized. It is therefore considered appropriate to include this, or similar basins contained within fold and thrust belts, in an analysis of inversion tectonics.

Acknowledgments: This project is funded by a Natural Environment Research Council Grant GR3/5528A to Dr K. R. McClay. Invaluable logistic support was provided by the British Columbia Department of Energy, Mines and Petroleum Resources, and by the Geological Survey of Canada through R. Campbell, H. Gabrielse, C. Findlay and D. Sangster. N. Way, J. Purcell, A. Scott and H. Beeley ably assisted in the field. We gratefully acknowledge valuable discussions with H. Gabrielse, D. Sangster, W. Goodfellow and R. Carne. M. Orchard is thanked for undertaking the conodont analyses.

References and further reading

BOND, G. C. & KOMINZ, A. 1984. Construction of tectonic subsidence curves for the early Palaeozoic miogeocline, southern Canadian Rocky Mountains: Implications for subsidence mechanisms, age of breakup and crustal thinning. *Bulletin of the Geological Society of America* **95**, 155–73.

CARNE, R. C. & CATHRO, R. J. 1982. Sedimentary exhalative (Sedex) Zinc-Silver-Lead deposits, northern Canadian Cordillera. *Bulletin of the Canadian Institute of Mining* **75**, 66–78.

COWARD, M. P. 1986. Heterogeneous stretching, simple shear and basin development. *Earth and Planetary Science Letters*, **80**, 325–336.

DEVLIN, W. J., BOND, G. C. & BRUECKNER, H. K. 1985. An assessment of the age and tectonic setting of volcanics near the base of the Windermere Supergroup in northeastern Washington: implications for latest Proterozoic–earliest Cambrian continental separation. *Canadian Journal of Earth Sciences*, **22**, 829–837.

EVANCHICK, C. A. PARISH, R. P. & GABRIELSE, H. 1984. Precambrian gneiss and late Proterozoic sedimentation in north-central British Columbia. *Geology*, **12**, 233–237.

FRITZ, W. H. 1980. Two Cambrian stratigraphic sections near Gataga River, northern Rocky Mountains, British Columbia. *Current Research, Part C. Geological Survey of Canada, Paper 80-C*, 113–119.

GABRIELSE, H. 1962. Kechika, British Columbia. *Geological Survey of Canada*, Map 42.

——, 1985. Major dextral transcurrent displacements along the northern Rocky Mountain Trench and related lineaments in north-central British Columbia. *Bulletin of the Geological Society of America*, **96**, 1–14.

GIBBS, A. D. 1984. Structural evolution of extensional basin margins. *Journal of the Geological Society of London*, **141**, 609–620.

GORDEY, S. P., ABBOTT, J. G. & ORCHARD, M. J. 1982. Devono–Mississippian Earn Group and younger strata in east-central Yukon. *Current Research, Part B, Geological Survey of Canada*, Paper 82-B, 92–100.

——, ——, TEMPELMAN-KLUIT, D. J. & GABRIELSE, H. 1987. Antler clastics in the Canadian Cordillera. *Geology*, **15**, 103–107.

LISTER, G. S., ETHERIDGE, M. A. & SYMONDS, P. A. 1986. Detachment faulting and the evolution of passive continental margins. *Geology*, **14**, 246–250.

MCCLAY, K. R. 1984. Tom barite-lead-zinc deposit, Yukon, Canada. *In*: MCCLAY, K. R. *Structural Geology of Stratiform Lead-Zinc deposits: Case histories. Geological Association of Canada, Cordilleran Section, Short Course No. 2 (Part II)*, 60–96.

—— Analogue models of inversion structures. *This volume.*

—— & INSLEY, M. W. 1986. Structure and mineralization of the Driftpile Creek area, northeastern British Columbia (94E/16, 94F/14, 94K/4, 94L/1), *British Columbia Ministry of Energy, Mines & Petroleum Resources, Paper 86–1*, 343–350.

——, ——, WAY, N. A. & ANDERTON, R. 1987. Stratigraphy and tectonics of the Gataga area northeastern British Columbia, (94E/16, 94F/14, 94K/4, 94L/1, 94L/7, 94L/9): *British Columbia Ministry of Energy, Mines and Petroleum Resources, Geological Fieldwork 1986, Paper 87–1*, 193–200.

——, ——, —— & —— 1988. Tectonics and mineralisation of the Kechika Trough, Gataga area, northeastern British Columbia. *In*: *Current Research, Part E, Geological Survey of Canada, Paper 88 1-E*, 1–12.

MACINTYRE, D. G. 1981. Geology of the Akie River Ba-Pb-Zn mineral district. *British Columbia Ministry of Energy, Mines and Pet. Res., Preliminary Map 41.*

—— 1983. Geology and stratiform barite sulphide deposits in northeastern British Columbia. *In*: Sangster, D. F. (ed.), *Sediment Hosted Stratiform Lead-Zinc Deposits, Mineralogical Association of Canada, Short Course Notes, No. 9*, 85–120.

MCKENZIE, D. 1978. Some remarks on the development of sedimentary basins. *Earth and Planetary Science Letters*, **40**, 25–32.

PRICE, R. A. 1986. The southeastern Canadian Cordillera: thrust faulting, tectonic wedging, and delamination of the lithosphere. *Journal of Structural Geology*, **9**, 239–254.

STRUIK, L. C. 1987. The ancient western North American margin: An Alpine rift model for the east-central Canadian Cordillera. *Geological Survey of Canada Paper 87–15*, 19 pp.

TAYLOR, G. C. & STOTT, D. F. 1973. Tuchodi Lakes map-area, British Columbia; *Geological Survey of Canada, Memoir*, **373**, 37 pp.

TEMPELMAN-KLUIT, D. J. 1979. Transported cataclasite, ophiolite and granodiorite in Yukon Territory: Evidence of arc-continent collision. *Geological Survey of Canada Paper 79–14*, 27 pp.

THOMPSON, R. I. 1981. The nature and significance of large 'blind' thrusts within the northern Rocky Mountains of Canada. *In*: MCCLAY, K. R. & PRICE, N. J. (eds) *Thrust and Nappe Tectonics. Special Publication of the Geological Society of London*, **9**, 449–462.

——, MERCIER, E. & ROOTS, C. 1987. Extension and its influence on Canadian Cordilleran passive-margin evolution. *In*: COWARD, M. P., DEWEY, J. F. & HANCOCK, P. L. (eds) *Continental Extension Tectonics. Special Publication of the Geological Society of London*, **28**, 409–417.

WERNICKE, B. 1985. Uniform-sense normal shear of the continental lithosphere. *Canadian Journal of Earth Science*, **22**, 108–125.

K. R. McCLAY, Department of Geology, Royal Hollway and Bedford New College, University of London. Egham, Surrey, TW2O OEX, UK.
M. W. INSLEY, Department of Geology, Royal Hollway and Bedford New College, University of London, Egham, Surrey, TW2O OEX, UK.
R. ANDERTON, Department of Applied Geology, University of Strathclyde, Glasgow GI IXJ, UK.

Basin inversion in the Osen–Røa thrust sheet, Southern Norway

C. K. Morley

SUMMARY: The Osen–Røa thrust sheet is the most external tectonic unit in the southern Norwegian Caledonides. This extensive thrust sheet may have originally covered over 100 000 km^2. It is largely composed of Late Precambrian and Lower Palaeozoic sedimentary rocks. Thickness variations in the Late Precambrian sedimentary package can be attributed to the passage from a rift basin about 3 km thick (the Hedmark Basin) into rift margin sediments a few hundred metres thick (Kumpulainen & Nystuen 1985). Sedimentary facies distribution in the Hedmark Basin indicates that frontal and lateral ramps in the sole thrust followed the rift margins. As the basin was thrust onto a footwall flat, inversion of the basin occurred and produced a large culmination. The short wavelength of the basin (approximately 70 km) indicates, that little downwarping occurred beneath the displaced Hedmark basin, if an elastic lithosphere is assumed. Hence, the culmination probably remained a positive feature and would have bowed up the overlying thrust sheets.

Introduction

The term basin inversion covers an extremely wide range of tectonic processes and structural styles. Mild forms of basin inversion may only involve localized compressional features associated with reversal of motion along a few extensional faults or strike–slip within a rift setting (e.g. Kingma 1958; Glennie & Boegner 1981). In contrast, an extreme form of basin inversion is seen in thrust belts where pre-thrusting or synthrusting sedimentary basins may be translated for tens or hundreds of kilometres, and portions of the original extensional, or strike–slip basin controlling fault system may be reactivated as thrust faults. The geometry of sedimentary packages on continental margins can vary considerably (e.g. Bernoulli 1981; Grow 1981) and such variations might be expected to exert a strong influence on the geometry of an evolving thrust belt.

In some thrust and fold belts, e.g. Appalachian and Southern Canadian Rocky Mountains (Hatcher 1981; Price 1981), the deformed (foreland) sedimentary sequence appears to restore to a relatively simple wedge-shaped template and the stratigraphy does not depart too severely from a layer cake shelf sequence. Other thrust belts, particularly those in the Alpine–Mediterranean region, display a Mesozoic stratigraphic sequence with marked stratigraphic and sedimentological changes related to extensional faults (and some strike–slip faults) developed in passive margin settings (e.g. Bosellini 1973; Bernouilli *et al.* 1979; Bernoulli 1981; Winterer & Bosellini 1981; Wildi 1983; Tricart & Lemoine 1986).

It has been noted by many authors that the pre-existing passive margin structural and sedimentary package geometry has influenced the geometry of superimposed thrust and fold belt deformation in a number of different ways. Extension faults have been re-activated as thrust faults (e.g. Helwig 1976; Jackson 1980; Dewey *et al.* 1986; Tricart & Lemoine 1986), salients and reentrants in thrust belts may be due to variations in sedimentary package geometry and thickness or the original passive margin configuration (e.g. Grubbs & van der Voo 1976; Thomas 1977; Powell 1979). Underlying passive margin trends may also influence the location of structures in the overriding thrust sheets (e.g. Laubscher 1977; Doglioni 1987; Morley 1987). Large faults, particularly those which juxtapose crystalline basement against sedimentary rocks may act as buttresses to compressional deformation (e.g. Collet 1927; Harberg *et al.* 1969; Tricart & Lemoine 1986; Morley 1986a).

The Scandinavian Caledonides originated from the closure of the Iapetus Ocean between the Laurentian and Baltoscandian continental plates (e.g. Phillips *et al.* 1976; Roberts & Gale 1978; Ziegler 1985). Collision of the continental plates produced a pile of structural units that originated from one of the following settings: the eastern margin of the Laurentian plate; the Iapetus Ocean and associated island arcs and back arc setting basins or the western margin of the Baltoscandian plate (e.g. Roberts & Gee 1985; Hossack & Cooper 1986). This paper illustrates how the original sedimentary basin configuration of a portion of the western Baltoscandian margin influenced the geometry of thrusting in the Osen Røa thrust sheet, south Norway.

From Cooper, M. A. & Williams, G. D. (eds), 1989, *Inversion Tectonics*
Geological Society Special Publications No. 44, 259–273.

The Late Precambrian to Devonian geological history of the Southwestern Baltoscandian margin

Rifting processes began to affect a supercontinent in the Late Precambrian and led to the formation of the Baltoscandian and Laurentian continents (e.g. Dewey 1969; Piper 1976; 1985; Scotese *et al.* 1979). Continental rifting commenced in the Middle to Late Riphean (1000–900 Ma) forming widespread rift basins and depressions in many parts of Baltica. The main site of rifting, which evolved into a passive margin, was probably more or less parallel to the present day Norwegian coast (NE–SW). The failed rift arms which extended into the craton, such as those observed along the Atlantic Ocean margins by Burke (1976) were oriented NW–SE relative to Norway today. Several NW–SE trending rift basins were filled with Upper Riphean sequences (Kumpulainen & Nystuen 1985) including the Risback, Hedmark, Valdres, Engerdalen and Tossasfjallet Basins. All of these basins were subsequently incorporated into Caledonian thrust sheets (Kumpulainen & Nystuen 1985). After the period of rift infilling, continental platform conditions developed during the Vendian and continued into the Lower Palaeozoic, prior to the collision with Laurentia. Onlap of Late Precambrian and Cambrian rocks onto Precambrian crystalline basement occurred over much of southern Norway. Following deposition of Lower Palaeozoic clastics and carbonates, part of the Hedmark Basin was incorporated into the Osen–Røa thrust sheet during Silurian and Devonian deformation phases (Nystuen 1982; Bockelie & Nystuen 1985). Palinspastic restorations of the late Precambrian rift basins to their depositional location are illustrated in Fig. 1. The restorations of the Valdres and Hedmark Basins are based upon shortening estimates from balanced cross-sections (Hossack *et al.* 1985; Morley 1986b).

The Osen–Røa thrust sheet

The Osen–Røa thrust sheet extends for about 270 km in the transport direction (NNW–SSE to NW–SE). It is the lowest tectonic unit in southern Norway and is predominantly composed of Late Precambrian to Lower Devonian sedimentary rocks (Nystuen 1981; Bockelie & Nystuen 1985). The sole thrust lies within shales of Cambrian age in the Oslo region, but to the NNW it cuts down section in the hanging wall via two hanging wall ramps into the Late Precambrian sedimentary rocks of the Mjøsa region (Morley 1986b, see Figs 1 & 2). In the footwall the thrust remains in the Lower Palaeozoic autochthon. The location of the largest hanging wall ramp is marked by the truncation of the Brøttum Formation (Fig. 1) and by an overturned syncline known as the Ringsaker inversion (Skjeseth 1963). This is a large ramp because the removal of the Brøttum Formation from the thrust sheet eliminates about 2000 m or more vertical thickness of section. The ramp can be traced as a large scale U-shaped trend in the map patterns of Precambrian rocks (Figs 1 & 3). The western portion of the trend is, however, better developed than the eastern side. This U-shaped trend suggests that the sole thrust has ramped both frontally and laterally. The area inside the 'U' displays a Riphean–Vendian age Precambrian stratigraphy that may be about 3–4 km thick (Nystuen 1982; 1983). To the west of the trend the Precambrian stratigraphy is less than 1 km thick (Hossack *et al.* 1985) and the lower part of the Precambrian stratigraphy (Riphean) is absent.

There are two possible explanations for the Hedmark basin geometry:

1 The location of the ramps in the sole thrust were dictated by the basin geometry (Fig. 4A). Hence most of the late Precambrian sedimentary basin has been incorporated into the hanging wall of the thrust and only small fragments of the basin remained behind in the footwall.
2 The sole thrust has cut up section frontally and laterally through an extensive sequence of Precambrian sedimentary rocks and left a substantial portion of them behind in the footwall (Fig. 4B). If the Hedmark Basin has been exhumed largely intact, then lateral sedimentary facies changes are to be expected approaching the basin margin, i.e. approaching the ramps in the sole thrust which followed the basin margins.

Sedimentology of the Hedmark Basin

The oldest sedimentary rocks of the Hedmark basin are of Late Riphean–Early Vendian age (Bjørlykke *et al.* 1976; Bockelie & Nystuen 1985). Two main formations (the Rendalen and Brøttum Formations) of Riphean age have been

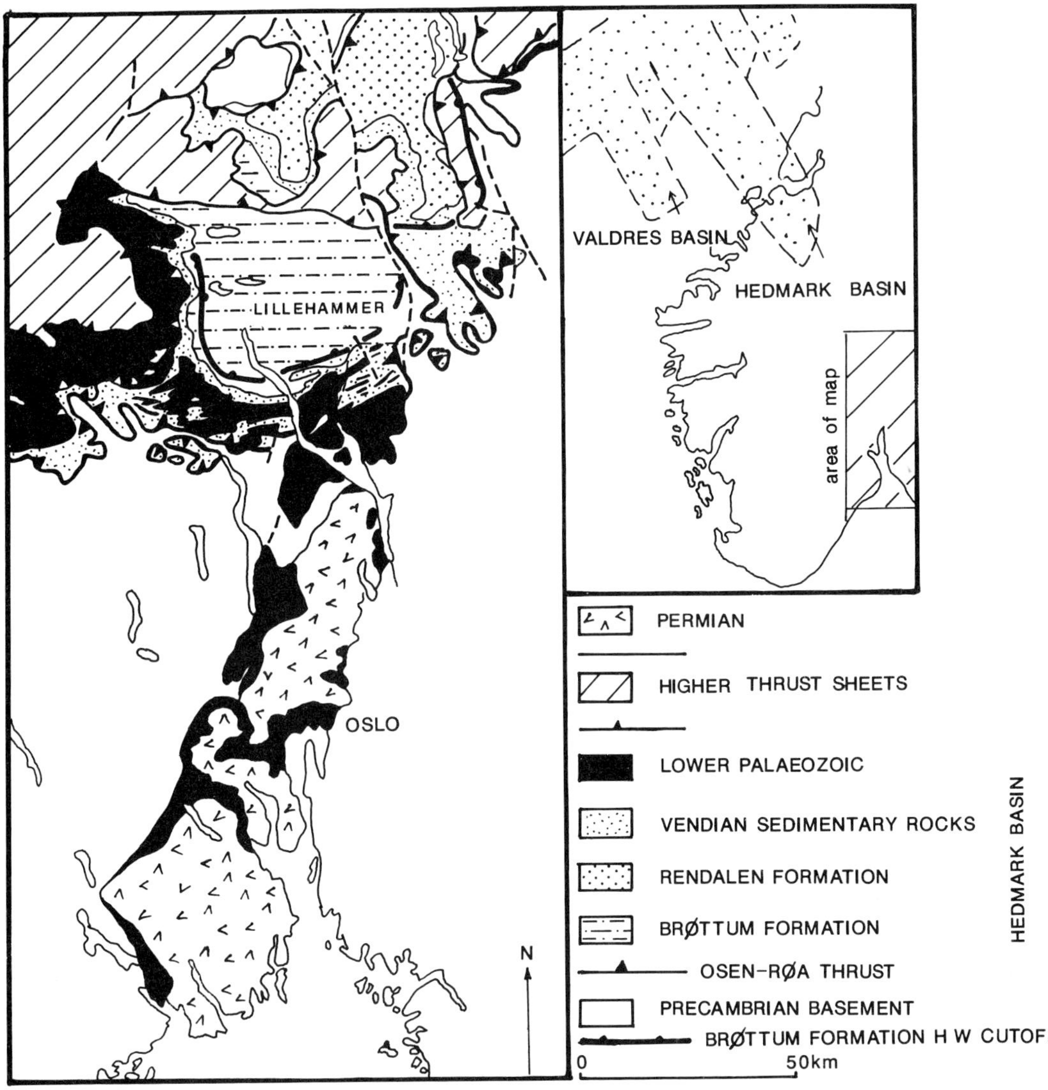

FIG. 1. Tectonostratigraphic map of the southeastern Scandinavian Caledonides (mostly after Bockelie & Nystuen 1985). Inset shows original location of late Precambrian rift basins relative to present day Norway (mostly after Kumpulainen & Nystuen 1985).

identified in the Hedmark basin (Fig. 5). The Rendalen Formation is at least 2000 m thick and is predominantly composed of braided stream deposits (Nystuen 1982; Kumpulainen & Nystuen 1985). These coarse arkoses and conglomerates grade northward into the shallow marine Atna quartzite and westward into a marine trough filled with sandstones and mudstones of the Brøttum Formation (Englund 1966; 1973; Bjørlykke *et al.* 1976). Generally palaeocurrent directions indicate transport toward the W and NW (Nystuen 1982). Extensional faults cut the Rendalen Formation but do not cause any facies changes (Nystuen 1981). However, variations in lithofacies are associated with basement highs in the basin which were truncated and incorporated within the Osen–Røa thrust sheet (Nystuen 1981).

The Brøttum Formation is composed of dark grey and black shales, interbedded with grey, mud-rich arkosic sandstones. The formation was deposited from turbidity currents and by pelagic

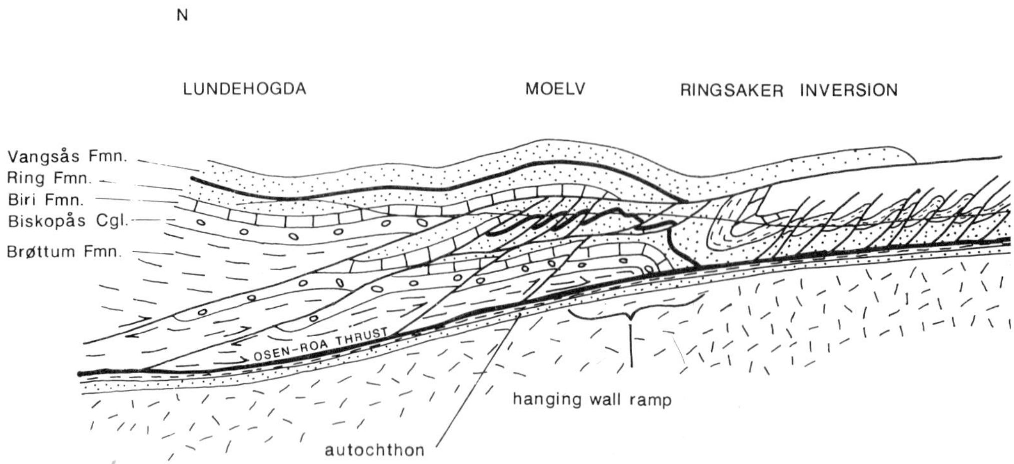

FIG. 2. Cross section through the Mjøsa area (after Morley 1986b), see Fig. 3 for location of section.

sedimentation (e.g. Bockelie & Nystuen 1985) and is thought to be of marine origin (Englund 1966; 1973; Bjørlykke *et al.* 1976; Kumpulainen & Nystuen 1985). The sedimentary facies is not conclusive because there are no palaeoecological indicators and turbidites are also found in lacustrine settings (e.g. the Newark Basin described by Manspeizer 1981).

A number of fan-shaped bodies are present within the rift basin (Biskopås Conglomerate, Osdalen, Imsdalen and Ring Formations). These sedimentary units represent conglomeratic alluvial fans and sub-aqueous gravity flow deposits that were deposited close to intrabasinal and basin-boundary faults (Sæther & Nystuen 1981, Nystuen 1982). They probably fringed the basin margin and pinched out basinward. Evidence for fault block-related topography is also inferred for the Biri Formation (Fig. 5). This formation is composed of dolomites, limestone breccias, dark shales and calcareous sandstones (Bjørlykke *et al.* 1976; Sæther & Nystuen 1981). It is thought to have been deposited in basins between eroded fault block highs (Nystuen 1982; Kumpulainen & Nystuen 1985). The Biskopås, Imsdalen and Ring fan delta and subaqueous fan deposits thin out into the Biri shales or Brøttum formation. The coarse fan deposits are thought to indicate a tectonic phase of rifting, which is marked further north by the extrusion of some local basalt flows over the Atna Quartzite (Sæther & Nystuen 1981).

Above the Precambrian sedimentary rocks described above are more widespread Precambrian sediments which were formed after the rift troughs were filled. The change between the Riphean to Vendian trough sedimentation, and the Vendian to early Cambrian predominantly shallow marine clastic sedimentation is marked by the Varanger ice age which occurred during the Vendian. The ice age is indicated by a massive diamictite interpreted as a basal till, which is called the Moelv Tillite in the Hedmark Group. Post-glacial deposits include the Ekre Shale, and the overlying Vangsås Formation. The latter is composed of prograding deltas, braided plains and shallow marine deposits.

The sedimentary facies distribution within the Hedmark Basin described above and by Kumpulainen and Nystuen (1985) summarizes the strong sedimentary evidence that the Osen–Røa thrust closely followed the Hedmark Basin margin. In fact, the strong resemblance of the Late Precambrian sedimentary distribution to a basin setting led earlier workers to consider that the region was parautochthonous and that the basin had undergone little transportation (e.g. Schiotz 1902; Skjeseth 1963; Ramberg & Englund 1969). Now it is thought that the southern basin margin has been transported approximately 150 km first to the SW then the SSW (Oftedahl 1943; Høy & Bjørlykke 1980; Nystuen 1981; Morley 1986b).

Passive margin geometry of the Hedmark Basin

A broad definition of basin inversion proposed by M. A. Cooper and G. D. Williams at the

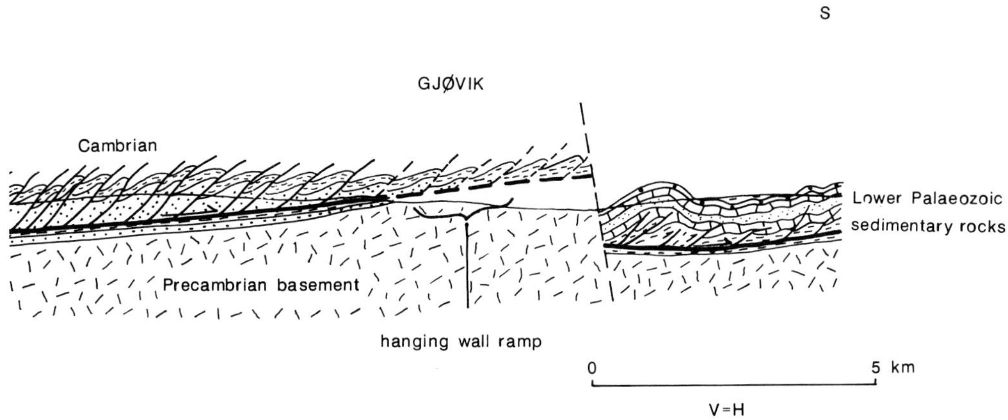

1987 Inversion Tectonics meeting at the Royal Society, London was 'reactivation of a fault or fault systems where the displacement vector varies from that of the earlier movement on the system.' It is most unlikely that all the faults in a basin, or the entire length of a large individual fault will be reactivated. More likely is the creation of new fault segments and reactivation of some segments of some old faults. For example, the low-angled detachment faults underlying a rift basin are unlikely to be reactivated by subsequent axis parallel wrenching. The upper portion of the rift where higher angled segments of faults are present is the area where reactivation of faults with strike–slip or oblique–slip motion is most likely. Similarly, when thrusting affects a basin, portions of the pre-existing faults are likely to be activated while other fault segments may remain unaffected. Thus new thrust fault segments link reactivated portions of older faults. In the example of the Osen Røa thrust sheet, the relatively low-angled portions of the thrust are probably related only to compressive deformation where the thrust propagated along or close to the basement–sedimentary rock interface. The role of older faults is most pronounced when the thrust ramped along old boundary faults. Thus, new flat fault segments link reactivated, originally steeply inclined fault segments.

The Hedmark Basin is about 70–90 km wide (ENE–WSW) and at least 100 km along the basin axis (NNW–SSE). Rifts are commonly of comparable width, but tend to be longer. For example, Lake Tanganyika in East Africa is 700 km long and up to 70 km wide (Fig. 6) and the Triassic–Jurassic rift basins of the Eastern USA are about 300 km long and 50–60 km wide (Van Houten 1977). Discrete rift segments such as Lake Tanganyika are not necessarily one long continuous sedimentary basin; high horst blocks can break up the rift up into several basins (Fig. 6). A thrust attempting to propagate through a rift segment may encounter such horst blocks. If there is a large mechanical contrast between the horst block and the surrounding rocks, such as crystalline basement horst flanked by sediment, then the propagating thrust may ramp out of the rift basin, rather than propagate through the horst. Consequently only part of the rift segment may be incorporated into the thrust sheet. The Hedmark Basin is small enough to suggest such a scenario. Although the entire width of the rift is probably preserved, it is unlikely that the entire length of the rift segment is preserved in the Osen–Røa thrust sheet. Hence other basins which originally comprised the rift segment to which the Hedmark Basin belonged may be buried beneath the hanging wall of the Osen–Røa thrust sheet, or thrust over the top.

The margins of rifts are largely defined by major extensional faults and monoclinal flexures (Fig. 6) (e.g. King 1978). The steep boundary faults are usually areas of maximum subsidence and sediment accumulation (Fig. 6). Commonly in east Africa and in other rift systems the geometry is that of a half graben (e.g. Chapman *et al.* 1978; King 1978; Desheng, 1980; Bosworth 1985; Chaoyuan 1985), where one boundary fault displays much more displacement than the opposing boundary fault (Fig. 7). In situations where boundary faults later influenced the location of thrust ramps, large thicknesses of

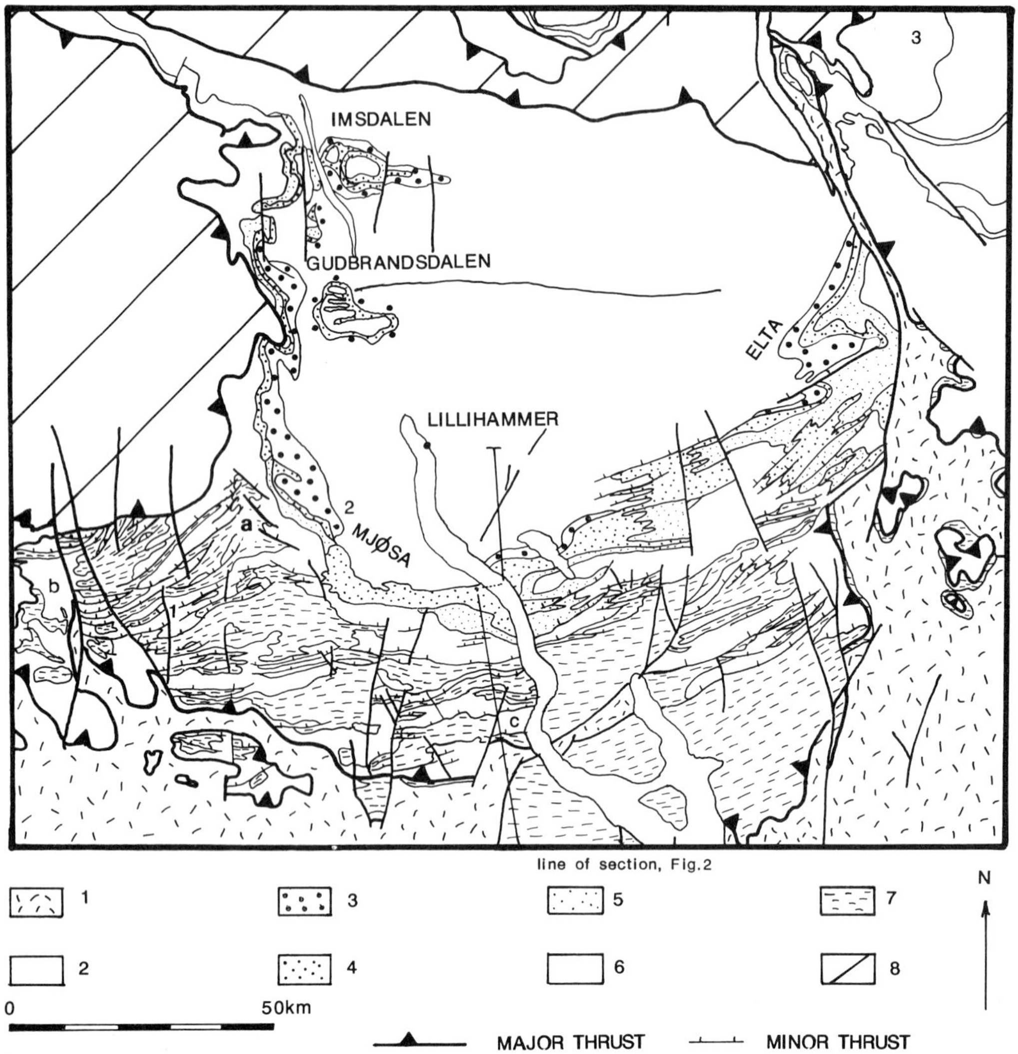

FIG. 3. Map of the Sparagmite region (mostly after Sigmond *et al.* 1984). 1, Precambrian crystalline basement (autochthonous); 2, turbidite facies (Brøttum Formation): 3, polymict conglomerate (Biskopås Conglomerate); 4, calcareous sandstone and shale (Biri Formation); 5, sandstone with conglomerate, tillite and other glacial beds, and shale (Ring Formation – Ekre shale); 6, shallow marine sandstone (Vangsås Formation); 7, Lower Palaeozoic shales, limestones and sandstones; 8, higher thrust sheets. 1, 2 and 3 locate the regions which the stratigraphic columns in Fig. 5 represent (a = Dokka, b = Aurdal, c = Gjøvik).

sediment (several kilometres) are likely to be cut out across hanging wall ramps. This is particularly true when a strong material anistropy exists across the rift boundary faults. Such was probably the case in the western Baltoscandian margin where clastic Precambrian sedimentary rocks in a rift setting would have been downthrown against Precambrian crystalline rocks. The southern and western margins of the Hedmark Basin display the most abrupt thickness changes across thrust ramps. Although, the eastern margin of the Hedmark Basin and the E oblique to lateral hanging wall ramp is less well defined, this is probably because less section is

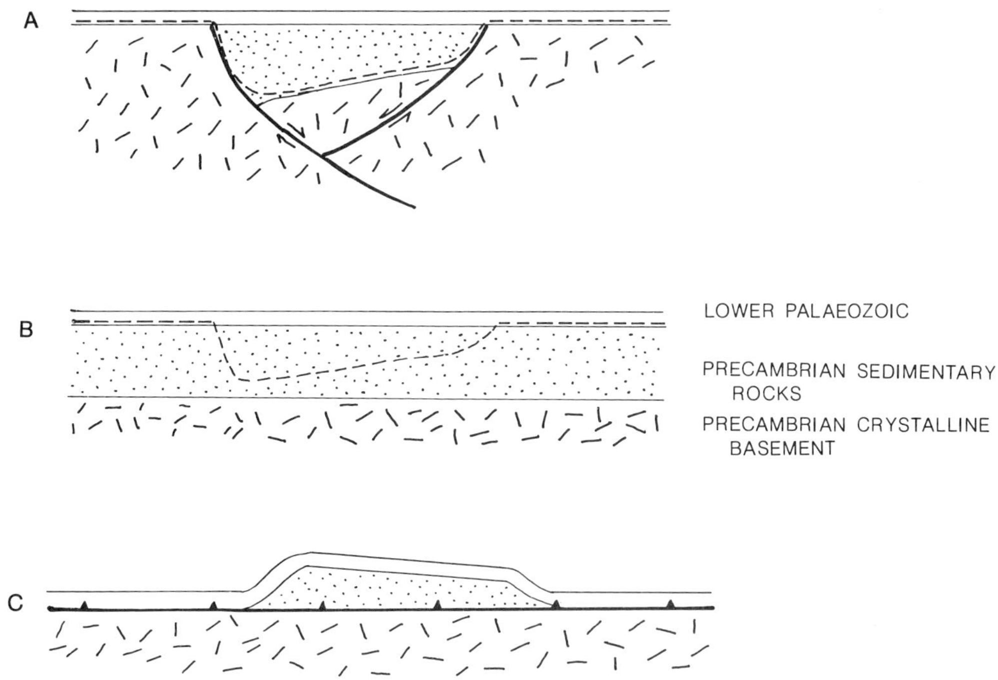

FIG. 4. Strike sections illustrating possible thrust-trajectory and sedimentary basin geometries for the Osen–Røa thrust sheet. A, Thrust follows rift basin geometry; B, lateral ramps generated within laterally extensive Precambrian sedimentary basin; C, inverted basin geometry after thrusting onto a footwall flat.

cut out across the ramp. The geometry observed above leads to the suggestion that the southern portion of the Hedmark Basin originally had the geometry of a half-graben, with the western side forming thickest portion of the basin. The southern margin of the basin may have developed by ramping of the sole thrust over a rift–oblique horst block such as those observed in Lake Tanganyika (Fig. 6).

The sedimentary facies distribution in the Hedmark basin is also in keeping with a half-graben, Newark-type basin (Manspeizer 1981) as illustrated in Fig. 7. The marginal piedmont facies of proximal fluvial–deltaic sandstones and conglomerates at the thinner end of the half graben sedimentary wedge is probably the equivalent of the Rendalen Formation. The central basin facies of lacustrine grey-black finely laminated shales and siltstones and turbidite sandstones is represented by the Brøttum Formation. The Ring Formation and Biskopås Conglomerate probably represent the alluvial fan and fan delta deposits associated with major boundary faults.

Although in detail the Hedmark basin may have resembled continental rifts, its plate setting was probably somewhat different because it was associated with a passive margin (e.g. Kumpulainen & Nystuen 1985). The continental margin of the eastern USA can be considered as an analogue to the Baltoscandian margin. Rift basins associated with the initial stages of continental break-up are not necessarily confined to the continent–oceanic crust transition zone (Fig. 8A). The E margin of the USA provides examples of rift basins on the craton side of the transition zone. Fig. 8 illustrates how the Brunswick Graben is located west of the main thinned section of continental crust in the Carolina through area. The Hedmark Basin may have occupied a similar position to the Brunswick Graben on the Baltoscandian margin.

Continental crust must be positioned on oceanward side of the Hedmark Basin because the Osen–Røa thrust sheet is overthrust by the Jotun thrust sheet, and the Jotun thrust sheet is thought to be derived from Baltoscandian continental crust (Hossack *et al.* 1985; Kumpulainen

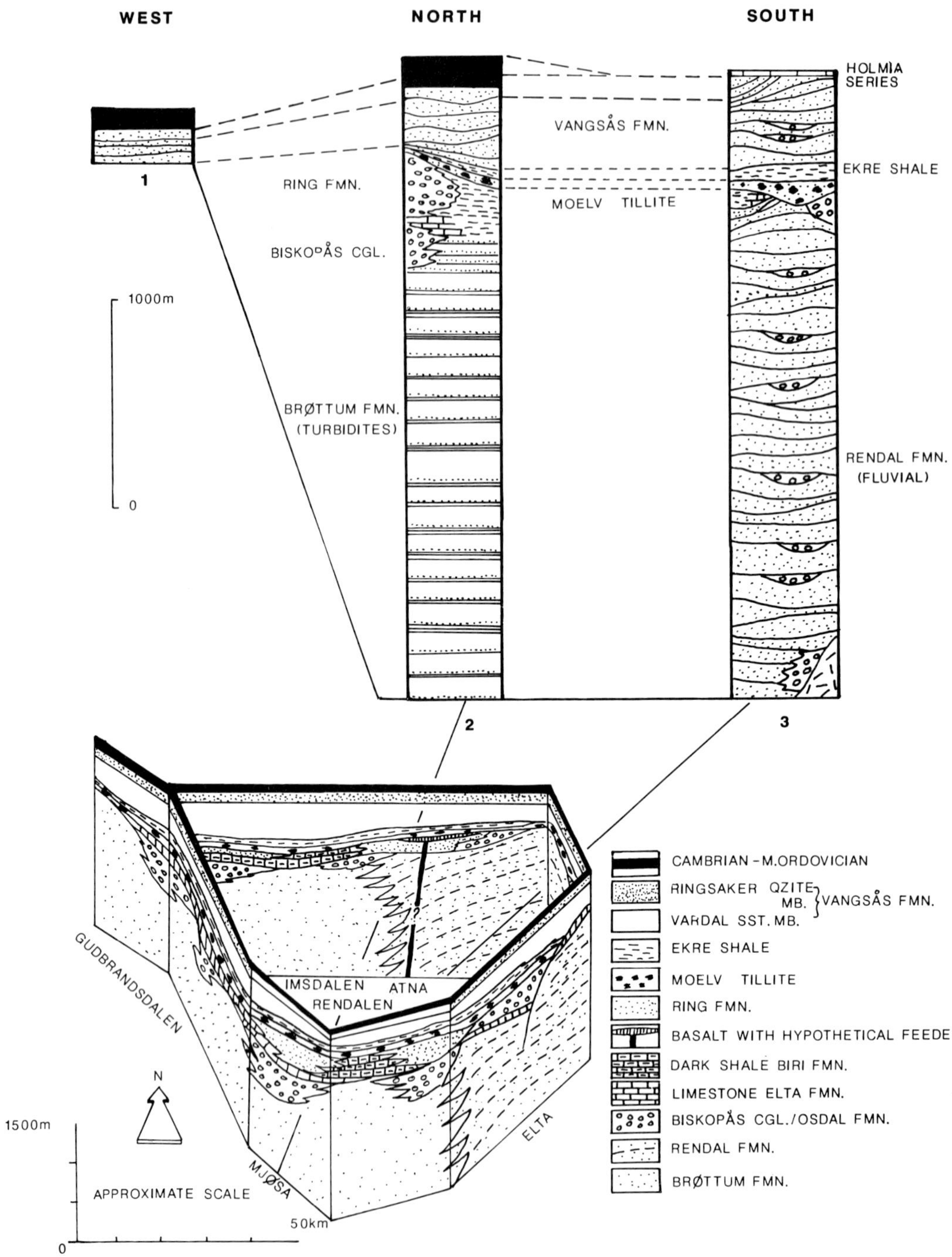

FIG. 5. Stratigraphic variations in the Osen–Røa thrust sheet (after Bockelie & Nystuen 1985; Hossack *et al.* 1985). Fence diagram after Nystuen (1981) illustrates facies changes present in the southern Hedmark Basin schematically. See Fig. 3 for location of the stratigraphic columns.

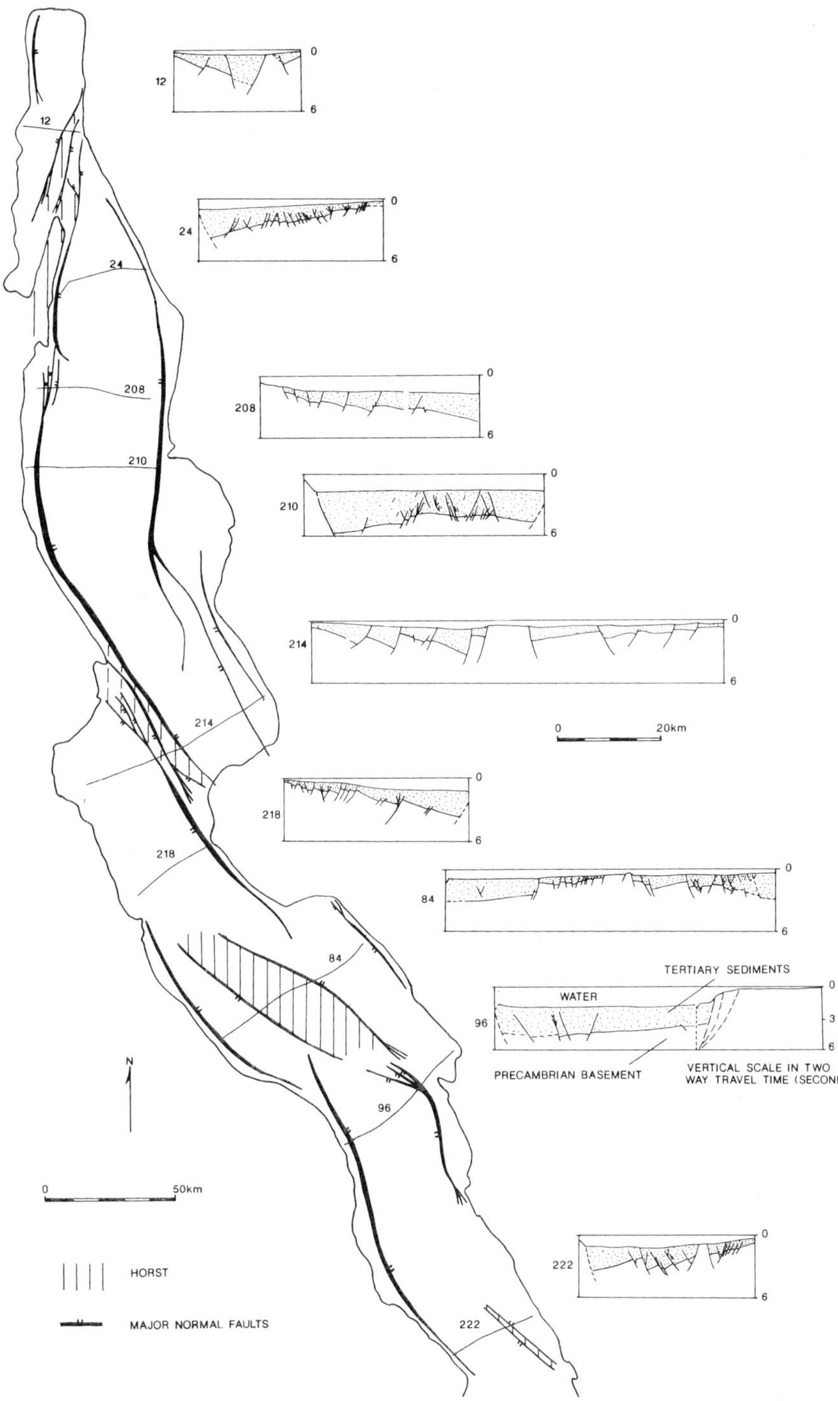

FIG. 6. Map of Lake Tanganyika showing the position of major boundary faults and horst trends. Cross-sections are simplified representations of the rift geometry seen on seismic lines shot by Duke University (Project PROBE) for a consortium of oil companies. For examples of the data see Rosendahl *et al.* (1986).

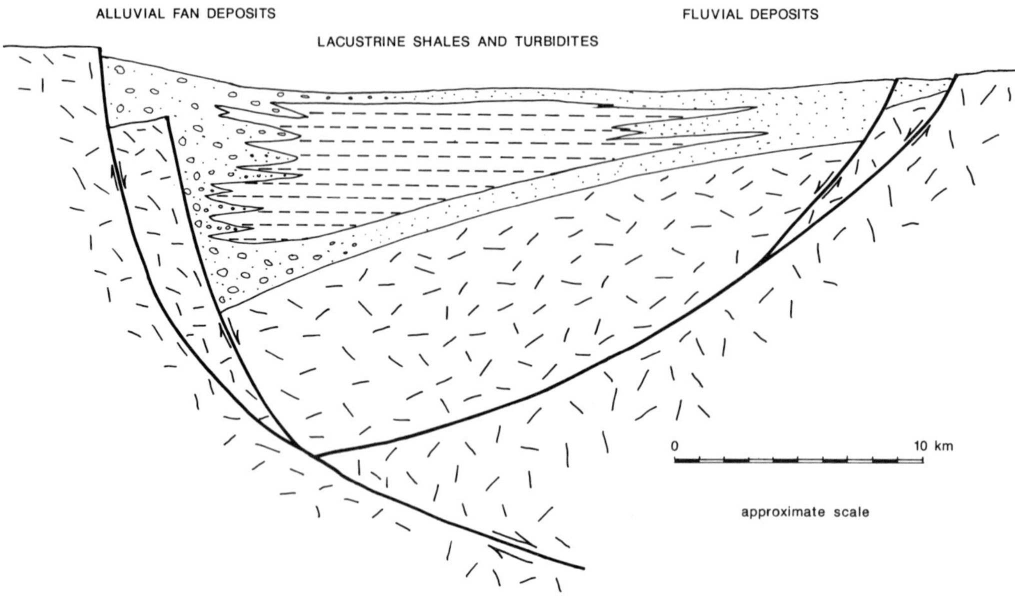

FIG. 7. Schematic half-graben geometry illustrating major sedimentary facies changes (after Manspeizer 1981).

& Nystuen 1985). Some examples of passive margins display over 300 km width of 20 km thick crystalline basement oceanward of the 'precursor' rift basins to continent break-up (Fig. 8C). Restorations of the southern Norwegian thrust sheets by Hossack *et al.* (1985) indicate that at least 280 km width of Baltoscandian basement and sedimentary rocks extended oceanward from the Osen–Røa thrust sheet. Hence an extensive transitional crust zone as illustrated by the Blake Plateau basin (Fig. 8C) probably existed on the Late Precambrian–Lower Palaeozoic Baltoscandian margin.

Consequences of inversion of the Hedmark Basin during thrusting

Uneven loading of the continental lithosphere

The Osen–Røa thrust appears to have closely tracked the base of sedimentary section (including the Hedmark Basin) which exhibits substantial lateral thickness variations (Fig. 5). During transport the Osen–Røa thrust sheet was emplaced onto a long footwall flat (Morley 1986a) composed of Cambrian shales above crystalline Precambrian basement. Consequently the movement of the Osen–Røa thrust sheet onto a footwall flat might be expected to have created an inversion of the original basin relief (Fig. 4). Hence the Hedmark Basin would have generated a very large culmination and the areas of relatively high basement and thin sedimentary cover would have formed relative lows. Any thrust sheet that rode 'piggy back' on the Osen–Røa thrust sheet are likely to have suffered extensional strains in the outer arc of the growing culmination.

The movement of a thrust sheet which contained large lateral variations in thickness (2–3 km in the Osen–Røa thrust sheet) would cause uneven loading of the crust. A consequence of the loading would be to decrease the amplitude of the culmination created by basin inversion resulting from flexural downwarping. The amplitude of the culmination is a function of the lithospheric response time to reaching Airy isostasy for a given load and the wavelength of the load. (Quinlan & Beaumont 1984; Turcotte & Schubert 1982). The rate that the load is moved by thrusting onto less isostatically compensated lithosphere.

If an elastic lithosphere is assumed, then the degree of compensation of the Hedmark Basin load, by the lithosphere (C) can be calculated. The wavelength of the load is a critical factor

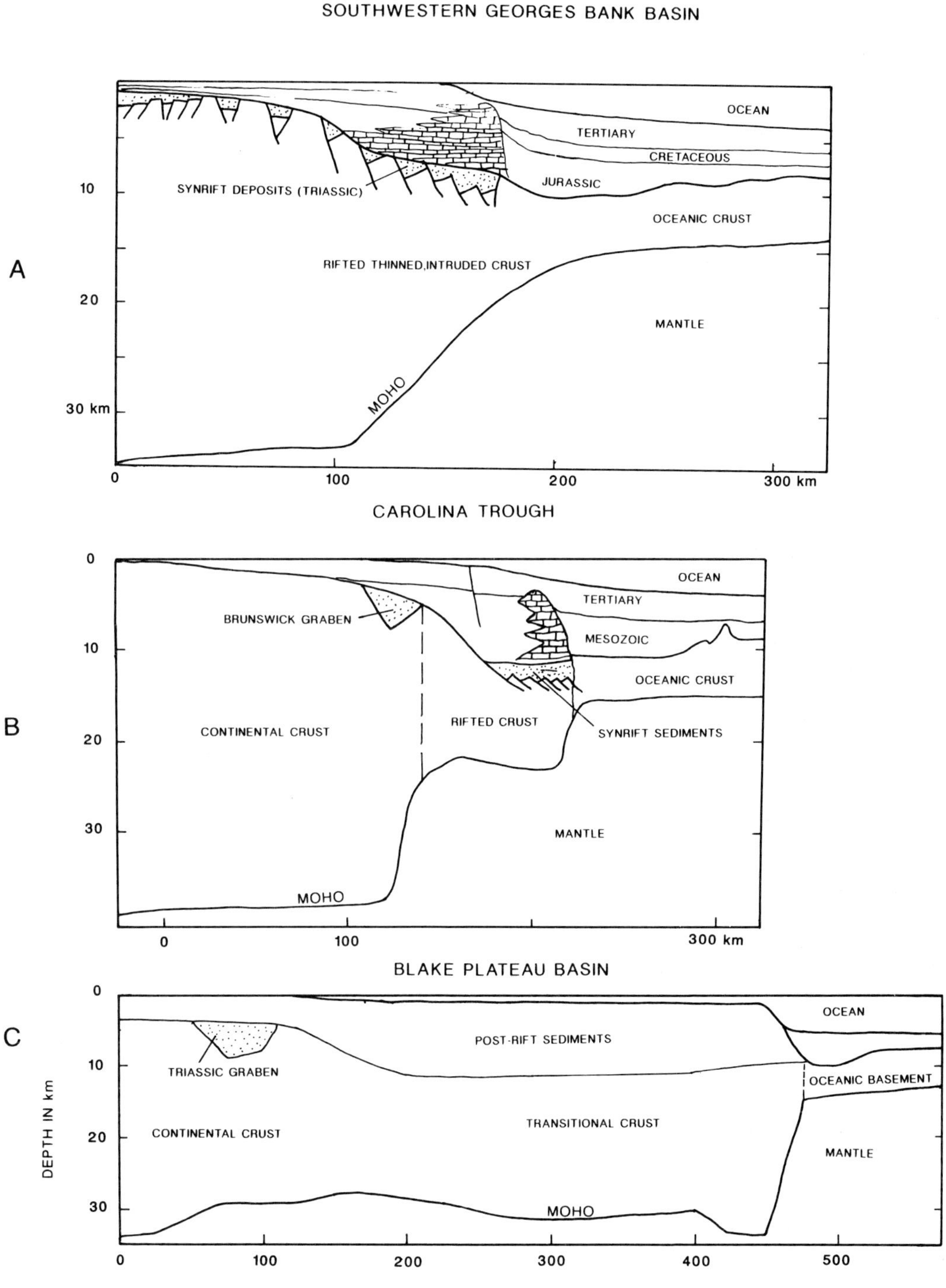

FIG. 8. Profiles through the western Atlantic passive margin, USA, illustrating variations in passive margin geometry (after Grow 1981).

when applying the following equation (after Turcott & Schubert 1982, pp. 115, 122, 123):

$$C = \frac{(P_m - P_c)}{P_m - P_c + \frac{D}{g}\left(\frac{2\pi}{\lambda}\right)^4}$$

Where P_m (density of the mantle) = 3300 kg m^{-3}; P_c (density of the crust) = 2800 kg m^{-3}; λ = 70 km and g is the gravitational constant. $D = Eh^3/12\ (1\text{-}V^2)$ where E (Youngs modulus) = 70 gPa, V (Poisson's ratio) = 0.25 and h (elastic thickness of the lithosphere) is assumed to be 25 km. For such a short wavelength load, the topography generated by inversion of the Hedmark Basin would be 10% compensated by the rigidity of the lithosphere. Hence the amplitude of the culmination was probably only reduced by approximately 10%.

Variations in deformation style related to basin inversion

The Hedmark Basin is deformed by imbricates and folds (Nystuen 1983), in some areas buckle folding appears to be the dominant deformation style (Morley 1986b). To the W (Aurdal–Dokka area) and S (Gjøvik area), the Vendian–Lower Palaeozoic sequence is deformed by much more closely spaced imbricates (Hossack *et al.* 1985; Bjørlykke 1979; Morley 1986b) (Fig. 3). The decrease in stratigraphic thickness and position of Vendian age sedimentary rocks at the base of the thrust sheet in the Aurdal–Dokka and Gjøvik areas is probably responsible for the intense imbrication style.

Passing eastward from the Aurdal area, the imbricated sequence appears to pass from ENE–WSW oriented bedding strikes into a NNW–SSE trending strip of Riphean and older age rocks that are not much affected by imbrication (Fig. 3). The latter form the W margin of the Hedmark Basin. As remarked earlier, this W margin probably represents a lateral ramp in the Osen–Røa thrust. There is a roughly triangular area or 'Christmas tree' shape of Lower Palaeozoic rocks in the vicinity of Dokka (Fig. 3) which lies between the Aurdal area and the western lateral ramp margin of the Hedmark Basin (Bjørlykke 1979; Sigmond *et al.* 1984). The W side of the triangle is composed of NE–SW striking imbricate slices, while the E side corresponds to NW–SE trending imbricate slices. The W side of the triangle follows the Aurdal area trends, while the E side corresponds more to the lateral ramp trends. Hence the Dokka area appears to be transitional between the structures which laterally terminate the Hedmark Basin area and the structures which permit shortening parallel to the tectonic transport direction in the Aurdal area. It corresponds to an area of interference between two different trending sets of imbricate faults generated within the same thrust sheet.

Conclusions

The original passive margin sedimentary basin geometry has influenced the development of the Osen–Røa thrust sheet in the following ways:

1 The sole thrust appears to have largely tracked the interface between late Precambrian sedimentary rocks and Precambrian crystalline basement. This involved following large thickness variations in the sedimentary sequence passing from rift basin to the rift margin. The largest ramps in the sole thrust, which cut out approximately 2–3 km of section, appears to correspond to the former rift margins.

2 During transport of the Hedmark Basin, stratigraphic thickness variations in the Osen–Røa thrust sheet probably created uneven loading of the crust, which generated minor downwarping of the lithosphere in the vicinity of the basin. The short wavelength of the Hedmark basin load indicates that the downwarping probably only compensated 10% of the thickness of the basin. The basin was therefore almost completely inverted during thrust transport and a large culmination corresponding to the basin shape would have been generated.

3 The deformation style of the thrust sheet changes passing from the thick Hedmark Basin sequence to the thin Vendian–Lower Palaeozoic sequence which flanks the basin. Very closely spaced imbricates are the dominant deformation style in the latter. Interference structures between the lateral ramp related structure on the W margin of the Hedmark Basin, and the Aurdal area (to the W) are seen in the Dokka area.

ACKNOWLEDGMENTS: I would like to thank Amoco Production Co. for permission to publish the paper and Jake Hossack and Mark Cooper whose helpful reviews improved the paper. The City of London Polytechnic and CNAA are gratefully acknowledged for funding fieldwork in Norway for my Ph.D. research.

References and further reading

BERNOULLI, D. 1981. Ancient continental margins of the Tethyian Ocean. In: Geology of Passive Continental Margins. *American Association of Petroleum Geologists Course Note Series*, No. 19, 36 pp.

——, KALIN, D. & PATACCA, E. 1979. A sunken continental margin of the Mesozoic Tethys. *Association of Sedimentologists of France Special Publication*, **1**, 197–210.

BJØRLYKKE, A. 1979. Gjøvik og Dokka. Beskrivelse til de berggrunnsgeologiske kart 18161 og 1816 IV (M. 1:50 000) *Norsk Geologisk Undersøkelse*, **344**, 48 pp.

——, ELVSBORG, A. & HØY, T. 1976. Late Precambrian sedimentation in the central sparagmite basin of South Norway. *Norsk Geologisk Tidsskrift*, **56**, 233–90.

BOCKELIE, J. F. & NYSTUEN, J. P. 1985. The south-eastern part of the Scandinavian Caledonides. *In*: GEE, D. G. & SURT, B. A. (eds) *The Caledonian Orogen – Scandinavian and Related Areas*. Wiley, Chichester, 69–88.

BOSELLINI, A. 1973. Modella geodynamico e paleotetitonico delle Alpi Meridionali durante il Giurassico-Cretacico. Sue possibili applicazioni agli Appenini. *In*: ACCORDI, B. *et al.* (eds) *Modern Vedute sulla Geologia del Appennina*. Accademia Nazionali Dei Lincei Quad, **183**, 163–205.

BOSWORTH, W. 1985. Geometry of propagating continental rifts. *Nature*, **316**, 625–7.

BURKE, K. 1976. Development of graben associated with the initial ruptures of the Atlantic Ocean. *Tectonophysics*, **36**, 93–111.

CHAPMAN, G. R., LIPPARD, S. J. & MARTYN, J. E. 1978. The stratigraphy and structure of the Kamasia Range, Kenya Rift Valley. *Journal of the Geological Society of London*, **135**, 265–281.

CHAOYUAN, H. 1985. Geological characteristics and oil exploration of small depressions in eastern China. *Geology*, **13**, 303–306.

COLLET, L. W. 1927. *The Structure of the Alps*. Edward Arnold, London, 289 pp.

DESHENG, L. 1980. Geological structure and hydrocarbon occurrence of the Bohai Gulf oil and gas basin (China). *In*: MASON, J. F. (ed.) *Petroleum Geology in China*. Penwell Books, Tulsa, 180–192.

DEWEY, J. F. 1969. Evolution of Appalachian/Caledonian Orogen. *Nature*, **222**, 124–129.

——, HEMPTON, M. R., KIDD, W. S. F., SAROGLY, F. & SENGOR, A. M. C. 1986. Shortening of continental lithosphere: the neotectonics of Eastern Anatolia–a young collision zone. *In*: COWARD, M. P. & RIES, A. C. (eds) *Collision Tectonics. Special Publication of the Geological Society of London*, **19**, 3–36.

DOGLIONI, C. 1987. Tectonics of the Dolomites (Southern Alps, Northern Italy). *Journal of Structural Geology*, **9**, 181–193.

ENGLUND, J. O. 1966. Studies on the latest Precambrian and Eocambrian rocks in Norway. No. 2. Sparagmittgruppens begarter ved favang, Gudbrandsdalen. En sedimentologisk og tektonisk undersøkelse. *Norges Geologisk Undersøkelse*, **238**, 55–103.

—— 1973. Stratigraphy and structure of the Ringebu–Vinstra district, Gudbrandsdalen; with a short analysis of the western part of the sparagmite region in southern Norway. *Norges Geologisk Undersøkelse*, **293**, 58 pp.

GLENNIE, K. W. & BOEGNER, P. L. E. 1981. Sole Pit inversion tectonics. *In*: ILLING, L. V. & HOBSON, G. D. (eds) *Petroleum Geology of the Continental shelf of North-west Europe*. Institute of Petroleum, London, 110–120.

GROW, J. A. 1981. Structure of the Atlantic margin of the United States, Geology of Passive continental margins, *American Association of Petroleum Geologists* Course Note Series, No. 19, 3–1–3–41.

GRUBBS, K. L. & VAN DER VOO, R. 1976. Structural deformation of the Idaho–Wyoming overthrust belt (USA) as determined by Triassic palaeomagnetism. *Tectonophysics*, **33**, 321–336.

HARBERG, L., NELSON, V. & CHURCH, V. 1949. Structural trends in central western Wyoming. *Geological Society of American Bulletin*, **60**, 183–216.

HATCHER, R. D. 1981. Thrusts and nappes in the North American Appalachian Orogen. *In*: MCCLAY K. R. & PRICE, N. J. (eds) *Thrust and Nape Tectonics. Special Publication of the Geological Society of London*, **9**, 491–499.

HELWIG, J. 1976. Shortening of continental crust in orogenic belts and plate tectonics. *Nature*, 266, 768–770.

HOSSACK, J. R. & COOPER, M. A. 1986. Collision tectonics in the Scandinavian Caledonides. *In*: COWARD, M. P. & RIES, A. C. (eds) *Collision Tectonics. Special Publication of the Geological Society of London*, **19**, 287–304.

——, GARTON, M. R. & NICKELSEN, R. P. 1985. The geological section from the foreland up to the Jotun thrust sheet in the Valdres area, south Norway. *In*: GEE, D. G. & STURT, B. A. (eds) *The Caledonian Orogen – Scandinavia and Related Areas*. Wiley, Chichester, 35–49.

HØY, T. & BJØRLYKKE, A. 1980. *Hamar, berggrunnsgeologisk Kart 1916 IV, M. 1 : 50 000*. Norsk geologisk Undersøkelse.

JACKSON, J. A. 1980. Reactivation of basement faults and crustal shortening in orogenic belts. *Nature*, **283**, 343–346.

KING, B. C. 1978. Structural and volcanic evolution of the Gregory Rift Valley. *In*: BISHOP, W. W. (ed.) *Geological Background to Fossil Man*. Scottish Academic Press Edinburgh, pp. 29–54.

KINGMA, J. T. 1958. Possible origin of piercement structures, local unconformities and basins in the Eastern Geosyncline, New Zealand. *New Zealand Journal of Geology and Geophysics*, **1**, 269–274.

KUMPULAINEN, R. & NYSTUEN, J. P. 1985. Late Proterozoic basin evolution and sedimentation in the westernmost part of the Baltoscandia. *In*: GEE D. G. & STURT, B. A. (eds) *The Caledonian Orogen–Scandinavia and Related Areas*. Wiley, Chichester, 213–232.

LAUBSCHER, H. P. 1977. Fold development in the Jura. *Tectonophysics*, **37**, 337–362.

MANSPEIZER, W. 1981. Early Mesozoic basins of the Central Atlantic passive margins. *In*: *Geology of Passive Continental Margins*. American Association of Petroleum Geologists Course Notes, No. 19, 4–1–4–60.

MORLEY, C. K. 1986a. A classification of thrust fronts. *Bulletin of the American Association of Petroleum Geologists*, **70(1)**, 12–25.

—— 1986b. The Caledonian thrust front and palinspastic restorations in the southern Norwegian Caledonides. *Journal of Structural Geology*, **8**, 753–765.

—— 1987. Origin of a major cross-element zone: Moroccan Rift. *Geology*, **15**, 761–764.

NYSTUEN, J. P. 1981. The Late Precambrian 'Sparagmites' of southern Norway: a major Caledonian allochthon—the Osen–Røa Nappe Complex. *American Journal of Science*, **281**, 69–84.

—— 1982. Late Proterozoic basin evolution on the Baltoscandian craton: the Hedmark Group, southern Norway. *Norges Geologisk Undersøkelse*, **375**, 74 pp.

—— 1983. Nappe and thrust structures in the Sparagmite Region, Southern Norway. *Norges Geologisk Undersøkelse*, **380**, 67–83.

OFTEDAHL, C. 1943. Overskyvninger i den norske fjellkjed. *Naturen*, **5**, 143–150.

PHILLIPS, W. E. A., STILLMAN, C. J. & MURPHY, T. 1976. A Caledonian plate tectonic model. *Journal of the Geological Society of London*, **132**, 579–609.

PIPER, J. D. A. 1976. Palaeomagnetic evidence for a Proterzoic supercontinent. *Philisophical Transactions of the Royal Society of London*, **280A**, 469–490.

—— 1985. Palaeomagnetism in the Caledonian–Appalachian orogen: a review. *In*: GEE D. G. & STURT, B. A. (eds) *The Caledonian Orogen–Scandinavia and Related Areas*. Wiley, Chichester, 35–49.

POWELL, C., MCA. 1979. A speculative tectonic history of Pakistan and surroundings: some constraints from the Indian Ocean. *In*: FARAH, A. & DEJONG, K. (eds) *Geodynamics of Pakistan, Geological Survey of Pakistan*. Geological Society of Pakistan, Quetta 5–24.

PRICE, R. A. 1973. Large-scale gravitational flow of supracrustal rocks, Southern Canadian Rockies. *In*: DEJONG, K. A. & SCHOLTEN, R. (eds) *Gravity and Tectonics*. Wiley, New York, 491–502.

—— 1981. The Cordilleran foreland thrust and fold belt in the Southern Canadian Rocky Mountains. *In*: MCCLAY, K. R. & PRICE, N. J. (eds) *Thrust and Nappe Tectonics. Special Publication of the Geological Society of London*, **9**, 33–39.

QUINLAN, G. M. & BEAUMONT, C. 1984. Appalachian thrusting, lithosphere flexure, and the Paleozoic stratigraphy of the Eastern Interior of North America. *Canadian Journal of Earth Science*, **21**, 973–996.

RAMBERG, I. B. & ENGLUND, J. O. 1969. The source rock of the Biskopås Conglomerate at Favang, and the western margin of the sedimentation basin of the Brøttum Formation at Favang-Vinstra, Southern Norway. *Norges Geologisk Undersøkelse*, **258**, 302–324.

ROBERTS, D. & GALE, G. H. 1978. The Caledonian–Appalachian Iapetus Ocean. *In*: TARLING, D. H. (ed.) *Evolution of the Earth's crust*. Academic Press, London, 255–347.

—— & GEE, D. B. 1985. An Introduction to the structure of the Scandinavian Caledonides. *In*: GEE, D. G. & STURT, B. A. (eds) *The Caledonide Orogen–Scandinavia and Related Areas*. Wiley, Chichester, 55–68.

ROSENDAHL, B. R., REYNOLDS, D. J., LORBER, P. M., BURGESS, C. F., MCGILL, J., SCOTT, D., LAMBIASE, J. J. & DERKSEN, S. J. 1986. Structural expressions of rifting: lessons from Lake Tanganyika, Africa. *In*: FROSTICK, L. E. RENAUT, R. W., REID, I. & TIERCELIN, J-J. (eds) *Sedimentation in the African Rifts. Special Publication of the Geological Society of London*, **25**, 29–43.

SÆTHER, T. & NYSTUEN, J. P. 1981. Tectonic framework, stratigraphy, sedimentation and volcanism of the Lake Precambrian Hedmark Group, Østerdalen, south Norway. *Norsk Geologisk Tidsskrift*, **61**, 193–211.

SCHIOTZ, O. E. 1902. Den sydostlige del au sparagmit—kvartsfjeldet i Norge. *Norges Geologisk Undersøkelse*, **35**, 131 pp.

SCOTESE, C. R., BAMBACK, R. K., BARTON, C., VAN DER VOO, R. & ZEIGLER, A. M. 1979. Palaeozoic base maps. *Journal of Geology*, **87**, 217–277.

SIGMOND, E. M. O., GUSTAVSON, M., & ROBERTS, D. 1984. *Berggrunnskart over Norge — M. 1:1 million*. Norges geologisk Undersøkelse.

SKJESETH, S. 1963. Contributions to the geology of the Mjøsa District and the classical sparagmite area in southern Norway. *Norges Geologisk Undersøkelse*, **220**, 1–126.

THOMAS, W. A. 1977. Evolution of Appalachian–Ouachita salients and recesses from reentrants and promontories in the continental margin. *American Journal of Science*, **277**, 1233–1278.

TRICART, P. & LEMOINE, M. 1986. From faulted blocks to megamullions and megaboudins: Tethyan heritage in the structure of the Western Alps. *Tectonics*, **5**, 95–118.

TURCOTTE, D. L. & SCHUBERT, G. 1982. *Geodynamics: Applications of Continuum Physics to Geological Problems*. Wiley, New York, 450 pp.

VAN HOUTEN, F. B. 1977. Triassic–Liassic deposits, Morocco and eastern North America: a comparison. *Bulletin of the American Association of Petroleum Geologists*, **61**, 79–99.

WILDI, W. 1983. La chaine tello-rifaine (Algerie, Maroc, Tunisie): structure, stratigraphie et evolution du Trias au Miocene. *Revue de Géographie Dynamique et de Géologie Physique*, **24**, 201–297.

WINTERER, E. J. & BOSELLINI, A. 1981. Subsidence

and sedimentation on a Jurassic passive continental margin (Southern Alps, Italy). *Bulletin of the American Association of Petroleum Geologists*, **65**, 394–421.
ZEIGLER, P. A. 1985. Late Caledonian framework of western and central Europe. *In*: GEE, D. G. & STURT, B. A. (eds) *The Caledonian Orogen – Scandinavia and Related Areas*. Wiley, Chichester, 3–17.

C. K. MORLEY, Amoco Production Co., P.O. Box 3092, Houston, Texas 77253, USA.

Devonian basins of Northern Scotland: extension and inversion related to Late Caledonian – Variscan tectonics

M. P. Coward, M. A. Enfield & M. W. Fischer

SUMMARY: The Devonian sedimentary basins of Northern Scotland illustrate the different processes of inversion along reactivated frontal and lateral thrust ramps, and the importance of strike–slip fault movements to late and post-Caledonian tectonics. The basins of the West Orkneys and West Shetland, west of the Great Glen Fault developed by collapse of thickened Caledonian crust during the Devonian. The basins extended in a NW–SE direction on dominantly SE-dipping faults, which curve and become parallel to the earlier Caledonian fabric at depth. The Caledonian structures developed a crustal anisotropy which was subsequently reworked during the later Palaeozoic and Mesozoic by extensional and compressional movements. Hence in the West Orkney and Minches basins, Caledonian structures influenced the kinematics of Early to Middle Old Red Sandstone extension, later (Devonian and Permo–Carboniferous) compression, Mesozoic extension and later Mesozoic–Tertiary compression. During basin extension, minor inversion of early listric normal faults occurred within the basin sediments, driven by gravitational gliding down tilted fault blocks. SE of the Great Glen Fault in eastern Shetland and possibly also the Inner Moray Firth, basins were affected by NE–SW extension during Middle to Upper Devonian times. This is possibly a result of sinistral strike–slip displacements on the Great Glen–Walls Boundary Fault system. This strike–slip episode may have been responsible for inversion of the earlier and more regionally widespread Lower and Middle Devonian basins formed by NW–SE extension. Localized zones of folding in Caithness and the Orkney Islands and wide zones of more intense folding in western Shetland are thought to be related to this deformation. Late Palaeozoic Variscan dextral strike–slip displacements juxtaposed basins along the Great Glen–Walls Boundary fault system which have distinct stratigraphies and structural evolutions and formed separately to the east and west of these faults. This strike–slip deformation inverted basins both to the east and west of the Great Glen Fault. In the East Shetland basins it produced local zones of overturned beds above oblique and frontal ramps. Further subsequent dextral movements in the Mesozoic produced pull-apart basins in the Moray Firth and East Orkneys.

The Devonian basins discussed in this paper occur along the coastal margins of Northern Scotland, the Orkney Isles and Shetland and offshore, west of the Orkney Islands (Fig. 1). Devonian basins of similar age are observed in western central Norway (Hossack 1984; Norton 1986; Seranne & Seguret 1987) and East Greenland (Haller 1971; Norton *et al.* in press) and probably occur beneath much of the Northern North Sea and Viking Graben (Beach *et al.* 1987).

The onland portions of the Northern Scottish Devonian basins are generally considered to belong to a large dominantly lacustrine basin which is commonly referred to as the Orcadian basin. The exposed Devonian sediments are largely correlated by common fish species found in laterally extensive marker beds. They are confined to the Middle Old Red Sandstone. The most important is the Achanarras–Sandwick horizon (see Traquair 1894a; 1894b; Rayner 1963; Allen & Marshall 1981; Mykura 1983). The aerial extent of the Old Red Sandstone basin has previously been confined to a small region including Caithness, the Orkney Isles, West Shetland and the Inner Moray Firth (for example, Watson 1985; Mykura 1983). However, from interpretation of the available seismic data we consider that the Orcadian basin is part of a much larger linked basin system, in which some sub-basins may have been locally isolated. We would extend the boundaries to the Devonian basin system considerably further to the west into the Outer Isles and Minches basins (Coward & Enfield 1987; Enfield & Coward 1987).

In this paper we aim to detail and synthesize the history of Devonian basin formation onshore and offshore of Northern Scotland and explain episodes of basin inversion produced both by Devonian and later tectonics. Some of the inversion structures within the basin are obviously fault controlled but some are of more enigmatic origin. The control of Caledonian pre-existing

From Cooper, M. A. & Williams, G. D. (eds), 1989, *Inversion Tectonics*
Geological Society Special Publications No. 44, 275–308.

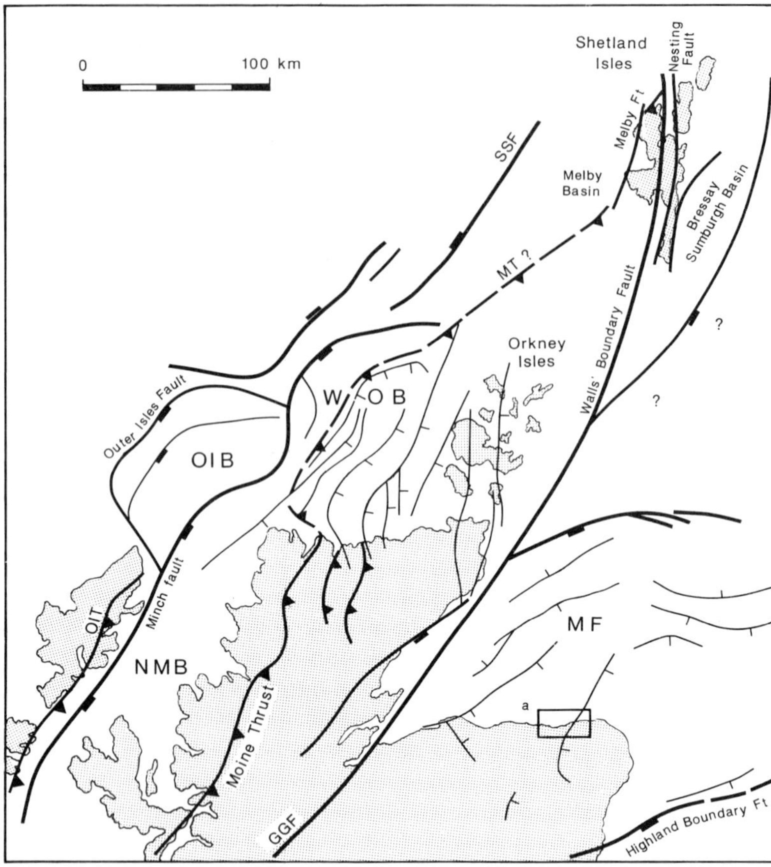

FIG. 1. Fault systems of the northern United Kingdom area. OIT, Outer Isles Thrust; GGF, Great Glen Fault; MT, Moine Thrust (possible offshore continuation); SSF, Shetland Spine Fault; WOB, West Orkney Basin; OIB, Outer Isles Basin; NMB, North Minch Basin. Major basin bounding faults are shown by heavy lines; other normal faults by thin lines. Ticks are on the hanging walls. Box A is shown in Fig. 14.

crustal structure and anisotropy on the geometry, distribution and evolution of the basins is discussed and an attempt made to quantify their effects on basin subsidence.

The data used in this project came from extensive fieldwork in the onland portions of the basin in Northern Sutherland, Caithness, the Moray Firth, the Orkneys and Shetland Isles. These data have been used in conjunction with commercial speculative seismic reflection surveys kindly provided by Western Geophysical (1982 WOB survey) GSI (reprocessed 1973 WO survey) and Jebco Seismic (1984 JS-MINCH survey), together with the published BIRPS lines from offshore N Scotland: MOIST, WINCH, DRUM, SHET (see Brewer & Smythe 1984; 1986; Cheadle *et al.* 1987; McGeary 1987; in press).

Regional geology

Pre-basin history

During the Caledonian orogeny Middle Proterozoic Moine meta-sediments were carried NW over a foreland of Archaean to Lower Proterozoic Lewisian gneiss and its upper Proterozoic (Torridonian) and Cambrian cover. The Moine Thrust forms the important tectonic boundary between rocks which were intensely deformed during the Caledonian and those of the essentially undeformed foreland. However, several Caledonian thrusts occur beneath the Moine Thrust, locally stacking foreland rocks in the Moine Thrust Zone. A major Caledonian thrust, the Outer Isles Thrust, occurs 80 km to the W of the Moine Thrust Zone. Minimum estimates of 50 km displacement for the WNW verging thrusts beneath the Moine Thrust have been made from constraints imposed by section restoration of the imbricated Cambro–Ordovician sediments (Butler & Coward 1984; Coward 1985). In Western Sutherland, the imbricates deform only the Cambrian cover sequences, indicating that Lewisian basement must continue in some form at least 50 km to the ESE from the present surface trace of the Moine Thrust Zone (Butler & Coward 1984).

The Moine meta-sediments, found in the hanging wall to the Moine Thrust, carry a strong Caledonian tectonic fabric developed parallel to bedding. The fabric generally dips eastward at approximately 15° near the Moine Thrust Zone but dips more steeply in the E (Barr *et al.* 1986). They have been imbricated and interfolded with a Lewisian-type basement gneiss to produce a pronounced lithological layering. To the E, the Great Glen Fault forms a major tectonic boundary across which it is difficult to correlate Moine stratigraphy and deformation (Harris *et al.* 1981).

The Great Glen Fault is not exposed along its length. However, there are several other strike–slip faults exposed in northern Scotland, e.g. the Strathconnan fault, on which sinistral strike–slip displacements can be demonstrated. On Shetland the Walls Boundary Fault has a similar character to the Great Glen Fault in that it separates Lewisian and older Moines in the W from younger Moines and Dalradian rocks in the E (Flinn *et al.* 1979; Flinn 1985).

Old Red Sandstone basins

Wherever observed, the Old Red Sandstone overlies Moine type basement with a landscape unconformity (see Peach & Horne 1914; Wilson *et al.* 1935; Blackbourn 1981). Traditionally the Old Red Sandstone is subdivided into three groups: the Lower, Middle and Upper Old Red Sandstone (Murchison 1859a; 1859b; Mykura 1983). This division is based essentially on structural grounds; each group is commonly separated by unconformities. A biostratigraphical subdivision of the groups is based on fish faunas (Traquair 1894a & b) and recent palynology (Allen & Marshall 1981; Marshall & Allen 1982; Marshall *et al.* 1985; Marshall in press, a & b). Old Red Sandstone sedimen tation is thought to have commenced by Lower Devonian (Emsian approximately 394 Ma) and to have continued into the Upper Devonian (Frasnian–Fammenian) and possibly into the Lower Carboniferous (Astin 1985).

The Lower Old Red Sandstone is predominantly composed of fanglomerate deposits, and has a characteristic association of coarse, compositionally immature, but commonly well rounded conglomerates with thin sandstone interdigitations (Mykura 1983). The depositional distribution is generally restricted and it was probably deposited in small basins.

The Middle Old Red Sandstone has a wider distribution. Its base is strongly diachronous, overstepping the Lower Old Red Sandstone to the W onto Moinian basement both on the Scottish Mainland and the northern isles of the Orkneys and Shetlands (Mykura 1983). Where the base of the Middle Old Red Sandstone sequence overlies the Lower Old Red Sandstone, an angular unconformity is commonly observed. This has a relief suggesting that sub-aerial erosion occurred between deposition of the Old Red Sandstone sequences. The lower part of the Middle Old Red Sandstone sequence is made up of thick lacustrine carbonate mudstone and sandstone. The section is up to 2 km in Orkney but approximately 4 km thick in Caithness and 0.8 km thick in Melby, W Shetland (Mykura, 1983). Towards the top of the Middle Old Red Sandstone the sequence shows a fluvio–lacustrine regressive episode And fluvial depositional facies become dominant.

On the Orkney Isles the Middle and Upper Old Red Sandstone sequences are separated by an erosional unconformity. This has been observed on Hoy (Wilson *et al.* 1935) and it has also been suggested to be present, but not been conclusively demonstrated, on the Scottish Mainland. Mykura (1983) has suggested that the unconformity may represent uplift and erosion of up to 1 km of the Middle Old Red Sandstone sequence and is associated with calc-alkaline lavas (Wilson *et al.* 1935; Halliday *et al.* 1977; Rock 1983).

Post-Caledonian extension: the West Orkneys, Outer Hebrides and North Minches Basins

Reflection seismic data from offshore in the West Orkney Basin (WOB), show a thick sequence of sediments dominated by gently to moderately NW-dipping reflectors. They are formed in a series of half-grabens bounded by SE-dipping faults (see for example, Fig. 2). In the deepest part of the basin, in the NW West Orkney Basin, depths to top of metamorphic basement reach 3.5 two-way-travel-time (TWT) (approximately 9 km, see Fig. 3). This and

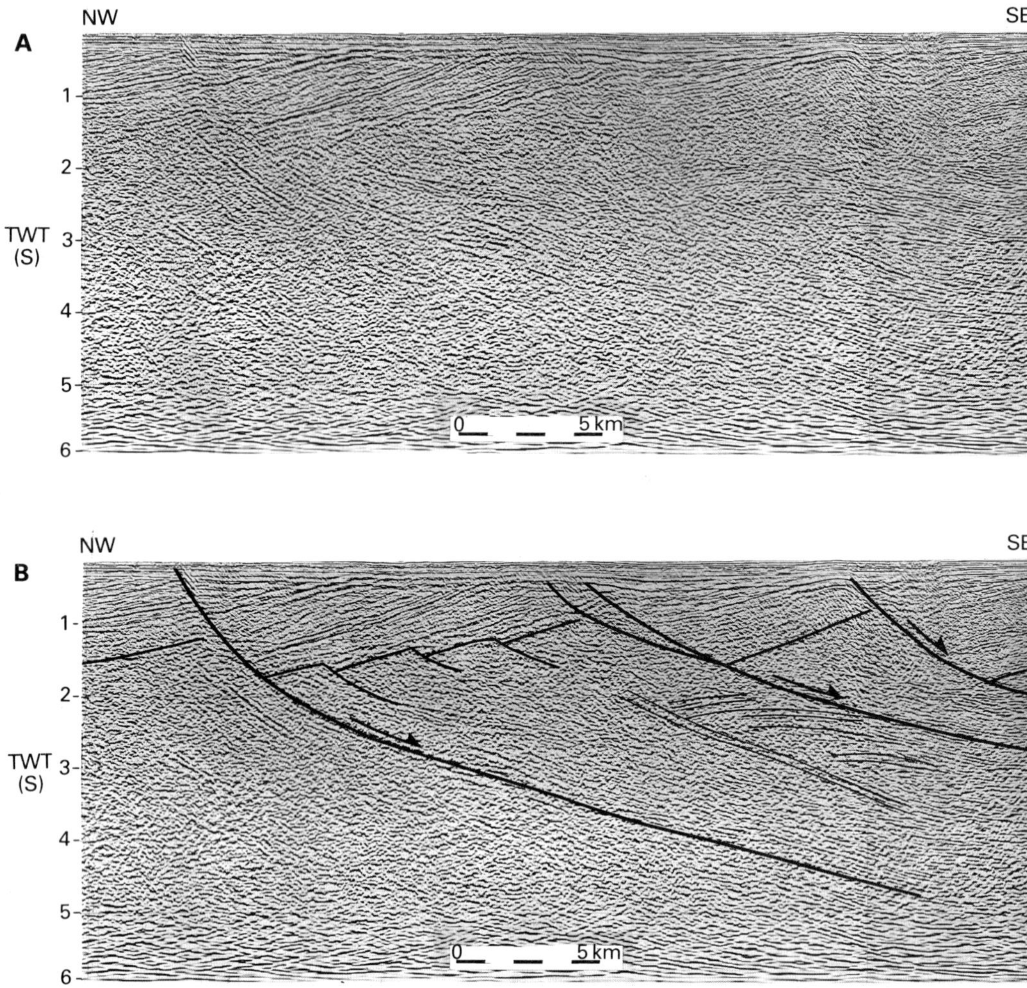

FIG. 2. A, Seismic profile from Western Geophysical, showing highly reflective basement structure continuing below the NE coastline of Caithness, Mainland Scotland, SE West Orkney Basin. See Fig. 4 for location of profile A. B, Structural interpretation of Fig. 2A.

other depth estimates are based on stacking velocities. However, over most of the basin, a more typical average maximum depth is 2.5 s TWT (roughly 3–4 km). Metamorphic basement occurs on the sea floor in several areas. Where exposed, as on the islands of Sule Sgear and Stack Skerry in the NW, the rock type is Lewisian gneiss (Kirton & Hitchen 1987).

Basement structure and fault geometry

Beneath the NW-dipping sediments of the West Orkney Basin reflectors which dip SE to ESE (Fig. 2) are considered to lie within metamorphic basement (Brewer & Smythe 1984; Enfield & Coward 1987). They are observed on all reflection seismic data, in the central and southern WOB including both shallow and deep profiles shot in this offshore N Scottish area. Within the West Orkneys, Outer Isles and North Minches basins, the faults, which bound the sediment-filled half-grabens, trend approximately parallel to the structure within the basement.

Two important sets of faults are present (Fig. 4). Faults with listric geometries, both in section and map plan, shallow toward a laterally extensive, but vertically thin, package of reflectors which dips between 10–15° to the ESE. This zone forms a decoupling zone for this fault

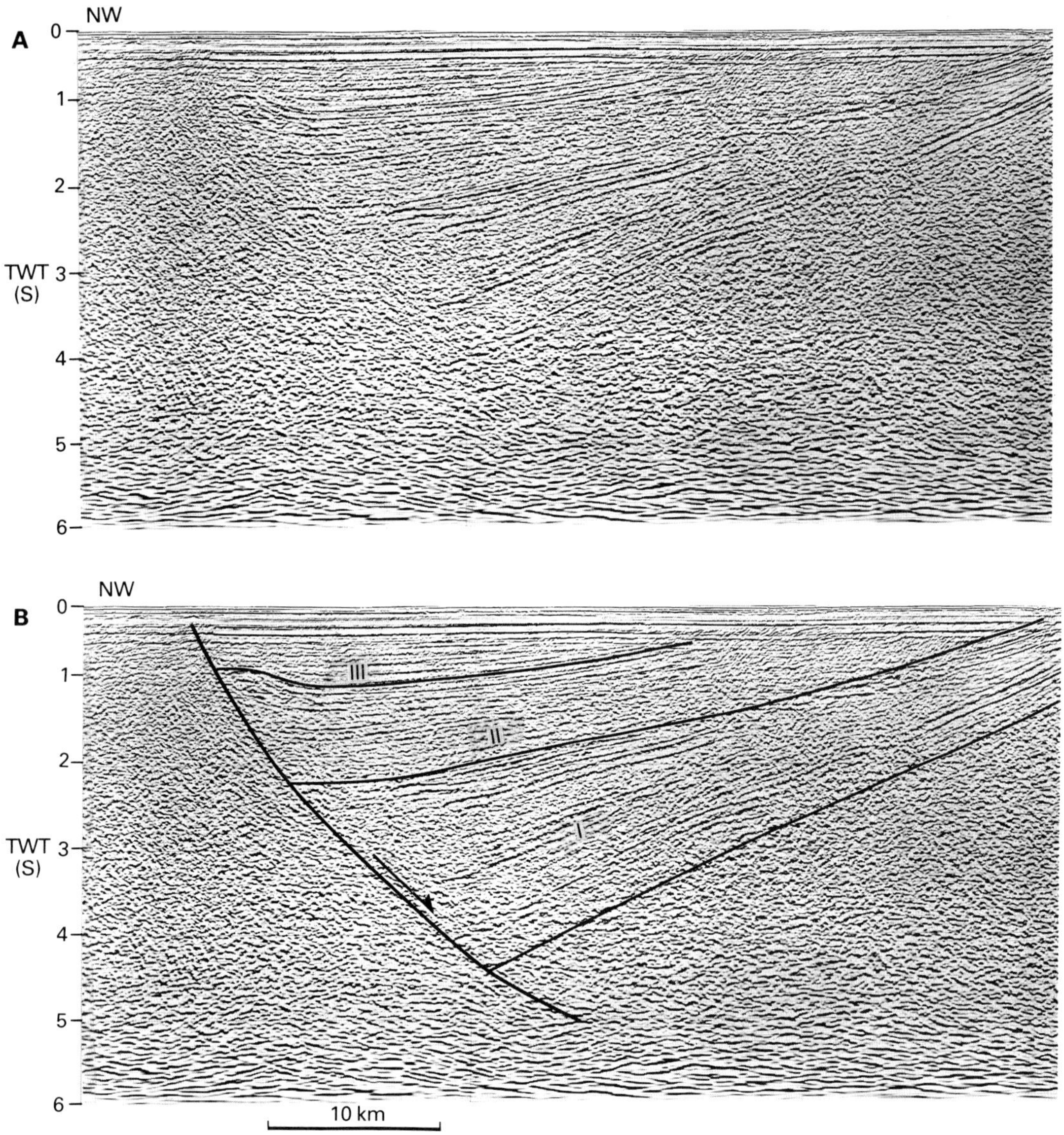

FIG. 3. A, Seismic profile from Western Geophysical showing NW bounding fault of the West Orkney Basin. Note the slightly curved fault plane geometry. B, Structural interpretation of Fig. 3A. See Fig. 4 for location of profile B.

set at a shallow depth of only a few kilometres (see Fig. 20). Other faults in the basin, including those best imaged on the deep seismic profiles, are straighter (Fig. 2) and appear to decouple at a deeper level into a zone of discontinuous reflectors within the lower crust (Cheadle *et al.* 1987).

The shallow decoupling zone forms the W and lower limit to the extensive SE-dipping set of basement reflectors, which are considered to represent structure within Moine type Caledonian basement (Brewer & Smythe 1984; Enfield & Coward 1987). The dominantly transparent upper-crust to the W is considered to represent older Lewisian gneisses of the Caledonian foreland (Brewer & Smythe 1984;

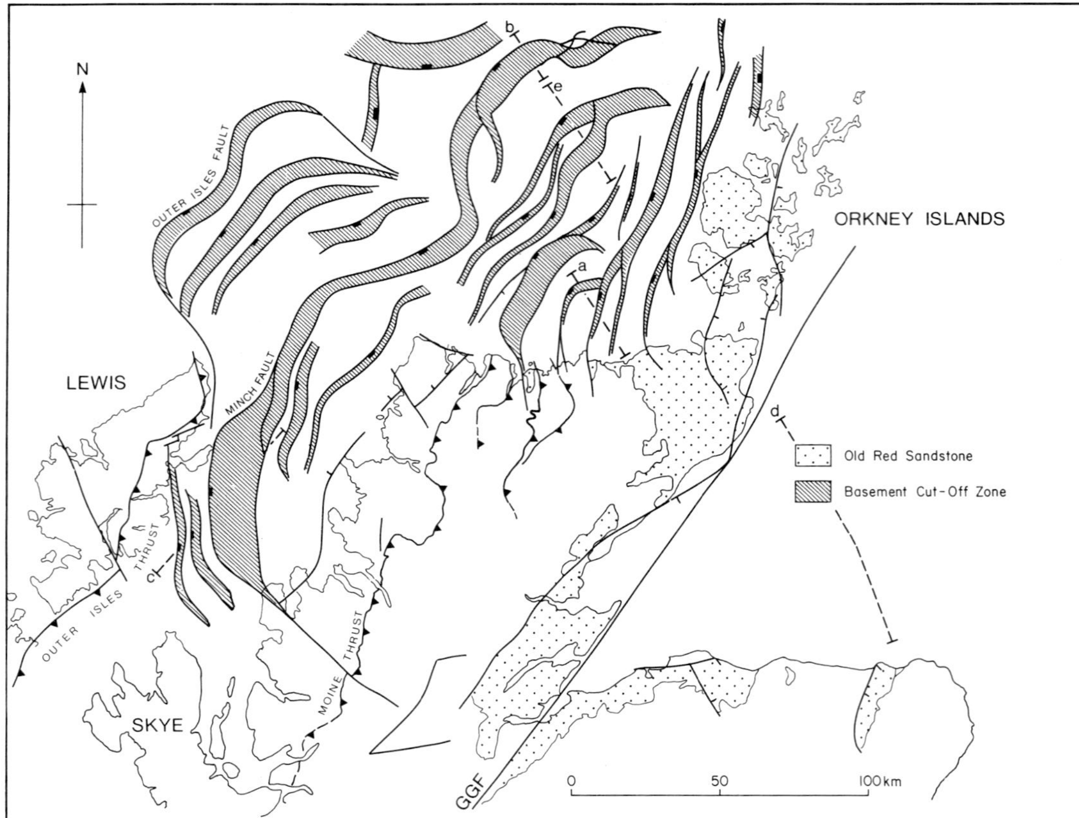

FIG. 4. Cut-off map of the fault system forming the West Orkneys, Outer Isles and North Minches Basins. Ticks are on the hanging walls at the upper cut-off lines or at the surface traces of the faults. Where the continuations of the faults to depth are not known, or where details of the seismic data are confidential, only the outcrop traces of the faults are given. The commercial speculative data used for this interpretation were provided by Western Geophysical, GSI and Jebco. Section line A refers to Fig. 2; B; to Fig. 3; C to Fig. 5; D to Fig. 15 and E to Fig. 20.

Enfield & Coward 1987). This structural boundary between these basement types is interpreted to be the offshore continuation of the Moine Thrust Zone (Enfield & Coward 1987). It is not known, however, whether this zone has the same character as that on the Scottish mainland, where deformed Cambro–Ordovician sediments are intensely imbricated with Lewisian basement, (cf. Brewer & Smythe 1984), or if it is simply the boundary of a moderately-dipping ductile shear zone, representing a continuation of the Moine Thrust at depth (cf. Coward 1986a), similar to the ductile shear which forms the boundary of Caledonian rocks in northern Shetland (Flinn *et al.* 1979).

The listric faults that flatten into the shallow decoupling zone form small half-grabens on their hanging walls which are limited by oblique to lateral ramps along strike (Fig. 4). These half-grabens demonstrate syntectonic sedimentation in their lower sequences, in that the sediment package thickens towards the fault (Fig. 1). The planar faults form parallel to the moderately SE-dipping reflectors present within the eastern basement (Fig. 2) but are also observed to breach and offset the detachment level to the shallower listric fault set in the centre of the basin. The bounding fault to the planar set is interpreted as the northern continuation of the Minch Fault (Figs 1 & 4; see also Coward & Enfield 1987). It is slightly listrict in section (Fig. 3) and strongly arcuate in map plan (Fig. 4). The fault penetrates basement to below 5 s TWT (approximately 15 km), and possibly flattens into a high amplitude gently SE-dipping reflector observed between 4.5 and 5 s TWT on several seismic profiles in the northern West Orkney Basin.

To the SW, the Minch Fault may either join or cut the fault set on which the Outer Isles Basin has formed (Fig. 4). The bounding fault to the Outer Isles Basin, the Outer Isles Fault (Fig. 1), is clearly visible on commercial and deep seismic data as a zone of generally SE-dipping reflectors on to which half-graben bounding normal faults detach. It is approximately planar in NW–SE section and can be traced to lower crustal levels. On the MOIST profile it is considered to merge with strong reflectors within the lower crust, perhaps to join the seismic Moho (Brewer & Smythe 1984; Cheadle *et al.* 1987). However, further to the N, below the West Orkney Basin, the continuation of the Outer Isles Fault shallows at a higher crustal level, to detach within the middle to lower crust.

Age of the offshore basins and inversion episodes

The interpretation of seismic sections which end close to the Orkney Isles (for example, Fig. 2), and maps contoured for depths to top of crystalline basement (see Coward & Enfield 1987) indicate that the sediments within the E part of the West Orkney Basin are structurally continuous with the Middle Old Red Sandstone outcropping onshore. Individual half-grabens observed on the seismic sections continue onshore into sedimentary sequences on Northwest Orkney mainland (Fig. 2 and Coward & Enfield 1987) and the N Scottish coast (Enfield & Coward 1987). These sequences are dated as Middle Old Red Sandstone (Peach & Horne 1914; Wilson *et al.* 1935; Blackbourn 1981). Thus the half-grabens which are developed on the hanging walls of faults that detach along reactivated Caledonian thrusts were initiated at least as early as the Middle Devonian, the age of the Middle Old Red Sandstone.

Although the lower sediments of the West Orkney Basin are confidently determined to be Devonian, there is more uncertainty regarding the ages of the upper sedimentary sequences within the West Orkney Basin and of the sequences within the Northern Minches and Outer Isles Basins. Shallow boreholes from within the West Orkney Basin intersected red beds at the sea bed. These beds are determined from their colour to be New Red Sandstone (Permo–Triassic) (Dunham 1973). The West Orkney Basin may therefore have a shallow capping of Permo–Triassic sediments (Kirton & Hitchen 1987).

To the W of the Solan Bank High a W-dipping and presumably Mesozoic fault system (Kirton & Hitchen 1987) is present. It extends to the NW or WNW and probably relates to Atlantic rifting. The sediments in this area and in the W part of the West Orkney Basin (sequence II, Fig. 3) comprise a thick sequence of parallel reflectors, down to approximately 3.5 s TWT. These are interpreted to be Permo–Triassic in age and were deposited prior to Jurassic faulting along the Atlantic margin.

Further SW, the North Minches Basin contains thick sequences of Mesozoic sediments (Binns *et al.* 1974; Chesher *et al.* 1983), with a thin capping of Tertiary (Roberts *this volume*). The age of the lowest sediments in the North Minches and the Outer Isles basins is not constrained by well data. They are considered by Brewer & Smythe (1984) and Kilenyi & Standley (1985) to be Torridonian.

However, on seismic sections along the E margins of the North Minches Basin, where

Torridonian rocks occur onshore, a distinction can be made between Torridonian sediments and the sedimentary fill (sequence I, Fig. 5) of the Minches Basin. The Torridonian is not imaged as a sedimentary succession but as acoustic basement. This is because the area of Torridonian contains no coherent reflections characteristic of a sedimentary sequence, and is separated from the sediments of the North Minches Basin by a W-dipping fault (Fig. 4). Therefore, we consider that the lower sediments observable on reflection seismic data within the North Minches Basin must be younger than Torridonian and probably Devonian in age.

The lower seismic sequence in the Minches Basin is separated from sequence II (Fig. 5), presumed Mesozoic (? Permo–Triassic and Jurassic) by a strong angular unconformity. If the lower sequence (I) is interpreted as Devonian, then this large unconformity, which represents several kilometres of uplift, can be interpreted as Permo–Carboniferous in age and related to Variscan fault movements and basin inversion. This unconformity is tentatively correlated with that observed on the West Orkney Basin seismic data in the half-graben developed on the hanging wall of the northern continuation of the Minch Fault (Fig. 3). Basin inversion and large amounts of uplift are known from further E from inversion structures seen within the Orcadian Basin. However, if the lower seismic sequence (I, Fig. 5) in the North Minches Basin is interpreted differently, say as Permo–Triassic, then the unconformity is more difficult to explain.

Extensional direction of the West Orkney Basin

The extension direction within the Devonian basin system has mainly been determined from the seismic data in the offshore area. Onshore, in the Orkneys and Caithness, extension direction indicators within fault zones have largely been obliterated by subsequent reactivation during episodes of tectonic inversion. The criteria used to determine the extension direction have included geometries of the fault systems and their hanging wall structures, together with a kinematic analysis of cut-off maps throughout the basin, (see below and Enfield & Coward 1987).

The mapped fault systems contain two main fault orientations. Short, steep to sub-vertical NW–SE-trending fault segments occur within a generally SE-dipping fault system, to give the faults bucket-shaped or arcuate map traces. These steep NW–SE-trending segments are interpreted as transfer or lateral structures and require an extension direction to the SE. The axes of rollover folds developed at both the top of basement and within the sedimentary sequence lie sub-parallel to the strike of the NE–SW faults, suggesting that these are frontal faults within the system. This provides further support for the interpretation of an SE extension direction.

A kinematic analysis was carried out on the fault map of Fig. 4. The cumulative displacements for all the faults within the basin, were examined by measuring the extension shown by the cut-offs at the top of basement level. For an assumed SE extension direction, the resulting cumulative displacement curve is smooth and approximately symmetrical (Fig. 6A) and does not require high shear strains in the basement related to changes in cumulative displacement. However, when the cumulative displacements are calculated in orientations other than parallel to the SE-trending transfer faults (Fig. 6B & C), the resultant curves are markedly asymmetric. If real they would require localized high shear strains in the basement to accommodate the differential extensions on the fault system (see also Enfield & Coward 1987). We prefer the interpretation that minimizes the predicted shear strain, i.e. one involving regional NW–SE extension (Fig. 6A).

Variations in extension magnitude

The total displacement measurable on faults that penetrate basement within the Devonian basins, and therefore extend and thin the crust beneath the basins, reaches a maximum of approximately 32 km (50% extension) in the centre of the West Orkney Basin, measured as the component of horizontal extension (heave) (Fig. 6). The measurable displacements on faults that form the Outer Isles Basin and faults to the SE up to the N Scottish coast, are of similar magnitude to that of the West Orkney Basin. They reach a maximum of approximately 30 km (45% extension). Displacements within the North Minches Basin are smaller and reach a maximum of approximately 25 km (60% extension) at the southern end, though in this basin 75–80% of the extension is accommodated on a single fault (the Minch Fault). These figures include Palaeozoic and Mesozoic extension in the North Minches Basin.

The faults and displacements observed on the seismic data within the West Orkney Basin die out to the NE within the area of seismic coverage (Fig. 4). However, because Devonian sediments are also known on Shetland and the West Shetland Platform (Fig. 1), we presume that

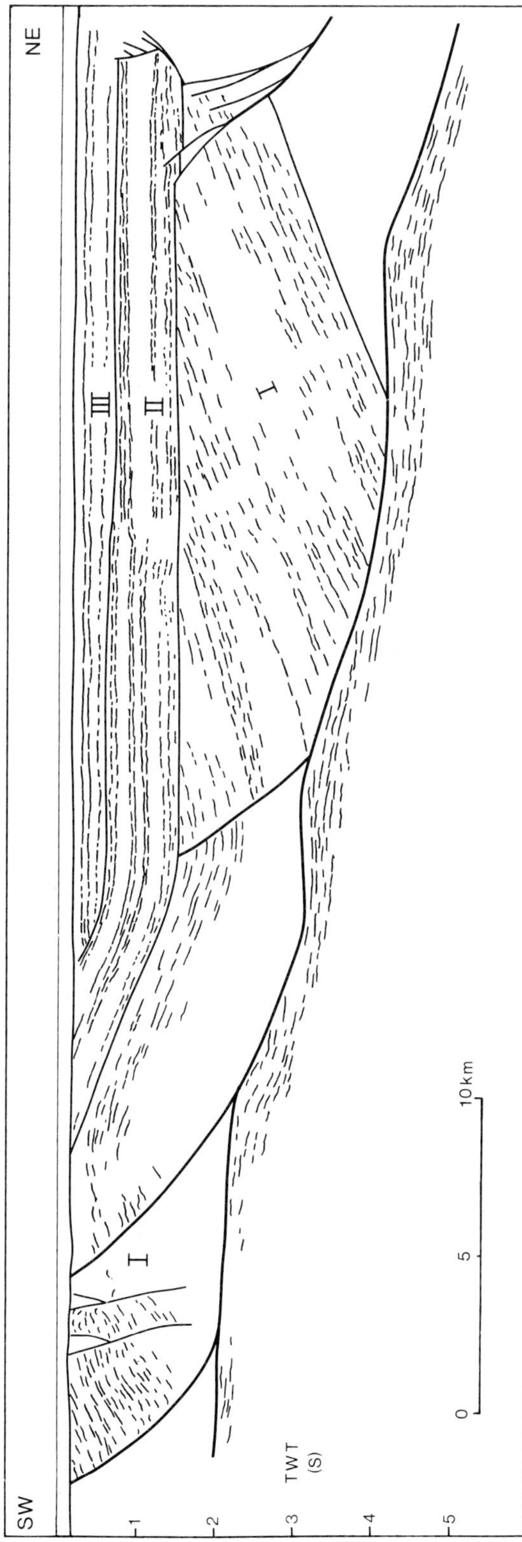

Fig. 5. Line drawing of 6 s (TWT) seismic profile from Jebco showing the North Minches Basin formed by extension on the Outer Isles and Minch Faults. The section is orientated perpendicular to the extension direction. The lower seismic sequnce (I) may be Devonian; sequence II is considered to be Mesozoic (Permo–Triassic and Jurassic?) and sequence III Cretaceous and/or Tertiary. See text for discussion. The unconformities (solid lines) are related to basin inversion episodes. See Fig. 4 for location of profile C.

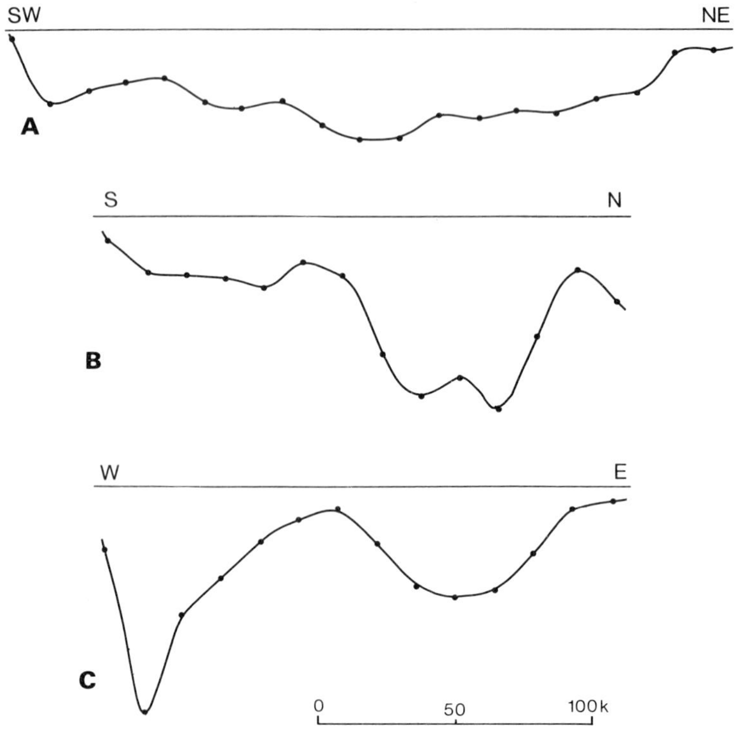

FIG. 6. Cumulative displacement curves from the West Orkney Basin. Curve A records the cumulative displacement measured assuming NW–SE extension. Curves B and C assume N–S and W–E extension, respectively.

the extension transfers to another fault system to the N. Displacements decrease rapidly to the S at the N Scottish coast. Displacements on faults within the Outer Isles Basin die out rapidly at NW–SE oriented lateral ramps. The extension presumably links with the fault systems of the West Orkneys and North Minches Basins.

Faults controlling small Devonian half-grabens observed along the N Scottish coast (Fig. 7) represent the southerly terminations of faults that in the central part of the West Orkney Basin (Fig. 4) are the major half-graben bounding faults. In this region their occurrence appears to be coincident with large early Caledonian thrust faults along the N Scottish coast (Fig. 7). Devonian basins are present to the SE where they are fault controlled along the Inner Moray Firth (see Norton *et al.* in press) though here we have insufficient data to plot cut-off maps and estimate displacement variations.

The Old Red Sandstone basins of the Scottish mainland and the Orkney Islands

On the Scottish mainland, the Old Red Sandstone deposits occur in two different facies associations. Around the basin margin in Sutherland and western Caithness small half-grabens with coarse clastic fills are developed on E-dipping faults. To the E, in Caithness, the Middle Old Red Sandstone is dominated by thick sequences of quiescent lacustrine sediments, with little evidence of fault control.

Along the N Scottish coast half-grabens at Coldbackie Bay, Kirtomy, Poulouriscaig and Strathy (Fig. 7) are developed on E to NE-dipping extensional faults. These faults penetrate into Moine-type metamorphic basement and can be correlated with major half-graben bounding faults offshore (Fig. 4). Estimation of the extension direction on these faults is difficult because no kinematic indicators are preserved on the master faults. However, at Kirtomy and Poulouriscaig the faults do appear

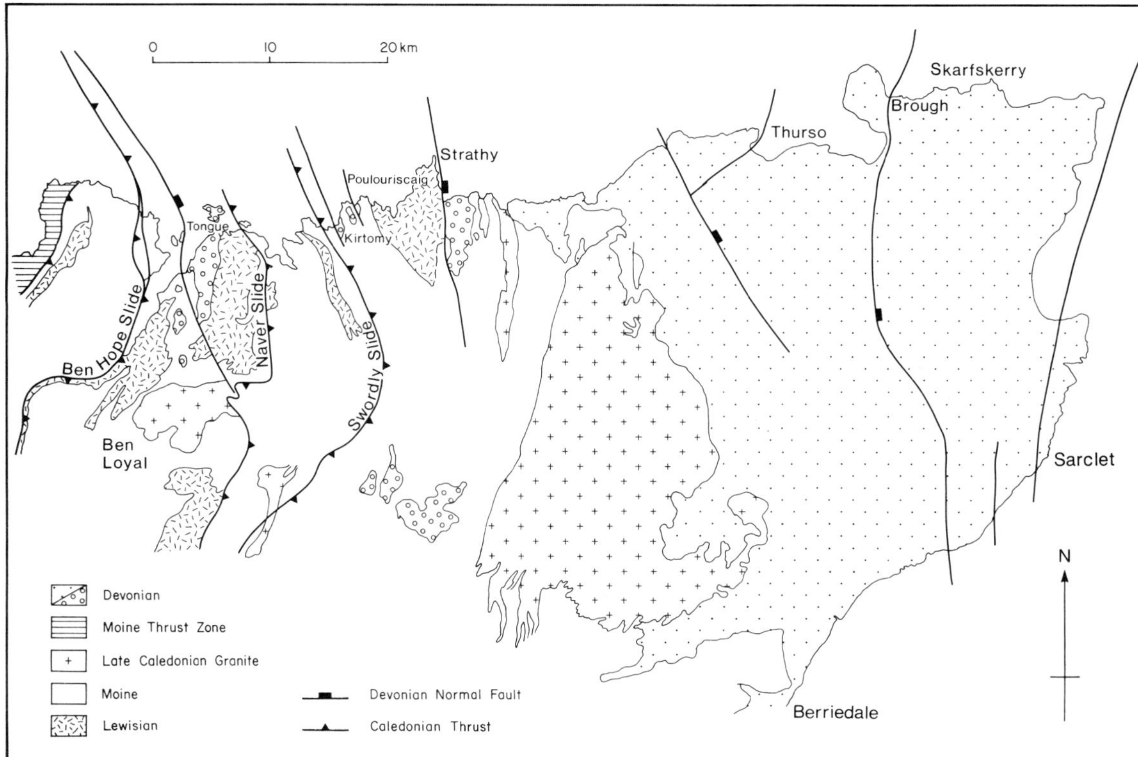

FIG. 7. Map of the north Scottish coast showing principal rock types, Caledonian and Devonian faults and the onshore distribution of Old Red Sandstone sediments.

to have components of oblique extension, as observed below.

The westernmost, NNW–SSE trending fault bounding the Kirtomy half-graben has small, steep WSW–ENE orientated faults developed on its hanging wall (Fig. 8 inset). This orientation would suggest ENE extension on the master fault if the small faults were interpreted as tear faults. These WSW–ENE orientated faults bound a small, conglomerate filled graben at the base of the sedimentary succession (Fig. 8 inset), which was active during Middle Old Red Sandstone sedimentation. Kinematic indicators on these WSW–ENE faults suggest frontal extension rather than strike–slip displacements. Subsidiary faults parallel to the main NNW–SSE half-graben-bounding fault have reverse offsets of sandstone beds, flower-like splays of minor faults and also disrupted conglomerate clasts. These features indicate dextral strike–slip displacements on this fault set. Thus the main NNW-trending half-graben bounding fault (Fig. 8) is the tear fault and the ENE-trending faults are minor frontal faults.

At Poulouriscaig (Fig. 7) the basin is bounded to the W by a sub-vertical, NNW–SSE trending fault. Small steep faults with reverse displacements are observed in the basal sediments. This indicates some strike–slip movement on the basin bounding fault, as at Kirtomy. Dips within the proximal sediments decrease upwards through the succession, suggesting that sedimentation occurred during faulting.

Thus, lateral facies changes and structures at Coldbackie Bay, Kirtomy and Poulouriscaig, indicate active extension on the half-graben bounding faults during Middle Old Red Sandstone sedimentation. Oblique displacements on NNW-trending faults bounding these half-graben are consistent with a dominantly NW–SE regional extension for the West Orkney Basin. All the small half-graben basin-fill sequences on the northern Scottish coast consist of coarse breccio-conglomerates in an arkosic matrix and include occasional welded clasts of sandstones and conglomerates. The breccio-conglomerate clasts are characteristically well rounded. This suggests a high degree of mechanical working. However, the matrix commonly contains a high proportion of large biotite crystals and clasts of characteristic locally derived rocks which suggests only short transport. The clasts are probably recycled from an earlier breccio-conglomerate basin-fill.

The sequence of Middle Old Red Sandstone lacustrine flagstones attains maximum thicknesses of up to 4 km in Caithness (Mykura 1976; 1983). The sequence thickness decreases in the

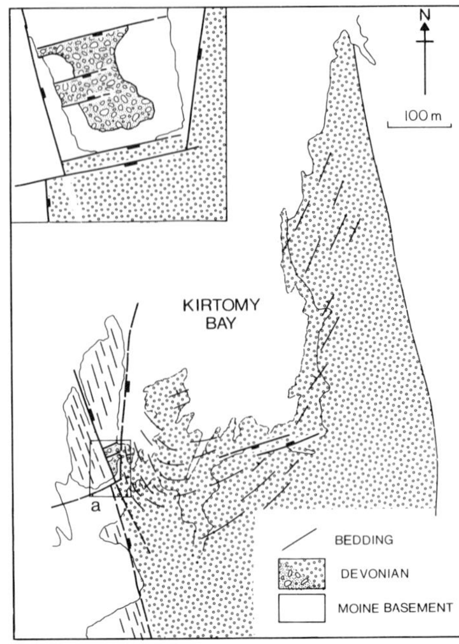

FIG. 8. Map of the Old Red Sandstone half-graben at Kirtomy. Inset shows the detail of region outlined by box A.

Orkney Isles, to the N, to approximately 2 km. The section also thins rapidly to the S, to pinch-out at the southern limit of Old Red Sandstone outcrop. Within the Middle Old Red Sandstone, facies changes along the basin margins are associated with synsedimentary faulting, for example along the East Scape Fault (Fig. 9) (Astin 1985) and the Thurso Bay Fault (Fig. 7). However, the majority of the Middle Old Red Sandstone basin sediments are parallel bedded, at least at outcrop scale. The general basin architecture and sedimentology are more typical of passive regional subsidence rather than fault-controlled growth sequences. However, the sediment thickness variations correlate well with variations in magnitudes of extension in the West Orkney Basin, where extension reaches a maximum in the basin centre and decreases laterally. The decrease is gradual to the N but more abrupt to the S (Figs 4 & 6).

The Old Red Sandstone basins of Shetland

On the Walls Peninsula of West Shetland, Middle Old Red Sandstone rocks occur in different structural environments, separated by the Melby Fault (Fig. 1). To the W, the sediments of the Melby Formation form part of the

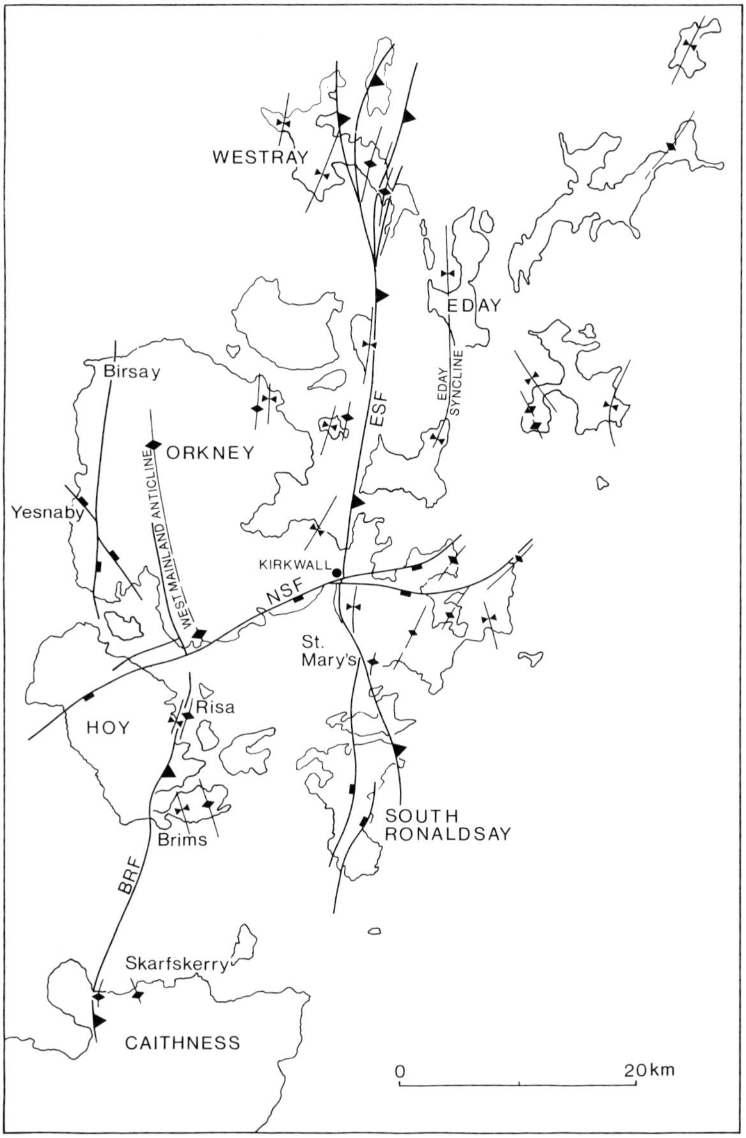

FIG. 9. Structural map of the Orkney Isles and NE Caithness, showing principal folds and faults. BRF, Brough–Brims Risa Fault; ESF, East Scapa Fault; NSF, North Scapa Fault.

West Shetland Platform and are cut by NW-dipping faults of unknown age, but which are possibly Mesozoic. These are related to the development of the West Shetland Basin.

Middle Old Red Sandstone age sediments of marginal lacustrine and alluvial facies are present to the E of the Melby Fault. They are locally bounded to the N by listric faults which curve to become parallel with a SE-dipping fabric in the underlying Caledonian basement (Figs 10 & 11). From our observations, the sediments on the hanging wall show a well developed rollover fold and locally dip moderately to steeply to the NW. Close by, Old Red Sandstone sediments are preserved in the footwalls to these faults. They rest unconformably on the Caledonian basement but have a similar SE dip. Thus the Devonian rocks of the Walls

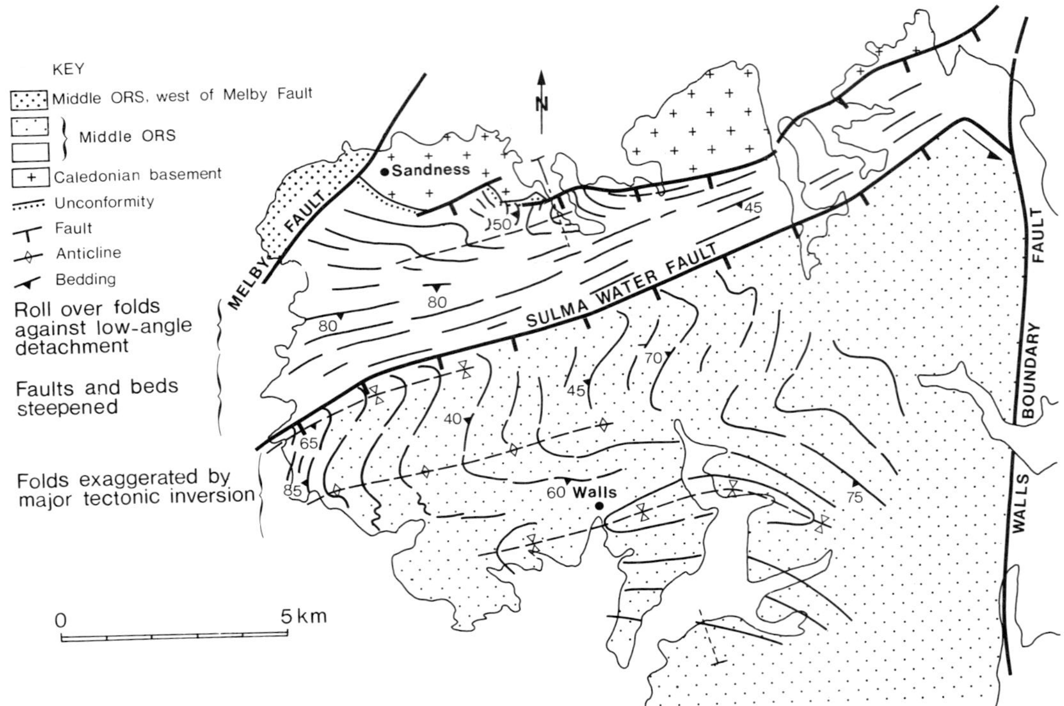

FIG. 10. Map of the Walls Basin, W Shetland (modified from Mykura 1976), situated between the Walls Boundary and Melby Faults. The bounding fault which locally joins or offsets the base Devonian unconformity E of Sandness is shown. The section line of Fig. 11 is also shown.

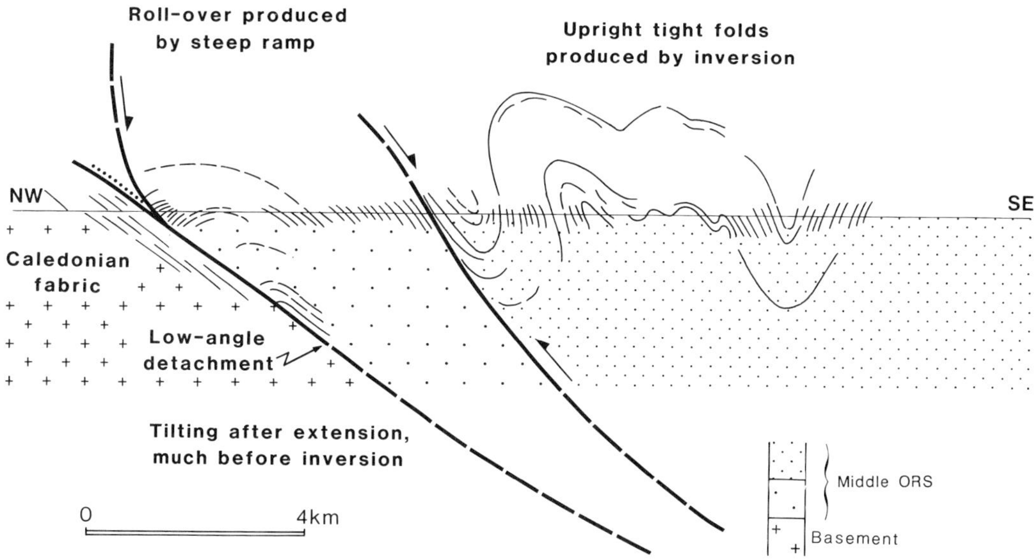

FIG. 11. Simplified NW–SE section through the Walls Basin, West Shetland, showing upright folds in the Middle Devonian sediments. See Fig. 10 for location. The tilting of the basal unconformity may be due to folding in the hanging wall of a deeper level Devonian fault. See text for discussion on the age of inversion.

Basin have been deformed by listric normal faults which flatten on to a low-angle detachment zone close to the Devonian unconformity, and have subsequently been tilted to dip at moderate angles to the SE (Fig. 11). The extension direction is taken to be NW–SE from the presence of small lateral ramps and tear faults which offset the normal faults and affect the sediments. Similar tear fault structures with the same orientation are also observed on Fair Isle (Mykura 1976). There the Devonian sediments are correlated with those of the Walls Basin (Astin 1982).

The normal faults along the NW boundary of the Walls Basin are not growth faults. In fact, they probably formed well after the main Old Red Sandstone sedimentation. Sedimentological studies by Astin (1982) suggest that stream flow directions were from the N. Thus the present boundary to the Walls Basin was not the original basin boundary. This may have occurred well to the NW, closer to the possible continuation of the Moine Thrust Zone (Fig. 1) in Shetland or to the SE where seismic data suggest that half grabens are bounded by W or NW dipping faults between Shetland and Fair Isle (cf. McGeary, in press). The present boundary to the Walls Peninsular Basin is formed by erosion of its tilted base (Fig. 11).

In East Shetland, Middle Old Red Sandstone rocks occur in the Bressay–Sumburgh Basin (Fig. 12), E of the Walls Boundary Fault. They form part of the East Shetland Platform. In places these sediments rest unconformably on basement rocks of Dalradian age. An irregular angular unconformity is inferred for the basin margin west of Sumburgh and is also clearly seen between Fladdabister and Quarff, about 10 km SW of Lerwick (Fig. 12). However, the basin is generally bounded by faults which drop down to the E. A steep fault occurs W of Lerwick and lies along a zone of tectonic melange, which probably initiated during Caledonian deformation (Flinn *et al.* 1979; Flinn 1985). This western boundary fault is sometimes vertical as near Sandwick. Slickensides suggest downthrow to the E, though a displacement direction may also be obtained from rollover geometry. Fig. 13 shows a detailed sketch map of part of the fault zone. Here normal faults trend N–S to NNW–SSE with a moderate ENE dip. The sediments on the hanging wall are tilted to the S and strike E–W to NE–SW. Although there may have been some reorientation of the beds during subsequent inversion, their strike and dip suggests that extension on this fault system occurred in a NE–SW direction rather than to the SE as in the other basins.

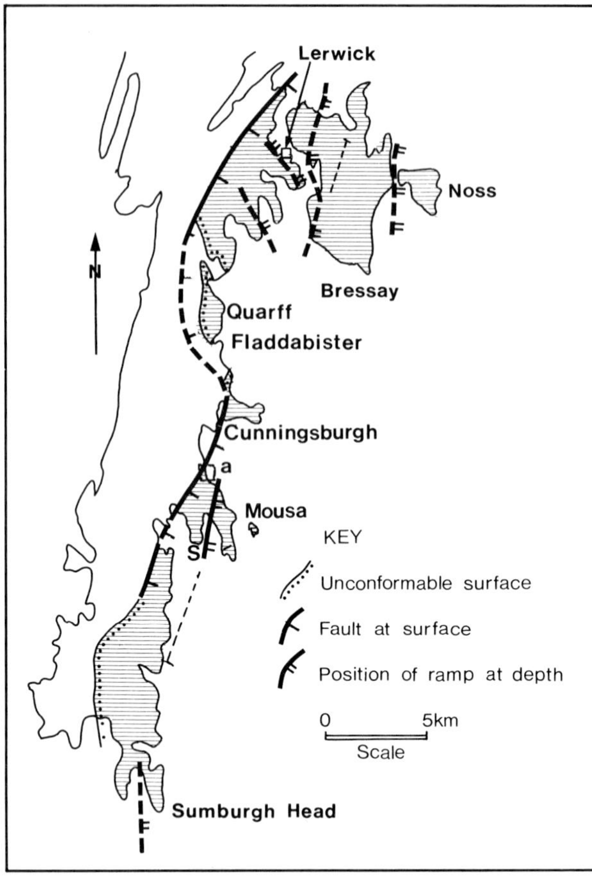

FIG. 12. Map of the Bressay–Sumburgh Basin in E Shetland, showing the main faults and the possible positions of ramps at depth, to explain folds in the Old Red Sandstone. a, location of Fig. 13; S, Sandwick.

The offshore continuation of this basin is uncertain. Commercial seismic data are not available for the East Shetland Platform SE of Bressay. On the BIRPS SHET deep seismic data (McGeary, in press) it is difficult to identify Devonian sediments with any certainty and difficult to identify the main basin bounding faults (S. McGeary, personal communication, 1987). Thus the importance of NE–SW extension relative to NW–SE extension is unknown.

The Moray Firth area

Most of the Old Red Sandstone sequences and structures of the Moray Firth, E of the Great Glen Fault, have been masked by Mesozoic sediments. However, Old Red Sandstone rocks outcrop along the S coast of the Moray Firth (Fig. 14A & B). Some indication of the nature of the Old Red Sandstone package and basement structure can be obtained from deep seismic lines (e.g. Fig. 15).

The Devonian section at Pennan on the Moray Firth, extends from Gamrie Bay in the W to Dundarg in the E. There is a further small occurrence at Quarry Head to the E of the main section (Fig. 14). The main section contains two fault-bounded blocks of Dalradian basement. The largest of these forms Troup Head (Fig. 14), and is separated from the Old Red Sandstone section by an ENE–WSW fault. This fault system, which occurs throughout the area, is considered to be later than the main Devonian faulting and related to extension on the Banff Fault as shown on the Moray Firth seismic line (Fig. 15).

The second Dalradian block occurs at Strahangles Point. Both the western and eastern

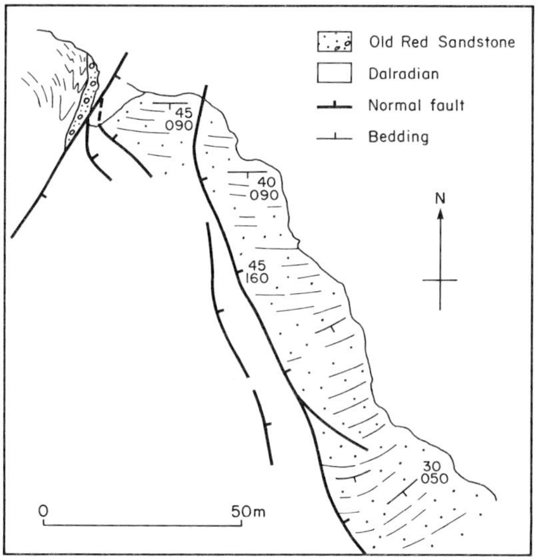

FIG. 13. Detailed map of part of the fault zone and basin fill in the Sumburgh Basin. See Fig. 12 for location.

margins are faulted. This block effectively separates the Old Red Sandstone basin into two parts (Fig. 14): The Pennan section to the W. and the Aberdour section to the E.

The sedimentology of the Aberdour section, which is exclusively Lower Old Red Sandstone is discussed by Sweet (1985). The section thickens dramatically towards the W and, if the section at Quarry Head is taken to be part of the same system, describes a half-graben geometry (Sweet 1985; Fig. 3), with a presumed boundary fault to the W. The fault at the W end of Aberdour beach appears to be a major normal fault with a sense of displacement down to the E. Flat-lying Dalradian fabrics are reorientated by the fault and metamorphic assemblages are strongly retrogressed. A minor fault occurs at the Castle of Dundary and has a similar down to the E displacement.

The Pennan section extends from the western end of Gamrie Bay to Strahangles Point. As in the Aberdour section, the western limit of the Devonian is marked by a steep normal fault

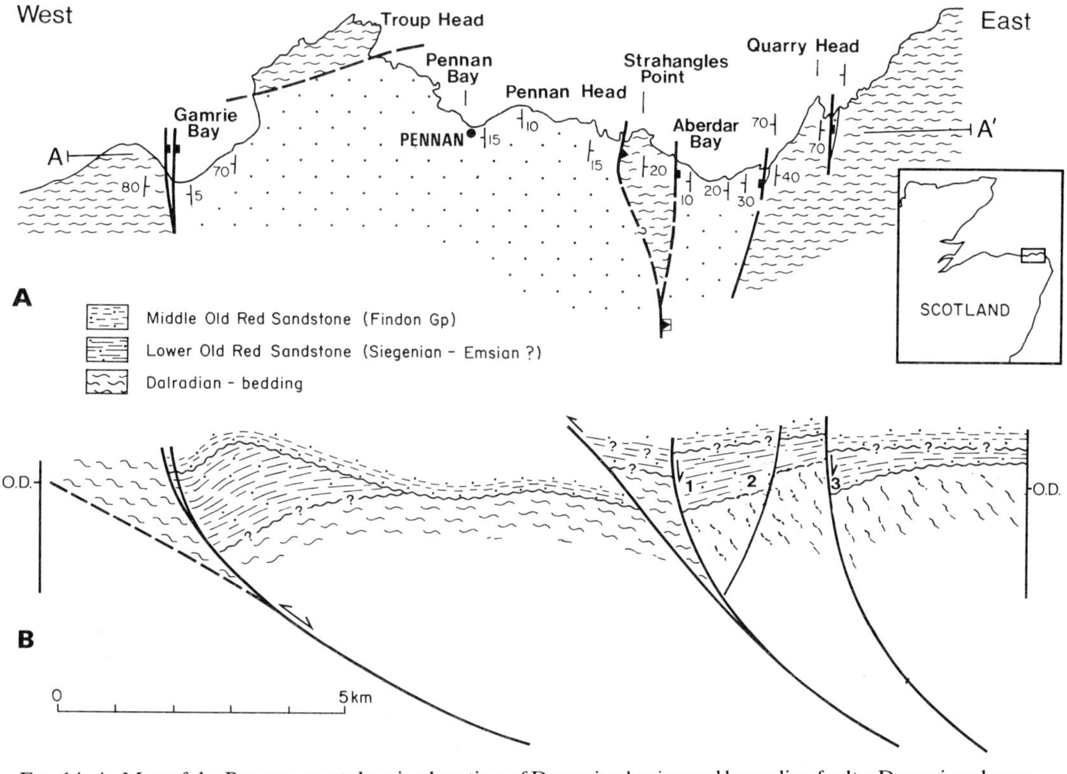

FIG. 14. A, Map of the Pennan coast showing location of Devonian basins and bounding faults. Devonian shown by coarse stipple. B, Section through the Pennan half-graben. The Lower Old Red Sandstone fill is a growth sequence, which is unconformably overlain by Middle Old Red Sandstone. The basin has been inverted by reactivation of the bounding faults, possibly in the Permo–Carboniferous (Variscan).

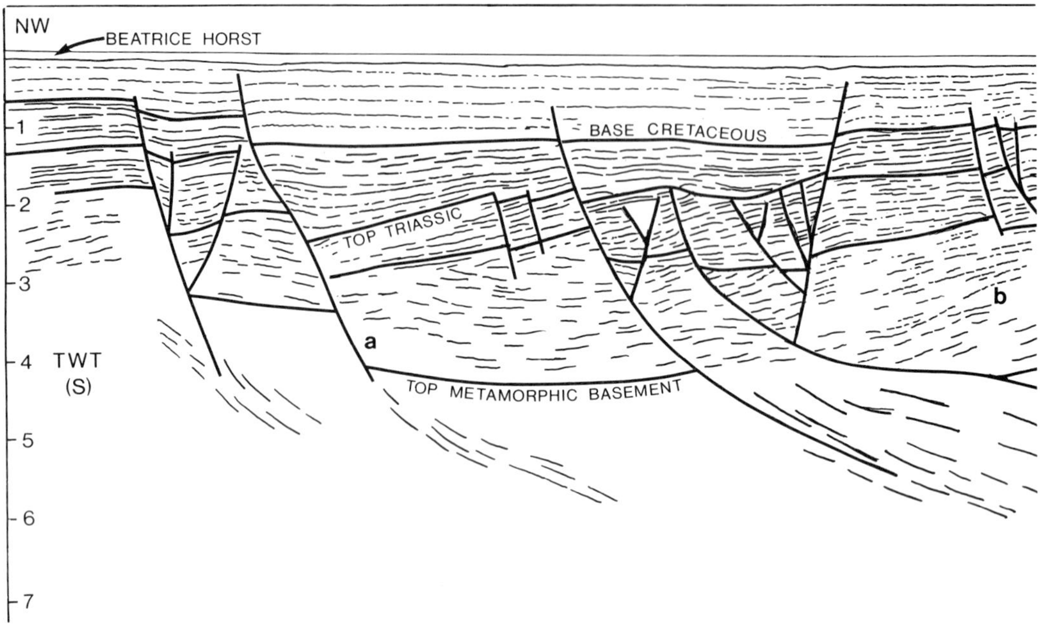

FIG. 15. Line drawing of the upper part of a 15 s (TWT) unmigrated seismic line through the Inner Moray Firth NMF 30, from Western Geophysical. Bold horizons represent the top of the metamorphic basement, top Palaeozoic (Devonian), top Triassic and base Cretaceous. The basin has been inverted so that Triassic sediments rest unconformably on an eroded Devonian sequence, where possibly over a kilometre of sediment was removed by erosion following Variscan (?) tectonics (location a). A Middle Devonian unconformity may exist in the sequence (e.g. location b, similar to that described from the Pennan coast. See Fig. 4 for location of profile D.

with down to the E sense of movement, along which the Dalradian metamorphic assemblage is strongly retrogressed. Slickenside lineations on the fault surface indicate a predominantly dip–slip movement. The fault juxtaposes Dalradian basement with probable Lower Old Red Sandstone, although to the E, at the W end of Pennan Bay, the unconformity between Lower Old Red Sandstone and Middle Old Red Sandstone can clearly be seen. It demonstrates the synrift nature of the earliest Middle Old Red Sandstone section (Fig. 14B). The Middle Old Red Sandstone is strongly erosive and is interpreted as having progressively cut across the Lower Old Red Sandstone to lie unconformably on top of the Dalradian in the vicinity of Pennan Head.

It would appear that the Pennan section represents a half-graben similar to that seen in Aberdour with a fault at its western end (Fig. 14). In the case of the Pennan section, however, the erosion level is such that Middle Old Red Sandstone is exposed.

The Western Geophysical 15 s (TWT) seismic line, NMF-30 (Fig. 15) shows a similar fault-bounded Devonian basin beneath Mesozoic cover south of the Beatrice Horst. Low angle faults accommodate the extension and appear to be formed parallel to a fabric within the basement, as in the West Orkney Basin.

Positive inversion

The sediments of the Orcadian, West Orkney and Walls Basins show several episodes of deformation, some of which can be considered to have been produced by basin inversion.

Unconformities exist both within the Lower Old Red Sandstone and between the Lower and Middle Old Red Sandstone, in Pennan (cf. Norton *et al.* in press), Struie Hill and Tongue (fieldwork by Enfield), and on the Orkneys (Coward & Enfield 1987). It has previously been suggested that these unconformities were formed by late Caledonian thrusting (see Armstrong 1977; Mykura 1983). However, we believe that there is no evidence for Lower Devonian thrusting and that uplift and erosion can be accounted for by extension during early basin tectonics.

Thrust faults and fault-associated folds are

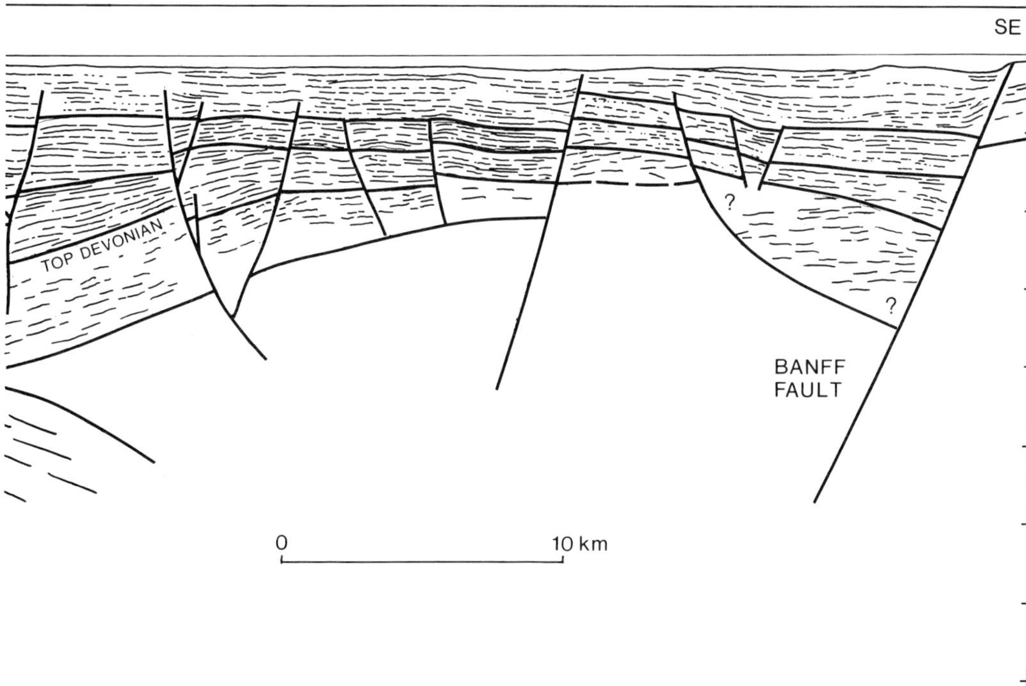

developed above detachments within the Middle Old Red Sandstone lacustrine sediments and are particularly well seen on Orkney mainland (Coward & Enfield 1987). These contractional structures are commonly localized around earlier extensional structures and exhibit excellent small scale examples of inversion geometries.

Within the West Orkney Basin, sediments interpreted as Middle Old Red Sandstone show unconformities related to tilting of major fault blocks and a reversal in the distribution of subsidence (see also Coward & Enfield 1987). This basin inversion event may correlate with a major, Middle Devonian, episode of inversion observed in the Walls Basin, West Shetland.

An episode of post-Upper Old Red Sandstone age deformation is demonstrated by strike–slip deformation on faults distributed within a broad zone along the line of the Great Glen Fault–Walls Boundary Fault system (Coward & Enfield 1987). This phase may be responsible for the uplift of the Devonian rocks to above sea level.

In the Outer Isles Basin and NW of the Hebrides, there has been major uplift associated with, or following Mesozoic extension, so that prerift sediments are eroded from the basement highs and Mesozoic sediments are tilted and onlapped by Tertiary sediments in the North Minches Basin. Presumably this uplift also affected much of Northern Scotland.

Lower to Middle Old Red Sandstone unconformities

Where the contact between the Lower and Middle Old Red Sandstones are observed at the basin margins, for example, at Pennan (Fig. 14). Yesnaby (Fig. 9) and possibly Coldbackie Bay, Kyle of Tongue (Fig. 7), an angular unconformity is present between the two series. However, this unconformity is apparently not present within the central parts of the basin. For example, at Sarclet (Fig. 7) there is a continuous transition from Lower Old Red Sandstone to the lower Middle Old Red Sandstone (Mykura 1983).

On Orkney mainland the Lower and Middle Old Red Sandstones are exposed at Yesnaby (Fig. 9) where an angular unconformity separates a lower sequence of sub-aerial sandstones and conglomeratic breccias from overlying finely-bedded lacustrine deposits. This unconformity is interpreted as the boundary between the Lower and Middle Old Red Sandstone sequences (see also Wilson *et al.* 1935). The dip of the lower, sub-aerial deposits varies between approximately 10 and 60°, with observed angular discordances between the sequences normally in the range of 7–10° but up to 45°. A systematic variation in the angularity of the unconformity in section from NW to SE suggests that the Middle Old Red Sandstone is resting on an eroded open fold that faces to the NW. There is no evidence for any contractional deformation in the lower sequence. The fold is considered to be a rollover fold developed on the hanging wall of an extensional listric fault.

At Coldbackie Bay, Tongue, on the North Sutherland coast (Fig. 7), a lower conglomerate sequence, resting on Moinian basement and dipping up to 45° to the NW, is overlain by an extensive sequence of sub-horizontally bedded Middle Old Red Sandstone conglomerates and sandstones. The bedding dips within sandy couplets in the lower conglomerate sequence are not original sedimentary dips, but have been tilted. This has occurred on extensional faults which possibly decouple along late or post-Caledonian structures present within the Moines (Holdsworth 1987) and which pre-date the main episode of deposition of Middle Old Red Sandstone conglomerates. It is suggested, here, that these lower conglomerates may be of Lower Old Red Sandstone age.

Thus large amounts of extension, on listric and/or planar faults produced rotation of the Lower Old Red Sandstone deposits. The unconformities discussed above were probably produced by sedimentation and basin formation during active extension in a sub-aerial environment, associated with regional uplift, rather than by tectonic inversion on faults.

Middle Devonian inversion on bedding parallel detachments

Within the Middle Old Red Sandstone lacustrine sediments, there are extensive bedding-parallel detachment horizons that are associated with both contractional and extensional structures (Fig. 16). Small folds and duplexes are produced locally by thrusts which form ramps from the bedding-parallel detachments (Fig. 17). Elsewhere listric normal faults, with less than 10 m displacement, flatten into these detachments (Fig. 18). These detachment zones have thus acted as the flats to small scale listric extensional fault systems (Fig. 18) and to thrusts (Fig. 17). Most of these verge NW towards the major Devonian normal faults. These structures are interpreted as growth faults (cf. Crans *et al.* 1980), generated by gravitational instability within dipping sediments in large half-grabens. The displacements occur on easy slip horizons. The dominant slip horizons in the Orkney Isles are overpressured horizons. These are often organic-rich carbonate mudstones and are commonly fish beds. We suggest that the contractional structures should match the amount of extension produced by listric extensional faults.

Locally, in Birsay, Northwest Orkney, earlier extensional structures were reactivated by thrusts, producing small-scale examples of tectonic inversion. We suggest that the contractional deformation developed where displacement on the detachment fault was buttressed by small pre-existing synthetic or antithetic faults within the sedimentary fill, as at Whitaloo Point (Fig. 17), Northwest Orkney, or by the major half-graben bounding fault, as at Birsay, Northwest Orkney (Fig. 9).

Middle Devonian tilting and folding

The majority of offshore basins show either evidence of syntectonic sedimentation, where the sedimentary sequence thickens towards the fault (for example, Fig. 3) and the sediments onlap away from the fault, or pre-tectonic sedimentation, where the orthogonal thickness of the sediment package is constant within the half-graben (Fig. 3, upper seismic sequence II). However, in the centre of the West Orkney Basin is a zone where the sediment package increases in thickness away from the fault with sedimentary onlap towards the fault (Figs 19 & 20). This suggests that there has been a reversal in the distribution of subsidence. This may be associated with a reversal in movement on the basin bounding fault which caused a reversal in rotation sense of the fault block. Fig. 19 shows the distribution of the syntectonic sedimentation in the West Orkney Basin and delineates the region where sedimentation occurred synchronously with tectonic inversion. The age of this inversion is thought to be Middle to Upper Devonian. This is equivalent to the Upper Middle Old Red Sandstone sediments of Orkney and Caithness.

On the Walls Peninsula of West Shetland, Old Red Sandstone sediments are more extremely deformed. The basal unconformity and listric faults are tilted by up to 45° (Fig. 11).

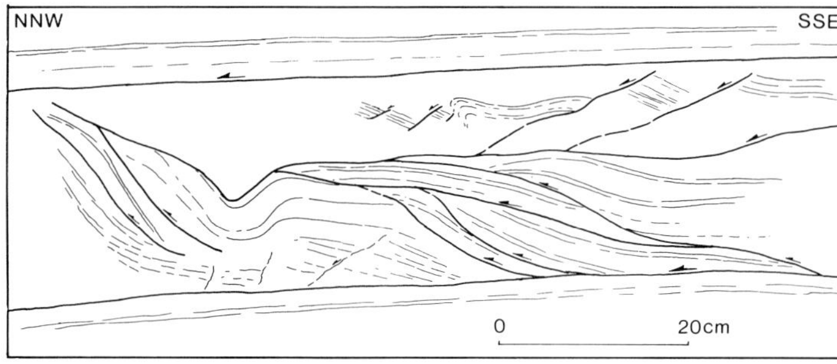

FIG. 16. Sketch of the structure developed within a detachment horizon utilized by listric normal faults at Yesnaby, W Orkney. Extensional shear bands are deformed by a contractional duplex. Both sets of structures were formed during extensional displacements on the detachment horizon.

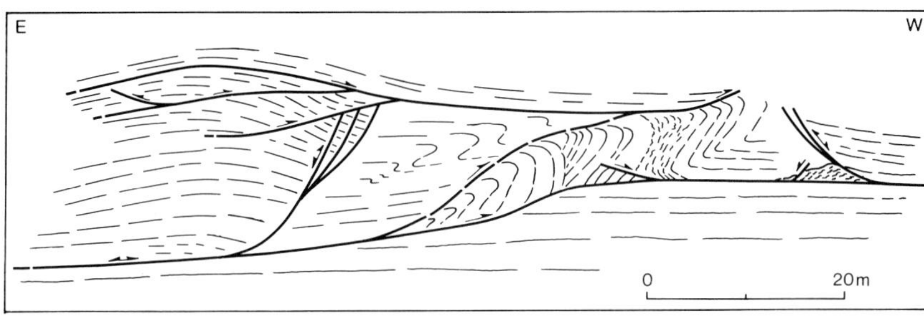

FIG. 17. Sketch illustrating relationships between early listric normal faults and inversion structures, from the coast at Birsay, NW Orkney mainland. Later thrusts utilize the pre-existing bedding-parallel detachments along which early extensional faults moved. The steeper normal faults commonly acted as buttresses to localize the folding during subsequent contraction.

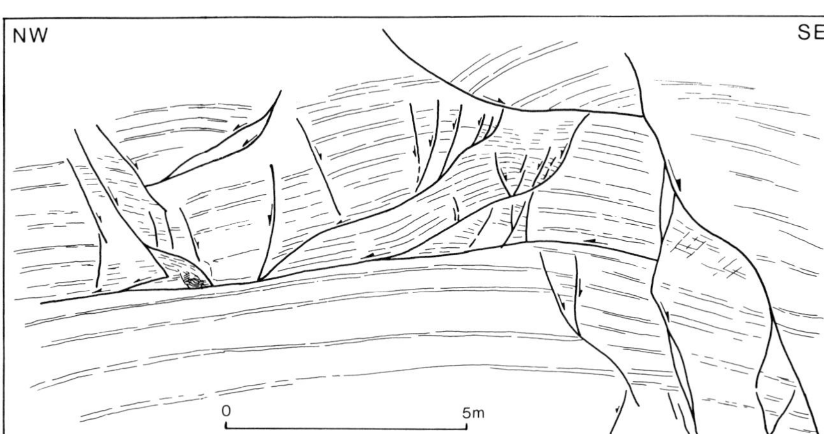

FIG. 18. Sketch of cliff section, Yesnaby, W. Orkney. Listric normal faults detach along bedding-parallel detachments in Middle Old Red Sandstone carbonates. See also Figs 16 & 17.

The tilting may be related to the reactivation of one of the basin-bounding faults to the N of the Walls Peninsula during Middle Old Red Sandstone times. Frank (1977) records a late antiformal fold in the Moines along the northern edge of the peninsula. We suggest that this antiform and the tilting of the Devonian unconformity originate by major tectonic inversion of

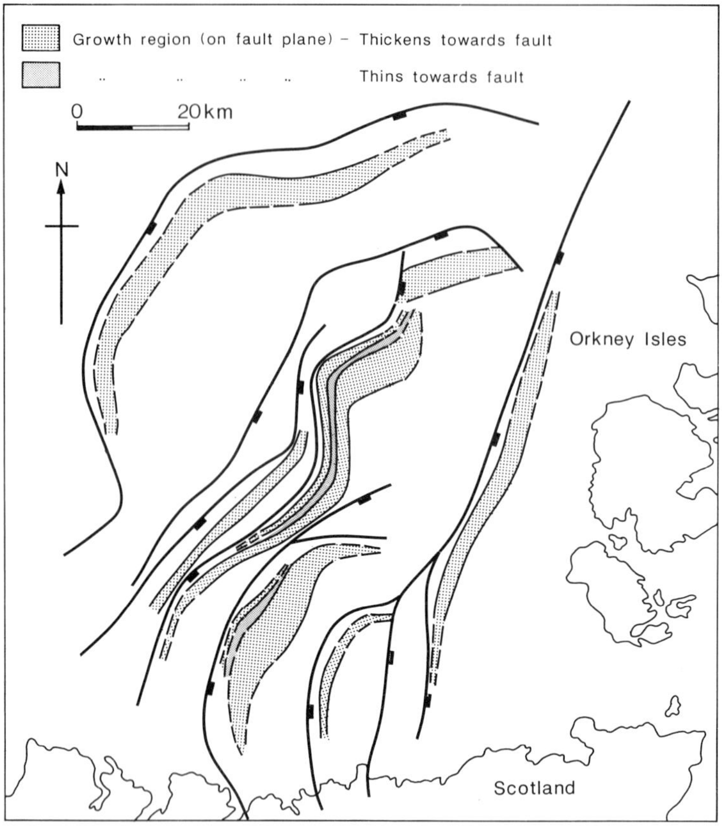

FIG. 19. Map showing the distribution of subsidence patterns throughout the West Orkney basin. Sediment packages that show growth either towards the half-graben bounding fault (see Fig. 3), or away from it (see Fig. 20), are mapped at their cut-offs with the bounding faults.

the original extensional system forming the Walls Basin.

The Devonian rocks of the Walls Basin are strongly folded into a large complex anticline (Fig. 11) trending ENE with an upright axial plane (Mykura 1976; Coward & Enfield 1987). The northern limb is steep to overturned adjacent to the Sulma Water Fault. This fault was originally considered to be an extensional fault because, it carries higher level rocks on its hanging wall. In detail the fold shape is complex (Fig. 10). Mykura 1983 interprets the structure as having been deformed by cross-folds. Some complications in strike and dip probably result from plunge variations during inversion, though in the east there is a later phase of minor folding. The rocks locally carry a pressure solution cleavage in sandstones which is approximately parallel to the major fold axial planes.

In the SE, steep to overturned beds are cut by the Sandsting Plutonic Complex which does not show evidence of the complex folding. Mykura (1983) considered the deformation of Old Red Sandstone sediments to be syn-intrusional. However, because of the present steep contacts and the discordant relationships we consider the intrusion to post-date the main folding. It is dated as 369 ± 10 and 360 ± 11 Ma (Mykura 1976). Therefore, the inversion phase was probably before the end of the Upper Devonian.

The East Shetland Basin shows no evidence of inversion at this time. The sediments have subsequently been tilted and folded, but on NW–SE trending axes, approximately perpendicular to the structures in the Walls Basin (Fig. 21).

Variscan deformation

On Shetland, the Walls Boundary Fault is a complex fault zone in which the shear criteria

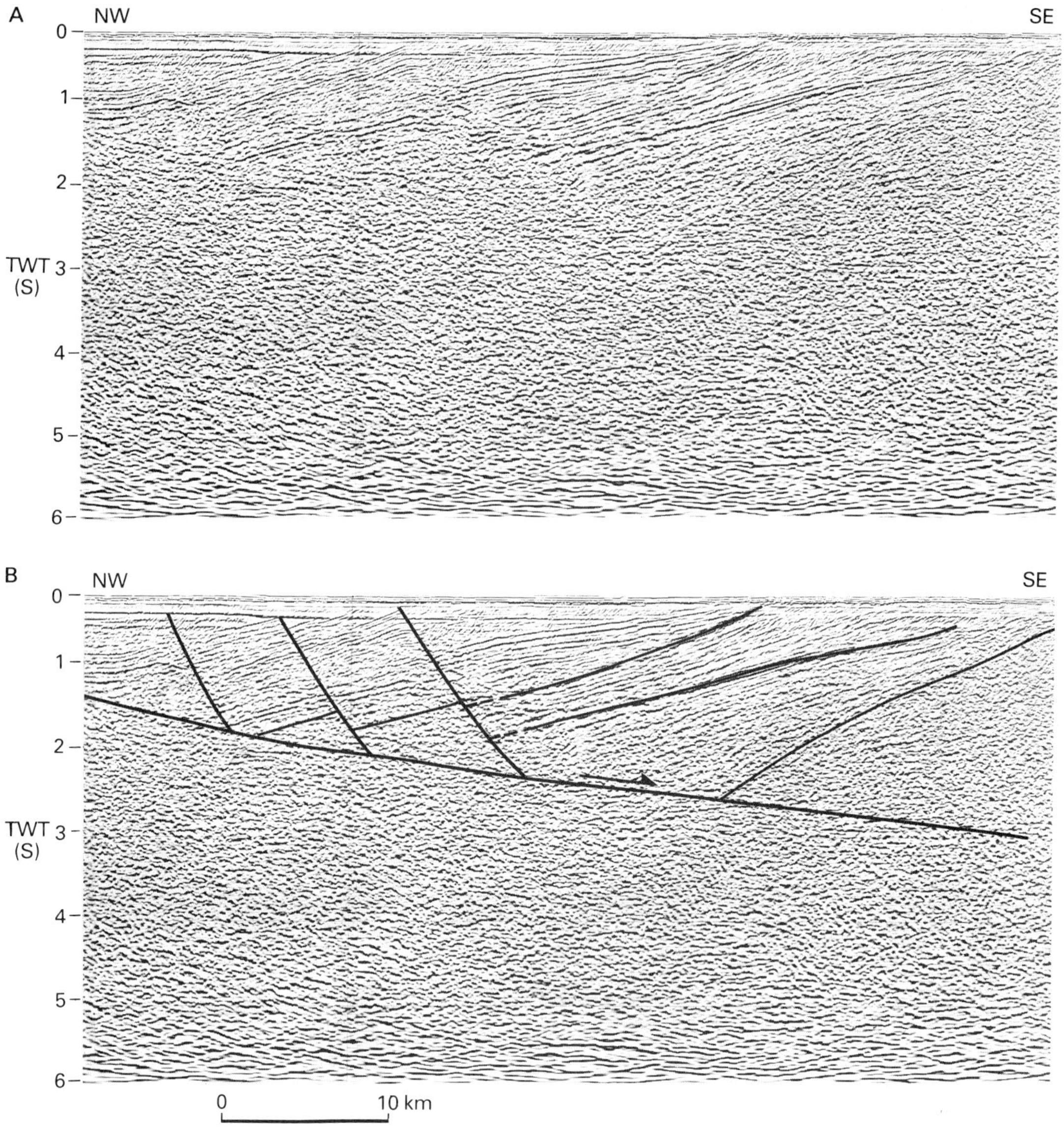

FIG. 20. A, Seismic profile from Western Geophysical, showing changing onlap directions through time, central West Orkney Basin (The line shown on Fig. 4 is profile E.) B, Structural interpretation of Fig. 20B.

are ambiguous. However, the dominant small-scale shear sense indicators suggest that the main displacement is dextral strike–slip. From the offset of the Graven igneous complex in Eastern Shetland, Miller & Flinn (1966) suggested that there was 16 km dextral displacement on the Nesting Fault which Flinn *et al.* (1979) consider to be synthetic to the Walls Boundary Fault. Flinn (1969; Flinn *et al.* 1979) suggests a dextral displacement of 65 km on the Walls Boundary Fault system based on correlation of magnetic anomalies around Fair Isle with those of the Sandsting area. Part of this displacement would include the 16 km dextral movement on the Nesting Fault. Astin (1982) considers the Devonian sediments of Fair Isle to have been deposited in a NE extension of the Walls Basin. This would correlate the plutonic rocks on Fair Isle with the Northmaven or Muckle plutons N of the Walls Peninsula. It

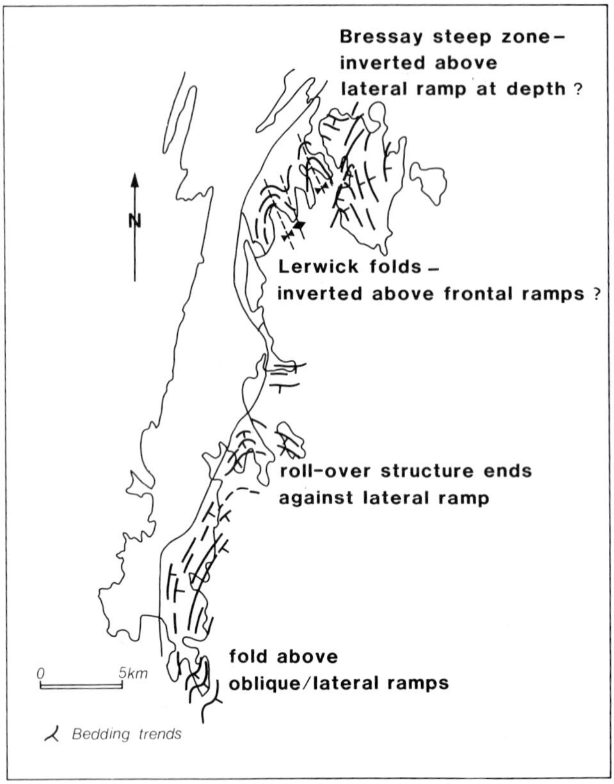

FIG. 21. Map of the Bressay–Sumburgh Basin, showing the distribution of inversion structures.

thus supports the case for over 60 km displacement on the dextral strike–slip faults of Shetland.

In East Shetland an inversion affects the Devonian rocks in the hanging wall to the NE-trending faults. Kinks and thrusts affect the peninsula near Sumburgh suggesting N–S compression. Also, large folds trend NNW–SSE to NW–SE across the region from Sumburgh to Lerwick. They may have been developed from reactivation of deeper normal faults. The inversion structures are dramatically seen on the island of Bressay where rocks are locally overturned (Fig. 21). The steeply dipping zones are the sites of intrusive breccias and agglomerate vents which cut through the overturned beds. From the trend of the folds and steeply tilted beds, this inversion episode involved either NE–SW compression or more likely, was associated with dextral strike–slip shear on the Shetland fault system.

In the Orkney Islands, NNE–SSW trending faults, some of which, such as the East Scapa Fault and perhaps the Brough–Brims–Risa Faults (Fig. 9), were active during Devonian sedimentation (Astin 1985). However, they exhibit dextral strike–slip kinematic indicators in the fault zones and are associated with folds that have orientations indicating dextral strike–slip displacements. This deformation pre-dates intrusion of late Permian dykes (Brown 1975; Rock 1983; Mitchell 1984), which intrude along the planes of these major faults (Coward & Enfield 1987).

Fig. 9 shows the positions and trends of the main folds on the Orkneys. The majority strike NNE–SSW and are spatially related, but slightly oblique to, the major East Scapa Fault. This fault changes in strike from NNW to NNE south of Kirkwall (Fig. 9). It may be related to an inherited irregular ramp shape during earlier Devonian extension. The fault is best exposed in the cliff sections close to St Mary's (Fig. 9) where there are numerous dextral shear indicators in the form of shear bands and second-order faults and oblique folds. From the changes in strike, dip and movement sense on the small-scale faults, we infer that dextral negative-flower structures were superimposed on earlier dextral positive-flower structures (*sensu* Wilcox *et al.* 1973). This local change from transpressive to transtensional displacement may reflect the change in strike of the main fault zone and the movement of rocks from a compressive bend to an extensional bend between Kirkwall and St Mary's.

Estimates of the magnitude of strike–slip displacements are difficult to assess on the Orkneys. However, in the western Orkney Isles a fault thought to be the westwards extension of the North Scapa Fault which is seen on Hoy and western Orkney mainland (Fig. 9) and which pre-dates the Middle to Upper Old Red Sandstone unconformity, appears to be offset dextrally by about 3 km on the Brims–Risa Fault (Fig. 9). In South Ronaldsay Lower Eday Sandstone units are obliquely offset by a dextral displacement of at least 4 km (see also Astin 1985).

The strike–slip faults of the Orkneys and Caithness are thought to link with the Helmsdale and Great Glen Faults to the S (Fig. 4). Along the Great Glen Fault in the Black Isle to the S, estimates of net dextral displacement since the Middle Devonian are approximately 25–29 km. (D. Rodgers, personal communication 1988). However, some of these faults must have been active during the Mesozoic because they form the lateral boundary to Mesozoic extension in the Moray Firth.

The W contact of the Dalradian block at Strahangles Point, S of the Moray Firth (Fig. 14) is not a simple normal fault. Dalradian rocks structurally overlie Middle Old Red Sandstone sediments in the cliff section (Fig. 14B) and the fault must be, by definition, a thrust. The Dalradian slice narrows towards the S, and the thrust at the W and the normal fault at the E merge, showing that at depth they must join at a branch line under Strahangles Point. This is a typical inversion geometry, where an inversion-related fold is developed in the hanging wall of the normal fault as in Gamrie Bay, while in the footwall a new thrust fault develops by cutting through the steep normal fault ramp. It is possible that a post-Devonian contractional fault also occurs to the W of the normal fault at Gamrie Bay, though complex structure within the Dalradian makes this difficult to identify. The age of this inversion is unclear. However by analogy with the Moray Firth and onshore Old Red Sandstone basins to the N, it is likely to be Permo–Carboniferous in age (Variscan) and is likely to have a strike–slip component. The offshore Devonian basins (Fig. 15) within the Inner Moray Firth are overlain unconformably by Permo–Triassic sediments. The unconformity, where observed immediately SE of the Beatrice horst, (above locality A, Fig. 15), appears to cut down stratigraphic section to the NW. This angular relationship suggests that fault-controlled inversion was responsible for the unconformity. The inversion may have also been related to Variscan fault movements.

Offshore deep seismic data (McGeary in press) record a step in the reflection seismic Moho along the line of the Walls Boundary fault, N of the Orkney Isles. This presumably juxtaposes sections of crust with different thicknesses. If correct, this requires strike–slip displacements on a steep fault which affect the whole crust and not strike–slip movement on a stepped fault as suggested by Gibbs (1987). We record no evidence to suggest that the Great Glen Fault or the Walls Boundary Fault were low angle structures. Indeed, they were probably originally steep and cut through the whole crust.

Mesozoic to Tertiary extension and inversion

In the North Minches Basin there is a thick sequence of Mesozoic sediments (Kilenyi & Stanley 1985) and lower Jurassic rocks outcrop in small fault graben onland. The area has been affected by later Mesozoic to Tertiary inversion (Fig. 5). N of the Hebrides, the Mesozoic fault system has been passively uplifted several kilometres. As the Triassic sediments predate the Jurassic extensional faulting, they must have had a much wider extent. After uplift and erosion they are now only preserved in the half-graben on the hanging walls of the faults. The age of the major uplift following Jurassic extension is uncertain. This uplift may have affected the West Orkney Basin.

Contractional structures of late Mesozoic to Tertiary age are present in the West Orkney, North Minches and Outer Isles Basins. The late Mesozoic to early Tertiary sediments show monoclinal folds and sometimes small contractional faults. Fig. 5 shows tilted fault blocks, involving what are interpreted to be Devonian sediments, eroded at their distal ends before the deposition of a Mesozoic (? Jurassic) growth sequence, which has itself been folded into a monocline of uncertain but possible Cretaceous or early Tertiary age. Good examples of Tertiary inversion occur as far W as the continental slope (Roberts *this volume*). Thus the uplift of mainland Scotland and the Devonian sediments of the West Orkney and Orcadian Basins may have been cumulative from the Carboniferous.

Discussion

Extension of thickened Caledonian crust

Precise estimates of the original shortening and thickening of the Caledonian crust cannot be made in Scotland. Estimates of displacement on the Moine Thrust Zone are of the order of 50–70 km (Elliott & Johnson 1980; Butler & Coward 1984). This is a minimum because it is unknown how much displacement occurred on the thick mylonites which form the Moine Thrust. The overall structure of the Moines can be considered in terms of a crustal scale duplex (Soper & Barber 1982), where maximum crustal thickening should occur above and behind the line where the Moine Thrust climbs the ramp from the lower detachment within basement into the Cambrian sediments. This must lie E of the outcrop of the Moine Thrust Zone (Butler & Coward 1984).

The displacement on the Moine Thrust continued for at least 50 km after the intrusion of the Loch Borrolan igneous complex (Coward 1985; Halliday *et al.* 1987), which is dated by van Breeman *et al.* (1979) as 430 ± 4 Ma. The earliest age assigned to the Lower Old Red Sandstone sediments is 394 Ma (Emsian). Thus, if we assume a displacement rate of 0.5 mm per year, then Moine thrust activity would have continued until about 420 Ma, leaving approximately 25+ My before the deposition of the earliest preserved post-orogenic sediments. During this period there may have been substantial uplift and erosion of the Caledonian belt.

The Lower Old Red Sandstone sediments were deposited sub-aerially. This required an elevated topography within the internal parts of the Caledonian mountain belt. For crust and lithosphere of normal density, under a normal geothermal gradient, the critical crustal thickness above which positive topography is formed is approximately 29–31 km. From Fig. 22 it can be seen that in a region of elevated topography associated with thickened crust or crust with lower than average density, large amounts of extension and thinning must occur for subsidence to lower the land surface to sea level. During the early extension, sub-aerial basins will form, which will remain elevated and therefore be subject to erosion. This results in their

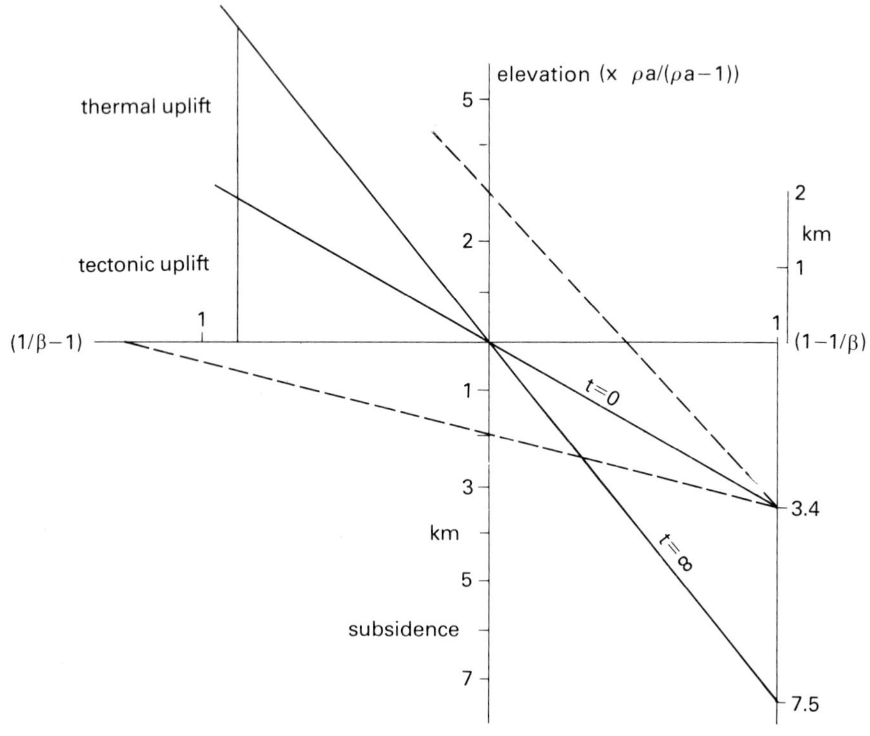

FIG. 22. Subsidence or elevation associated with homogeneous deformation of the lithosphere. See text for discussion.

cannibalization to fill subsequent basins. There will be a lack of any record for early extensional phases of basin formation. This may lead to a late age being estimated for the initiation of extension.

Examples

Fig. 23 shows the subsidence for heterogeneous stretched crust (see Appendix for details). If the lithosphere had previously been thickened during compressional tectonics, but had not thermally equilibrated, then the subsidence curves will differ from those of Figs 22 and 23. Homogeneous thickening of the lithosphere generates an initial phase of uplift, followed by thermal uplift as the lithosphere temperature profile equilibrates. Thinning causes tectonic and thermal subsidence but the final elevation relative to sea level depends on the amount of thermal equilibration before extension and the amount of erosion of the thickened crust. Thus it depends on the changes in average lithospheric density and the crust to lithosphere thickness ratio. In some circumstances, the uplift associated with thermal equilibration after thickening may strongly counteract subsidence related to extension. Certainly the basins developed above previously thickened crust would differ from those produced by homogeneous stretching of the normal crust. The differences are described below.

Much of the stretching may have occurred while the topographic surface lay above sea level so that early inter-montaine basin deposits were cannibalized by later basin development. Thus in the Old Red Sandstone basins, Marshall (in press) has observed that reworked Upper Silurian spores are observed in Devonian sandstones on Papa Stour. Angular unconformities within the early basin sediments (the Lower Old Red Sandstone) may have formed during this extension period. There may have been considerable extension without the preservation of rift phase sediments. However, if extension was heterogeneous and confined to the upper part of the lithosphere, then local deep half-grabens may develop. There may also have been relatively little thermal subsidence compared to the extension observed from the fault geometry.

We can assume that the process of crustal shortening during the Caledonian orogeny produced an abnormally thick and cool lithosphere which then began to thermally equilibrate. The thickest crust would have occurred above and on the hinterland side of the major crustal scale

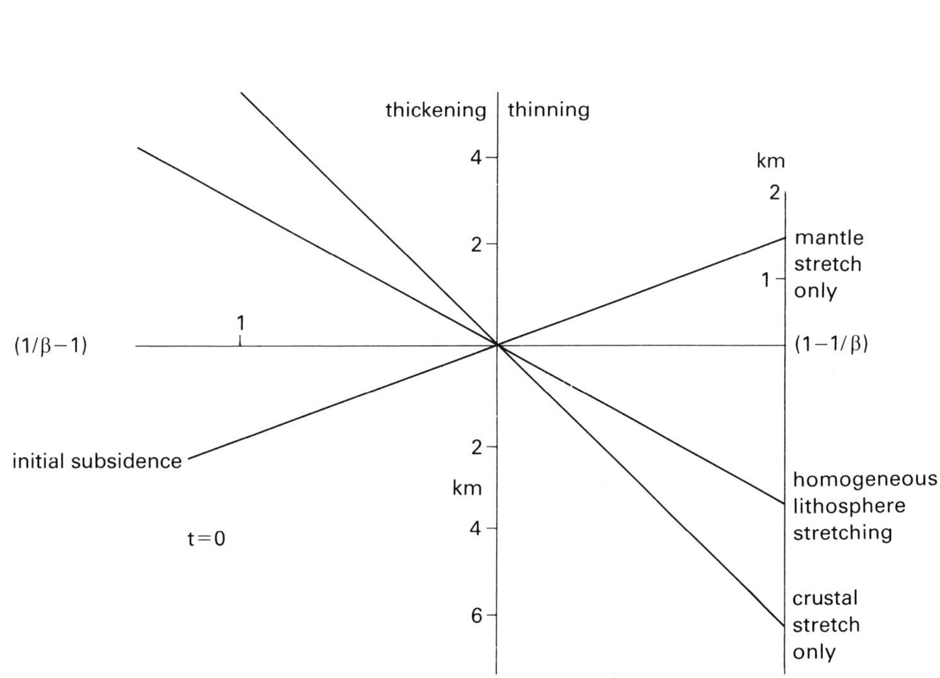

FIG. 23. Subsidence or uplift associated with heterogeneous deformation of the lithosphere. Thinning of the crust alone causes excessive subsidence, while thinning of the mantle produces initial elevation, followed by a thermal subsidence phase.

ramps, such as the Outer Isles Thrust and the deeper parts of the Moine Thrust. Sonder *et al.* (1987) calculate that continental crust will fail in extension and begin to spread under its own mass when the Moho achieves approximately 750°C. In the Caledonides the time lag between the end of the principal compressional deformation and the onset of spreading appears to be a maximum of 25–30 My. The uncertainty is due to lack of knowledge about the age of cessation of thrusting and the lack of preservation of the earliest sediments.

Once the lower crust achieved the critical conditions for spreading, it would be unable to support contractional deformation. Thrusting affecting the upper crust would cease and brittle extensional faulting, driven by the spreading, would reactivate Caledonian structures. This would cause negative inversion of the orogen. Compressional deformation could still occur on the foreland side of the thickened and now weakened crust (cf. Houseman & England 1986; Sonder *et al.* 1987). The Outer Isles and Flannan structures, with their deep crustal to lithospheric shears, may have formed late in the Caledonian history as deformation propagated into the foreland.

Much of the early extension generating the Old Red Sandstone basins developed on faults that sole into gently dipping detachments. These are interpreted here as the Moine Thrust and as major shear zones in the Moines. Half-grabens observed offshore on seismic data also occur onshore on the north Scottish coast (Figs 2 & 4). In each case they seem to coincide with the position of major ductile thrusts (Barr *et al.* 1986; Holdsworth 1987). The extensional faults occur a short distance into the hanging walls of the thrusts (Fig. 7). Subsequent extension occurred on deeper level decoupling zones in the lower crust beneath the West Orkney Basin or along the Moho beneath the Outer Isles Basin. This indicates a gradual cooling and strengthening of the crust.

Middle Devonian basin inversion

The major effects of basin inversion during Middle Old Red Sandstone times appear to be concentrated around the Devonian faults in Orkney and the Shetland Isles, though some of the fault blocks in the central part of the West Orkney Basin also show inversion, probably during this time. The inversion may be due to the end effects of the Caledonian compression, in that initial extension may have strengthened the crust sufficiently for the transmission of compressive deformation from the latest phases of the Caledonian orogeny. However, a mechanism involving inversion due to strike–slip movements is preferred, possibly on an early equivalent of the Great Glen Fault. From the trend of folds on W Shetland and structures within fault zones in the Orkney Isles, the displacement along the line of a Devonian Walls Boundary Fault–Great Glen Fault system would be sinistral. From the relative position of the Caledonian fronts on Shetland and the Scottish mainland, Flinn (1985) suggests that 100–200 km of late Caledonian sinistral offset may have occurred along the Walls Boundary Fault. Note that sinistral strike–slip deformation would not only generate inversion in a NW–SE direction, but could also generate NE–SW extension and hence be the cause of the anomalous extension directions observed in the Bressay–Sumburgh Basin. The East Shetland Basin could have been growing in one region during the period of Middle Old Red Sandstone deposition while the Walls Basin was being deformed elsewhere.

There are arguments for Devonian strike–slip displacement along other major Caledonian fault zones in Scotland, such as within the Midland Valley (Gibbs 1987). However the amount of sinistral displacement along the Great Glen Fault must remain speculative. It could be related to a late transpressive phase of Caledonian deformation (Soper & Hutton 1984) or may be related to the effects of Acadian deformation further S in the Caledonian belt. This could cause indentation tectonics similar to those described by Tapponier *et al.* (1986) from Tibet and West China. The Great Glen Fault would then be equivalent to a structure such as the Tien Shan Fault in Tibet, and related to the localized lateral expulsion of part of the Caledonian belt.

Variscan and later inversions

The Variscan deformation and inversion episodes affect a narrow region along the line of the Great Glen Fault and Walls Boundary Fault system, an area which is dominantly above sea level today. The orientations of folds formed during inversion of the East Shetland Basin by strike–slip displacements on reactivated Devonian faults in the Orkney Isles indicates a dextral strike–slip sense. The Old Red Sandstone basins of Shetland were juxtaposed during this time by displacements of tens of kilometres on the Walls Boundary Fault–Great Glen Fault systems.

Major dextral strike–slip shear zones occur

within the northern edge of the Variscan belt in NW Europe (Badham 1982) and the northern margin of the Variscan in Britain has dextral oblique movements (Coward & Smallwood 1984). These zones trend WNW–ESE parallel to the Hercynian closure direction. It is not certain how the NE–SW trending Great Glen Fault would fit with this system, though within the Appalachians of Canada and the USA there are similarly trending dextral strike–slip zones.

It is probable that the Moho step across the Walls Boundary Fault, proposed from the BIRPS SHET data (McGeary in press) is also produced by strike–slip displacements at this time. This deeply penetrating fault would produce an offset of the lower crustal decoupling level of the West Orkney Basin fault systems, to act as a buttress to any further displacement and thus lock the fault system.

This may explain the presence of an area of thick, stable crust east of the Great Glen Fault–Walls Boundary Fault systems (Klemperer 1987) and may explain the distribution of the Mesozoic extension and basin development (Fig. 24). Thick Mesozoic basins are developed east of the Great Glen Fault–Walls Boundary Fault trend in the Inner Moray Firth and E Shetland Basins, and also west of the Outer Isles and West Orkney Basins on the Atlantic margin. Fault controlled extension in the Minches and western part of the West Orkney Basin (Fig. 24) has formed thinner Mesozoic basins in these areas.

Conclusions

Small Devonian half-grabens situated along the N coast of Scotland are formed on the hanging wall of major Caledonian thrusts. These can be correlated offshore with large half-grabens developed along major thrusts in the Caledonian

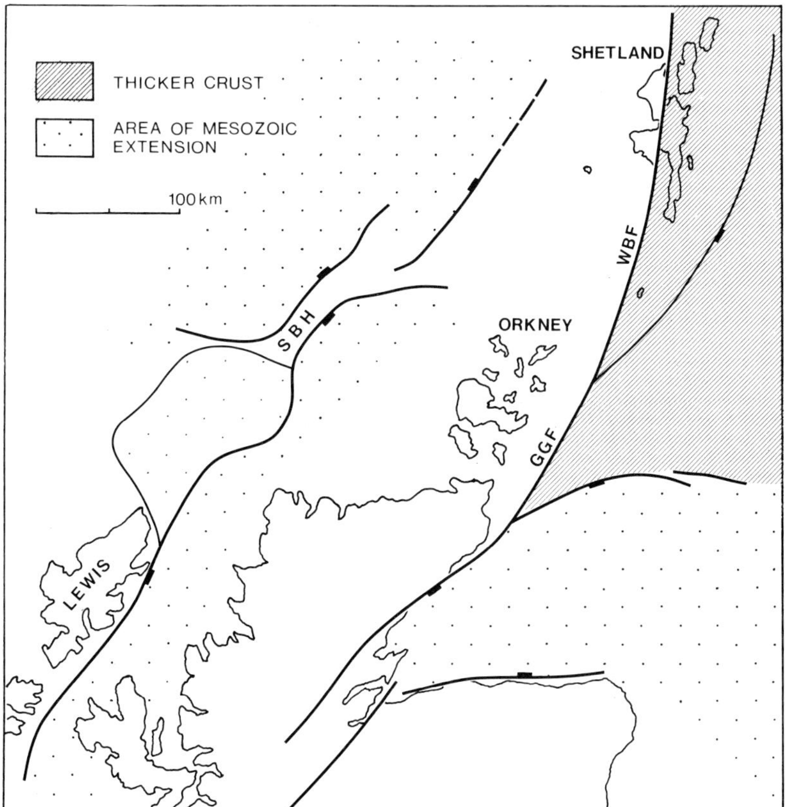

FIG. 24. Distribution of Mesozoic extension in the offshore northern Scottish area, showing regions of thick crust and the main basin bounding faults. SBH, Solan Bank High; GGF, Great Glen Fault; WBF, Walls Boundary Fault.

basement that were reactivated during the Devonian extension. The geometry of the early extensional fault systems suggests that the over-thickened Caledonian crust was either extended or spread by gravitational collapse along the reactivated thrust faults on which it had been thickened, failing under its own mass (Coward, unpublished abstract Tectonic Studies Group, 1987). The thinning was presumably aided by large amounts of erosion that preceded extension.

Early basins developed sub-aerially. They were elevated above sea level by isostatic compensation of the thickened crust. Until extension took the basins to near sea level a cycle of erosion and deposition resulted in the cannibalization and recycling of the basins formed early.

Tectonic inversion occurred during the Middle Devonian and may be related to sinistral strike–slip displacements on basement structures, which caused reactivation of Devonian growth faults. The Walls Basin was intensely deformed during this inversion episode. Fault reactivation in the West Orkney Basin further S caused changes in subsidence patterns resulting in heterogeneous subsidence within the basin.

During the Variscan orogeny, structures along the line of the Great Glen Fault–Walls Boundary Fault system were reactivated with a dextral strike–slip sense of displacement. This caused several kilometres of uplift along this trend and juxtaposition of different parts of the Devonian basin system. This fault line offsets the Moho and apparently juxtaposes crusts of different thickness. The effect of this inversion was to breach the decoupling zone of the extensional fault system beneath the Devonian basins, thus producing a stable block of thick crust which was unaffected by later extension during the Mesozoic. To the W, on the Atlantic margin, there was localized extension and subsidence along the western edge of the Devonian basin system during the Mesozoic. This formed the grabens related to the North Minch system and these were themselves subsequently inverted during the later Mesozoic–Tertiary.

ACKNOWLEDGMENTS: Seismic data and a great deal of assistance were given by Western Geophysical, Geoservices International and Jebco Seismic and we wish to thank them for their support. M.A.E. was funded by a NERC Studentship, No. GT4/84/GS/128.

References and further reading

ALLEN, P. A. & MARSHALL, J. E. 1981. Depositional environments and palynology of the Devonian, South-east Shetland Basin. *Scottish Journal of Geology*, **17**, 257–273.

ANDERTON, R., BRIDGES, P. H., LEEDER, M. R. & SELLWOOD, B. W. 1979. *A Dynamic Stratigraphy of the British Isles*. George Allen & Unwin, London, 301 pp.

ARMSTRONG, M. 1977. The Old Red Sandstone of the Easter Ross and the Black Isle. *In*: GILL, G. (ed.) *The Moray Firth Area*. Geological Studies of the Inverness Field Club. John G. Eccles, Inverness.

ASTIN, T. 1982. *The Devonian Geology of the Walls Peninsula, Shetland*. PhD Thesis, University of Cambridge, UK.

—— 1985. The Palaeogeography of the Middle Devonian Lower Eday Sandstone, Orkney. *Scottish Journal of Geology*, **21**, 353–375.

BADHAM, J. P. N. 1982. Strike–slip orogens; an explanation for the Hercynides. *Journal of the Geological Society of London*, **139**, 493–504.

BARR, D. 1987. Lithospheric stretching, detached normal faulting and footwall uplift. *In*: COWARD, M. P., DEWEY, J. F. & HANCOCK, P. (eds) *Continental Extension Tectonics. Special Publication of the Geological Society of London*, **28**, 75–94.

——, HOLDSWORTH, R. E. & ROBERTS, A. M. 1986. Caledonian ductile thrusting in a Precambrian metamorphic complex: The Moine of north-western Scotland. *Bulletin of the Geological Society of America*, **97**, 754–764.

BAXTER, A. N. & MITCHELL, J. G. 1984. Camptonite – monchiquite dyke swarms of Northern Scotland; age relationships and their implications. *Scottish Journal of Geology*, **20**, 297–308.

BEACH, A. 1985. Some comments on sedimentary basin development in the Northern North Sea. *Scottish Journal of Geology*, **21**, 493–512.

——BIRD, T. & GIBBS, A. 1987. Extensional tectonics and crustal structure: deep seismic reflection data from the Northern North Sea Viking Graben. *In*: COWARD, M. P., DEWEY, J. F. & HANCOCK, P. (eds) *Continental Extensional Tectonics. Special Publication of the Geological Society of London*, **28**, 467–476.

BINNS, P. E., MCQUILLIN, R. & KENOLTY, N. 1974. *The geology of the Sea of the Hebrides. Report of the Institute of Geological Sciences*, **73/4**, 433 pp.

BLACKBOURN, G. 1981. Probable Old Red Sandstone conglomerates around Tongue and adjacent areas, north Sutherland. *Scottish Journal of Geology*, **17**, 103–118.

BREWER, J. A. & SMYTHE, D. K. 1984. MOIST and the continuity of crustal reflector geometry along the Caledonian–Appalachian orogeny. *Journal of the Geological Society of London*, **141**, 105–120.

—— & —— 1986. Deep Structure of the Foreland to the Caledonian Orogen, NW Scotland: Results from the BIRPS WINCH Profile. *Tectonics*, **5**, 171–194.

BROWN, J. F. 1975. Potassium–argon evidence of a Permian age for the camptonite dykes: Orkney. *Scottish Journal of Geology*, **11**, 259–262.

BUTLER, R. W. H. & COWARD, M. P. 1984. Geological constraints, structural evolution and deep geology of the northwest Scottish Caledonides. *Tectonics*, **3**, 347–365.

CHEADLE, M. J., MCGEARY, S., WARNER, M. R. & MATTHEWS, D. H. 1987. Extensional structures on the western UK continental shelf: a review of evidence from deep seismic profiling. *In*: COWARD, M. P., DEWEY, J. F. & HANCOCK, P. (eds) *Continental Extensional Tectonics. Special Publication of the Geological Society of London*, **28**, 445–465.

CHESHER, J. A., SMYTHE, D. K. & BISHOP, P. 1983. *The geology of the Minches, Inner Sound and Sound of Raasay. Report of the Institute of Geological Sciences*, **83/6**.

COWARD, M. P. 1985. The thrust structures of southern Assynt, Moine thrust zone. *Geological Magazine*, **122**, 595–607.

—— 1986a. Moine thrust structures. *In*: HARRIS, A. L. & FETTES, D. (eds) *The Caledonides*. NATO ASI, **175**, 259–280.

—— 1986b. Heterogeneous stretching and basin development. *Earth and Planetary Science Letters*, **80**, 325–336.

—— & ENFIELD, M. A. 1987. The Structure of the West Orkney and Adjacent Basins. *In*: BROOKS, J. & GLENNIE, K. (eds) *Proceedings of the Third Conference on the Petroleum Geology of NW Europe*. Graham A. Titman, London, 687–696.

—— & SMALLWOOD, S. 1984. An interpretation of the Variscan tectonics of SW Britain. *In*: HUTTON D. H. W. & SANDERSON, D. J. (eds) *Variscan Tectonics of the North Atlantic Region. Special Publication of the Geological Society of London*, **14**, 89–102.

CRANS, W., MANDL, G. & HAREMBOURE, J. 1980. A geomechanical delta model based on gravity sliding. *Journal of Petroleum Geology*, **2**, 265–307.

DAHLSTROM, C. D. A. 1969. Balanced cross sections. *Canadian Journal of Earth Sciences*, **6**, 743–757.

DEWEY, J. & PANKHURST, R. J. 1970. The evolution of the Scottish Caledonides in relation to their isotopic age pattern. *Transactions of the Royal Society, Edinburgh*, **68**, 361–388.

DUNHAM, K. C. 1973. *Annual Report of the Institute of Geological Sciences 1972*.

ELLIOTT, D. & JOHNSTON, M. R. 1980. Structural evolution of the northern part of the Moine thrust belt, NW Scotland. *Transactions of the Royal Society of Edinburgh*, **71**, 69–96.

ENFIELD, M. A. & COWARD, M. P. 1987. The West Orkney Basin, northern Scotland. *Journal of the Geological Society of London*, **144**, 871–884.

FLINN, D. 1969. A geological interpretation of aeromagnetic maps of the continental shelf around Orkney and Shetland. *Geological Journal*, **6**, 279–292.

—— 1977. Transcurrent faults and associated cataclasis in Shetland. *Journal of the Geological Society of London*, **133**, 231–248.

—— 1985. The Caledonides of Shetland. *In*: GEE, D. G. & STURT, B. A. (eds) *The Caledonide Orogen–Scandinavia and Related Areas*. Wiley, Chichester 1159–1172.

——, FRANK, P. L., BROOK, M. & PRINGLE, I. R. 1979. Basement-cover relationships in Shetland. *In*: HARRIS, A. L., HOLLAND, C. H. & LEAKE, B. E. (eds) *The British Caledonides Reviewed. Special Publication of the Geological Society of London*, **8**, 109–116.

FOSTER, R. J. 1972. *The solid geology of Northeast Caithness*. PhD Thesis, University of Newscastle, UK.

FRANCIS, P. W. & SIBSON, R. N. 1973. The Outer Hebrides Thrust. *In*: PARK, G. & TARNEY, J. *The early Precambrian of Scotland and related rocks of Greenland*. University of Keele, 95–104.

FRANK, P. L. 1977. *The structure and metamorphism of the Walls Metamorphic Series, Shetland*. PhD Thesis, University of Liverpool, UK.

GIBBS, A. 1987. Development of extension and mixed-mode sedimentary basins. *In*: COWARD, M. P., DEWEY, J. F. & HANCOCK, P. L. (eds) *Continental Extensional Tectonics. Special Publication of the Geological Society of London*, **28**, 19–33.

HALLER, J. 1971. *Geology of East Greenland Caledonides*. Wiley, Chichester.

HALLIDAY, A. N., MCALPINE, A. & MITCHELL, J. G. 1977. The age of the Hoy lavas, Orkney. *Scottish Journal of Geology*, **13**, 43–52.

HARRIS, A. L., PARSON, L. M., HIGHTON, A. J. & SMITH, D. I. 1981. New/Old Moine relationship between Fort Augustus and Inverness. *Journal of Structural Geology*, **3**, 187–8.

HOLDSWORTH, R. 1987. *Basement cover relationships, reworking and Caledonian ductile thrust tectonics of northern Moine, NW Scotland*. PhD Thesis, University of Leeds, UK.

HOSSACK, J. 1979. The use of balanced cross sections in the calculation of orogenic contraction: a review. *Journal of the Geological Society of London*, **136**, 705–711.

—— 1984. The geometry of listric growth faults in the Devonian basins of Sunnfjord, W Norway. *Journal of the Geological Society of London*, **141**, 629–637.

HOUSEMAN G. A., & ENGLAND, P. C. 1986. A dynamical model for lithosphere extension and sedimentary basin formation. *Journal of Geophysical Research*, **91**, 719–729.

JOHNSON, R. J. & DINGWALL, R. G. 1981. The Caledonides: their influence on the stratigraphy of the North West European continental shelf. *In*: ILLING, L. V. & HOBSON, G. D. (eds) *Petroleum geology of the continental shelf of Northwest Europe*. Heyden & Sons., London, 85–97.

JOHNSTONE, G. S. 1975. The Moine Succession. *In*: HARRIS, A. L., SHACKLETON, R. M., WATSON, J., HARLAND, W. B. & MOORBATH, S. (eds) *A correlation of Precambrian rocks in the British Isles. Special Report of the Geological Society of London*, **6**, 30–42.

KILENYI, T. & STANLEY, R. 1985. Petroleum prospects in the northwestern seaboard of Scotland. *Oil & Gas Journal*, **10–7**, 100–108.

KIRTON, S. R. & HITCHEN, K. 1987. Timing and style of crustal extension north of the Scottish mainland. *In*: COWARD, M. P., DEWEY, J. F. & HANCOCK, P. (eds) *Continental Extensional Tectonics. Special Publication of the Geological Society of London*, **28**, 501–510.

KLEMPERER, S. L. Crustal thinning and nature of extension in the northern North Sea from deep seismic reflection profiling. *Tectonics* (in press).

LE PICHON, & SIBUET, J-C. 1981. Passive margins: a model of formation. *Journal of Geophysical Research*, **86**, 3708–3720.

——, ANGELLIER, J. & SIBUET, J-C. 1982. Plate boundaries and extensional tectonics. *Tectonophysics*, **81**, 239–256.

MARSHALL, J. E. A. Devonian miospores from the Walls Sandstone, Shetland. *Transactions of the Royal Society of Edinburgh* (in press a)

—— Devonian miospores from Papa Stour, Shetland. *Transactions of the Royal Society of Edinburgh* (in press b).

—— & ALLEN, K. C. 1982. Devonian miospore assemblages from Fair Isle, Shetland. *Palaeontology*, **25**, 277–312.

——, BROWN, J. F. & HINDMARSH, S. 1985. Hydrocarbon source rock potential of the Devonian rocks of the Orcadian Basin. *Scottish Journal of Geology*, **21**, 301–320.

MCGEARY, S. 1987. Non-typical BIRPS on the margin of the northern North Sea. *Geophysical Journal of the Royal Astronomical Society*, (Special issue) **89**, 231–238.

—— Profiling a major continental strike–slip fault: The SHET surrey. *Journal of the Geological Society of London.* (in press).

MCKENZIE, D. 1978. Some remarks on the development of sedimentary basins. *Earth and Planetary Science Letters*, **40**, 25–32.

MILLER, J. A. & FLINN, D. 1966. A survey of the age relations of Shetland rocks. *Geology*, **5**, 95–116.

MITCHELL, A. H. G. 1984. The British Caledonides: interpretation from Cenozoic analogues. *Geological Magazine*, **121**, 35–46.

MORTON, W. H. & BLACK, R. 1975. Crustal attenuation in Afar. *In*: PILGER, A. & ROSSLER, A. (eds) *Afar Depression of Ethiopia*. Inter-Union Commission on Geodynamics, E. Schweizerbartsche Verlagbuchhandlung, Stuttgart, 53–65.

MURCHISON, R. I. 1859a. On the succession of the older rocks in the northernmost counties of Scotland. *Quarterly Journal of the Geological Society of London*, **17**, 145–151.

—— 1859b. On the succession of the older rocks in the northernmost counties of Scotland; with some observations on the Orkney and Shetland Isles. *Quarterly Journal of the Geological Society of London*, **15**, 353–418.

MYKURA, W. 1976. *Orkney and Shetland. British Regional Geology*, Institute of Geological Sciences, HMSO, Edinburgh, 149 pp.

—— 1983. Old Red Sandstone. *In*: CRAIG, G. Y. (ed.) *Geology of Scotland*. Academic Press, London, 205–251.

NORTON, M. G. 1986. Late Caledonide extension in western Norway; a response to extreme crustal thickening. *Tectonics*, **5**, 195–204.

——, MCCLAY, K. R. & WAY, T. A. Tectonic evolution of Devonian basins in northern Scotland and Southern Norway. *Norsk Geologisk Tidsskrift* (in press).

PEACH, B. N. & HORNE, J. 1914. Outliers of Old Red Sandstone between Bighouse Bay and Kyle of Tongue. *In*: CRAMPTON, C. B. & CARRUTHERS, R. G. (eds) *The Geology of Caithness. Memoir of the Geological Survey, Scotland.*

——, ——, GUNN, W., CLOUGH, C. T., HINXMAN, L. W. & TEALL, J. J. H. 1907. The geological structure of the north-west Highlands of Scotland. *Memoir of the Geological Survey United Kingdom*.

PEDDY, C. P. 1984. Displacement of the Moho by the Outer Isles thrust shown by seismic modelling. *Nature*, **312**, 628–630.

PIASECKI, M. A. J. 1980. New light on the Moine rocks of the central Highlands of Scotland. *Journal of the Geological Society of London*, **137**, 41–59.

RAYNER, D. H. 1963. The Achanarras Limestone of the middle Old Red Sandstone, Caithness, Scotland. *Proceedings of the Yorkshire Geological Society*, **34**, 117–138.

ROCK, N. M. S. 1983. The Permo–Carboniferous camptonite–monchiquite dyke suite of the Scottish Highlands and Islands: distribution, field and petrological aspects. *Report of the Institute of Geological Sciences*, **82/14**.

SERANNE, M. & SEGURET, M. 1987. The Devonian basins of western Norway, tectonics and kinematics of an extending crust. *In*: COWARD, M. P., DEWEY, J. F. & HANCOCK, P. (eds) *Continental Extensional Tectonics*. Special Publication of the Geological Society of London, **28**, 537–548.

SMYTHE, D. K., DOBINSON, A., MCQUILLAN, R., BREWER, J. A., MATTHEWS, D. H., BLUNDELL, D. J. & KELK, B. 1982. Deep structure of the Scottish Caledonides revealed by the MOIST reflection profile, *Nature*, **299**, 338–340.

SONDER, L. J., ENGLAND, P. C., WERNICKE, B. P. & CHRISTIANSEN, R. L. 1987. A physical model for Cenozoic extension of western North America. *In*: COWARD, M. P., DEWEY, J. F. & HANCOCK, P. (eds) *Continental Extensional Tectonics. Special Publication of the Geological Society of London*, **28**, 187–201.

SOPER, N. J. & BARBER, A. J. 1982. A model for the deep structure of the Moine Thrust Zone. *Journal of the Geological Society of London*, **134**, 41–44.

—— & HUTTON, D. H. W. 1984. Late Caledonian sinistral displacements in Britain: implications for a three plate collisional model. *Tectonics*, **3**, 781–794.

STEEL, R. J. & WILSON A. C. 1975. Sedimentation and tectonism (?Permo–Triassic) on the margin of the North Minch Basin, Lewis. *Journal of the Geological Society of London*, **131**, 183–202.

SWEET, I. P. 1985. The sedimentology of the Lower Old Red Sandstone near New Aberdour, Grampian Region. *Scottish Journal of Geology*, **21**, 239–259.

TAPPONNIER, P., PELTZER, G. & ARMIJO, R. (eds) 1986. *On the Mechanics of the Collision Between India and Asia*. Special Publication of the Geological Society of London, **19**, 115–157.

TRAQUAIR, R. H. 1894a. Achanarras Revisited. *Proceedings of the Royal Physical Society, Edinburgh*, **12**, 279–286.

—— 1894b. A Monograph of the Fish of the Old Red Sandstone of Britain. *Monograph of the Palaeontological Society*, **1**, Part ii.

VAN BREEMAN, O., AFTALION, M. & JOHNSON, M. R. W. 1979. Age of the Loch Borralon complex, Assynt and late movement on the Moine thrust. *Journal of the Geological Society of London*, **136**, 489–496.

WATSON, J. 1985. Northern Scotland as an Atlantic–North Sea Divide. *Journal of the Geological Society of London*, **142**, 221–243.

WERNICKE, B. 1981. Low-angle normal faults in the Basin and Range province: nappe tectonics in an expanding orogen. *Nature*, **291**, 645–648.

—— 1985. Uniform-sense normal simple shear of the continental lithosphere. *Canadian Journal of Earth Science*, **22**, 8–25.

WILCOX, R. E., HARDING, T. P. & SEELY, D. R. 1973. Basic wrench tectonics. *Bulletin of the American Association of Petroleum Geologists*, **57**, 57–96.

WILSON, G. V., EDWARDS, W., KNOX, J., JONES, R. C. B. & STEPHENS, J. V. 1935. *The Geology of the Orkneys. Memoirs of the Geological Survey, United Kingdom*, **117–122**.

M. P. COWARD, Department of Geology, Imperial College of Science and Technology, London SW7 2BP, UK.

M. A. ENFIELD, Department of Geology, Imperial College of Science and Technology, London SW7 2BP, UK. Present address: Maxus Energy Corporation, 717 North Harwood, Dallas, Texas 75201, USA.

M. W. FISCHER, BP Petroleum Development, Farburn Industrial Estate, Aberdeen, UK.

Appendix

The effects of stretching or thickening

The effects of stretching or thickening of an anomalous crust can be demonstrated using the method of Le Pichon *et al.* (1982) where subsidence (S_i) is related to stretching by the relationship:

$$S_i = C(1 - 1/\beta)$$

where β is the stretching factor (McKenzie 1978) and C is the constant related to crustal lithospheric densities and thicknesses. For infinite stretching $S_i = C$ that is, C represents the level to which the hot asthenosphere mantle would rise after infinite instantaneous stretching of the lithosphere. For typical lithospheric densities and thermal conditions, C lies between 3.2 km and 3.65 km and averages about 3.4 km. This level is known as the mantle geoid. After thermal equilibrium the total subsidence (S_t) is given by:

$$S_t = C_t\,(1 - 1/\beta)$$

where C_t is about 7.5 and the thermal subsidence $S_{th} = S_t - S_i$.

If the lithosphere is more buoyant, that is, if it contains more material of lower density, such as thick crust, and if the lithosphere is stretched homogeneously, the average density does not change but the thickness changes to $L \times 1/\beta$, where L is the initial thickness and the elevation of the lithosphere (E) above the mantle geoid is given by:

$$E = \frac{L/\beta\,(\rho_a - \rho_l)}{\rho_a - \rho_w}$$

when under water and:

$$E = [L/\beta\,(\rho_a - \rho_l) + C\rho_w]\,\rho_l/\rho_a$$

when above water, where ρ_a is the average density of the asthenosphere, ρ_l the average density of the lithosphere, ρ_w the density of sea water and C the depth of the mantle geoid.

The subsidence associated with different topographies is shown in Fig. 22. A change in scale is used above sea level to convert virtual water depth to actual altitude by multiplying by a factor $= \rho_a\,(\rho_a - \rho_w)$. The curves have a simple relationship:

$$S_i = S_0 + (1 - 1/\beta)\,(C - S_0)$$
$$S_t = S_0 + (1 - 1/\beta)\,(C_t - S_0)$$

for initial (S_0) and final depths (S_i, S_t) below sea level and

$$e_i = e_0\left(\frac{\rho_a}{\rho_a - \rho_w}\right) - \left[e_0\left(\frac{\rho_a}{\rho_a - \rho_w}\right) + C\right]\left(1 - \frac{1}{\beta}\right)$$

$$e_t = e_0\left(\frac{\rho_a}{\rho_a - \rho_w}\right) - \left[e_0\left(\frac{\rho_a}{\rho_a - \rho_w}\right) + C_t\right]\left(1 - \frac{1}{\beta}\right)$$

for an initial elevation (e_0) above sea level, where e_i and e_T are the final elevations for initial and total subsidence. These curves assume a sediment starved basin. The final depths can be modified by the addition of sediment of different densities. Thus, for an initial topographic elevation of 2 km, an extension value of about $\beta = 2$ is required before the rocks subside to below sea level. These relationships can be used to find the depths to which the top corners of rotated fault blocks will subside (cf. Barr 1987) and hence predict the amount of erosion associated with faults with particular spacing.

The above discussion assumes that ρ_l does not change with extension or with time. However, ρ_l will change if extension is heterogeneous, following the Wernicke model (Wernicke 1981; 1985) or its modification (Coward 1986b). Different subsidence plots will be produced depending on the levels of decoupling and density variations in the crust, as well as the degree of heterogeneity of stretching. The elevation of heterogeneously stretched crust above the mantle geoid is given by

$$E = \left[\frac{L_u}{\beta}\;(\rho_a - \rho_{ul}) + L_z\,(\rho_a - \rho_{zz})\right]/(\rho_a - \rho_w)$$

where L_u is the thickness of the upper stretched lithosphere of average density ρ_{ul} and L_z the thickness of the lower unstretched lithosphere, of average density ρ_{zz}. Simple curves are given in Fig. 23 to show the subsidence histories of heterogeneously stretched lithosphere, assuming that stretching was confined to an upper plate, with a detachment 11 km above the Moho and stretching was confined to the lower plate, using the average densities given by Coward (1986b). For an initial topographic elevation, less stretching
is required before the rocks subside below sea level if stretching was confined to upper crustal levels, rather than homogeneous throughout the lithosphere. Note that these relationships can be inversely applied to compressional tectonics of initially thin crust, that is to positive inversion of a stretched basin.

Rift basin evolution in Africa: the influence of reactivated steep basement shear zones

M. C. Daly, J. Chorowicz & J. D. Fairhead

SUMMARY: Phanerozoic rift basin evolution in Africa has been strongly influenced by the pre-existing basement structure inherited from Precambrian tectonics. Two modes of basement reactivation may be identified; the largely dip–slip reactivation of gently dipping shear zone and thrust structures and the largely strike–slip reactivation of steep shear zones and faults. In particular, crustal scale steep basement shear zones have exercised a profound control on rift basin formation on a regional scale. Such shear zone structures are seen to govern the location, trend and structural style of rifts and appear to constrain the orientation of displacements generating the rifts. The shear zones are relatively narrow, high strain zones formed during ductile deformation. They occur as steep, crustal scale ramps that define a pronounced mechanical anisotropy within the African continental crust. These structures are difficult to reactivate by dip–slip movement and thus tend to act as major strike–slip transfer zones during later extensional events. This paper discusses examples of the steep shear zone–rift relationship from three of the rift systems of Africa: the Upper Palaeozoic (Karoo) rifts of Central Africa, the Mesozoic rifts of the Sudan and the Cenozoic rifts of East Africa. These discussions indicate that steep shear zones are sites of rift nucleation and play a fundamental role in the mechanisms of rift propagation and subsequent continental breakup.

Introduction

The reactivation of pre-existing structures has long been invoked as a major influence on the evolution of geological structures (Sykes 1978). Experimental rock deformation studies demonstrate that under conditions of brittle failure, individual faults may have a lower cohesive strength and frictional sliding threshold than the rock surrounding the fault. Hence, given an appropriate stress orientation, failure is likely to occur along a pre-existing fault (Donath & Cranwell 1981). For rocks characteristic of more ductile deformation conditions the reactivation of a pre-existing fabric defining a strong anisotropy appears mechanically likely. Watterson (1975) recorded a complicated history of reactivation on crustal scale ductile shear zones and suggested that the repeated usage of these structures was a reflection of the finer grain size of the mylonitic shear zones in comparison to the country rock. Such structures have a high 'reactivation potential' given the appropriate stress conditions. Sykes (1978) supported this reasoning by pointing out that continental intraplate earthquake zones are commonly concentrated along pre-existing continental fault zones. This identifies these as weaknesses within the continental crust.

From the above discussion it seems a compelling argument that pre-existing faults and shear zones represent a mechanical weakness both at upper and lower crustal levels. However, not all fault and shear zones reactivate or have any apparent influence on later structural developments. When structures do reactivate, the manner of the reactivation and its precise influences on later structural development is often poorly defined.

Critical in the reactivation of a structure is its attitude to an imposed stress. A steep structure will be difficult to reactivate in a reverse dip–slip sense, but relatively easy to reactivate in a strike–slip sense (Etheridge 1986). In the continental crust the crustal scale faults and shear zones that bound orogenic belts have a variable geometry corresponding crudely to the ramp–flat trajectories widely described in thrust belts (Coward & Daly 1984; Ando *et al.* 1984). Such ramps develop parallel (lateral), obliquely (oblique) and perpendicular (frontal) to the regional tectonic transport direction and in doing so define a complex 3-D shear zone geometry (Daly 1986). It is observed that the lateral ramps tend to be relatively steep dipping structures (>60°) and the frontal ramps tend to be relatively shallow dipping. Given that such geometries exist, two modes of extensional reactivation may be identified (Fig. 1):

1 Dip–slip reactivation of gently dipping fabrics and structures.

From Cooper, M. A. & Williams, G. D. (eds), 1989, *Inversion Tectonics*
Geological Society Special Publications No. 44, 309–334.

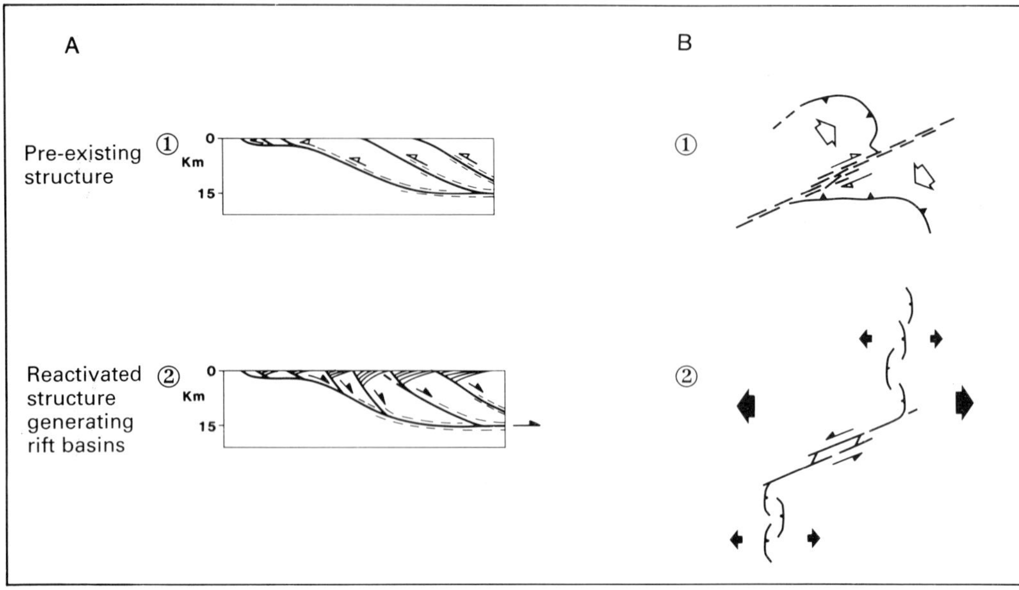

FIG. 1. Modes of basement reactivation during extension. A, Reactivation of a gently dipping fabric, eg. thrusts. B, Reactivation of a steeply dipping fabric, eg. strike–slip shear zones.

2 Strike–slip reactivation of steeply dipping fabrics and structures.

The study of the first mode of reactivation is best approached from the view of the deep reflection seismic profile (Cheadle *et al.* 1987) and is not discussed here. The second mode is perhaps most clearly demonstrated in map view and is the subject of this paper. We discuss three rift systems from Africa where we believe a major control on the rift basin location, structure and orientation has been exercised by steep pre-existing shear zones in the continental crust. In each example a brief description of the rift system is followed by a summary of the major basement structure underlying the rift zone. The relationship between the two is then highlighted and the rift systems placed in their regional tectonic setting. The implications of the observed relationships for rift propagation and continental breakup are then discussed.

The African continental crust and reactivation

For some time it has been observed that Phanerozoic rift systems avoid cratonic areas. They tend to 'wrap' around them, much in the way the Proterozoic orogenic belts do. This relationship is clearly observed on a geological map of Africa (Fig. 2). It is an undisputed form of basement control on rift location (King 1970). The same pattern is revealed by the location of earthquakes in Africa (Fig. 3). The gross distribution of earthquakes clearly outlines the major cratonic areas (Fairhead & Henderson 1977). Similarly, heat flow data (Ballard *et al.* 1987) show fundamental differences in the thermal structure between cratons and orogenic belts in Southern Africa.

The craton-bounding orogenic belts comprise a number of major lateral ramp structures, seen as ductile shear zones, that developed parallel to estimated tectonic transport direction (Coward & Daly 1984; Daly 1988). These shear zones tend to be steep and relatively narrow structures and are of paramount importance in rift basin formation in Africa.

Several accounts of rift tectonics make reference to a basement influence on the rifts (Rosendahl *et al.* 1986) but do not identify the nature of this influence or the basement structures involved. A rather contentious view of the basement control on rift systems in Africa has been proposed by McConnell (1980). He claimed the East African rift system followed a Precambrian lithospheric lineament extending 5000 km from Afar to the Orange river of South Africa. McConnell (1980) defined this

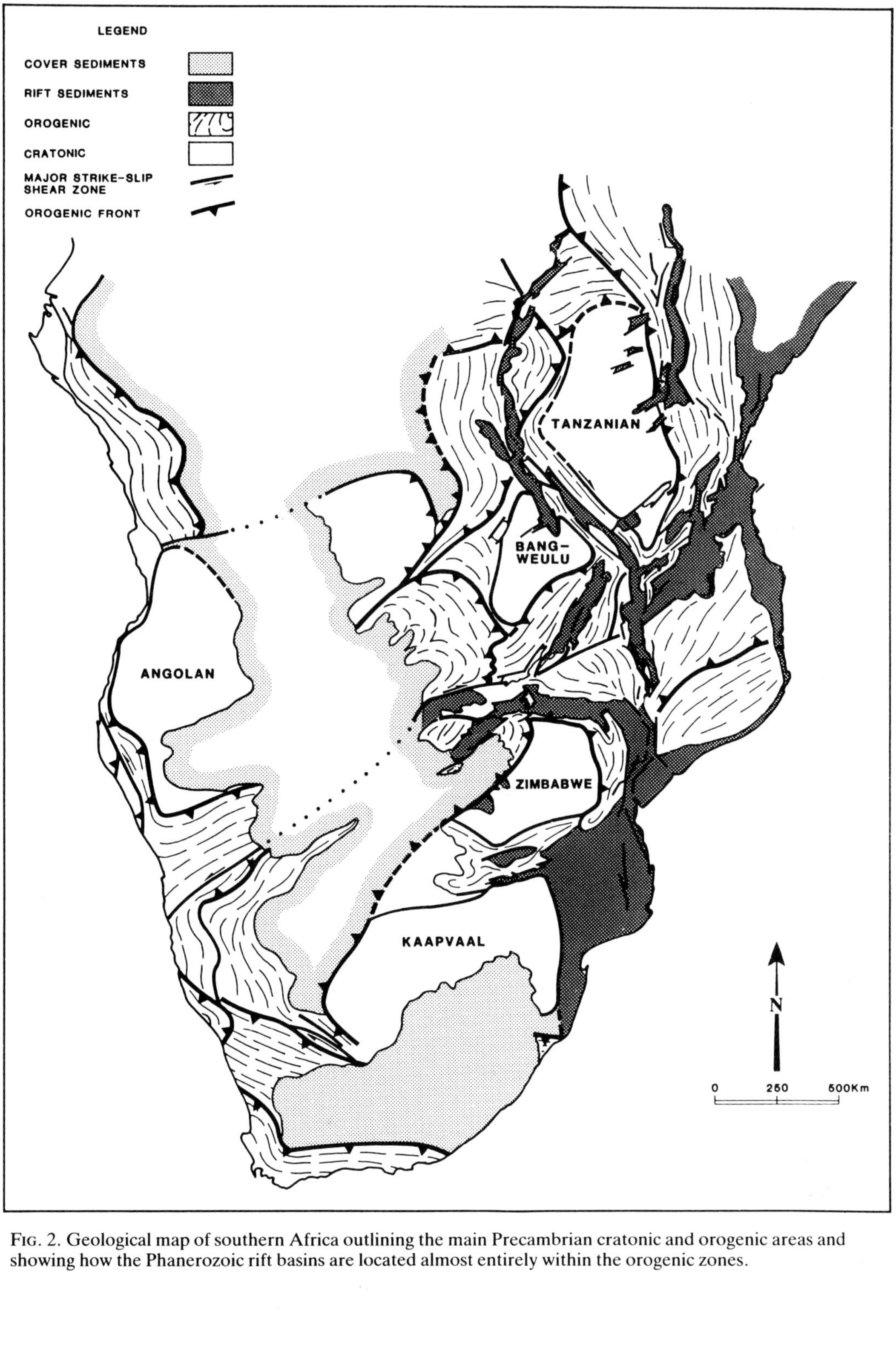

FIG. 2. Geological map of southern Africa outlining the main Precambrian cratonic and orogenic areas and showing how the Phanerozoic rift basins are located almost entirely within the orogenic zones.

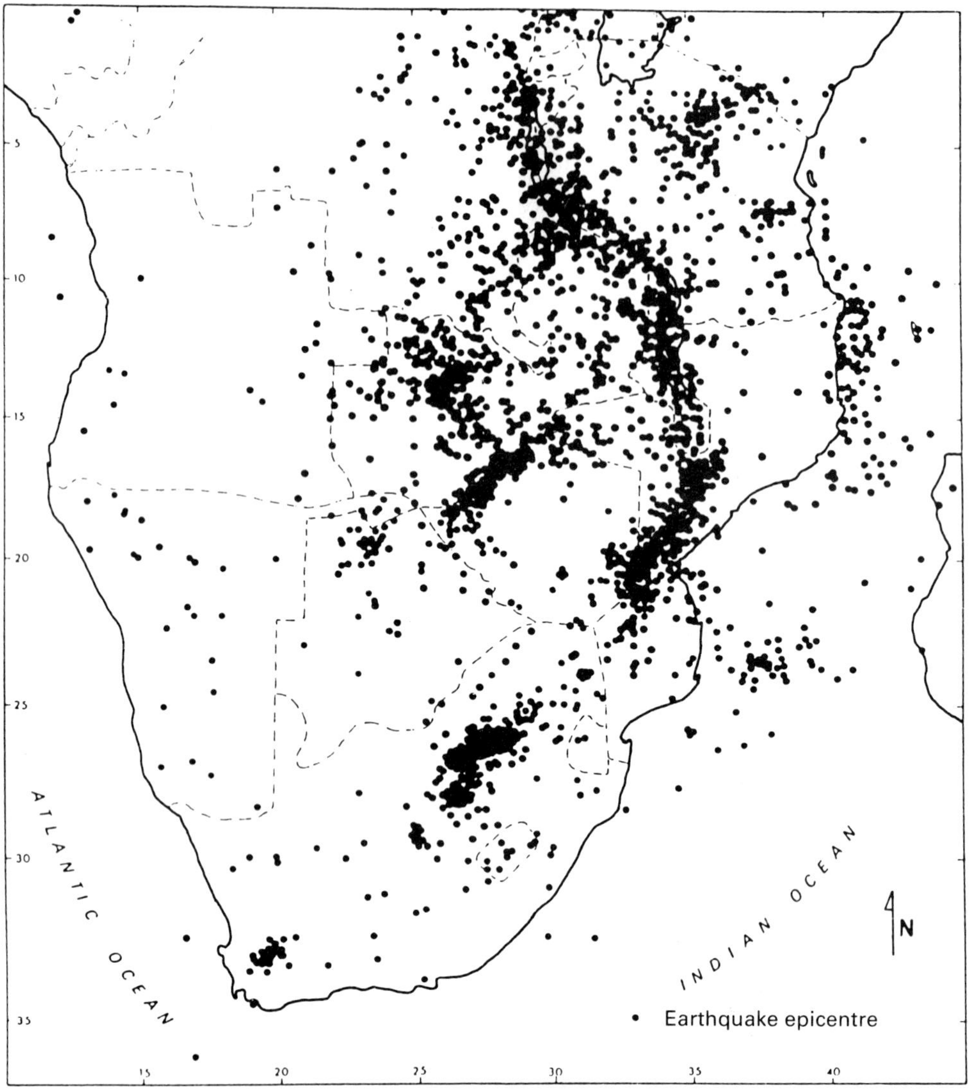

FIG. 3. Map showing the distribution of historical seismic events in Africa. Note how the seismicity is also confined to the orogenic zones.

lineament by linking such diverse geological features as the Great Dyke, the Bushveldt igneous complex and the Gregory Rift. The existence of the lineament is not supported by the present pattern of African seismicity nor by a consideration of the Late Precambrian tectonic evolution of Africa generally. Other discussions of reactivation in Africa have concentrated on potential basement control on the location of Jurassic and Cretaceous volcanism (Cox 1970) and the superimposition of Karoo sediments in the Proterozoic age. Waterberg Graben (Green *et al.* 1980).

The East African rift system

The East African rift system defines a diffuse plate boundary within Africa (McKenzie *et al.* 1970) and is often regarded as an example of the early stage of continental breakup. The rift system comprises two main branches: an eastern branch running from Afar to northern Tanzania and a western branch from Lake Albert to the Mozambique coastal plane (King 1970; Baker & Wohlenberg 1971). The two branches are linked by a poorly defined zone of faulting to the east and NE of Lake Albert. This broad zone parallels the Proterozoic Aswa shear zone of Uganda and incorporates the Nandi fault of Kenya (Fig. 4). The present discussion will restrict itself to the W branch of the rift.

Geometry of the Western Rift

The Western Rift has an overall sigmoidal shape made up of three segments (Fig. 4): a N segment from Lake Albert to Lake Tanganyika; a central segment from Lake Tanganyika to Lake Malawi and a S segment from Lake Malawi to the Mozambique coastal plain.

The *N* segment includes the Lakes Albert, Mobutu, Kivu and North Tanganyika. Lake Albert has a NE–SW trend that swings southwards to become N–S in the region of Lake Tanganyika. The northernmost Albert Nile graben is bordered to the N by the Aswa zone, a Late Proterozoic ductile shear zone. The Aswa zone comprises a series of small, Neogene, *en échelon* fault scarps that appear to link into the Nandi fault zone east of Mount Elgon. The ill-defined Aswa zone and the Mesozoic rifts to the north act as a transfer zone between the Western and Eastern Rifts.

The individual rift basins that make up the Western Rift have an asymmetrical, half-graben geometry. They are bounded on one side by a major border fault and on the other side by antithetic faults or a monoclinal flexure.

This N segment of the Western Rift comprises eight separate basins, each defined by a NE–SW to N–S trending border fault which terminates in transverse ridges of basement. These basement ridges define the transfer zones that link the basins. The ridges characteristically have a NW–SE strike. The well defined border fault segments vary from 60 to 120 km in length and form the major morphological features of the region.

The central segment of the Western Rift comprises the central part of Lake Tanganyika, Lake Rukwa and the N part of Lake Malawi (Figs 4 & 5). The major border faults which define this segment of the rift have a NW–SE trend, are linear in outcrop, and are continuous for 300–500 km along strike (Fig. 5). These major faults have an en-echelon arrangement and define a deformation zone 150 km wide and 900 km long. A number of smaller faults of NE–SW strike occur at the N end of Lake Rukwa and SE Lake Tanganyika. All these smaller faults link into the large NW–SE faults and appear to be part of the same fault system. At the N end of Lake Malawi the NW–SE trending Livingstone fault curves progressively southwards into a N–S fault (Figs 4, 5 & 6). This curvature and the faults at the NW end of Lake Rukwa define a narrow, deep, rhomb-shaped basin of characteristic pull apart geometry (Aydin & Nur 1982).

The S segment comprises the Lake Malawi rift and the smaller basins of Mozambique (Fig. 6). The Malawi rift basins comprise a succession of half-grabens separated by NW–SE and NE–SW striking transfer zones seen as narrow basement highs between the basins. These highs are defined by NW–SE striking faults that link into the major border faults. The Malawi rift comprises nine border faults that define the polarity of the individual basins (Ebinger *et al.* 1986). This border fault repetition occurs at an average wavelength of about 70 km and is comparable in structural style to the N segment of the Western Rift described above.

To the south of Lake Malawi the Tertiary rift is poorly defined and consists of a N–S valley along which the Shire presently flows. There has been no significant subsidence in this area. The Shire valley is abruptly cut by the NW–SE striking Mwanza fault in the Lower Zambezi area. This major fault defines a regional transfer zone that links displacement to the Urema and Dombe graben of Mozambique (Fig. 4). These ill-defined basins appear to fade southwards and have little expression on the Mozambique coastal plain.

The above discussion highlights three important features of the Western Rift system:

1. The northern and southern segments of the Western Rift comprise a series of half-graben basins defined by arcuate border faults, often of alternating polarity. Individual border fault segments have a length of 60–120 km and define an across rift deformation zone of 50 km.
2. The transfer zones linking these border faults are dominantly NW–SE-trending in the northern and southern segments and are characteristically defined by fault bounded upstanding basement blocks defining inter-basinal highs.
3. The central segment of the Western Rift differs markedly in structural style and trend from the northern and southern segments. The major rift border faults are linear in outcrop. They are between 300 and 500 km long and define a broad deformation zone of 150 km. The polarity along these linear faults does change along strike although a regular periodicity is difficult to discern due to the poor definition of maximum vertical displacements.

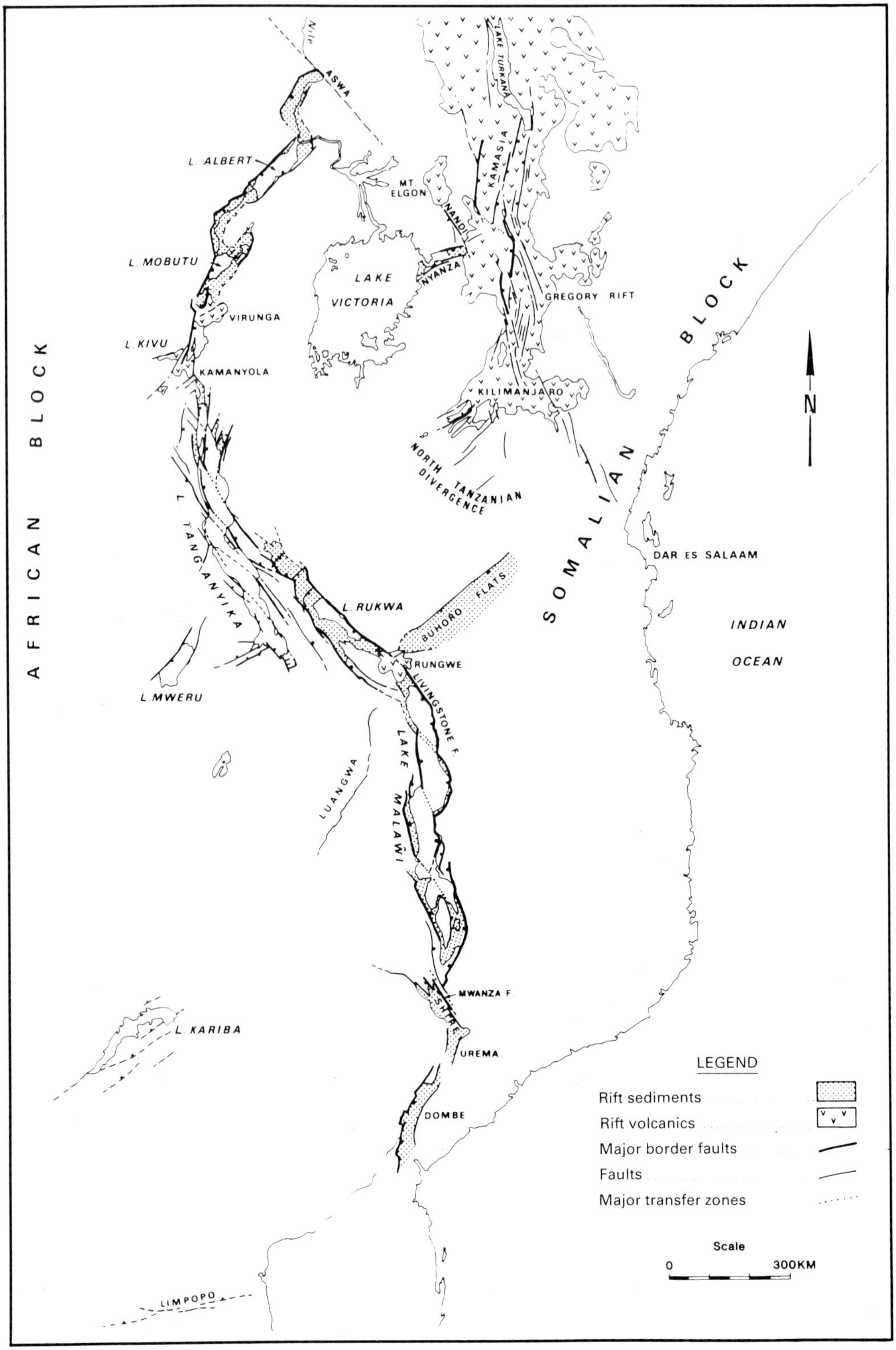

FIG. 4. Regional structural sketch map of the African Tertiary rift system. Note the major kink in the central part of the Western Rift. It divides into northern, central and southern segments.

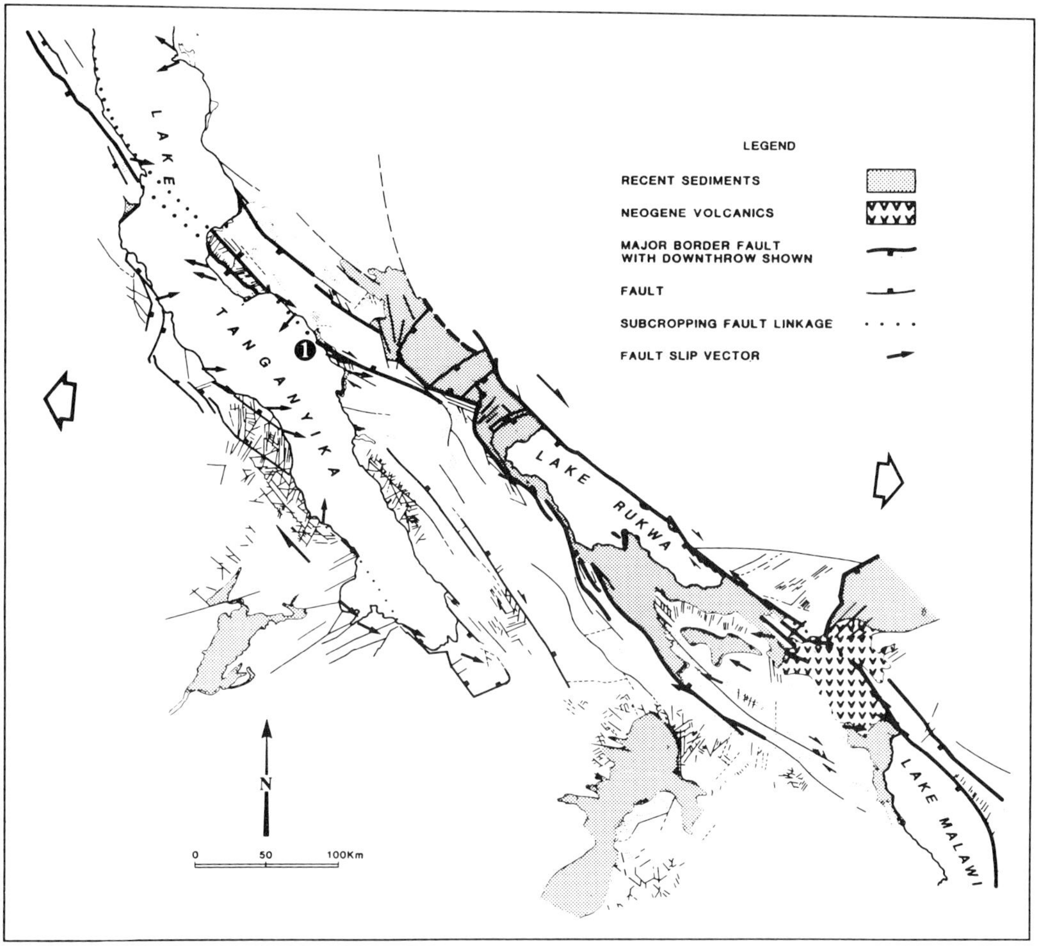

FIG. 5. Structural sketch map of the central segment of the Western Rift, Lake Rukwa area. Note the long, straight border faults, the broad deformation zone (*circa* 250 km) and the oblong nature of the rift basin. 1, marks the location of Karema Mission.

Kinematics of the Western Rift

An analysis of fault surface slip indicators, slickensides, and striations, together with outcrop scale fracture patterns has enabled the identification of the direction and sense of slip along the major faults of the rift system (Chorowicz *et al.* 1987). These data have been interpreted in terms of the local slip vector for a particular fault (Figs 5 & 6).

The Neogene age of the fault plane structures is constrained by several features. These include observations made in Neogene sediments or volcanics, faults that displace a Neogene erosional surface and fault scarps bordering the rift. It is common that the faults studied only show one set of brittle features and that dilatant fractures within these zones are not mineralized. They are instead empty or filled with recent sediment.

In the Lake Malawi region, the N–S trending rift structures comprise dominantly normal faults with an oblique to dip–slip displacement of normal sense (Fig. 6). This implies an approximately WNW–ESE to E–W extension direction associated with the main phase of fault movement. In contrast NW–SE striking faults of the Malawi area, such as the Mwanza and Livingstone faults (Fig. 6) show consistantly sub-horizontal slip vectors with a dextral shear sense.

In the central segment of the Western Rift the major faults strike NW–SE. The slip vector is commonly horizontal and indicates dextral, strike–slip displacements (Fig. 5). These major

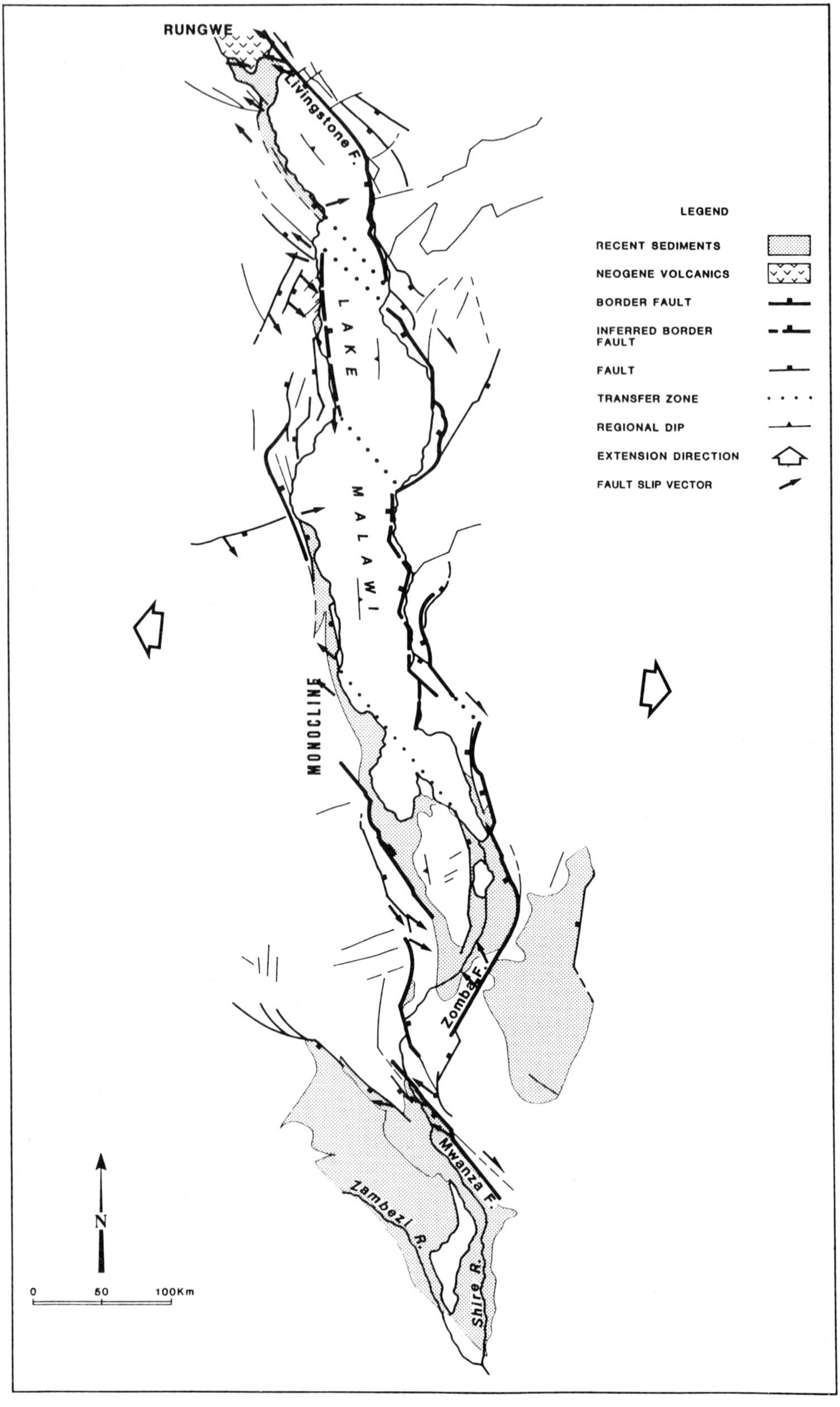

FIG. 6. Structural sketch map of the S segment of the Western rift, the Lake Malawi area. Note the frequently changing polarity of the curved major border faults, and the narrow width of the deformed zone (*c.* 100 km).

faults have subsidary fault sets striking NNW–SSE with small sinistral off-sets and NE–SW trending tension gashes. These relationships are displayed in the Karema Mission area of Lake Tanganyika (Fig. 5) where the Ufipa Fault comes onshore from the lake. The NNW–SSE subsidary fractures are interpreted as Reidel shears associated with regional NW–SE dextral strike–slip displacement on the Ufipa Fault.

From both the central and southern rift segments the fault mechanisms deduced from fault surface slip vectors are compatible with WNW–ESE extensional displacements with a dextral shear sense on NW–SE striking faults. A similar WNW–ESE extension direction is indicated for the northern segment of the Western Rift.

The fault kinematics and geometry of the Western Rift highlights the changing structural environment along the rift system. The NW–SE striking faults, dominant in the central segment and southern Malawi are essentially strike–slip zones with a minor dip–slip component. The N–S and NE–SW striking faults, dominant in the northern and southern segments, are dip–slip to oblique–slip normal faults. This arrangement of structures is compatible with a regional WNW–ESE to E–W extension. It is also compatible with the active displacements indicated by earthquake focal mechanism solutions from the Western Rift (Shudofsky 1985).

In plate tectonic terms the large NW–SE striking zones of the rift system may be regarded as continental 'leaky' transforms. The diffuse northern transform boundary of the Western Rift includes the Aswa zone and the reactivated faults of the Mesozoic rift basins of southern Sudan. This boundary links the Western and Eastern Rift systems. The central rift segment of the Lake Rukwa area defines the transform zone between the northern and southern segments of the Western Rift. The Mwanza fault of southern Malawi is similarly a dextral transform zone between the Malawi rift and the Urema and Dombe graben.

It is noteworthy that the Western Rift appears to die out to the N and S of the central segment. It is suggested that the rift nucleated in this central strike–slip section and propagated both N and S from it.

Basement structure of the Western Rift

The basement structure underlying the Western Rift system is extremely variable in orientation, type and origin. In the northern (Lake Albert to Lake Tanganyika) and southern (Lake Malawi) segments of the Western Rift the regional trend of the underlying Proterozoic orogenic belts is often at a high angle to the rift boundary faults (McConnell 1980). In Malawi the basement Mozambique belt gneisses show no suggestion of controlling the orientation or location of the major border faults (Fig. 7). Similarly, there is little suggestion of the preferential development of faults to 'avoid' relatively strong lithologies (Ebinger *et al.* 1986).

Our observations of the basement structure–rift border fault relationship for the northern and southern parts of the Western Rift agree with Crossley & Crow (1980), i.e. that no obvious control is exercised on the location and geometry of the border fault systems. This lack of a control is reflected in the periodicity recorded by their repeated change in polarity approximately every 80 km (Rosendahl *et al.* 1986). This periodicity apparently reflects a more fundamental aspect of rift systems (Bosworth 1985), yet to be understood. However, it is also apparent that the rift transfer zones are generally sub-parallel to the tectonic transport direction of the basement orogenic belt on which the rift system lies.

The central Western Rift, between northern Lake Tanganyika and northern Lake Malawi is underlain by rocks of the early Proterozoic Ubendian Belt. The Ubendian Belt comprises a number of geologically distinct fault or shear zone-bounded blocks or terranes (Fig. 8) (McConnell 1972; Daly 1988). These terranes are lithologically and structurally distinct. They are strongly elongated in a NW–SE direction and are laterally persistant for up to 400 km. Fig. 8 outlines the distribution of the terranes. Internally the terranes are heterogeneously deformed with extensive tracts of foliated, mylonitic and ultramylonitic gneisses imparting a strong anisotropy to the individual terranes.

The Ubendian terranes are bounded by a series of major, steep to vertical shear zones which are persistant throughout the belt for up to 600 km along strike. The shear zones are rectilinear in outcrop. This, together with their steep attitude at the present day erosion level, suggests a significant vertical dimension. The shear zones comprise a mixture of ductile and brittle fault rocks which vary from ultramylonites to gouge. They define a major mechanical anisotropy in the basement of the central segment of the Western Rift.

Basement and rift structure

Consideration of Figs 5 & 8 highlights the control on rift structure exercised by the terrane-bounding shear zones. The change in rift trend

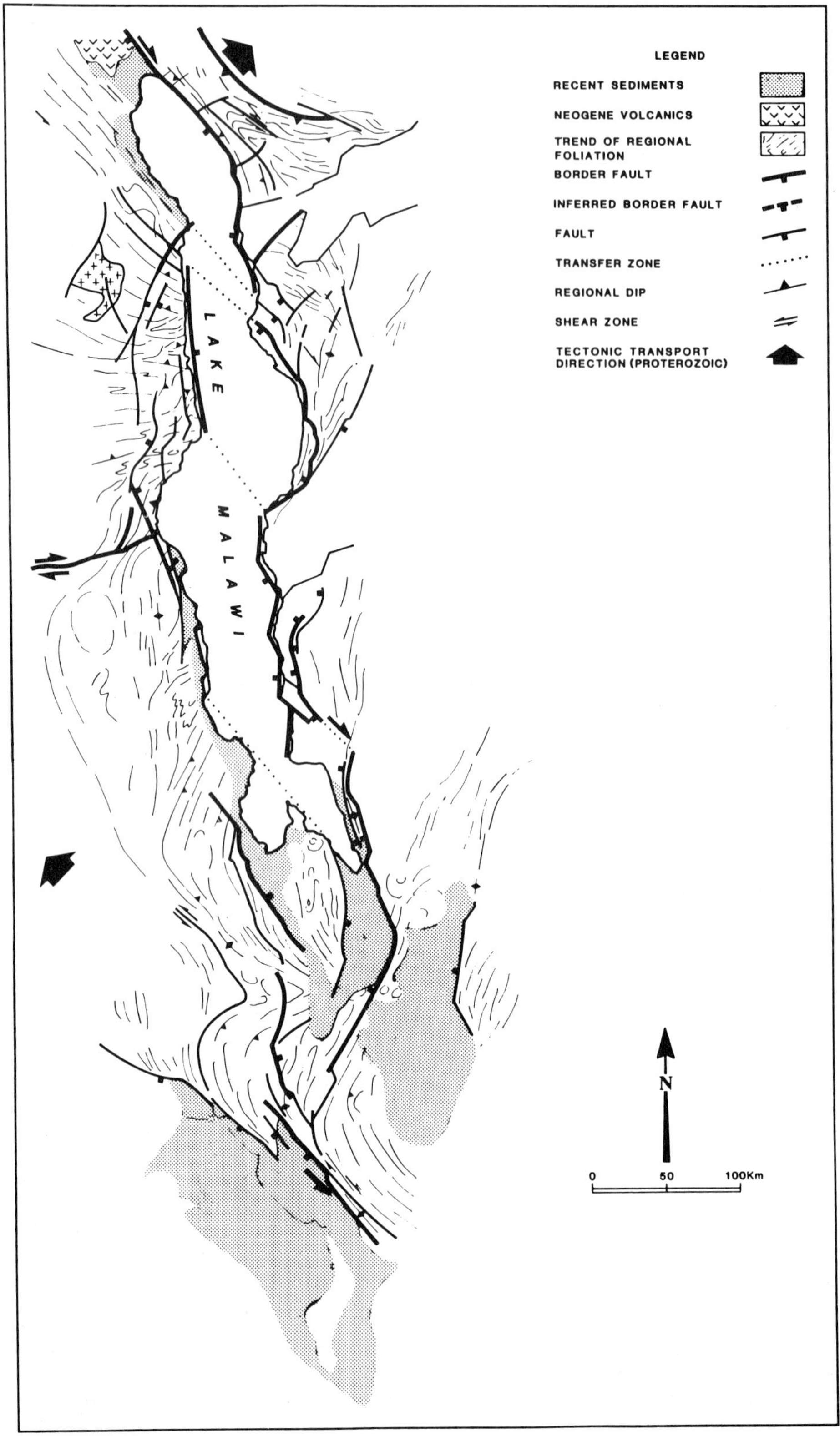

FIG. 7. Sketch map of the basement structure around the Lake Malawi rift. Note the variable strike of the basement foliations and the lack of parallelism between basement structural trend and rift border faults in the central part of the rift.

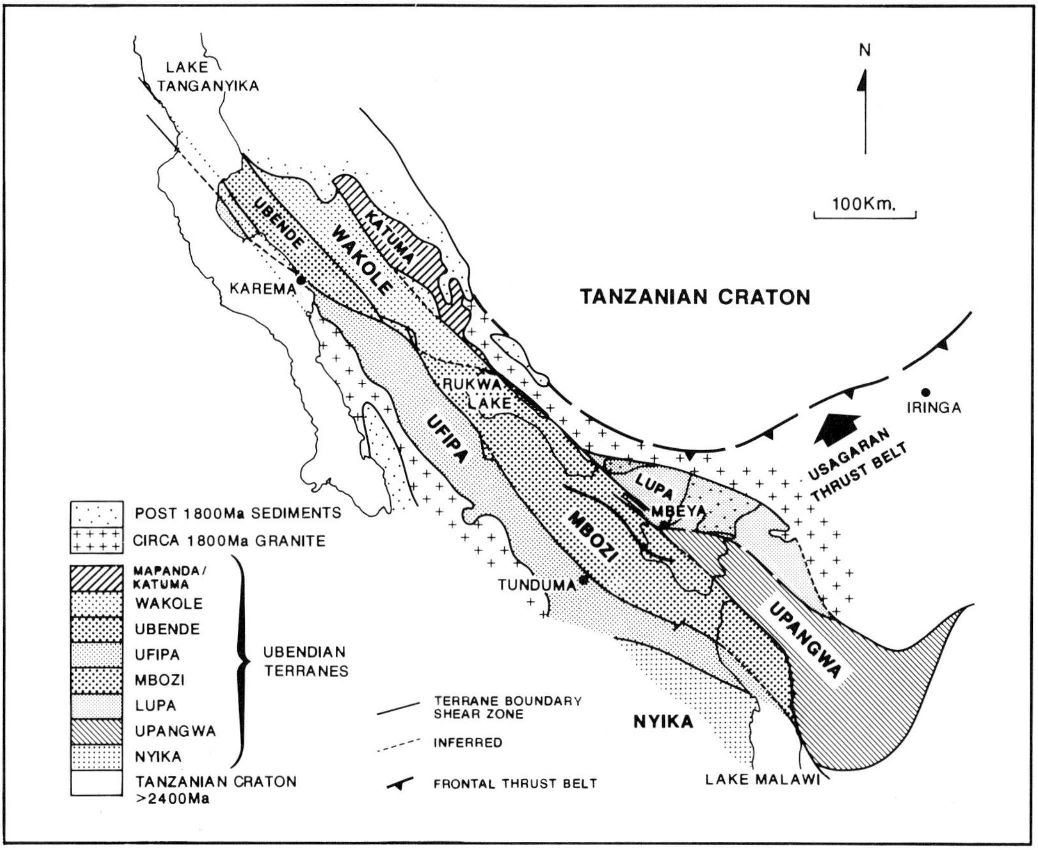

FIG. 8. Sketch map of the main terranes of the Ubendian belt, SW Tanzania. The terrane boundaries are defined by steep brittle–ductile shear zones. To the east the terranes pass into a NW-directed thrust belt known as the Usagaran.

and structural style in the central segment of the Western Rift is a direct result of the reactivation of the shear zones as brittle rift basin border faults. This results in the change of trend of the rift and the loss of the characteristic 80 km border fault length.

The southern border fault of the Rukwa Rift lies within a terrane-bounding shear zone (Mugesse shear zone). Rosendahl *et al.* (1986), by use of seismic profiling, show this same structure to define the northern margin of a basement high in Lake Tanganyika. Along strike to the NW, a mylonitic shear zone and accompanying fault occur on the lake shore of Zaire, a distance of 600 km.

The northern border fault of Lake Rukwa is defined by the shear zone between the Upangwa and Mbozi terranes. The present day expression of this structure is the Livingstone Fault (Figs 5 & 6). This also has a long rectilinear outcrop trace of 500 km. Together with the southern border fault it defines the oblong shape of the Rukwa rift.

The discussion of the fault kinematics of the central segment of the Western Rift demonstrated a dominantly NW–SE dextral strike–slip displacement pattern along the rift faults. This parallels the inferred displacement on the ductile shear zones bounding the Ubendian terranes (Daly 1988), and indicates that the reactivation of these structures has been subparallel to their original displacement direction.

This discussion demonstrates the profound influence of the steep basement shear zones of the Ubendian belt of Tanzania on the central segment of the Western Rift system. This influence is exhibited by the 45° difference in orientation of the rift system; the change in structural style from a series of short (80 km) overlapping, curvilinear border faults to long (300 km) rectilinear border faults and the change in rift kinematics from dominantly dip–slip (northern and southern segments) to dominantly strike–slip (central segment).

The Upper Palaeozoic basins of Southern and Central Africa

The Upper Palaeozoic Karoo basins of southern and central Africa (Fig. 9) occur in two main tectonic settings:

1 The Karoo foredeep basin of South Africa which developed and deformed in front of the advancing Cape fold belt.

2 The localized and fault-bounded rifts of Central and East Africa which appear related to intra-continental tectonics.

In spite of the two different tectonic settings and structural styles, a broad correlation of the stratigraphic units between South and Central Africa is possible. Their broad time equivalence is supported by fossil evidence (Taverner-Smith 1962). The deposition of the Karoo sediments terminated with the extrusion of a vast volume of basaltic lava of Early to Mid-Jurassic age.

The Karoo basins of Central Africa comprise a number of narrow and discrete half-graben structures that define a complex rift system (Fig. 10). The Mid and Lower Zambezi rifts and the Luano, Lukasashi, and Luangwa rifts are the two most extensive systems. The remaining rifts are the Maniamba rift of

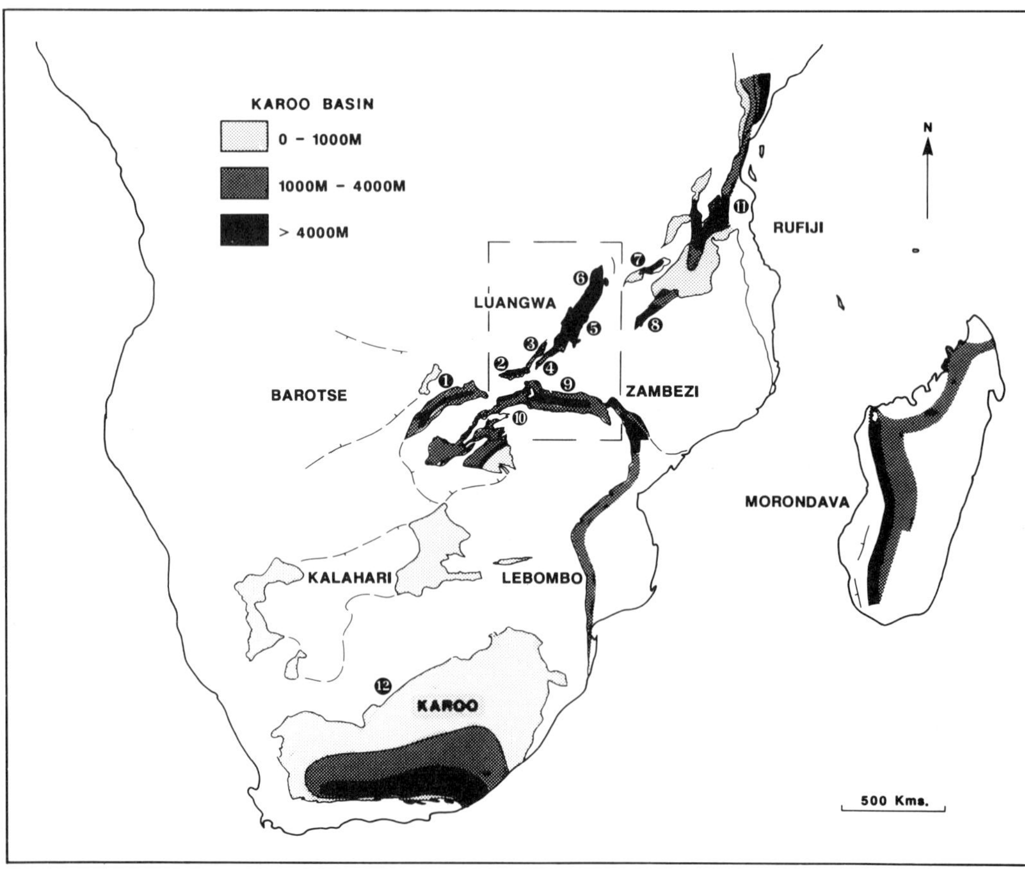

FIG. 9. Sketch map showing the distribution of the main Karoo basins of southern Africa and an indication of their depth. Karoo basins discussed in the text: 1, Kafue; 2, Luano; 3, Lukusashi; 4, South Luangwa; 5, Mid Luangwa; 6, North Luangwa; 7, Ruhuhu; 8, Maniamba; 9, Lower Zambezi; 10, Mid-Zambezi; 11, Rufiji and Coastal basins; 12, Karoo Foreland basin. The box indicates the location of Fig. 10.

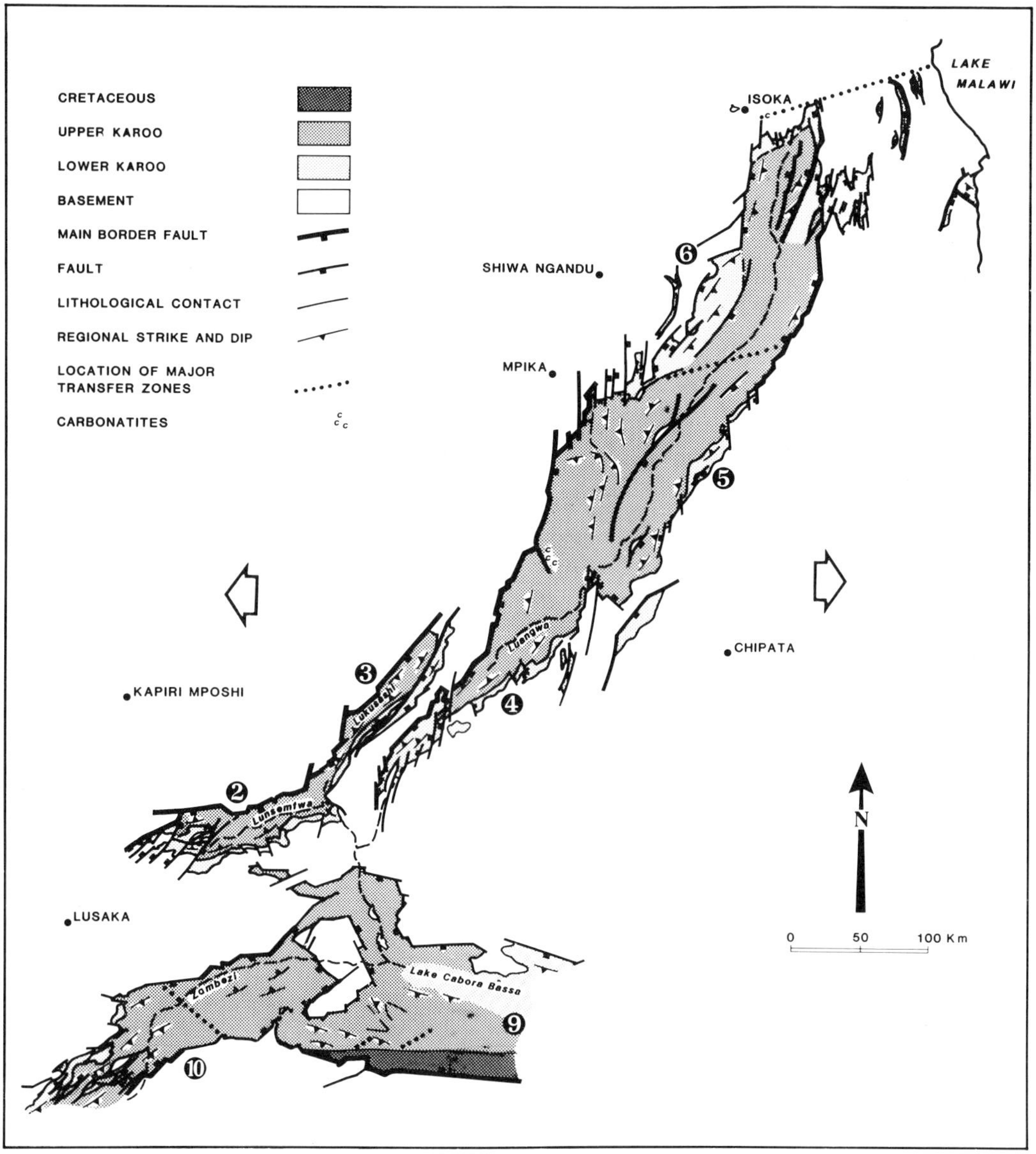

FIG. 10. Structural sketch map of the Karoo basins of eastern Zambia, the Luangwa, Luano and Zambezi. The location and key to numbers is shown on Fig. 9.

Mozambique and the Ruhuhu rift of Tanzania. The latter links eastwards into the coastal Karoo basins (Fig. 9).

Fault geometries in the various rift basins indicate a complex and variable extension history, much complicated by Tertiary reactivation that deforms a regional Miocene erosion surface (Gair 1959). However, an insight into the tectonic evolution of the rift system can be gained by reference to the basic fault pattern and inferred displacement directions.

Rift basin fault patterns

The elongate, ENE trending Kafue, Luano, Ruhuhu and Maniamba rift basins are almost entirely fault bounded. They exhibit an asymmetry defined by the major border faults. This gives them a half-graben geometry. The location of the major fault is variable, lying both N (Luano) and S (Ruhuhu) of the basins. It is almost always accompanied by antithetic faulting on the other margins (Fig. 11). The E and W terminations of the basins are also commonly fault bounded.

The faults within the basins are characteristically oblique to the basin border faults and link with them at the basin margins. This feature is seen clearly in the Luano, Ruhuhu and Maniamba rifts. The Karoo sediments in the hanging wall to the internal faults dip towards the fault plane. Such fault and basin geometry is characteristic of pull-apart basins developed along strike–slip fault systems.

Aeromagnetic data clearly outline the Kafue, Luano and Ruhuhu basins by their absence of the complex magnetic signature characteristic of the surrounding basement. This feature indicates thicknesses of non-magnetic Karoo sediments to be a minimum of 5 km. Estimates of 2 km of Karoo sediments in the Luano rift of Zambia have been made from field mapping and shallow drilling (Reichwalder & Brandon 1978).

The isolated, elongate fault-bounded basins are interpreted as pull-apart basins developed during a period of ENE–WSW, sinistral strike–slip displacement. Within the individual basins a similar structural style and stratigraphic sequence occurs. The stratigraphic sequence characteristically commences with localized coal bearing shales and sandstones and passes up through a mudstone sequence into coarse clastic sandstones correlated with the Stormberg of South Africa. The major basin forming event appears to be marked by the onset of coarse clastic sedimentation which had a more localized, within-rift, distribution compared to the preceeding, more regionally developed, lacustrine facies.

Oblique to the ENE-trending rifts lie the Lukusashi, Luangwa and Zambezi rifts. The Lukusashi–Luangwa rifts comprise a series of four *en échelon* half-graben basins. The Lukusashi and South Luangwa grabens are much smaller than the Mid- and North Luangwa graben (Fig. 10). The major border faults of these basins trend NNE to N–S and generally downthrow to the SE. This polarity changes in the North Luangwa graben where the major border fault lies to the E and downthrows to the west. Basement inliers within the rift system outline the interbasin highs and define the four *en échelon* rifts (Fig. 10). Transfer zones within the rifts are ill-defined but appear to trend ENE–WSW.

Both the Lukusashi and Luangwa rifts terminate along a line extending from the S margin of the Luano rift (Fig. 10). The stratigraphies of the basins are similar but the sedimentary

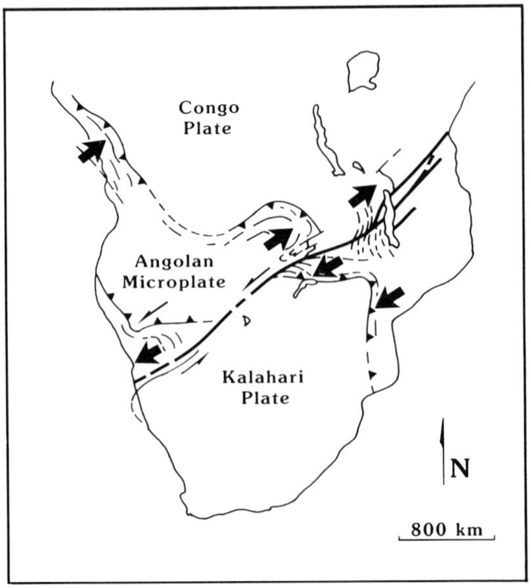

FIG. 11. Sketch map showing the Late Proterozoic tectonic setting of Central Africa. This comprises a pair of opposed thrust belts separated by the continental scale Mwembeshi shear zone.

thicknesses present in the Lukusashi and Luangwa rifts are uncertain. Aeromagnetic data reveals a near total suppression of the basement magnetic signature. This indicates a minimum thickness of the order of 5 km. The NNE to N–S *en échelon* half-grabens of the Lukusashi–Luangwa rifts are compatible with the sinistral strike–slip tectonic setting proposed for the associated ENE-trending basins. They are interpreted as a major splay from the strike–slip system and were generated by oblique, E–W extension.

The Lower Zambezi rift, defined by a major E–W trending border fault in the south, has sediments dipping both to the SW–WSW and NE–ENE. The folds defined by these dip reversals plunge to the ESE and are progressively overlain by Cretaceous sediments. This dates the folding as pre-Cretaceous. The fold structures may indicate a period of contractional tectonics which post-dates the original rift formation. In contrast the Mid-Zambezi rift has an internal fault geometry indicative of WNW–ESE extension (Fig. 10) that clearly deforms the supposedly Miocene erosional surface (Gair 1959). This extensive neotectonic imprint makes it difficult to interpret the Karoo aged fault geometry. However, the sedimentology of the Escarpment Grit Karoo Formation shows evidence of syndepositional faulting (Gair 1959; Taverner-Smith 1962) and palaeocurrent data from the same formation indicates a axial drainage system flowing from NE to SW along the rift. This strongly suggests the present rift developed syntectonically with sedimentation.

Our interpretation of the Central African Karoo basins indicates they developed as a sequence of pull-apart and oblique extensional basins associated with ENE–WSW sinistral strike–slip displacements. These displacements may have instigated the clastic phase of Karoo sedimentation, thought to be Triassic (Stormberg) in age.

Basement structure

The Central African Karoo basins lie between the Zimbabwe and Tanzania cratons in a complex network of orogenic belts (Fig. 2). The major tectonic features of this region are a series of Late Proterozoic thrust belts, the Irumide belt, the Lufilian arc and Zambezi belt, and an intervening steep ductile shear zone, the Mwembeshi zone (Fig. 11) (Daly 1986). The Lufilian arc and Zambezi thrust belts are coeval and opposed in their dominant thrust direction, i.e. the Lufilian arc thrusting was ENE-directed and the Zambezi belt was WSW-directed (Fig. 11). The Mwembeshi shear zone accomodates this opposed tectonic transport and also has real ENE–WSW dextral displacement associated with it. The Mwembeshi shear zone has associated eclogite intrusives (Vrana *et al.* 1975) and is thought to link into the Damaran belt of Namibia to the SW and a major shear zone in Malawi to the NE. As such it forms a continental scale structure and has been proposed as a plate boundary between the Kalahari and Congo plates (Coward & Daly 1984).

The basement–rift relationship

The Central African Karroo basins are located within or adjacent to the Mwembeshi shear zone and its splays (Fig. 9). The Kafue and Luano rifts have their N and S border faults respectively, lying within the main shear zone. The Lukusashi and South Luangwa graben terminate against the Mwembeshi shear zone, whilst the South and North Luangwa sub-basins are separated by a major splay of the Mwembeshi that can be traced northward, across the Luangwa, into Zambezi age structures in the northern Irumide belt (Fig. 12).

The eastward continuation of the Mwembeshi into the Chimaliro shear zone and its bifurcation before crossing Lake Malawi (Cannon *et al.* 1969) correspond precisely with the location of the Ruhuhu and Maniamba rifts of Tanzania and Mozambique. It is suggested that a similar relationship exists here, and that the rifts represent a reactivation of the eastward continuation of the Mwembeshi shear system.

In detail, the rift faults within the shear zone show no consistent relationship to an individual lithology but merely seem to exploit the steep strongly foliated mylonitic gneisses. The oblique Lukasashi and Luangwa rifts lie sub-parallel to the regional trend of the Irumide orogenic belt but show no significant relationship either to the trend of the Irumide foliation or the major discontinuities within it. However, the mid-basin high that separates the North and South Luangwa basins is a splay of the Mwembeshi shear zone and as such exercises a major influence on the basin morphology.

The localization of the Karoo basins along the Mwembeshi zone, the pull-apart geometry of the basins parallel to the Mwembeshi shear zone and the kinematic association of the oblique basins to the zone support the contention that it exercised a major controlling influence during their development. This control occurred by the reactivation of a steep, ductile strike–slip shear zone system by dominantly

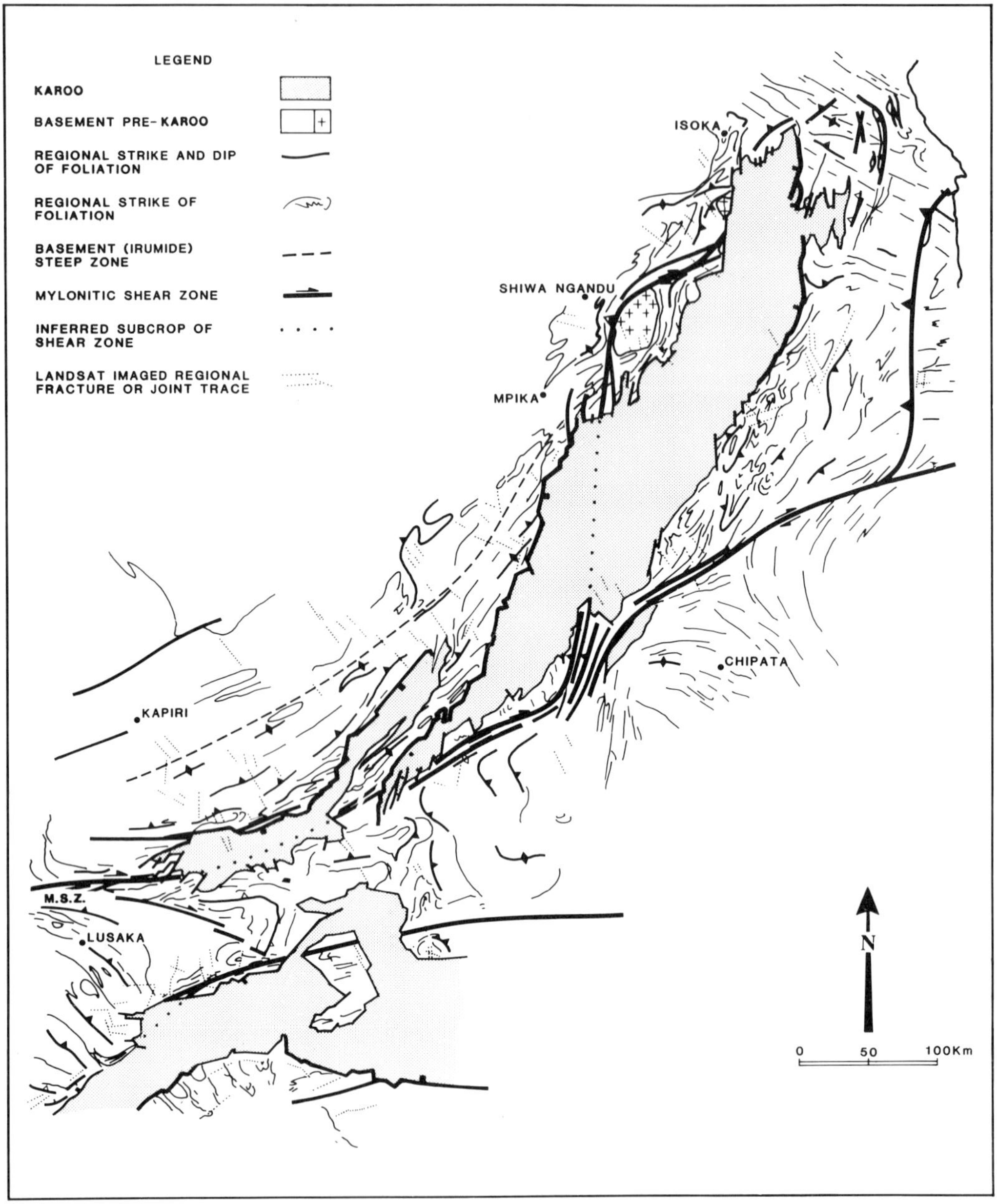

FIG. 12. Sketch map showing the basement structure of eastern Zambia and its relationship to the rift basins. In the main part of the Luangwa rift there is no consistant control exercised by the basement structure which is often at a high angle to the border faults. However, in the south the Luangwa rift terminates against the major Mwembeshi shear zone.

brittle, sinistral strike–slip displacements during the Permo–Trias.

Tectonic setting of the Karoo rifts

The plate tectonic control on the evolution of the Central African Karoo basins may be viewed in terms of the compressional active margin tectonics of Gondwana during the Permo–Trias. A major foredeep basin developed in front of the northward directed Cape fold belt. Far field compressional reactivation of the foreland to the Cape fold belt exploited a major pre-existing vertical weakness in the African plate, the Mwembeshi shear zone. Oriented sub-parallel to the maximum shear stress, albeit some 2000 km to the N, the Mwembeshi shear zone reactivated as a sinistral strike–slip zone and resulted in the formation of the Central African Karoo basins (Fig. 13).

The Cretaceous basins of West and Central Africa

Introduction

A major Cretaceous rift system occurs in West and Central Africa. This system comprises a complex array of linked rift basins that straddle the continent from Nigeria to Kenya (Fig. 14) (Fairhead 1986). Large parts of the rift system have no present-day surface expression and were discovered largely by regional geophysical studies (Browne & Fairhead 1983). The basic geometry of the rift system is that of a series of major fault zones diverging from the Gulf of Guinea into Africa. The rift system comprises two parts:

1 The Central African fault zone that extends from the Gulf of Guinea in an ENE direction through Southern Chad and the Central African Republic into the Western Sudan. In the Sudan the fault is linked to an orthogonal system of extensional basins striking NW–SE through the Sudan and Northern Kenya.

2 The Benue Trough and Gongola Rift, which together form a series of basins that pass to the NE into the Lake Chad region and link into a series of NNE-trending basins from

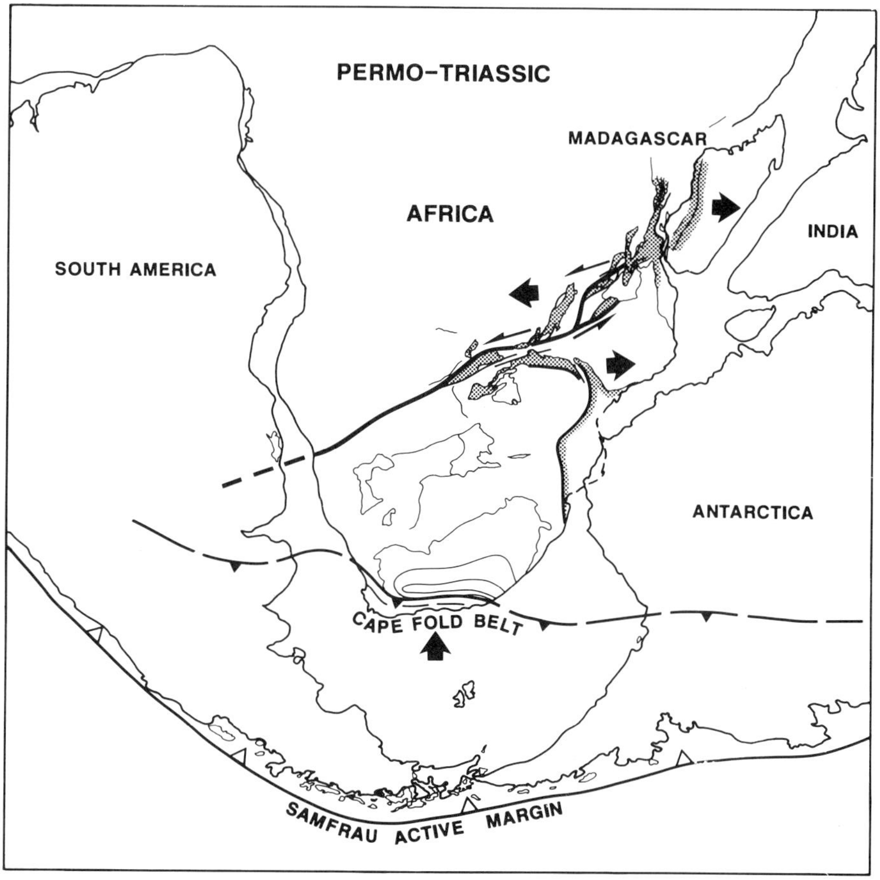

FIG. 13. Tectonic model to explain the Karoo rift basins of Central Africa developed contemporaneously with the foreland Karoo basin and Cape fold and thrust belt of South Africa.

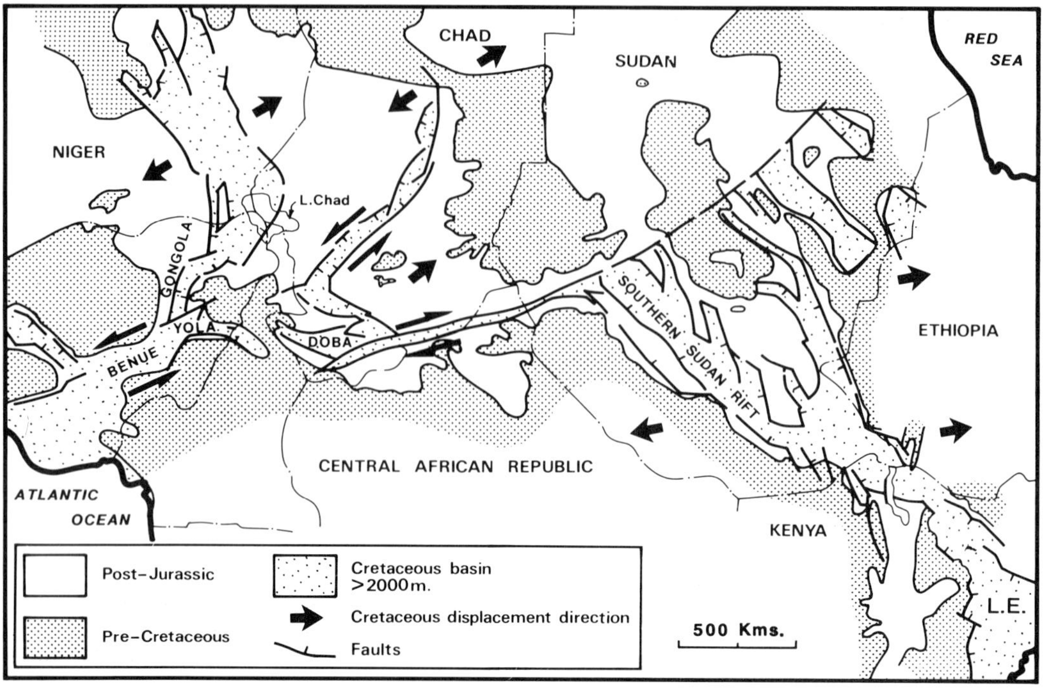

FIG. 14. Sketch map outlining the basin systems of West Africa and the Sudan. The evolution of these basins appears to be linked in time. Note the abrupt termination of the Sudanese basins along a postulated ENE trending fault zone. L. E., Lama Embayment.

Chad, through Eastern Niger into southernmost Algeria.

Geophysical studies of the Cretaceous rifts show them to be characterized by major syn- and postrift subsidence. This contrasts markedly with the Karroo rifts which show no postrift subsidence and the East African rift system which is presently associated with uplift.

The Central African fault zone and the rift basins of Sudan and Kenya

The Central African rifts of the Cameroon are developed along an ENE–trending fault zone generated during periods of dextral strike–slip displacement in the Late Cretaceous. Such displacement is indicated by the numerous pull-apart basins and flower structures developed along the length of the fault zone and characterized by Cretaceous sedimentation (Conoco & Esso Petroleum companies, personal communication). Geological studies along the zone estimated at least 40 km dextral displacement (Cornacchia & Dars 1983).

The fault zone is a major crustal structure which is clearly visible on regional gravity and magnetic maps (Cornacchia & Dars 1983). In the Sudan the fault zone appears to terminate and its displacement is transferred into dip–slip extension and the formation of the SE-trending rifts of Sudan and Kenya. Of these, the Southern Sudan and White Nile rifts are the largest (Browne *et al.* 1985). The multiple rift complex narrows to the SE in southern Sudan and continues as an important SE-trending basin through Northern Kenya into the Lamu Embayment on the Indian Ocean coast (Fig. 14). In Northern Kenya the Cretaceous extension appears to have reactivated the Jurassic Anza graben, developed during the Callovian rifting of Madagascar from Africa (Segoufin & Patriat 1979). Geophysical studies in the Kenya section of the rift show a similar crustal character to the Sudan rifts which is indicative of lithospheric extension (Reeves *et al.* 1987).

The striking feature of this Cretaceous rift system is the abrupt termination of the wide zone of Sudan basins against the Central African fault zone to the NW. Across this zone there is no evidence of a continuation of the rift system. It is therefore a geometrical necessity that the displacement that generated the Sudan basins be accommodated by the fault zone. The fault zone acts as a major lateral ramp with respect to the Sudan and Kenya Cretaceous rifts.

Basement structure

The basement structure in the area of the termination of the Sudan rift system is unexposed. However, the Central African fault zone, apparently responsible for the termination, is traceable geophysically to the WSW (as discussed above) where it is exposed in the Central Cameroon. Here the Cretaceous fault zone is seen to follow the trace of the steep, Late Proterozoic Foumban shear zone system.

The Foumban shear zone comprises an ENE striking sequence of anastomosing mylonite zones. On plate tectonic reconstructions the Foumban shear zone appears to be a continuation of the Pernambuco shear zone of NE Brazil (Fig. 15). In the latter the shear zones are associated with WSW-directed thrusts of Late Proterozoic age (Caby & Arthaud 1986). To the south of the Foumban shear zone the Late Proterozoic structure swings southerly into the ENE directed thrusts of the West Congolian thrust belt of Gabon and Congo Brazzaville (Fig. 15). These thrust and shear zones appear to form a linked and opposed set of Late Proterozoic structures that were reactivated during the formation of the intra-continental rifts.

The major Cretaceous fault system that traverses Cameroon and acts as a regional transfer zone to the Sudan rifts is superimposed on the Pernambuco–Foumban Proterozoic shear zone system. This is clear in the Cameroon but difficult to prove for the Sudan due to the lack of outcrop of the fault zone for some hundred kilometres. However, by analogy with the other two rift–steep basement shear zone relationships outlined in this paper, we suggest this is another example where the gross geometry of a rift system has been profoundly controlled by a pre-existing crustal weakness. The basement structure is responsible for the abrupt termination of the Sudan rifts and allowed the transfer of displacement between dip–slip structures of the Sudan and the strike–slip structures of the Cameroons.

Geodynamics of the West and Central African rift system (WCARS)

The evolution of the WCARS can be directly related to the tectonic evolution of the Atlantic Ocean, and in particular the equatorial fracture zone. This provides a convincing example of how plate tectonic processes of sea floor spreading may result in major zones of continental deformation. In the classical case of a ridge–transform–ridge offset there is no displacement along the fracture zones other than that due to differential cooling across the fracture zone.

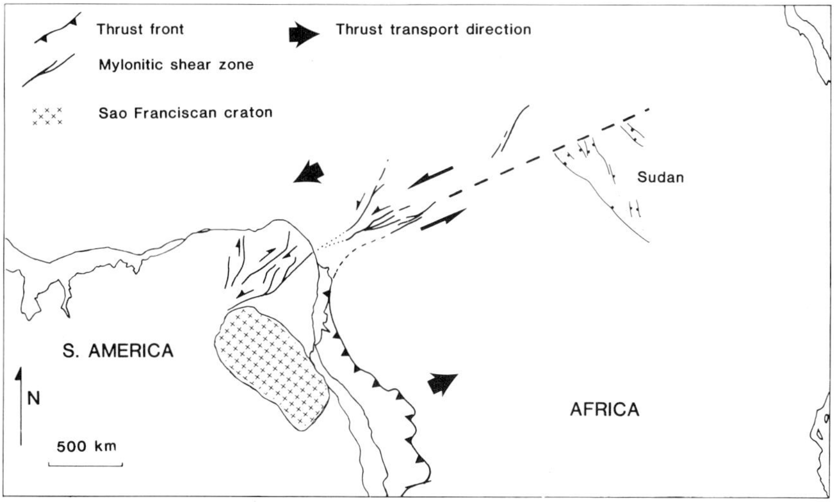

FIG. 15. Sketch map showing a plate tectonic reconstruction of South America and Africa and the major Late Proterozoic structures of the area. A regionally developed ENE-trending shear zone system (Pernambuco–Foumban shear zone) occurs in NE Brazil and the Cameroon and may be extrapolated eastwards to define the northern termination of the Sudanese basins. Note that the displacement on the Shear Zone was dextral during Cretaceous–Tertiary times.

However, if the ridges do not share the same sea floor spreading rate or direction, displacement along the fracture zones is possible.

The Central and South Atlantic Oceans comprised two distinct sea floor spreading systems from the Early Cretaceous to present. The plate boundary between these two spreading systems in the Late Cretaceous was located in the equatorial fracture zone. Differential relative plate motion across this zone was likely and the nature of such displacement can be estimated by consideration of the rotation parameters of both the Central and South Atlantic opening histories (Cande *et al.* 1986; Klitgord & Schouten 1986).

From the inception of the equatorial Atlantic at about 105–74 Ma the differential displacement favoured a sinistral displacement along the equatorial fracture zones. However, between 74 Ma and 65 Ma differential opening of the equatorial Atlantic implies a dextral shear along the fracture zones (Fairhead 1988). This period of dextral shear coincides with a contractional deformation of the Benue trough, previously opened by sinistral strike–slip motion during Atlantic rifting. It also coincides with renewed dextral displacement along the Central African fault system and the related extension within the basins of the Sudan. Thus we have a well constrained example of a plate tectonic derived stress causing failure of the African continental crust. This occurred by the reactivation of a pre-existing weakness provided by a steep basement shear zone.

Conclusions

Basement–rift relationship

The pattern of present day seismicity (Fig. 3) and the continent wide distribution of Palaeozoic, Mesozoic and Cenozoic rift systems in Africa shows a clear preference for rifting in the non-cratonic continental crust (Fig. 2). This is partially due to the age structure of the cratonic lithosphere. The fact that it is old, implies it is also thick and relatively strong. However, a second factor is the presence of crustal-scale mechanical anisotropies in the orogenic zones, some of which are Late Archean and Early Proterozoic in age. As such they are as old as many of the apparently inert cratonic units.

Within the orogenic zones large, steep shear zones developed during Proterozoic orogeny. These steep basement shear zones have been reactivated since the Precambrian. Reactivation was dependant on the regional stress pattern. The mechanical weakness afforded by these zones controlled several features of the rift systems discussed above (Fig. 16). These include:

1 The location of the three rift systems.
2 The change in structural style, border fault geometry and rift basin orientation from the N (Lake Albert to Lake Tanganyika) and N (Lake Malawi) segments of the Western rift to the central (Lake Rukwa) segment (Fig. 16A).
3 The variable orientation of the Karoo rifts of Southern Central Africa, depending on whether they are located within the Mwembeshi shear zone system or adjacent to it (Fig. 16B).
4 The abrupt northerly termination of the Sudan rifts against the ENE fault zone that links them to the Equatorial fracture zone of the Atlantic (Fig. 16C).

Rift kinematics

In all of the basement-rift examples there is a notable sub-parallelism between the displacement direction of the Proterozoic shear zone concerned and the later rifting. We have already discussed the difficulty involved in the reactivation of steep structures under certain displacement patterns. This problem is highlighted in Fig. 17, which shows the reactivation potential of a schematic shear zone surface with a steep ramp. The presence of such steep structures will facilitate reactivation at a low angle to the steep ramps. This implies that the steep ramps may exercise a kinematic control on the formation of rift basins by allowing ramp-parallel displacement in preference to displacement at a high angle to the ramp.

The length of the lateral ramps discussed here is very large. Their vertical, sub-surface expression is, however, uncertain. The great linear extent of the structures, often in excess of several hundred kilometres, suggests they are likely to be 'significant' crustal structures in the vertical dimension. It has been suggested that both the Mwembeshi and Ubendian shear zones represent ancient plate suture zones. As such they may be expected to have vertical lithospheric dimensions of 100–200 km. Such deep, steep structures have been modelled along cratonic margins in Africa (Briden *et al.* 1981) emphasizing the profound and persistent significance of craton–orogenic margins.

If the vertical expression of steep structures is lithospheric and roughly vertically below their outcrop trace, therein may lie the key to their

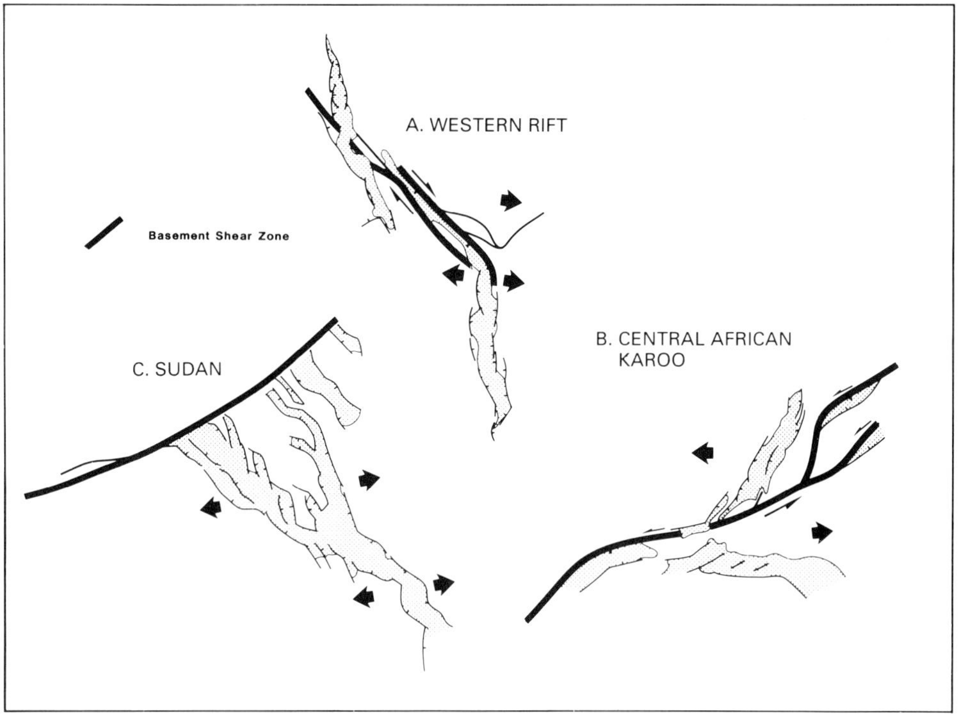

FIG. 16. Summary diagram showing the three rift systems discussed and the control exercised by steep basement shear zones: A, The steep shear zones of the Ubendian belt cause a Karoo-lateral step in the Western Rift and change the structural style of the rift basins from extensional, north and south of the shear zones, to pull apart within them. B, The Central African Karoo basins are located about the steep Mwembeshi shear zone. The structural style of the rift basins changes depending on whether they lie within the shear zone (strike–slip pull-apart basins) or adjacent to it (oblique extensional basins). C, The Sudan basins terminate abruptly against the proposed continuation of the Pernambuco–Foumban shear zone system, exploiting the steep shear zone as a lateral ramp or transfer zone.

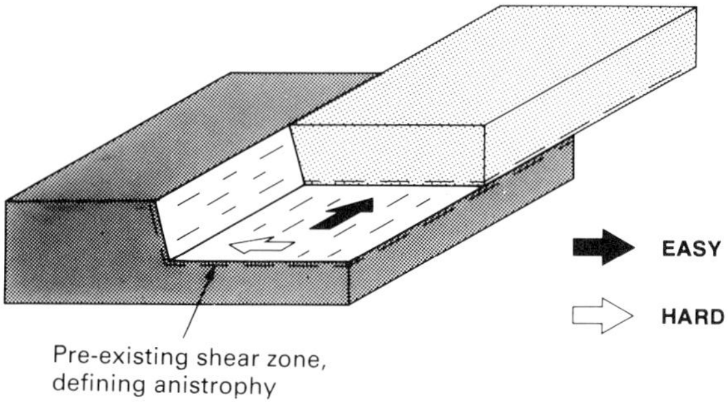

FIG. 17. The pre-existing anisotropy will be difficult to reactivate by displacements perpendicular to steep structures but easy parallel or sub-parallel to them.

apparently profound importance. Consideration of lithospheric strength, based on the location of intra-plate earthquakes and assumed rheological models (Chen & Molnar 1983; Dewey *et al.* 1986), implies that the strongest parts of the lithosphere are the mid-crust and upper mantle. For areas of very low heat flow values, as may be expected in Proterozoic and Archean crust, the upper mantle is by far the stronger of these two layers (Fig. 18B). Therefore, the presence of an anisotropy at upper mantle levels would have a profound effect on localizing lithospheric failure. Fig. 18A shows the Mohr failure envelope for isotropic and anisotropic situations. The anisotropic material fails under a much reduced differential stress or under a much greater variety of fault angles for the same stress. Taking this analysis to the lithospheric scale we suggest an anisotropy at the upper mantle level, inherited from a plate suturing event, will act as a major 'stress guide' during rift nucleation and propagation, due to the significantly reduced lithospheric strength.

In general, the reactivation of the steep shear zones initially occurs when the zone is oriented close to the orientation of maximum shear stress (Figs 18 & 19). However, in some instances regional rift displacements appear to be parallel to the pre-existing shear zones, and highly oblique to the maximum shear stress. In the latter case it is suggested that initiation of the reactivation occurred whilst the structure was in a maximum shear stress orientation (Fig. 19) but after initiation displacement became constrained parallel to it (Fig. 20). Thus the steep shear zone structure may exercise a kinematic control on the rift basin evolution.

The role of basement structure in propagating rifts and continental breakup

Courtillot (1982) pointed out that the generation of rifts in a heterogeneous lithosphere will be uneven, with the development of 'locked' zones

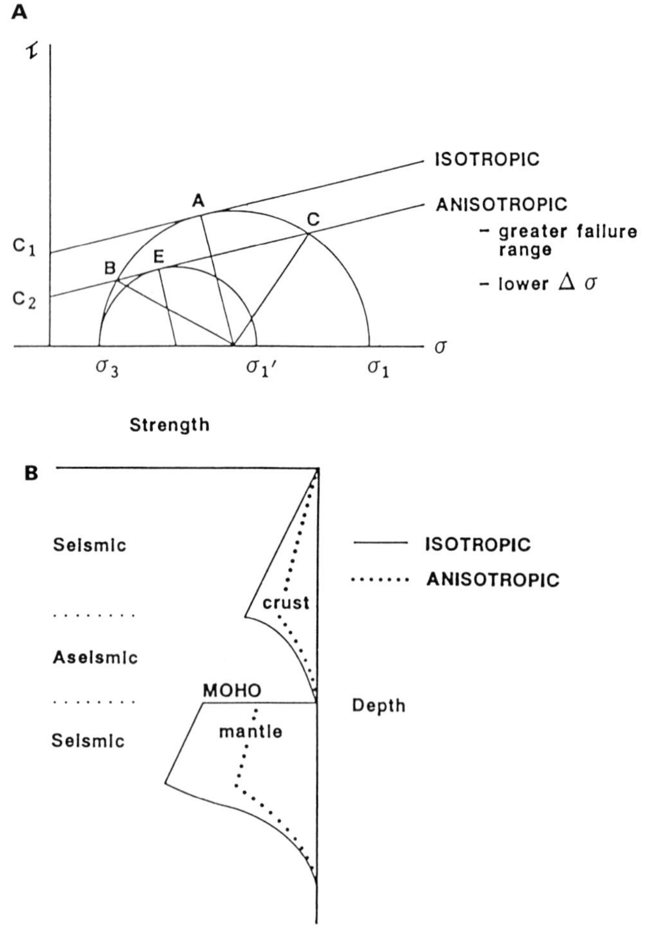

FIG. 18. Mechanical considerations of failure and anisotropies. A, Mohr failure envelopes for isotropic and anisotropic materials. The anisotropic material fails under a wider range of angles with respect to a differential stress (B–C) or at a much lower differential stress. B, The same principles applied to the lithosphere.

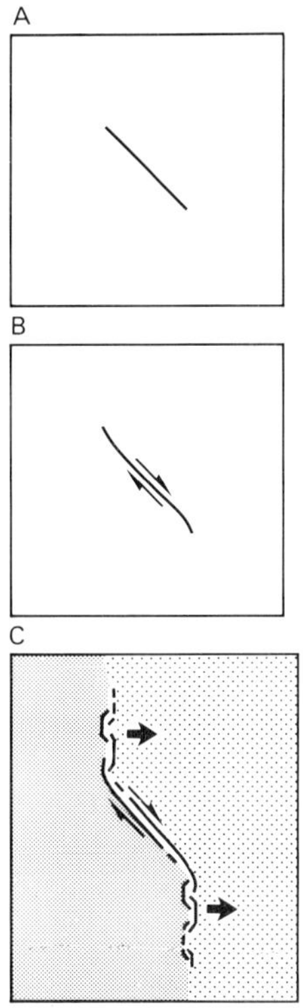

FIG. 19. Rift nucleation and propagation model. A, Continental crust with a steep basement shear zone. B, Same crust subjected to an E–W extension. First failure occurs by dextral shear along the shear zone anisotropy. C, Continued extension causes rift propagation to N and S of the anisotropy, away from zone of nucleation.

between areas actively rifting. The location of the active rifts is likely to be controlled by the presence of anisotropies such as the major shear zones outlined above. This may result in markedly diachronous rift basin development within a rift system. The oldest basins would be developed above favourably oriented basement structures, whilst the younger basins developed on isotropic crust or crust containing unfavourably oriented structures.

Both the Tertiary and Palaeozoic rifts discussed here appear to die out away from steep basement structures where rift development was largely by strike–slip displacements. We suggest that the rifts nucleated on the basement structures by strike–slip movements and propagated away from them by dominantly extensional displacements (Fig. 19). This implies, in the case of the Tertiary Western rift, that the central (Lake Rukwa) segment is the oldest part of the rift and rifting has propagated both N and S away from this strike–slip zone. A similar arguement can be made for the Palaeozoic rifts.

The suggestion that steep shear zones exercise a kinematic constraint of rift formation has important implications for the reconstruction of the synrift and early drift phases of continental breakup. In passive margins with a broad and deep thermal sag basin obscuring the rift phase, constraints on rift kinematics can be gained by understanding the early sea floor spreading directions. From the present discussion it is suggested that an equally important constraint on this early rift direction is likely to be the basement structure, and in particular the orientation of crustal-scale, steep shear zones.

Fig. 20 shows a postulated orogenic front

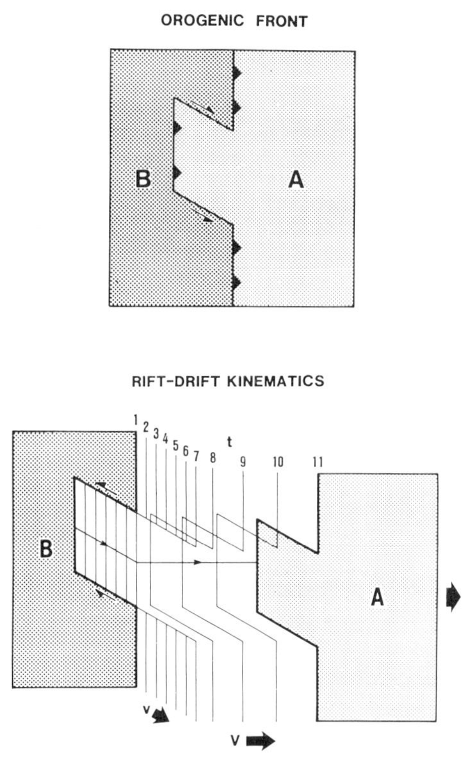

FIG. 20. Cartoon showing re-entrant salient (crustal scale shear zone, thrust zone) developed in an orogenic structure and its effect on later rifting and drift. A, The re-entrant showing the frontal ramp (gentle dips) and the lateral ramps (steep dips) of an orogenic belt. B, Later rifting reactivates that pre-existing geometry which constrains the early rift kinematics. Once the salient clears the re-entrant and the continent–continent contact is gone the main control on the kinematics of the system will be global plate motions.

defined by major lateral and frontal shear zones resulting in a marked re-entrant feature. From the above discussion, extensional reactivation of this shear zone system would preferentially occur by displacement sub-parallel to the lateral ramps. As continental rifting proceeds to sea floor spreading a time will come when the two plates are no longer in contact. This point depends on the irregularities of the plate-to-plate contact. Once free of this contact the spreading system will be free to readjust to an orientation dictated solely by the plate tectonic driving mechanism (Fig. 20).

This implies that significant plate readjustments will occur during the early stages of continental drift when the width of new ocean crust exceeds the largest re-entrant feature on the plate boundary. Such readjustments are observed, notably in the early evolution of the South and Central Atlantics where major ridge jump phases, early in the spreading history (Rabinowitz & LaBrecque 1979; Klitgord & Schouten 1986) mark plate adjustments possibly related to the freeing of re-entrant features. Examples of possible large scale re-entrant geometries occur on the NW continental shelf of Australia. In the Cuvier basin–Exmouth Plateau area a major salient–re-entrant pair are expressed by the bathymetry. They were generated by continental breakup during the Early Cretaceous (Fig. 21). This geometry clearly reflects the breakup kinematics of this part of Gondwana.

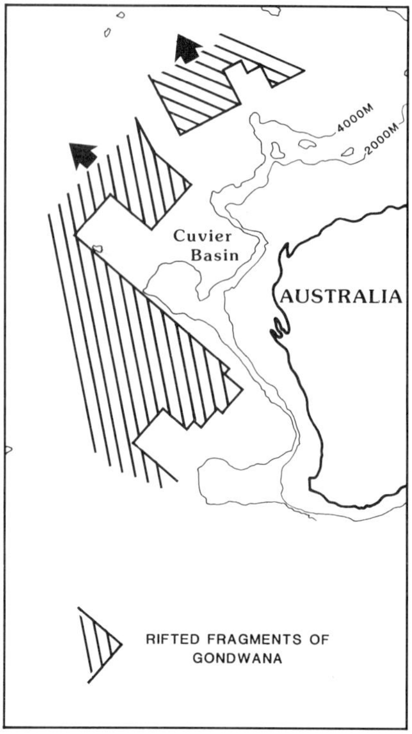

FIG. 21. Sketch map of the bathymetry of the NW shelf of Australia where several salient–re-entrant features exist related to the rifting history of Gondwana.

ACKNOWLEDGMENTS: Part of this contribution arises from the participation of M.C.D. and J.C. in the UNESCO/IUGS sponsored Geological Applications of Remote Sensing (GARS) project. This paper is published by kind permission of the British Petroleum Company.

References and further reading

ANDO, C. J., CZUCHRA, B. L., KLEMPERER, S. L., BROWN, L. D., CHEADLE, M. J., COOK, F. A., OLIVER, J. E., KAUFMAN, S., WALSH, T., THOMPSON, J. B., LYONS, J. B. & ROSENFELD, J. L. 1984. Crustal profile of mountain belt: COCORP deep seismic reflection in New England Applachians and implications for architecture of convergent mountain chains. *Bulletin of the American Association of Petroleum Geologists*, **68**, 819–837.

AYDIN, A. & NUR, A. 1982. Evolution of pull-apart basins and their scale independance. *Tectonics*, **6**, 91–105.

BAKER, B. H. & WOHLENBERG, J. 1971. Structure and evolution of the Kenya Rift Valley. *Nature*, **229**, 538–542.

BALLARD, S., POLLACK, H. N. & SKINNER, N. J. 1987. Terrestial heat flow in Botswana and Namibia. *Journal of Geophysical Research*, **92**, 6291–6300.

BOSWORTH, W. 1985. Geometry of propagating rifts. *Nature*, **316**, 625–762.

BRIDEN, J. C., WHITCOMBE, D. N., STUART, G. W., FAIRHEAD, J. D., DORBATH, C. & DORBATH, L. 1981. Depth of geological contrast across the West African craton margin. *Nature*, **292**, 123–128.

BROWNE, S. E. & FAIRHEAD, J. D. 1983. Gravity study of the Central African Rift System: a model of continental disruption 1. The Ngaundere and Abu gabra rifts. *Tectonophysics*, **94**, 187–203.

——,—— & MOHAMED, I. I. 1985. Gravity Study of the White Nile Rift, Sudan, and its regional tectonic setting. *Tectonophysics*, **113**, 123–137.

CABY, R. & ARTHAUD, M. 1986. Major Precambrian

nappes of the Brazilian belt, Ceara, northeast Brazil. *Geology*, **14**, 871–874.

CANDE, S. C., LABRECQUE, J. L. & HAXBY, W. F. 1988. Plate Kinematics of the South Atlantic: chron 34 to present. *Journal of Geophysical Research*, **93**, 13479–13492.

CANNON, R. T., HOPKINS, D. A., THATCHER, E. C., PETERS, E. R., KEMP, J., GASKELL, J. L. & RAY, G. E. 1969. Polyphase deformation in the Mocambique belt, northern Malawi. *Bulletin of the Geological Society of America*, **80**, 2615–2622.

CHEADLE, M. J., MCGEARY, S., WARNER, M. R. & MATTHEWS, D. H. 1987. Extensional structures on the western UK continental shelf: a review of evidence from deep seismic profiling. *In*: COWARD, M. P., DEWEY, J. F. & HANCOCK, P. (eds) *Continental Extensional Tectonics. Special Publication of the Geological Society of London*, **28**, 445–465.

CHEN, W-P. & MOLNAR, P. 1983. Focal depths of intra continental and intra-plate earthquakes and their implications for the thermal and mechanical properties of the lithosphere. *Journal of Geophysical Research*, **88**, 4183–41214.

CHOROWICZ, J., LE FOURNIER, J. & VIDAL, G. 1987. A model for rift development in Eastern Africa. *Geological Journal*, **22**, 495–513.

CORNACCHIA, M. & DARS, R. 1983. Un trait structural majeur du continent Africain. Les lineaments centrafricains du Cameroun au golfe d'Aden. *Bulletin de la Sociéte Géologique de France*, **25**, 101–109.

COURTILLOT, V. 1982. Propagating rifts and continental break up. *Tectonics*, **1**, 239–250.

COWARD, M. P. & DALY, M. C. 1984. Crustal lineaments and shear zones in Africa: their relationship to plate movements. *Precambrian Research*, **24**, 27–45.

COX, K. G. 1970. Tectonics and vulcanism of the Karoo period and their bearing of the postulated fragmentation of Gondwanaland. *In*: CLIFFORD, T. N. & GASS, I. G. (eds) *African Magmatism and Tectonics*. Oliver & Boyd, Edinburgh, 211–236.

CROSSLEY, R. & CROW, M. J. 1980. The Malawi rift. *In*: *Geodynamic Evolution of the Afro-Arabian Rift System*. Accademia nazionale dei Lincei, **7**, 77–87.

DALY, M. C. 1986. Crustal shear zones and thrust belts: their geometry and continuity in Central Africa. *Philosopical Transactions of the Royal Society of London*, **A317**, 111–128.

—— 1988. Crustal shear zones in Central Africa: A kinematic approach to Proterozoic tectonics. *Episodes*, **11**, 5–11.

DEWEY, J. F., HEMPTON, M. R., KIDD, W. S. F., SAROGLU, F. & SENGOR, A. M. C. 1986. Shortening of continental lithosphere: the neotectonics of Eastern Anatolia—a young collision zone. *In*: COWARD, M. P. & RIES, A. C. (eds) *Collision Tectonics. Special Publication of the Geological Society of London*, **19**, 3–36.

DONATH, F. A. & CRANWELL, J. 1981. *In*: CARTER, N. L., FRIEDMAN, M. & LOGAN, J. M. (eds) *Mechanical Behaviour of Crustal Rocks*. American Geophysical Union, Washington, DC, 231–241.

DRYSDALL, A. R. & WELLER, R. K. 1966. Karoo sedimentation in Northern Rhodesia. *Transactions of the Geological Society of South Africa*, **69**, 39–70.

EBINGER, C., ROSENDAHL, B. R. & REYNOLDS, D. J. 1986. Tectonic model of the Malawi rift, Africa. *In*: BEN-AVRAHAM, Z. (ed) *Sedimentary Basins within the Dead Sea and other Rift Zones. Tectonophysics*, **141**, 215–235.

ETHERIDGE, M. A. 1986. On the reactivation of extensional fault systems. *Philosophical Transactions of the Royal Society of London*, **A317**, 179–194.

FAIRHEAD, J. D. 1986. Geophysical controls on sedimentation in the African Rift System. *In*: FROSTICK, L. E., RENAUT, R. W., REID, I. & TIERCELIN, J. J. (eds) *Sedimentation in the African Rifts. Special Publication of the Geological Society of London*, **25**, 15–23.

—— 1988. Mesozoic plate tectonic reconstructions of the Central-South Atlantic Ocean: the Role of the West and Central African Rift System. *Tectonophysics*, **155**, 181–191.

—— & HENDERSON, N. B. 1977. The seismicity of southern Africa and incipient rifting. *Tectonophysics*, **41**, 19–26.

GAIR, H. S. 1959. The Karoo system and coal resources of the Gwembe District, north-east section. *Bulletin of Geological Survey of Northern Rhodesia*, **1**, 88 pp.

GREEN, D., CROCKETT, R. N. & JONES, M. T. 1980. Tectonic control of Karoo sedimentation in Mid-Eastern Botswana. *Transactions of the Geological Society of South Africa*, **83**, 213–219.

KING, B. C. 1970. Vulcanicity and rift tectonics in East Africa. *In*: CLIFFORD, T. N. & GASS, I. G. (eds) *African Magmatism and Tectonics*. Oliver & Boyd, Edinburgh, 263–284.

KLITGORD, K. D. & SCHOUTEN, H. 1986. Plate kinematics of the Central Atlantic. *In*: TUCHOLKE, B. E. & VOGT, P. P. (eds) *The Western North Atlantic Region. Geological Society of America Decade of North American Geology, Series 1*, 351–377.

MCCONNELL, R. B. 1972. Geological development of the rift system of Eastern Africa. *Bulletin of the Geological Society of America*, **83**, 2549–572.

—— 1980. A resurgent taphrogenic lineament of Precambrian origin in Eastern Africa. *Journal of the Geological Society*, **137**, 483–489.

MCKENZIE, D., DAVIES, D. & MOLNAR, P. 1970. Plate tectonics of the Red Sea and East Africa. *Nature*, **226**, 243–248.

RABINOWITZ, P. D. & LA BRECQUE, J. 1979. The Mesozoic South Atlantic Ocean and evolution of its continental margins. *Journal of Geophysical Research*, **84**, 5973–6002.

REEVES, C. V., KARANJA, F. M. & MACLEOD, I. N.1986. Geophysical evidence for a Jurassic triple junction in Kenya. *Earth and Planetary Science Letters*, **81**, 299–311.

REEVES, C. V., KARANJA, F. M. & MACLEOD, I. N.

1987. Geophysical evidence for a failed Jurassic rift and triple junction in Kenya. *Earth and Planetary Science Letters*, **81**, 299–311.

REICHWALDER, P. & BRANDON, A. 1978. The geology of the Luano Valley area; explanation of degree sheet 1429, SW and SE quarters. *Report of the Geological Survey of Zambia*, **61**, 28–85.

ROSENDAHL, B. R., REYNOLDS, P. M., LORBER, C. F., BURGESS, C. F. MCGILL, J., SCOTT, D., LAMBIASE, J. J. & DERKSEN, S. J. 1986. Structural expressions of rifting: lessons from Lake Tanganyika, Africa. *In*: FROSTICK, L. E., RENAUT, R. W., REID, I. & TIERCELIN, J. J. (eds) *Sedimentation in the African Rifts. Special Publication of the Geological Society of London*, **25**, 29–43.

SEGOUFIN, J. & PATRIAT, P. 1979. Existence d'anomalies mesozoiques dans le bassin de Somalie. Implications pour les relations Afrique–Antarctique–Madagascar. *Comptes Rendus de l'Académie des Sciences, Paris, Serié B*, **291**, 85–88.

SHUDOFSKY, G. N. 1985. Source mechanisms and focal depths of East African earthquakes using Raleigh-wave inversion and body wave modelling. *Geophysical Journal of the Royal Astronomical Society*, **83**, 563–614.

SYKES, L. R. 1978. Intraplate seismicity, reactivation of preexisting zones of weakness, alkaline magmatism, and other tectonism postdating continental fragmentation. *Review of Geophysics & Space Physics* **16**, 621–687.

TAVERNER-SMITH, R. 1962. Karoo sedimentation in a part of the mid-Zambezi valley. *Transactions of the Geological Society of South Africa*, **65(1)**, 43–74.

VRANA, S., PRASAD, R. & FEDIUKOVA, E. 1975. Metamorphic kyanite eclogites in the Lufilian arc of Zambia. *Contributions to Mineralogy and Petrology*, **51**, 139–160.

WATTERSON, J. 1975. Mechanisms for the persistence of tectonic lineaments. *Nature*, **253**, 520–522.

M. C. DALY, B. P. Petroleum Development Ltd., Britannic House, Moor Lane, London EC2Y 9BU, UK.

J. CHOROWICZ, Département de Géotectonique, Université Pierre-et-Marie Curie, Paris, France.

J. D. FAIRHEAD, Department of Earth Sciences, The University, Leeds, UK.

Inversion tectonics – a discussion

M. A. Cooper, G. D. Williams, P. C. de Graciansky, R. W. Murphy, T. Needham, D. de Paor, R. Stoneley, S. P. Todd, J. P. Turner & P. A. Ziegler

Introduction

The term 'inversion' to describe an inverted basin was first used by Glennie & Boegner (1981) although inverted basins had been recognized many years before e.g. Lamplugh (1920) and Stille (1924). During this meeting it became apparent that the application of the term had broadened to such an extent that the understanding of 'inversion' in the petroleum industry was incompatible with much of the current usage. The discussion that follows illustrates many of the points of disagreement, perhaps the most contentious of which is the use of the term 'negative inversion' although this was also introduced by Glennie & Boegner (1981). Most of the discussion was presented verbally at the meeting and was recorded, transcribed and returned to speakers for their corrections. In addition, a number of written contributions were received. All contributions to the discussion have been edited as gently as possible so as to retain the exact meaning intended by the contributor. All contributors are included as co-authors in this discussion article but clearly this does not mean that individuals necessarily accept all the points made by other contributors. The editors have identified individual contributions. The discussion commenced with some proposals by the editors which are briefly reproduced here. This discussion article concludes with a considered revision of the proposals on nomenclature which aims to satisfy some of the shortcomings identified in the discussions.

Recognizing inversion structures (M. A. Cooper & G. D. Williams)

Our initial definition of inversion relied on the concept of regional elevation. The regional elevation of a marker horizon is the structural elevation of the horizon if the effects of faulting or folding are removed. It is not necessarily a horizontal planar surface but can be deformed by isostatic, thermal and flexural processes affecting the lithosphere, usually of a significantly larger wavelength than the fault-related structuring. Contractional deformation will elevate beds above regional elevation and extension will lower them below regional elevation (Fig. 1).

We can therefore recognize inversion when in the hanging wall of the fault or fault system controlling deformation two marker horizons display different elevations above and/or below their respective regionals. In the case of the classical inverted half-graben model of Bally (1984) the upper part of the fault causes elevation above regional whilst on the lower part of the fault beds are still beneath their regional elevation (Fig. 2). We term the change-over point (where beds are at regional in the hanging wall) the null point (Williams *et al. this volume*).

One key feature that is a consequence of the geometric criterion that we propose is that it must be possible to identify a synrift (or passive infill) sequence in all inversion structures. If the synrift package cannot be positively identified then inversion cannot be unequivocally recognized. Note that we are using the term synrift to describe sedimentation synchronous with extension which may not necessarily produce a rift system.

The postrift sequence is longer (Fig. 2, length *Y*) than the prerift sequence (Fig. 2, length *X*) due to the intervening phase of extension and on inversion must shorten whilst the prerift remains in net extension. This creates the asymmetric anticline in the postrift sequence which is characteristic of an inversion structure which faces towards the footwall (Fig. 2).

Note that this geometric criterion does not require that extension be followed by contraction – positive inversion. It is equally applicable to contraction followed by extension – negative inversion. What the criterion does necessitate is that the regional *stress* regime would undergo a fundamental change such that the *dip–slip* component of deformation is reversed.

From Cooper, M. A. & Williams, G. D. (eds), 1989, *Inversion Tectonics*
Geological Society Special Publications No. 44, 335–347.

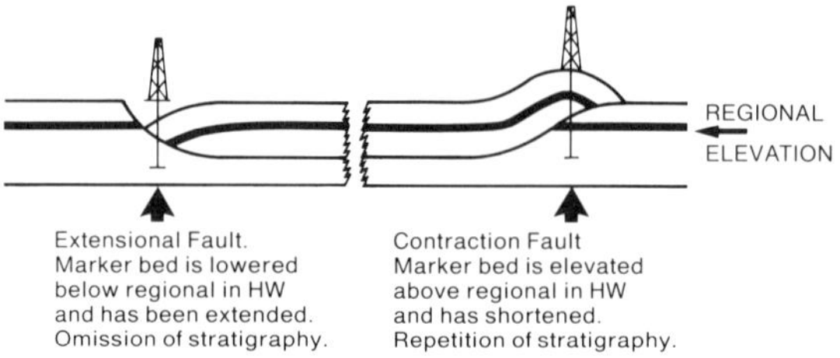

FIG. 1. An illustration of the regional concept in a simple tectonic model.

We did not include non-fault related uplift and subsidence e.g. thermal subsidence following extensional deformation because we considered that this would broaden the concept of inversion too far. Such mechanisms tend to affect areas beyond the limits of basins, i.e. they are often more regional in their effect.

Open discussion

The discussion was initiated by Dr Murphy who described the concept of inversion as currently used in the petroleum industry.

Petroleum industry usage (R. W. Murphy)

For a number of years the concept of inversion tectonics has been current within the petroleum industry. We saw an increasing number of examples on regional seismic lines of basins in which the depositional thicks, located at or near the basin centres, had been uplifted by late-stage compressive forces. A good case in point is provided by the Malay Basin (Bubb & Hayash 1980; Fig. 3). The Malay basin profile was first shown publicly by Jack Armitage at a Petroleum Symposium of the Geological Society of Malaysia. The basin centred over SP 600–700 and the half-grabens at SP 1000 are separated by a basement ridge which was topographically high during deposition up to about end Miocene time. Some 1200 m of section were eroded from the main basin prior to the deposition of lower Pliocene beds. Some compression continued, as the Pliocene beds are arched upward, but movement ceased in more recent times, since the youngest beds are flat-lying over the basin. Similarly, the half-graben underwent inversion, although to a lesser degree. In the main basin the inversion appears to have taken place by reversal of offset along the major faults, whereas compression within the half-graben is expressed

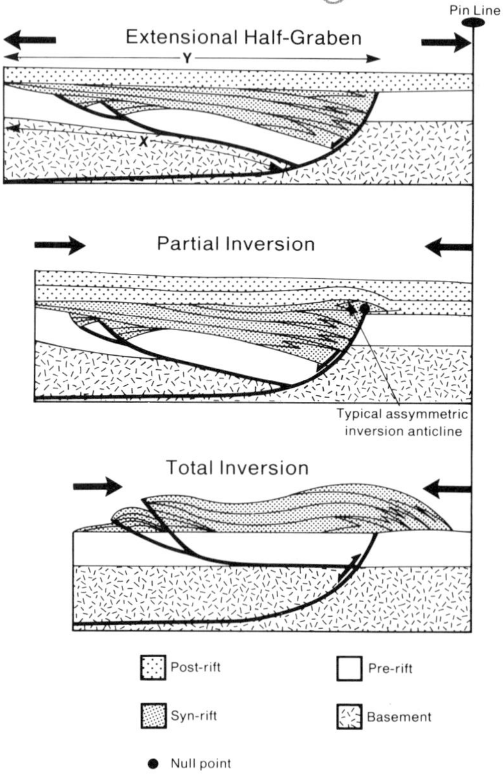

FIG. 2. The development of an inversion structure (modified from Bally 1984). Regional is defined from the footwall of the half-graben.

on this seismic profile by folding without apparent faulting.

A three-stage model of evolution is suggested: 1. Basin formation by normal faulting of Cretaceous basement; 2. rapid sedimentation decreasing through time as activity along the faults diminishes to zero; 3. a short, rapid compressive phase ending in quiescence.

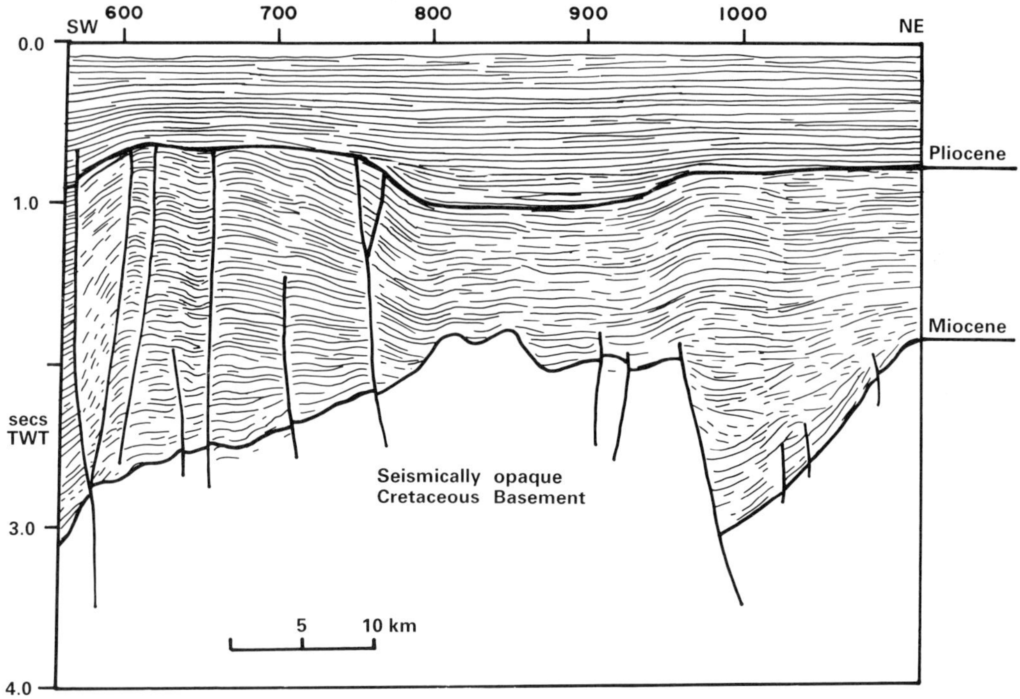

FIG. 3. Line drawing of an example seismic section of an inversion structure from the Malay basin. Vertical scaling is approximately 8 × horizontal.

From regional mapping we believe the compressional phase to be associated with throughgoing wrench faults in an essentially transpressive tectonic pattern.

In oil company practice inversion tectonics are restricted to situations which are similar to that which is exemplified by the Malay Basin: developed on continental crust, initial basin formation usually along normal faults, and a three-stage development: extensional–quiescent–compressional. Inverted basins are not the result of collision tectonics (except perhaps very indirectly). They do not lie at continental margins. They are not minor elements within a tectonic regime that is overwhelmingly of a different structural style.

These comments do not offer a definition of inversion tectonics. They are designed to convey to the scientific world at large what the concept means within the context of the petroleum industry. Since the existence of 'inversion tectonics' as a separate style was first documented on regional seismic profiles (if we designate the earlier usage of 'Germano-type' tectonics as a broader antecedent concept of which inversion is a special case) I feel there is merit, when attempting to define the term, to research its origins.

The next contributor, Dr Ziegler, supported this view whilst recognizing the value of applying the concepts of basin inversion to examples other than those from intra-cratonic settings. However both Ziegler and Murphy imply that fault reactivation is not a necessary requirement, although in most cases the transtensional–extensional basin development phase would involve fault systems.

Classical basin inversion (P. A. Ziegler)

I feel that the concept of 'basin inversion' as developed by the petroleum industry, on the basis of extensive seismic surveys calibrated by well data, should be retained and should not be rendered meaningless by the introduction of such terms as 'positive inversion' and 'negative inversion'.

An archetype of an inverted basin has been demonstrated in this discussion by Dr Murphy (for other examples see P. A. Ziegler 1987). The basic concept is illustrated by Fig. 3 (Bally 1984).

Inverted basins have been recognized for a long time and have been discussed amongst others by Lamplugh (1920), Stille (1924), Prouvost (1930) and Voigt (1963). Inverted basins occur most commonly in intra-cratonic areas (e.g. Northwest Europe) and in back-arc areas (e.g. Sunda arc), but have also been recognized within mega-sutures (e.g. Southern Altiplano of Bolivia, Devonian basins superimposed on the Arctic–North Atlantic Caledonides) and locally also on passive margins (e.g. Mid Norway Basin, Bukowicz & Ziegler 1985). The structural style of inverted basins indicates that their deformation was governed by compressional/transpressional stress regimes. Empirically it can be established that grabens and troughs, which evolved under tensional–transtensional stress regimes, are characterized by attenuated crust that is prone to inversion under tangential stresses.

The degree of basin inversion can vary greatly and can range from partial, mild inversion (configuration of the original basin at depth still retained whilst its shallower parts are deformed into an anticlinorium) to total inversion whereby movements along extensional faults controlling the original basin configuration are reversed to the null point or even beyond it (Bally 1984).

In the following discussion some arguments are given in support of the recommendation presented in the opening paragraph of this contribution.

The term 'negative inversion' is essentially identical to basin subsidence. The principal mechanisms governing the subsidence of sedimentary basins, as discussed in an ever expanding literature, involve: 1. crustal and lithospheric thinning related to rifting processes or wrench induced transtentional deformations; 2. deflection of the lithosphere under tectonic loads; 3. cooling and contraction of a thermally destabilized lithosphere and 4. water and sedimentary loading of the lithosphere. The term 'negative inversion' does not contribute to the understanding of basin subsidence and is therefore redundant.

The term 'positive inversion', corresponding essentially to uplift and basin destruction, has been used during this conference as encompassing the orogenic deformation of passive margin sedimentary prisms and of successor basins in the interior of major orogens, the compressional–transpressional deformation of intra-cratonic basins and thermal uplift associated with rifting processes.

Of these phenomena only the compressional–transpressional deformation of grabens and troughs located in the interior of major lithospheric plates is compatible with the term 'basin inversion' as used by the petroleum industry. Such deformation can be related to changes in the regional stress regime which may be induced either by plate margin collisional processes (e.g. latest Cretaceous and Tertiary basin inversion in NW Europe) or by the inception of partial abandonment of multidirectional rift systems (e.g. Late Jurassic–Early Cretaceous development of NW Europe, Ziegler, 1982).

Thermal uplift of a rift zone, involving the elevation of the rift shoulders, and under certain conditions also of the axial graben, as well as hot spot induced regional arching accompanied by only minor crustal extension, differs, fundamentally from the concept of 'basin inversion' as used by the petroleum industry from a geodynamic point of view. It is therefore recommended to exclude hot spot and rift induced thermal uplift from the term 'inversion'.

The collision-related orogenic deformation of continental margin prisms, and of foredeeps and success or basins, involves the development of thin- and thick-skinned thrust sheets and nappes which account for major crustal shortening. In this respect the structural style and deformation patterns of such 'Alpino-type' plate margin fold belts (megasutures) differs greatly from the structural style of intra-cratonic grabens and troughs which became inverted in response to the build-up of intraplate compressional stresses. As noted above, such stresses may be related to collisional processes affecting one or more of the margins of the affected plate (e.g. Cenozoic inversion of the Celtiberian trough of Iberia, Ziegler 1988) or may be of an anorogenic nature (e.g. Palaeogene inversion of the Doba–Doseo trough of Chad and the Central African Republic). In view of this it is suggested to distinguish between intraplate inversion tectonics and 'Alpinotype' orogenic plate margin deformations. In particular, the term 'inversion' should not be applied to the scooping out of sedimentary ramps by imbricate thrust sheets, for instance in the Canadian Rocky Mountains. Yet, the orogenic deformation of passive margin sedimentary prisms may involve, as in the case of the Western Alps (Gillcrist *et al. this volume*), the deformation of grabens and half-grabens which developed during the rifting stage preceding crustal separation and the subsequent development of the respective passive margin. The compressional deformation of such grabens and troughs, mainly during early orogenic phases and in part at considerable distances from the contemporaneous orogenic thrust front, is compatible with the concept of intra-plate basin inversion.

In some cases a clear boundary between compressional intra-plate and plate margin deformation cannot be defined if the foreland is transsected by inverted grabens which extend into the orogenic belt (e.g. Dauphinois, Western Alps).

'Inverted basins' are a classical form of intra-plate compressional deformation. Another expression of such deformation are basement blocks which are uplifted along steep reverse, trust and/or wrench faults (e.g. Wind River Mountains, USA; Harz Mountain , Germany). Such basement uplifts develop by compressional–transpressional reactivation of pre-existing fracture systems and may or may not involve the inversion of basins and troughs.

In summary, it is recommended to restrict the terms 'inverted basin' and 'inversion tectonics' to the intra-plate compressional–transpressional deformation of basins which had developed under a largely tensional–transtensional setting. This definition retains the concept developed by the petroleum industry who use the term 'inverted basin' on a routine basis. It is therefore suggested that if authors wish to use the term 'inversion' in any other sense they should specify in a few words what in their usage this term means.

Professor de Graciansky outlined the reasons for preferring a less restrictive application of inversion. The use of inversion concepts in the W Alps has revolutionized the models of evolution for the orogen and provides a powerful tool for interpreting similar orogenic belts, notably those in the Tethyan realm.

Inversion in orogenic belts (P. C. de Graciansky)

During the discussion on the definition of inversion tectonics, we noted a reluctance amongst oil geologists to apply concepts of inversion to mountain belts, i.e. to severely deformed areas.

Experience in the Alps has shown that some of the most demonstrable examples of inversion can be found in the sub-Alpine ranges which have undergone relatively mild tectonism.

Moreover, the concepts of inversion help to explain the geometry of the more severely deformed zones. This can be documented by: 1. the gradient in the intensity of deformation from external to internal zones; 2. the comparison of structural styles between internal and external zones; 3. the reconstruction of the geometry of the early stages of compression in every zone.

For example, the initiation of megamullions in the Belledonne–Pelvoux area and of megaboudins as in the Penninic zone, which were both inherited from tilted extensional fault blocks, could not be understood without considering early Alpine inversion of Tethyan extensional faults. This is also true for the Helvetic and Digne nappes which derived from the Tethyan half-grabens.

We fully agree that if compression is in the order of magnitude observed in the internal zones of folded belts, for example, in the metamorphic areas, this should not be considered to be inversion.

However, where extension and compression are of the same order of intensity, the concepts of inversion is extremely useful for structural reconstruction *whatever the geodynamic environment*, including intra-cratonic as well as oceanic basins, in epeiric basins as well as in marginal environments, and in low relief areas as well as in mountain belts.

Within mountain belts it seems reasonable to restrict the application of inversion concepts to external zones of mild tectonism. However, it must be appreciated that even in the internal zones the early phases of compression may well have produced inversion structures subsequently overprinted.

A definition that is too rigid is inconvenient, particularly when several extensional phases are interspersed with compressional events as observed in Alpine and North European basins.

Professor Stoneley introduced a note of caution raising the important issue of the origins of basins. A particularly pertinent point is that basins related to compressional tectonics e.g. foreland basins, cannot be subsequently considered to have been inverted by continuing compression. The implication is that we must look for a change in the regional stress field to identify inversion.

The importance of understanding basin forming process (R. Stoneley)

The proposed definitions involving fault movements are imposing conditions on Nature which probably do not exist. The idea that faults have to be involved in order to get inversion, implies that Nature is going to make all her vertical displacements on faults. This can be achieved in other ways, for example as a result of flexing or as gentle basement uplift. We should not impose any such requirement on Nature through the definitions.

To define negative inversion–positive inversion as a result of extension following compression or vice versa denies the scientific method. Basins can form as a result of extension, and most of the ones described at the meeting apparently did. Other basins, however, can form from compression, and a change in the sense of vertical movement may or may not involve a change from extension to compression. Furthermore, there are some intra-cratonic basins which are circular in shape, such as the Moscow Basin and the Michigan Basin, which do not seem to involve any compression or extension in their development. If we exclude such basins from our definitions, by referring only to compression and extension, then we are implying that this sort of basin never inverts, whether in fact they do or not.

We have seen interpretation encroaching on definition in one or two papers during this meeting, for example in referring to unconformities as thermal relaxation unconformities or crustal thickening unconformities. This is just an interpretation.

I hope that students here will not be misled by this rolling bandwagon of enthusiasm. We understand a lot about how basins are formed and how they invert, but we must continue to describe basins objectively and then try to interpret how they developed, rather than force interpretations into a model at the outset.

Finally, to define basin inversion, we must consider what we understand by a basin. Is, for example, any rollover depression on a listric normal fault considered to be a basin?

The following discussion illustrates many of the conceptual points raised by Professor Stoneley but concludes that the definition of inversion *should include* cases where the regional stress regime remains constant, in direct conflict to the comments above.

Inversion in the Pyrenees (S. P. Todd & J. P. Turner)

In the W part of the South Pyrenean foreland basin system, flexural subsidence was generated by thrust sheet loading along the basin margin during the late Eocene. The Jaca thrust tip basin, a detached compartment at the N margin of the main foreland basins, was created by further propagation of the thrust front southwards. The presence of growth folds and numerous intraformational unconformities in the Jaca basin indicate that shortening and sedimentation were contemporaneous during the Oligocene–Miocene. As the thrust front propagated through the basin, various sub-areas experienced a transition from being zones of subsidence and net sedimentation to uplift and net erosion. Thus the timing of this "inversion" was highly diachronous across the basin, ranging from early Oligocene in the E to early Miocene in the W.

Several points emerge from this potted basin history which are pertinent to the discussion.

1 The transition from subsidence to uplift was achieved without a significant change in the regional stress regime.
2 Basin uplift was achieved without reactivation of pre-existing faults.
3 The timing of inversion was highly diachronous across the basin.

We therefore suggest that the term 'inversion' be applied in the following broad sense: 'the transition within a basin from subsidence and net deposition to uplift and net erosion'. This definition has the advantages of encompassing both tectonic and thermal uplift whilst not necessarily implying fault reactivation.

Two contributors suggested introducing the term 'reactivation' to avoid confusion with 'inversion' for overturning of strata or structures. One (D de Paor) discusses the potential value of null points in describing reactivated faults.

Reactivation of faults and the null point (D. de Paor)

I suggest that we need to make a clear distinction between basin inversion and fault reversal. An inverted fault is one that has been rotated through the vertical so that its hanging wall and footwall are interchanged, as opposed to a reactivated fault whose sense of slip has undergone reversal. Regarding the latter, two categories of dip–slip reversal may be defined, reversed normal faults (RN) and normalized reverse faults (NR). If a fault is entirely post–sedimentary, there is no stratigraphic record of slip increments, so only the net slip vector may be determined. However, during synsedimentary faulting, growth sedimentation creates a vernier scale against which subsequent slip reversals may be measured. As long as sedimentation keeps pace with fault slip, a point of no offset, or null point exists at the synorogenic surface.

When a fault (or either RN or NR type) reverses its sense of slip, this null point migrates down the plane at a rate that depends on the ratio of the fault slip to sediment growth rates. If syntectonic sedimentation persists, a new null point is generated at the surface. Subsequent reversal of slip leads to RNR or NRN type faults with two migrating null points, one of which travels down the fault plane, the other up. After a certain amount of slip, these two points can meet and annihilate one another in Pacman style. All that then remains of the incremental slip history is a variation in net slip vector with stratigraphic level.

Reactivation not inversion (T. Needham)

The term 'inversion', prefixed by a variety of other terms, has been used to describe reversal of the sense of motion along a given fault or within a fault system. When describing the re-use of a pre-existing fault, the well known term 'reactivation' is readily applicable. It is then easy to describe a fault in terms such as, 'contractional reactivation of an extensional fault'. Whilst this is longer and somewhat more cumbersome than, for instance, 'positive inversion' it is unambiguous and requires the use of no new terminology. The term is also non-genetic. It describes a feature based on geometrical observations and is not restricted to particular tectonic settings. This also side-steps the possibly contentious issues of what is inversion and what is orogenesis, as well as the separation of plate margin from intra-plate deformation. Use of 'reactivation' also leaves the term 'inversion' free to be used in the more general sense which many of the conference participants seemed to prefer.

Secondly, usage of the term 'regional inversion' may also lead to confusion, especially in that 'regional' is often used as an abbreviated form of regional elevation which itself is used as the datum relative to which inversion is measured.

Finally, uplift, when not obviously fault related, was described as thermal uplift by some speakers. Surely this is placing too much interpretation on a simple observation. Perhaps we should be more cautious and refer to non-fault related uplift for any given situation until further data allows the distinction to be made between, for example, thermal or isostatic uplift.

Inversion structures – recognition and characteristics (M. A. Cooper & G. D. Williams)

The discussion on 'inversion' reproduced above indicates the extent of the disagreements on what constitutes an inversion structure. We propose that our initial geometric criterion for recognition be retained with minor modifications.

Recognition on geometric criteria

Inversion produces variation in the structural elevation, relative to regional elevation of marker beds depending on their location on the fault or fault system controlling deformation. The variation should be systematic; a zone of depression below regional will be associated with one portion of the fault separated by the null point from a zone of elevation above regional (Fig. 2).

This should normally be generated by a fundamental change in the regional stress field from extensional to compressional. The converse sequence, compression followed by extension, produces 'negative inversion', a less widely accepted concept.

Exclusions

These criteria exclude such effects as the commonly observed truncation of extensional fault blocks prior to the deposition of the postrift sequence, e.g. the Kimmerian phase in the North Sea. These effects are thermally and isostatically driven due to mantle re-equilibration and we do not therefore consider such regionally extensive changes in structural elevation inversion.

We would also exclude inversion-type structures produced by movements on strike–slip faults. The complex array of faults that typify many strike slip faults at shallow depths, e.g. the San Andreas Fault facilitate localized transpressional and transtensional regimes which can migrate with time (Fig. 4). Thus, in such systems alternations of localized compressional and extensional tectonics will be common but the regional stress regime will be unchanged.

Basin processes

The processes of basin formation require some consideration in this discussion. Basins which

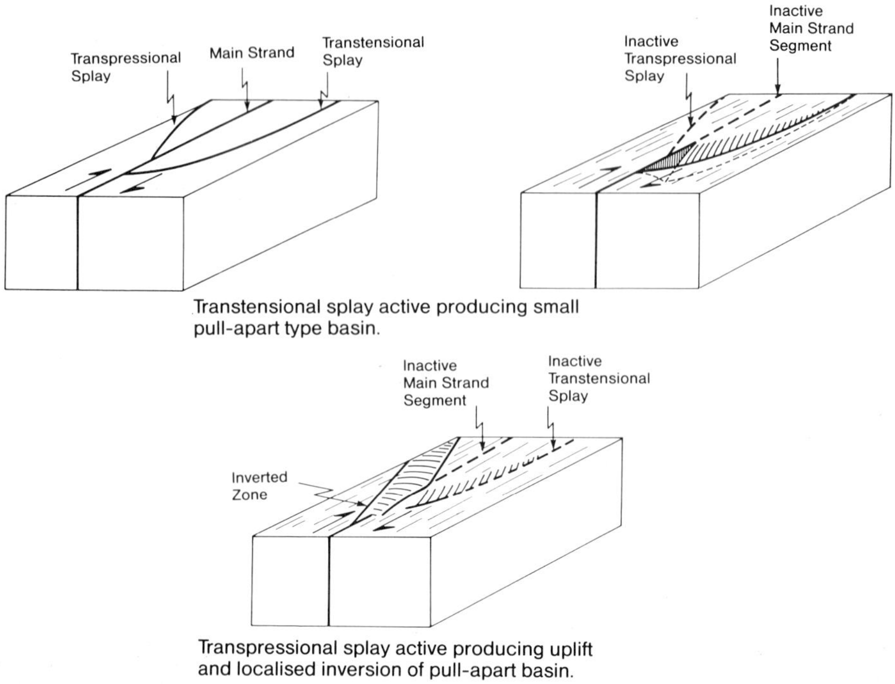

FIG. 4. The development of inversion-type structures in a transcurrent fault zone.

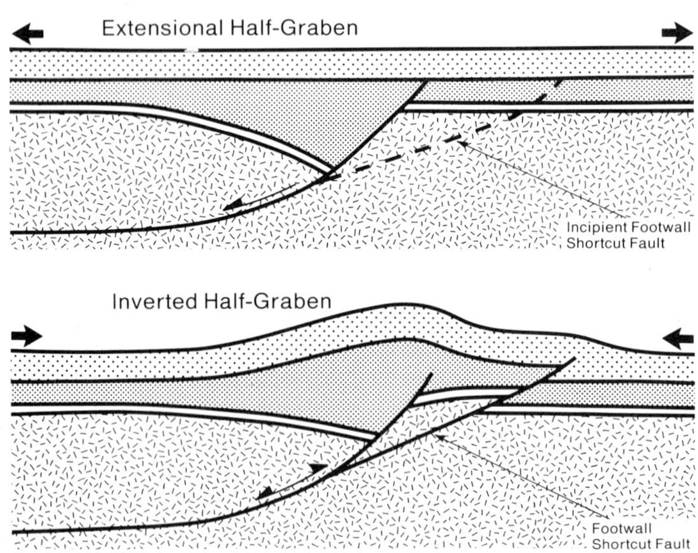

FIG. 5. The development of a footwall shortcut fault with a shallower dip during inversion of a steeply dipping extension fault.

develop at divergent plate boundaries and/or passive margins, e.g. the East African Rift Valley (Rosendal *et al.* 1986), are clearly related to fault movements which control basin development. Fault control is also apparent for basins associated with transform plate boundaries

(Mann *et al.* 1983) and failed rifts e.g. the North Sea (Ziegler 1983). This is also true for some basins associated with convergent plate boundaries, notably back-arc basins. However, a major basin forming mechanism at convergent plate boundaries is loading and flexure of the lithosphere, e.g. trenches and foreland basins. Such basins may contain minor extensional faults but these are secondary and do not control basin development. These basins are controlled by the nature and distribution of the load on the lithosphere which is indirectly controlled by the thrust system. There are other basins, e.g. the Michigan Basin whose origins are obscure and for which no fault control is evident. We would suggest that the concept of inversion be restricted to basins whose development is clearly controlled by fault systems. Note that this does not preclude the concept of negative inversion provided the compressionally deformed area subsequently developed fault-controlled extensional basins. This does however preclude foreland basins from becoming 'inverted basins' due to continuing advance of the thrust system as advocated for the Pyrenees (Todd & Turner in the previous section.)

Restriction to intracratonic environments

We do not favour the restriction of the term inversion to intra-cratonic basins (as suggested by Murphy and Ziegler in this paper) because of the difficulties in determining which basins are truly intracratonic, particularly in SE Asia where numerous microplates fragment the lithosphere. It is also useful (P. C. de Graciansky this paper) to apply concepts of inversion to the deformation of the old Tethyan passive margin in the Alps. We consider that the inversion process may be an important effect of orogenic deformation which may either precede or postdate thin-skinned deformation above a shallow detachment. To recognize this thicker skinned element of orogenic contraction the inversion concept is useful and should be used. Thus we believe that basins can be inverted by orogenesis and are sometimes incorporated into orogenic belts, but that these basins must have demonstrably existed before the compression rather than being generated as a result of the orogeny, e.g. foreland basins.

Reactivation of earlier fault systems

Another factor in the recognition of inversion is the extent to which the fault system controlling basin development is re-used during inversion. It appears that when an extensional half-graben is inverted, footwall shortcut faults (Fig. 5) will commonly be produced in order to generate a more gently inclined fault trajectory favourable to a sub-horizontal compressional stress (Gillcrist *et al.* 1987; Hayward & Graham *this volume*). Steep faults are highly efficient at producing structural elevation but achieve only minor shortening. The shallower dipping footwall shortcut creates an efficient trajectory for shortening to occur. In negative inversion the hanging wall shortcuts to produce a steeper fault gradient more efficient in extension (Powell & Williams *this volume*). Clearly there is the possibility that the old fault system is bypassed during the later deformation phase and a new fault system is developed to accommodate the movements (Fig. 6). This is an end member of a continuous spectrum through to total re-use. We suggest that substantial re-use of the preexisting fault system should be a prerequisite of recognizing an inversion structure. In some

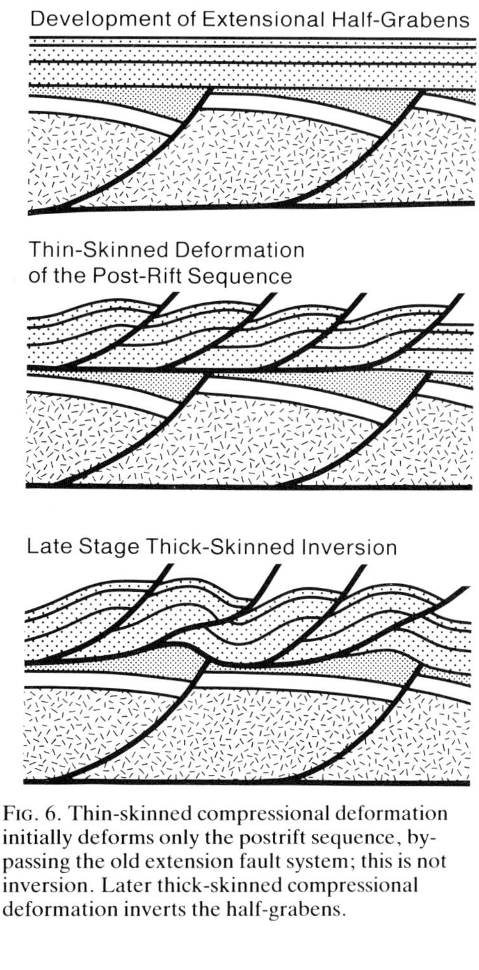

FIG. 6. Thin-skinned compressional deformation initially deforms only the postrift sequence, bypassing the old extension fault system; this is not inversion. Later thick-skinned compressional deformation inverts the half-grabens.

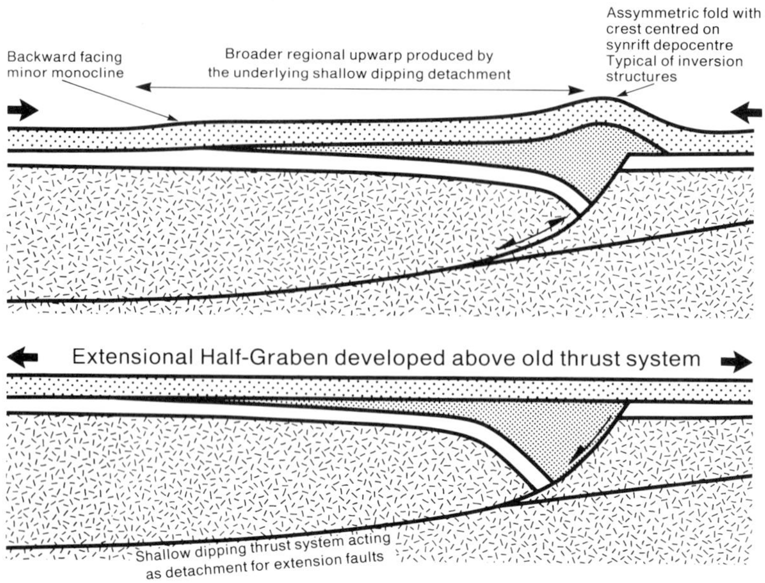

FIG. 7. The development of a broad regional uplift by inversion of a half-graben which utilized a shallow dipping rather than horizontal detachment, e.g. an older thrust system.

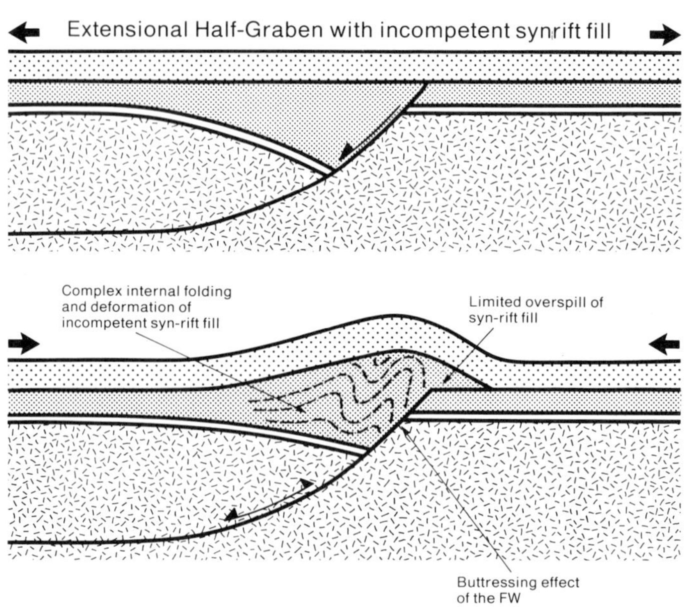

FIG. 8. Buttressing by the rigid basement footwall generates internal folding to shorten the synrift sequence.

cases, e.g. the Wessex Basin (Simpson *this volume*) the inversion is multiple. Mesozoic extension utilized the Variscan thrust system (Chadwick *et al.* 1983) which was placed into compression again during the Tertiary. This suggests that once a throughgoing deep crustal fault system has been established it will be re-used to some extent by all subsequent phases of deformation.

Positive inversion appears to be favoured by

a short time interval between the extensional and compressional phases (Professor, M. P. Coward, personal communication). This is probably because the attenuated lithophase has not thermally re-equilibrated and being relatively weak is prone to reactivation under any compressional stress (Ziegler *this volume*). The fault system established during extension will also be relatively weak and easily reactivated.

The geometry of positive inversion structures is highly dependent on the geometry of the original basin-controlling fault system. In the simple case of inverting a half-graben, the inversion fold produced will be an asymmetric monocline facing the footwall with the crest located above the synrift depocentre (Figs 3 & 4). However, if a more complex fault geometry occurs, e.g. as in Southern England where the extensional faults re-use Variscan thrusts, then this relationship will not necessarily hold (Fig. 7). Above such a shallow dipping detachment a broad, regionally developed inversion anticline can be produced with low amplitude which is superimposed on the higher amplitude asymmetric anticline located above the synrift depocentre.

The polarity of the extension faults which are being inverted is also significant. If the polarity is dipping towards the direction of compression then the fault may act as a buttress generating backthrusts on the roll-over crest (Fig. 2) or complex folding of the synrift fill (Hayward & Graham 1988; Gillcrist *et al.* 1987; de Graciansky *et al. this volume* (Fig. 8). Ultimately, however, the fault will behave as a forward directed thrust perhaps by shortcutting the footwall. When the polarity of the extension faults is such that they dip away from the compressional stress then there is potential for the development of backthrusts, perhaps with related footwall shortcuts, or thin-skinned fore-thrusts on the crest of the rollover. Another possibility here is that the fault may be rotated through the vertical by simple shear to a fore-thrust orientation during the compressional deformation (McClay *et al. this volume*).

Previous 'definitions' of inversion

There are three definitions of inversion which we feel it is useful to include here.

1 *Glennie & Boegner 1981*: 'Structural inversion . . . involves conversion of a basin area into a structural high. The converse is also possible so that inversion can be considered in both positive (uplift) and negative (subsidence) senses relative to the immediately preceeding history'.

2 *Bally 1983*: '. . . inverted extensional features. The main thought here is that basins or half-grabens or graben systems are to varying degrees turned inside out by compressional forces that reverse the deformation along pre-existing normal faults'.

3 *Harding 1985*: 'Positive structural inversion—a change in polarity of structural relief, from a previous low to a high as applied to a

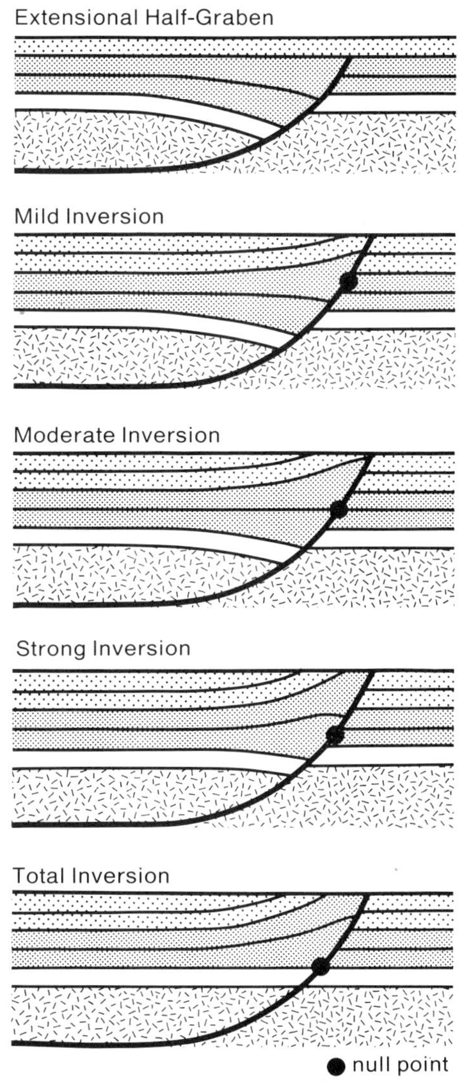

FIG. 9. Degrees of inversion illustrated by the progressive inversion of a half-graben.

specific structure. On a large scale this type of deformation is referred to as basin inversion'.

Conclusions

The requirements for inversion are:

1 Development of a basin actively controlled by faults such that a synrift or passive infill stratigraphic sequence can be recognized.

2 A change in the regional stress system resulting in the extensive re-use of the pre-existing fault (system), with the uplift affecting the hanging wall rather than the footwall.

The term positive inversion is unnecessary because inversion is normally assumed to be positive. The descriptor negative can be prefixed in cases of compression followed by extension.

Thus we can define an inversion structure or an inversion as: a pre-existing extensional–transtensional fault controlling a hanging wall basin containing a synrift or passive fill sequence that has subsequently undergone compression–transpression producing uplift and partial extrusion of the basin.

Basin inversion can be defined as: a basin controlled by a fault system that has been subsequently compressed–transpressed producing uplift and partial extrusion of the basin fill.

The degree of inversion can be indicated by the inclusion of a descriptor such as mild, moderate or strong depending on the location(s) of the null point(s) or total when no null point exists due to complete extrusion (Fig. 9).

Basins developed by processes other than those involving substantial crustal extension by faults should not be described as inverted even if subsequently uplifted.

Inversion structures can readily be recognized in cross-sections and on geological maps using the geometric criteria and the null point concept described earlier in this discussion.

References and further reading

Bally, A. W., 1984. Tectogénèse et sismique réflexion. *Bulletin Société Géologique de France*, **(7)XXIV(2)**: 279–285.

Bubb, J. N. & Hayash, J. M. 1980. Techniques in a seismic interpretation for hydrocarbon assessment: Assessment of Undiscovered Oil and Gas, *Committee for the Co-ordination of Joint Prosperity for Mineral Resources in Asian Offshore Areas. Technical Publications*, **10**, Bangkok.

Bukowicz, C. & Ziegler, P. A. 1985. Tectonic development of Mid-Norway continental margin. *Marine and Petroleum Geology*, **2(1)**, 12–22.

Chadwick, R. A., Kenolty, N. & Whittaker, A. 1983. Crustal structure beneath southern England from deep seismic reflection profiles. *Journal of the Geological Society of London*, **140**, 893–911.

de Graciansky, P. C., Dardeau, G., Lemoine, M., Tricart, P. The inverted margin of the French Alps and forehand basin inversion. *This volume*.

Gillchrist, R., Coward, M. P., & Mugnier, J. L. 1987 Structural inversion examples from the Alpine Foreland and the French Alps. *Geodinamica Acta*, **1**, 5–34.

Glennie, K. W. & Boegner, P. L. E. 1981. Sole Pit Inversion Tectonics. *In*: Illing, L. V. & Hobson, G. D. (eds) *Petroleum Geology of the Continental Shelf of Northwest Europe*. Institute of Petroleum, London, 110–120.

Harding, T. P. 1985. Seismic characteristics and identification of negative flower structures, positive flower structures, and positive structural inversion. *Bulletin of the American Association of Petroleum Geologists*, **69**, 582–600.

Hayward, A. B. & Graham, R. H. Some geometrical characteristics of inversion. *This volume*.

Lamplugh, G. W. 1920. Structure of the Weald and analogues tracts. *Quarterly Journal Geological Society* **75**: LXXIII–XCV (Anniversary Address of the President).

Mann, P., Hempton, M. R., Bradley, D. C. & Burke, K. 1983. Development of pull-apart basins. *Journal of Geology*, **91**, 529–554.

McClay, K. R., Insley, M. & Anderson, R. Inversion of the Kechika Trough, Northeastern British Columbia, Canada. *This volume*.

Powell, C. M. & Williams, G. D. 1988. The Lewis thrust Rocky Mountain trench system in Northwest Montana USA: an example of negative inversion tectonics? *This volume*.

Prouvost, P. 1930. Sédimentation et subsidence, *Centenaire de la Société Géologique de France Livre Jubilaire, 1820–1990*, Vol. II. Paris, 545–564.

Rosendahl, B. R., Reynolds, P. M., Lorber, C. F., Burgess, C. F., McGill, J., Scott, D., Lambiase, J. J. & Derksen, S. J. 1986. Structural expressions of rifting: lesson from Lake Tanganyika, Africa. *In*: Frostick, L. E. *et al.* (eds) *Sedimentation in the African Rifts. Special Publication of the Geological Society of London*, **25**, 29–43.

Simpson, I. R., Gravestock, M., Ham, D., Leach, H. & Thompson, S. D. Notes and cross-sections illustrating inversion tectonics of the Wessex Basin. *This volume*.

Stille, H., 1924. *Grundfragen der Vergleichenden*

Tektonik. Brontrager, Berlin, 443 pp.
VOIGT, E. 1963. Uber Randtroge vor Schollenrandern und ihre Bedeutung im Gebiet der mitteleuropaischen Senke und angrenzender Gebiete *Zeitschrift der Deutschen Geologischen Gesellschaft*, **114(2)**, 378–418.
WILLIAMS, G. D., POWELL, C. M. & COOPER, M. A. 1988. Geometry and kinematics of inversion tectonics. *This volume*.
ZIEGLER, P. A. 1983. Inverted Basins in the Alpine Foreland. *In*: Bally, A. W. (ed.) *Seismic Expression of Structural Styles—A picture and work atlas. American Association of Petroleum Geologists, Studies in Geology. Series No. 15*, Vol. 3 3.3.3.–3.3.12.
—— 1987. Compressional intra-plate tectonics in the Alpine foreland. *Tectonophysics*, **137**, 420 pp.
—— 1988. Evolution of the Arctic North Atlantic and the Western Tethys *American Association of Petroleum Geologists Memoir*, **43**, 198 pp.
—— Geodynamic model for Alpine intra-plate compressional deformation in Western and Central Europe. *This volume*.

M. A. COOPER, B.P. Canada Inc., Oil and Gas Division, 333 Fifth Ave SW Calgary, Alberta, T2P 3B6, Canada.
G. D. WILLIAMS, Department of Geology, University of Keele, Keele, Staffordshire, ST5 5BG, UK.
P. C. DE GRACIANSKY, Ecole Nationale Supérieure des Mines, 60 Blvd St Michel, 75272 Paris, Cédex 061 France.
R. W. MURPHY, 34 Bridge St, Walton-on-Thames, Surrey, KT12 1AJ, UK.
T. NEEDHAM, Department of Geology, Goldsmiths College, London, UK.
D. DE PAOR, Department of Earth Sciences, Johns Hopkins University, Baltimore, USA.
R. STONELEY, Department of Geology, Imperial College of Science and Technology, London SW7 2BP, UK.
S. P. TODD, Department of Geology, University of Bristol, Bristol, UK.
J. P. TURNER, Department of Geology, University of Bristol, Bristol, UK.
P. A. ZIEGLER, Shell International Petroleum Mij. B. V., PO Box 162, 2501 AN The Hague, The Netherlands.

Abstracts

Inversion of circum-Pacific marginal basins

The initial phase of compression (mid-Cretaceous) in the history of Andean orogenesis involved inversion of a composite back-arc basin along at least 7500 km of the Pacific margin. Uplift of segments of basin floor as ophiolitic complexes appears to have taken place along reactivated normal faults. In the most intensely deformed region stacking of the passive continental margin of the basin during arc–continent collision may have been controlled by basin bounding faults. Present day analogues for the Cretaceous basin in the Andes can be recognized elsewhere in the circum-Pacific, for example the Bransfield Trough off the Antarctic Peninsula. Incipient inversion is suggested by first motion studies of earthquakes bordering the Japan Sea. Marginal basin formation and subsequent inversion has probably played an important role in orogenesis throughout the tectonic history of the Earth.

I. W. D. Dalziel, Institute for Geophysics, University of Texas at Austin, 4920 North IH35, Austin, Texas 78751, USA.

Kinematics and dynamics of basin inversion

The scale, style and degree of basin inversion varies from the collisional inversion of continental margin rift complexes to the subtle and gentle inversion of intracontinental rifts and from gentle whole basin uplift to major structural inversions with substantial thrust remobilization of earlier extensional structures. Some inversion mechanisms are intrinsic to the existence and lithospheric structure of the basin. Others, especially those involving simple basin uplift are coincidental. The causes of inversion are many and may include:

1 Regional temporal variations in stress patterns within plates, resulting from forces caused by changes in plate boundary configuration, the blocking of subduction zones by buoyant crust (collision) and changes in relative motion at nearby plate boundaries.

2 Global, episodic, intraplate, stress changes from deviatoric compression during the collisional assembly of Pangaea-type continental configurations to deviatoric tension in assembled configurations to deviatoric compression in dispersed configurations. Inversions are commonly associated with rifted margin breakup unconformities caused by the transition from rift–phase tension to ridge push compression.

3 Local inversion in strike–slip rhombohedral pull-aparts as a natural consequence of alternating phases of transtension with negative flower structure and transpression with positive flower structure.

4 Complex alternating phases of extension and shortening at the margins of upper crustal rotational flakes in strike–slip zones.

5 Progressive diminution of the ratio of crustal to lithospheric thickness during slow extension causing whole basin uplift.

6 Uplift on lithospheric flexural arches and hotspots.

7 Body force mechanisms including salt mud diapirism, salt/mud decollement tectonics, gravity spreading, sliding and consequent relaying of heel extension to toe thrusting.

8 Inversion of strong negative isostatic gravity anomalies in confined deep basins caused by upper crustal stretching. Critical to whether a basin becomes inverted are, also, the timing of a compression phase relative to the initial basin-forming extensional event and the extensional strain rate. Short extension–compression intervals and high extensional strain rates facilitate basin inversion. Inversion may be prevented by long intervals and low strain rates.

The way in which extension is balanced through the lithosphere determines the style of subsequent shortening inversion. Upper crustal extensional basins lead to long distance compression relaying along mid-crustal detachments while preserving their negative gravity anomalies. Whole lithospheric extension with substantial crustal thinning and Moho pull-up leads to mantle involvement in thrust complexes and positive gravity anomalies in inverted basins.

J. F. DEWEY, Department of Earth Sciences, Parks Road, Oxford OX1 3PR, UK.

Extensional tectonics in a convergent intra-plate setting: linked inversions on oblique and frontal ramps

A model of a sub-horizontal detachment with ramps and flats and thrusts climbing section in the hanging wall, is commonplace in the interpretation of thrust belts and generally used in two dimensions. The existence of a regional detachment beneath the foreland of a convergent zone is less well known. The coexistence of areas of both compressional uplift and extensional subsidence in the regional setting of a convergent foreland can be described by one simple geometric model. In a system of sub-horizontal (thrusting) displacement of a hanging wall, movement over footwall ramps will give rise to deformation of the hanging wall. In particular, oblique ramps that allow right lateral transfer of the thrust displacement will give rise to extensional deformation, and those that allow left lateral transfer will give rise to compressional deformation.

These geometric relationships are seen in the Alpine foreland in France. The general NW convergence of this segment of the Alps passes into the foreland, where regional uplifts (Jura, Morvan, Vosges, Massif Central, Ardennes, etc.) are separated by zones of extensional subsidence (Dauphine, Bresse, Clermont, Rhone, etc. basins) in a linked system of regional hanging wall deformation. The model provides a consistent geometric explanation for their origin, as well as a predictive framework for the evaluation of subsurface structures.

A. Gibbs, Midland Valley Exploration, 14 Park Circus, Glasgow G3 6AX, UK.
A. Beach, Midland Valley, Exploration, 14 Park Circus, Glasgow G3 6AX, UK. Present address: Alastair Beach Associates, Lockend Farm, Beith, Ayrshire, KA15 2LN, UK.

Structural inversion in the external French Alps

The classical thrust rules presently used to interpret thrust-mountain belts were derived from thrust belts which shorten and telescope a non-tectonized layer cake shelf sequence. The indiscriminate use of these rules to interpret the parts of thrust belts which involve oceanic and passive margin sedimentary sequences affected by numerous growth faults will produce erroneous results.

Stratigraphic, sedimentological and structural relationships in the external French Alps provide evidence of various Mesozoic extensional phases, which culminate in the formation of a complex faulted passive margin. With the onset of Alpine contraction the passive margin fault system was subsequently inverted forming a fold–thrust belt so that the resultant thrust sequences, geometries, variations in finite shortening direction, and arcuate shape of the Alpine thrust belt are all controlled by the geometry and the inversion kinematics of the early extensional fault system. Characteristic inversion geometries have been produced on a large scale. These include both hanging wall and footwall shortcuts, large scale pop-ups, complex back fold–thrust systems and buckling of the half graben fills where the steeper pre-existing normal fault scarps act as butresses to the contractional displacements. Further complications arise from the polyphase nature of the extension and compression during the Cretaceous to Pliocene tectonics.

A re-examination of the basic assumptions currently used to constrain balanced cross-section construction in such regions, with a greater emphasis on the initial template construction, is needed. Where the amount of orogenic contraction is large, the early extensional geometries may be difficult to recognize. Here the final 'stacking' geometry derived from cross-sections, which need not necessarily balance when using a simple 2-D plane strain model, will provide more information on the geometry of the inverted extensional system, than a 'forced' precision balanced cross-section constructed in the classical way.

R. Gillcrist, Department of Geology, Imperial College of Science and Technology, London SW7 2BP, UK.
M. P. Coward, Department of Geology, Imperial College of Science and Technology, London SW7 2BP, UK.
B. Trudgill, Department of Geology, Imperial College of Science and Technology, London SW7 2BP, UK.
A. Pecher, Ecole des Mines, Nancy, France.
J. L. Mugnier, Institute Dolemieu, Grenoble, France.

Geometric, thermal and isostatic constraints on basin inversion

Many extensional basins show inversion generated by the compressional reactivations of the same major low angle fault systems originally responsible for the formation of the basin. The geometric, thermal and isostatic consequences of the inversion of extensional basins by compressional reactivation of low angle faults and detachments have been investigated using a numerical model. The model assumes simple shear in the upper lithosphere and pure shear at depth. The isostatic response of the lithosphere during inversion is driven by crustal thinning by both simple shear and pure shear, associated perturbation of the geothermal gradients, erosion of the uplifted extensional basin and sediment deposition on adjacent lithosphere. Model predictions are compared with observations of basin inversion for the Mesozoic basins of Southern Britain. For these basins, the geometries of the associated major low angle faults have been imaged by deep seismic reflection profiling, providing a powerful constraint in modelling both the formation of the basin and their inversion.

N. J. Kusznir, Department of Geology and Geophysics University of Liverpool, Brownlow Street, P.O. Box 147, Liverpool L69 3BX, UK.
G. D. Williams, Department of Geology, University of Keele, Keele, Staffordshire, ST5 5BG, UK.

Structural evolution, timing and tectonic style of the Sole Pit inversion

The NW–SE oriented Sole Pit High, located in the western half of the Southern North Sea Basin, is the result of three different structural processes initiated during the Late Palaeozoic. Basin subsidence, probably related to Late Hercynian pull-apart movements led to continuous sedimentation from the Permian to the Middle Jurassic. Mid-Jurassic doming of its eastern flank, followed by erosion, transtensional movements and differential subsidence until the Early Cretaceous reflect the Mid and Late Kimmerian crustal distension and fragmentation of the North Sea Rift. Basin-wide inversion occurred between the Turonian and the Campanian and was repeated in the Oligocene. This resulted in a total uplift of 1500 m.

The present structural style is represented by a complex network of transpressional and transtensional faults at Permian Rotliegend level, which are thought to represent the upward termination of a series of wrench-induced flower structures with a right-lateral sense of displacement. These faults are detached by Zechstein and Triassic salt from more orderly transtensional fault systems at shallower levels. In areas where salt is absent Late Kimmerian normal faults became reactivated in a reverse sense during inversion. However, the relationship of these shallow mappable tectonic components to the deeper pre-Permian structure of the Sole Pit High and to the underlying crustal configuration remains speculative.

B. van Hoorn, Shell Expro, Shell-Mex House, Strand, London WC2R 0DX, UK.

Index

Note: Page numbers in *italic* refer to figures and those in **bold** refer to tables.